D1187123

I G Macdonald

1990

A TREATISE ON THE CIRCLE AND THE SPHERE

A TREATISE ON THE CIRCLE AND THE SPHERE

BY

JULIAN LOWELL COOLIDGE, Ph.D.

ASSISTANT PROFESSOR OF MATHEMATICS IN
HARVARD UNIVERSITY

CHELSEA PUBLISHING COMPANY
BRONX, NEW YORK

THE PRESENT, 1971, EDITION IS A REPRINT, TEXTUALLY UNALTERED EXCEPT FOR THE INCORPORATION INTO THE TEXT OF THE AUTHOR'S NOTES AND THE CORRECTION OF ERRATA, OF A WORK FIRST PUBLISHED IN 1916 AT OXFORD

IT IS PRINTED ON LONG-LIFE ALKALINE PAPER

INTERNATIONAL STANDARD BOOK NUMBER 0-8284-0236-1

LIBRARY OF CONGRESS CATALOG CARD NUMBER 78-128872

LIBRARY OF CONGRESS CLASSIFICATION NUMBER QA484

DEWEY DECIMAL CLASSIFICATION NUMBER 516'.2

PRINTED IN THE UNITED STATES OF AMERICA

PREFACE

Every beginner in the science of geometry knows that the circle and the sphere have always played a central rôle, yet few people realize that the reasons for this are many and various. Attention was first called to these figures by their mechanical simplicity and importance, and the fortunate position thus won was further strengthened by the Euclidean tradition of limiting geometry, on the constructive side, to those operations which can be carried out with the aid of naught but ruler and compass. Yet these facts are far from sufficient to account for the commanding position which the circle and the sphere occupy to-day.

To begin with, there would seem no *a priori* reason why those curves which are the simplest from the mechanical point of view should have the greatest wealth of beautiful properties. Had Euclid started, not with the usual parallel postulate, but with the different assumption either of Lobachevski or Riemann, he would have been unable to prove that all angles inscribed in the same circular arc are equal, and a large proportion of our best elementary theorems about the circle would have been lacking. Again, there is no *a priori* reason why a curve with attractive geometric properties should be blessed with a peculiarly simple cartesian equation; the cycloid is particularly unmanageable in pure cartesian form. The circle and sphere have simple equations and depend respectively on four and five independent homogeneous parameters. Thus, the geometry of circles is closely related to the projective geometry of three-dimensional space, while the totality of spheres gives our best example of a four-dimensional projective continuum. Still further, who could have predicted that circles would play a central rôle in the theory of linear functions of a complex variable, or that every conformal transformation of space would carry spheres into spheres? These are but examples of the way in which circles and spheres force themselves upon our notice in all parts of geometrical science.

The result of all this is that there is a colossal mass of literature dealing with circles and spheres, the various parts of which have been developed with little reference to one another. The elementary geometry of the circle was carried to a high degree of perfection by the ancient Greeks, but by no means completed, for in comparatively recent times there have been notable contributions from mathematicians of no mean standing, Steiner and Feuerbach, Chasles and Lemoine, Casey and Neuberg, and a countless following host. The relation between circle geometry and projective geometry has been thoroughly studied by Reye, Fiedler, Loria, and their pupils. Every text-book of the theory of functions of a complex variable discusses the relation of circles to the linear function, while the general theory of circle transformations has had such distinguished exponents as Möbius and von Weber. The circle and sphere with positive or negative radius have been the subject of admirable studies by Laguerre and Lie, algebraic systems of circles in space have been studied by Stephanos, Koenigs, Castelnuovo, and Cosserat, while circle congruences in general have received no little attention from recent writers on differential geometry, notably Ribaucour, Darboux, and Guichard.

The present work is an attempt, perhaps the first, to present a consistent and systematic account of these various theories. The greatest difficulty in any such undertaking is obviously that of selection. This is particularly the case in the early part of the subject. A complete account of all known elementary theorems regarding the circle would be far beyond the strength of any writer, or reader. The natural temptation is to go to the other extreme, and omit entirely the elementary portions; yet this would be equally fatal. How could one write at length on the geometry of the circle without discussing the Apollonian problem and the nine-point circle? But if we include the circle of Feuerbach, why should we exclude the circles of Lemoine, Tucker, and Brocard? Where does the geometry of the circle end, and that of the triangle begin? Clearly any principle of choice must be largely arbitrary and illogical.

In the present treatise preference is shown to those theorems which are unaltered by inversion, and to those which are as general as possible in their scope. The author has tried to say something about every circle that is known by a recognized name, but the vast subject of the geometry of the triangle is treated only in a superficial manner. Similarly, only a small number of the most famous problems in construction have been discussed, but these have been treated at some length.

When we pass from the elementary to the more advanced portions of the subject, we find a tolerably clear line of demarcation running through the geometry of the circle and the sphere, namely, the separation of those theorems which involve the centre or radius from those which do not. Otherwise stated, we have those theorems which are invariant under the group of conformal collineations, and those which are invariant for inversion. An attempt is made to keep these two classes as far separate as practicable. For this reason, distinction is drawn between cartesian space which is supposed to have been rendered a perfect continuum by the adjunction of a plane at infinity, and pentaspherical space where the finite region is defined, in the real domain, as a single point. Among the cartesian theorems there is a sharp sub-division between those where the radius is looked upon as essentially signless and those where a positive or negative radius is allowable. The circle and the oriented circle should be considered as essentially dissimilar figures; the former is a locus of points, the latter, in the plane, is best handled as an envelope of oriented lines, and considered under a totally different group. In the present work the oriented circle and sphere are discussed in three chapters entirely devoted to them.

Every writer knows that the pleasantest part of his task consists in writing the preface, for here he has a chance to express his gratitude to the generous friends who have helped him with suggestion and counsel. The present author would especially mention his colleague Professor Maxime Bôcher, who kindly read the proof of Ch. VIII, and his former pupil Dr. David Barrow, who not only supplied much of the material in Ch. XIV but also did yeoman service in unearthing mistakes in various parts of the work. Another pupil, Dr. Roger Johnson, has kindly suggested a number of minor corrections, mostly of a bibliographical nature. Yet the greatest debt is not to any one of these.

The present work went to press in the spring of 1914. During the two years which have intervened, the Delegates of the Clarendon Press, despite the fact that their country was passing through the most severe trial in her history, have yet seen fit to continue the publication of a book which dealt with a subject utterly remote from all that occupied men's thoughts, and which was not even written by one of their countrymen. Let the author's last word be one of gratitude to them for this great kindness, as signal as it is undeserved.

CAMBRIDGE, U.S.A.
July, 1916.

CONTENTS

CHAPTER I

THE CIRCLE IN ELEMENTARY PLANE GEOMETRY

§ 1. Fundamental Definitions and Notation

§ 2. Inversion

§ 3. Mutually Tangent Circles

§ 4. Circles related to a Triangle

§ 5. The Brocard Figures

§ 6. Concurrent Circles and Concyclic Points

§ 7. Coaxal Circles

CHAPTER II

THE CIRCLE IN CARTESIAN PLANE GEOMETRY

§ 1. The Circle studied by means of Trilinear Coordinates

§ 2. Fundamental Relations, Special Tetracyclic Coordinates

§ 3. The Identity of Darboux and Frobenius

§ 4. Analytic Systems of Circles

CHAPTER III

FAMOUS PROBLEMS IN CONSTRUCTION

CHAPTER IV

THE TETRACYCLIC PLANE

§ 1. Fundamental Theorems and Definitions

§ 2. Cyclics

CHAPTER V

THE SPHERE IN ELEMENTARY GEOMETRY

§ 1. Miscellaneous Elementary Theorems

§ 2. Coaxal Systems

CHAPTER VI

THE SPHERE IN CARTESIAN GEOMETRY

§ 1. Coordinate Systems

§ 2. Identity of Darboux and Frobenius

§ 3. Analytic Systems of Spheres

CHAPTER VII

PENTASPHERICAL SPACE

§ 1. Fundamental Definitions and Theorems

§ 2. Cyclides

CHAPTER VIII

CIRCLE TRANSFORMATIONS

§ 1. General Theory

§ 2. **Analytic Treatment**

§ 3. **Continuous Groups of Transformations**

CHAPTER IX

SPHERE TRANSFORMATIONS

§ 1. **General Theory**

§ 2. **Continuous Groups**

CHAPTER X

THE ORIENTED CIRCLE

§ 1. Elementary Geometrical Theory

§ 2. Analytic Treatment

§ 3. Laguerre Transformations

§ 4. Continuous Groups

§ 5. Hypercyclics

§ 6. The Oriented Circle Treated Directly

CHAPTER XI

THE ORIENTED SPHERE

§ 1. Elementary Geometrical Theorems

§ 2. Analytic Treatment

§ 3. The Hypercyclide

§ 4. The Oriented Sphere Treated Directly

§ 5. Line-sphere Transformation

§ 6. Complexes of Oriented Spheres

CHAPTER XII

CIRCLES ORTHOGONAL TO ONE SPHERE

§ 1. Relations of two Circles

§ 2. Circles Orthogonal to one Sphere

§ 3. Systems of Circle Crosses

CHAPTER XIII

CIRCLES IN SPACE, ALGEBRAIC SYSTEMS

§ 1. Coordinates and Identities

§ 2. Linear Systems of Circles

§ 3. Other Simple Systems

CHAPTER XIV

THE ORIENTED CIRCLE IN SPACE

§ 1. Fundamental Relations

§ 2. Linear Systems

§ 3. The Laguerre Method of Representing Imaginary Points

CHAPTER XV

DIFFERENTIAL GEOMETRY OF CIRCLE SYSTEMS

§ 1. Differential Geometry of the S_6^5 of all Circles

§ 2. Parametric Method for Circle Congruences

§ 3. The Kummer Method

§ 4. Complexes of Circles

CHAPTER I

THE CIRCLE IN ELEMENTARY PLANE GEOMETRY

§ 1. **Fundamental Definitions and Notation.**

ALL figures discussed in the present chapter are supposed to exist in the real and finite domain of the Euclidean plane; the domain of elementary plane geometry. As fundamental objects, we shall take *points, lines*, and *circles*. We shall make no attempt to define a *point*. By *line* we shall mean a *straight line*; a class of points uniquely determined by any two of its members. It extends to an infinite distance on either side of any of its points. That portion of a line which is on either side of any point shall be called a *half-line*; the portion which includes two points and all between them shall be a *segment*. If two half-lines be given which are not collinear, but are bounded by a common point, that portion of the plane which includes all segments whose extremities are on the given half-lines shall be called their *interior angle*, or, more shortly, their *angle*. The remainder of the plane shall be their *exterior angle*. These definitions may be easily extended to include null and straight angles. Three non-collinear points will determine three segments forming together a *triangle*. The given points and segments are the *vertices* and *sides* respectively, the lines whereon the segments lie shall be called the *side-lines*.* The three angles, each of which is bounded by two half-lines including two sides of the triangle, shall be called the *angles of the triangle*,

* This term suggests football rather than geometry. It is, however, proper to distinguish between the side of a triangle, and the line whereon that side lies.

their supplements its *exterior angles*. A line through a vertex perpendicular to the opposite side-line shall be called an *altitude line*, its intersection with the side-line its *foot*, and the *segment* bounded by the foot and the opposite vertex, the *altitude*.

We shall mean by a *circle* the locus of points at a given distance from a fixed point called the *centre*. A segment bounded by two points of a circle shall be called a *chord*, its line a *secant*. The limiting position of a secant as the two points of the circle approach one another shall be a *tangent*. A segment bounded by the centre and a point of the circle is a *radius*, that which is made up of two collinear radii a *diameter*.

Let us pass from these definitions to establishing certain conventions as to notations. Points shall be denoted by large *italic* letters as $A\ B\ P_i$. The segment bounded by A and B, or the distance of these points, shall be written (AB). When a question of algebraic sign arises, or a segment is looked upon as measured in a particular sense, we shall superpose an arrow pointing to the right, to indicate that the segment is measured from the point denoted by the first letter to that denoted by the second, thus

$$(\overrightarrow{AB}) = -(\overrightarrow{BA}).$$

The line determined by the points A and B shall be indicated AB. It is often convenient to indicate a line by a single small *italic* letter as a, l_i. The angle of the half-lines which include the segments $(AB)(AC)$, when considered as a quantity bereft of sign, shall be indicated $\measuredangle BAC$. When the sense of description is essential we shall introduce a right-pointing arrow, as

$$\measuredangle \overrightarrow{BAC} = -\measuredangle \overrightarrow{CAB}.$$

When we wish one of the lesser angles determined by two lines, *including its sense of description*,* we shall use the notation $\angle BAC$ or $\angle l_1 l_2$. Parallelism shall be denoted by $\parallel$, perpendicularity by $\perp$. The distinction in meaning between

* There is, of course, a slight ambiguity when the lines are mutually perpendicular; it does not, however, cause any practical inconvenience.

our various symbols will appear from the following familiar equations:

If $l_1 \parallel l_1'$ and $l_2 \parallel l_2'$, then $\angle l_1 l_2 = \angle l_1' l_2'$.
If $l_1 \perp l_1'$ and $l_2 \perp l_2'$, then $\angle l_1 l_2 = l_1' l_2'$.
If $ABCD$ be concyclic,

$$\angle ABC = \angle ADC,$$
$$\measuredangle ABC = \measuredangle ADC \text{ or } \pi - \measuredangle ADC,$$
$$\measuredangle \overrightarrow{ABC} = \measuredangle \overrightarrow{ADC} \text{ or } \pm(\pi + \measuredangle \overrightarrow{ADC}).$$

If $ABCD$ be any four coplanar points,

$$\angle ADB + \angle BDC + \angle CDA \equiv 0 \pmod{\pi},$$
$$\measuredangle \overrightarrow{ADB} + \measuredangle \overrightarrow{BDC} + \measuredangle \overrightarrow{CDA} \equiv 0 \pmod{2\pi}.$$

A triangle where vertices are ABC shall be indicated ΔABC.

It is useful to make certain further conventions for the study of a single triangle. The vertices shall be $A_1 A_2 A_3$, this order of letters corresponding to a circuit of the triangle in a counter-clockwise or positive sense. If the letters i, j, k indicate a circular permutation of the numbers 1, 2, 3,

$$\measuredangle A_j A_i A_k = \measuredangle A_k A_i A_j = \measuredangle A_i \quad (A_j A_k) = a_i \quad \Sigma a_i = 2s.$$

If P be any other point of the plane, the line $A_i P$ shall meet $A_j A_k$ in P_i; a line through $P \perp A_j A_k$ shall meet $A_j A_k$ in Pa_i. The middle point of $A_j A_k$ shall be M_i; the centre of gravity of the triangle is thus M. The centre of the circumscribed circle shall be O, the orthocentre, the point of concurrence of the altitude lines, shall be H. We have thus, incidentally, $H_i = Ha_i$. The area of this triangle shall be Δ, the radius of the circumscribed circle shall have the length r. A theorem shall be referred to as x] or y] while an equation is (p) or (q).

§ 2. Inversion.

A truce to these preliminaries! Suppose that we have given a circle whose centre is O and radius has the length $r \neq 0$. Let P and P' be any two points collinear with O such that

$$(\overrightarrow{OP}) \times (\overrightarrow{OP'}) = r^2. \tag{1}$$

The relation between the two is perfectly symmetrical, each is said to be the *inverse* of the other with regard to that circle, and the transformation from one to the other is called an *inversion*. The point O is called the *centre of inversion*, the given circle the *circle of inversion*, and its radius the *radius of inversion*.*

Theorem 1.] *Every point other than the centre of inversion has a single inverse.*

Theorem 2.] *The circle of inversion is the locus of points which are their own inverses.*

Theorem 3.] *Points within the circle of inversion other than the centre will invert into points without, points without will always invert into points within.*

Another transformation similar to inversion is found by taking S and S' collinear with O so that

$$(\overrightarrow{OS}) \times (\overrightarrow{OS'}) = -r^2.$$

This is seen immediately to be the product of an inversion and a reflection in the centre, though algebraically it is an inversion in a circle of imaginary radius. We shall make but little use of this transformation in the present chapter. Returning to the direct study of inversion, let the reader show that if P be without the circle, P' is the intersection of OP with the chord of contact of tangents from P to this circle, i.e. with the polar of P. We notice further that if OP meet the circle in H and K, H lying between O and P,

$$\frac{(\overrightarrow{HP})}{(\overrightarrow{HP'})} = \frac{(\overrightarrow{OH}) - (\overrightarrow{OP})}{(\overrightarrow{OH}) - (\overrightarrow{OP'})}; \quad \frac{(\overrightarrow{KP})}{(\overrightarrow{KP'})} = \frac{(\overrightarrow{OK}) - (\overrightarrow{OP})}{(\overrightarrow{OK}) - (\overrightarrow{OP'})},$$

* This transformation is usually credited to Plücker. See his *Analytisch-geometrische Aphorismen*, Crelle, vol. xi, 1836. It was rediscovered a decade later by Sir William Thompson, *Principe des images électriques*, Liouville, vol. x, 1845. The most recent view, however, seems to be that the method was found some time previous by Steiner. Cf. Bützberger, *Ueber bizentrische Polygone*, Leipzig, 1913, pp. 50-5. The inversion of a small region can be effected mechanically by link works invented by Peaucellier, Hart, Kempe, and others.

$$\frac{(\overrightarrow{HP})(\overrightarrow{KP'})}{(\overrightarrow{HP'})(\overrightarrow{KP})}=\frac{-r^2+r^2-[(\overrightarrow{OP})(\overrightarrow{OK})+(\overrightarrow{OP'})(\overrightarrow{OH})]}{-r^2+r^2-[(\overrightarrow{OP})(\overrightarrow{OH})+(\overrightarrow{OP'})(\overrightarrow{OK})]}=-1.$$

We thus reach a theorem slightly beyond the limits of elementary geometry strictly construed.

Theorem 4.] *Mutually inverse points are harmonically separated by the intersections of their line with the circle of inversion.*

If P' and Q' be the inverses of P and Q respectively, we have

$$(\overrightarrow{OP})(\overrightarrow{OP'})=(\overrightarrow{OQ})(\overrightarrow{OQ'}),\quad \frac{(OP)}{(OQ')}=\frac{(OP')}{(OQ)}. \tag{2}$$

$\Delta\, OPQ$ and $\Delta\, OQ'P'$ are similar.

$$(P'Q')=(PQ)\frac{(OP')}{(OQ)}=(PQ)\frac{r^2}{(OP)(OQ)}. \tag{3}$$

If $PQRS$ be four points whose inverse are $P'Q'R'S'$,

$$\frac{(\overrightarrow{P'Q'})(\overrightarrow{R'S'})}{(\overrightarrow{S'P'})(\overrightarrow{Q'R'})}=\frac{(\overrightarrow{PQ})(\overrightarrow{RS})}{(\overrightarrow{SP})(\overrightarrow{QR})}. \tag{4}$$

We shall make great use of this equation subsequently. For the moment we merely draw therefrom an extension of the previous proposition.

Theorem 5.] *The cross ratio of four points collinear with the centre of inversion, but distinct therefrom, is equal to that of their inverses.*

We now assume specifically that P and Q are not collinear with O. We see from (2) that $\Delta\, OPQ$ and $\Delta\, OQ'P'$ are similar, hence

$$\measuredangle\, \overrightarrow{OPQ}=\measuredangle\, \overrightarrow{OQ'P'}.$$

If R be a fourth point in general position,

$$\measuredangle\, \overrightarrow{OPR}=\measuredangle\, \overrightarrow{OR'P'}.$$

We substitute for each angle on the right its equivalent in terms of the other two angles of the triangle whose vertices are thereby designated, then subtract;

$$\measuredangle\, \overrightarrow{RPQ}+\measuredangle\, \overrightarrow{R'P'Q'}=\measuredangle\, \overrightarrow{ROQ}=\measuredangle\, \overrightarrow{R'OQ'}.$$

Theorem 6.] *The algebraic sum of the corresponding angles of two mutually inverse triangles is equal to the angle subtended at the centre of inversion by the sides opposite these angles.*

Theorem 7.] *If two opposite angles of a quadrilateral be measured in such a way that the two initial sides and the two terminal sides meet respectively in vertices of the quadrilateral, their algebraic difference is numerically equal to the corresponding difference for the inverse quadrilateral.*

Of course, when we say that two triangles or quadrilaterals are mutually inverse, we merely mean that this is true of their corresponding vertices. We next let Q approach P as a limit, so that PQ and $P'Q'$ approach tangency in two mutually inverse curves.

Theorem 8.] *The angle made at any point by a curve with a line from there to the centre of inversion is numerically the supplement of the corresponding angle for the inverse curve at the inverse point.*

Theorem 9.] *An angle at which two curves intersect at any point other than the centre of inversion is the negative of the corresponding angle made by the inverse curves at the inverse point.*

Theorem 10.] *Curves which intersect at right angles not at the centre of inversion will invert into curves intersecting at right angles.*

Any curve which is its own inverse is said to be *anallagmatic.**

Theorem 11.] *If the circle of inversion intersect an anallagmatic curve at any point which is a simple point for the latter, the two will intersect at right angles.*

Theorem 12.] *A line through the centre of inversion is anallagmatic.*

Theorem 13.] *A circle through a pair of inverse points is anallagmatic.*

* This curious word seems to be due to Moutard.

We see, in fact, that if we consider any pair of points on such a circle collinear with the centre of inversion, the product of their distances therefrom is the square of the radius of inversion. Let the reader show that

Theorem 14.] *A circle which cuts the circle of inversion at right angles is anallagmatic.*

Theorem 15.] *If two intersecting circles cut a third at right angles, their intersections are inverse in the third circle.*

This last theorem leads to another way of looking at anallagmatic curves. If we have a system of circles moving continuously yet always orthogonal to a fixed circle, we see that the intersections of infinitely near circles are inverse in the fixed circle, i. e. the envelope is anallagmatic. Conversely, if an anallagmatic curve be given, a circle through two inverse points and tangent at one, will be tangent at the other; the curve is the envelope of circles orthogonal to the circle of inversion. The locus of the centres of the moving circles shall be called the *deferrent.*

If a circle orthogonal to the circle of inversion be anallagmatic, what is the inverse of a circle in general position?

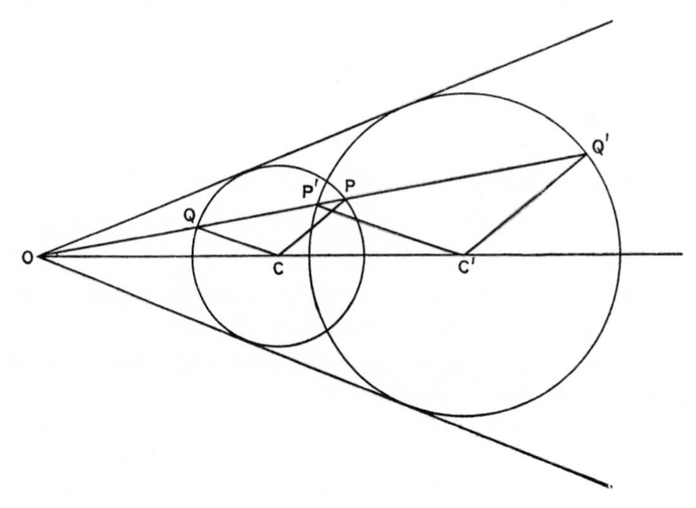

Fig. 1.

Let C be the centre of such a circle, ρ the length of the radius; PQ shall be two points of the circle collinear with O the centre of inversion, P' and Q' their inverses. We assume for the moment that our given circle does not pass through the centre of inversion. A line through $P' \parallel QC$ shall meet OC in C'. Now $(\overrightarrow{OP}) \times (\overrightarrow{OQ})$ and $(\overrightarrow{OP}) \times (\overrightarrow{OP'})$ have constant values, hence

$$\frac{(\overrightarrow{OP'})}{(\overrightarrow{OQ})} = \text{const.} = \frac{(\overrightarrow{OC'})}{(\overrightarrow{OC})} = \frac{(\overrightarrow{C'P'})}{(\overrightarrow{CQ})}.$$

The locus of P' is thus a circle of centre C' and radius

$$\rho' = \rho \frac{(\overrightarrow{OC'})}{(\overrightarrow{OC})}.$$

Theorem 16.] *The inverse of a circle not passing through the centre of inversion is a circle of the same sort.*

The reasoning above is inapplicable when the given circle passes through the centre of inversion. In this case Q coincides with O. Let R be diametrically opposed to this point, R' its inverse. Then since $\Delta\, OPR$ is similar to $\Delta\, OR'P'$

$$\measuredangle OR'P' = \frac{\pi}{2}.$$

Theorem 17.] *The inverse of a circle passing through the centre of inversion is a line not passing through that centre.*

Theorem 18.] *The inverse of a line not passing through the centre of inversion is a circle through that point.*

Theorem 19.] *Parallel lines invert into circles tangent to one another at the centre of inversion.*

Theorem 20.] *If two figures be mutually inverse with regard to a circle, their inverses in a second circle whose centre does not lie on the first are mutually inverse in the inverse of the first circle with regard to the second.*

Suppose, in fact, that P and P' are inverse in a circle C_1. Every circle through them will, by 13], cut C_1 at right angles. The inverses of these circles with regard to a second circle C_2 will cut the inverse of C_1 at right angles, and the two points common to them will be mutually inverse in that inverse of C_1.

Theorem 21.] *If a circle be inverted into a straight line, a pair of points inverse with regard to the circle will become a point and its reflection in the line.*

Theorem 22.] *If a curve be anallagmatic with regard to two circles, it is anallagmatic with regard to every circle that can be obtained by successively inverting one circle of inversion in another.**

We saw in the reasoning which led up to 16] that mutually inverse circles are similar figures radially situated. If two figures be similar we may clearly adjoin to the one and the other as many points as we please, getting more comprehensive figures which are still similar with the same ratio of similitude, and include the originals as parts of themselves. If there be a point which corresponds to itself in two such similar figures, it is called a *double* or *self-corresponding* point. When the figures are radially situated, corresponding points are collinear with the double point, and their distances therefrom bear to one another a ratio fixed in magnitude and sign. The double point is called the *centre of similitude*, and the fixed ratio the *ratio of similitude.*

Theorem 23.] *If two circles be mutually inverse, the centre of inversion is a centre of similitude for them while the ratio of similitude is numerically that of their radii. If this centre lie outside of one circle it is outside of the other, and is the point of intersection of their direct common tangents.*

Suppose, conversely, that we have two circles which are neither concentric nor of equal radius. Let us divide the segment bounded by their centres in two parts proportional to the radii, and find the harmonic conjugate of this point with regard to those centres (loosely called dividing the

* Cf. Möbius, *Collected Works*, vol. ii, p. 610; also Finsterbusch, *Die Geometrie ebener Kreissysteme*, Werdau, 1893, p. 68. For the conditions that an algebraic curve should be anallagmatic see Picquet, *Sur les courbes et surfaces anallagmatiques*, Comptes rendus de l'Association française pour l'avancement des sciences, Session of 1878 at Paris.

segment externally in that ratio). These points are the internal and external centres of similitude respectively, and are the points of intersection of such common tangents as the circles may have. Let O be one of these points and let a line through it meet one circle in PQ and the other in $Q'P'$. Then

$$\frac{(\overrightarrow{OP})}{(\overrightarrow{OQ'})} = \frac{(\overrightarrow{OQ})}{(\overrightarrow{OP'})} = \pm\frac{\rho}{\rho'},$$

$$(\overrightarrow{OP}) \times (\overrightarrow{OP'}) = (\overrightarrow{OQ}) \times (\overrightarrow{OQ'}) = k.$$

We easily find that k will be positive in the case of one point when the circles do not intersect, and in the case of both when they do. They are thus certainly mutually inverse in one circle of radius $\sqrt{k}$.

Theorem 24.] *Any two circles of different centres and unequal radii are mutually inverse in at least one circle whose centre is one of their centres of similitude.*

The circle or circles in which the given circles are mutually inverse are called their *circles of antisimilitude*; that on the segment bounded by the centres of similitude as diameter is their *circle of similitude.*

Theorem 25.] *If two circles of unequal radius lie outside of one another, their common tangents intersect at their centres of similitude and at four points of the circle whose diameter is the segment bounded by their centres.*

Let us define as a *tangential segment* of a point with regard to a circle a segment bounded by that point and the point of contact of a tangent to the circle which passes through the point. The common tangential segments of two circles will be segments lying on common tangents and bounded by the points of contact. Let us find the locus of a point whose tangential segments to two circles are proportional to their radii. The circles being $c_1 c_2$, their centres $C_1 C_2$, while the radii have the lengths $r_1 r_2$, if P be a point of the locus while t_i is the tangential segment from there to c_i

$$\frac{t_1^2}{t_2^2} = \frac{r_1^2}{r_2^2} = \frac{t_1^2 + r_1^2}{t_2^2 + r_2^2} = \frac{(PC_1)^2}{(PC_2)^2}, \quad \frac{(PC_1)}{(PC_2)} = \frac{r_1}{r_2}.$$

We have, thus, by a familiar theorem of elementary geometry,

Theorem 26.] *The locus of points whence the tangential segments to two non-concentric circles of unequal radius are proportional to the radii is so much of the circle of similitude as lies without the circles.*

Theorem 27.] *The distances from a point of the circle of similitude of two given circles to their centres are proportional to the respective radii.*

Theorem 28.] *The circle of similitude of two given circles includes all points whereat equal angles are determined by the pairs of tangents to the two.*

We find at once from Menelaus's theorem

Theorem 29.] *If three circles be given, no two concentric nor of equal radius, a line connecting a centre of similitude of one pair with a centre of similitude of a second pair will pass through a centre of similitude of the third pair.*

If two circles touch one another, their point of contact is a centre of similitude.

Theorem 30.] *If a circle touch two others of unequal radius, the line connecting the points of contact will pass through a centre of similitude of the two.*

Theorem 31.] *The centres of similitude determined by three circles whereof no two are concentric or of equal radius lie by threes on the sides of a complete quadrilateral, whose diagonal lines connect the pairs of centres of the circles.**

We find at once from the theorem of Ceva

Theorem 32.] *If three circles be given, no two being concentric or of equal radius, the lines connecting each centre with the centres of similitude of the other two are the side-lines of a complete quadrangle whose diagonal points are the centres of the given circles.*

Let us return to the point of view where we regarded the two circles as inverse in a circle of antisimilitude. If their radii be ρ and ρ', the radius of inversion

$$\frac{\rho}{\rho'}=\frac{(OP)}{(OQ')},\quad \rho'=\frac{(OQ')\,\rho}{(OP)}=\rho\,\frac{r^2}{(OP)\,(OQ)}\,. \tag{5}$$

* Chasles, *Traité de géométrie supérieure*, Paris, 1852, p. 539.

If we define as the *power* of a point with regard to a circle the product of its oriented distances to any two points of the circle collinear with it (the square of the tangential segment when the point lies without) we have

Theorem 33.] *The radius of the inverse of a given circle not through the centre of inversion is equal to the radius of the given circle multiplied by the square of the length of the radius of inversion, and divided by the absolute value of the power of the centre of inversion with regard to the given circle.*

Let us next follow the fate of the centre of the given circle. This point has the property that all straight lines through it cut the given circle at right angles. These lines invert into circles through the centre of inversion, whence by 15]

Theorem 34.] *The inverse of the centre of a circle which does not pass through the centre of inversion is the inverse of that centre in the inverse of the given circle. The inverse of the centre of a circle through the centre of inversion is the reflection of that centre in the line which is the inverse of the given circle.*

If two circles be given which do not intersect, either they lie outside of one another, or the one includes the other. In the first case we may easily find a point of the segment bounded by their centres which has the same positive power with regard to the two. This will be the centre of a circle cutting the two at right angles, and intersecting the line of centres in two points inverse in both circles. In the second case, if a point move off indefinitely on the line of centres from that intersection with the outer circle which is nearer to the centre of the inner one, its inverse in the outer circle will trace a segment which includes in itself the segment which is the locus of its inverse in the inner circle. In each case we can find a pair of points which are inverse in both circles. If we take either as centre of inversion we find:

Theorem 35.] *Any two circles which do not intersect may be inverted into concentric circles.*

§ 3. Mutually Tangent Circles.

The last theorem enables us to solve a problem very dear to Jakob Steiner.* Suppose that we have given two non-intersecting circles. What relations must exist between their radii and the distances of their centres in order that there should be a finite succession of circles all tangent to the given two, and each tangent to its two neighbours in the ring? Let us imagine that there are n circles in the ring, and that they make m complete circuits. These numbers will be invariant when we invert the given circles into two concentric circles of radii r_1 and r_2 respectively. If the common radius of circles of the new ring be r,

$$\tan \tfrac{1}{2}\left(\frac{2m\pi}{n}\right) = \frac{r}{\sqrt{(r+r_1)^2 - r^2}},$$

$$r + r_1 = \tfrac{1}{2}(r_1 + r_2),$$

$$\tan^2 \frac{m\pi}{n} = \frac{r^2}{r_1 r_2}.$$

Next, let any line through the common centre of the two meet them in $P_1'Q_1'$ and $P_2'Q_2'$.

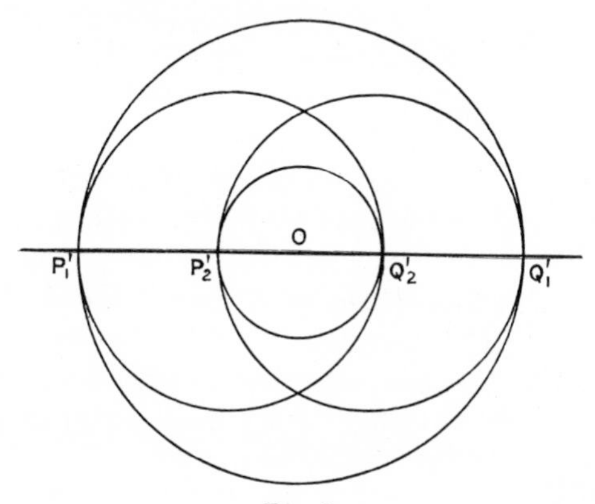

Fig. 2.

* See his Collected Works, vol. i, pp. 43 and 135. The resulting systems of circles are described by English writers as 'poristic'. See H. M. Taylor, 'Porism on the ring of circles touching two circles', *Messenger*

To be definite, we assume that the former pair includes the latter, and that $P_1'P_2'$ are on one side of the centre while $Q_1'Q_2'$ are on the other side. Then

$$\frac{r^2}{r_1 r_2} = \frac{(\overrightarrow{P_1'P_2'}) \times (\overrightarrow{Q_2'Q_1'})}{(\overrightarrow{P_1'Q_1'}) \times (\overrightarrow{P_2'Q_2'})}.$$

We saw, however, in equation (4) that the right-hand side of this is invariant for inversion, the centre of inversion being on the line of centres of the given circles. If, thus, this line meet the original circles in P_1Q_1 and P_2Q_2

$$\tan^2\frac{m\pi}{n} = \frac{(\overrightarrow{P_1P_2}) \times (\overrightarrow{Q_2Q_1})}{(\overrightarrow{P_1Q_1}) \times (\overrightarrow{P_2Q_2})}.$$

This equation has a simple geometric meaning. Reverting to the concentric case, let us construct circles on $(P_1'Q_2')$ and $(P_2'Q_1')$ as diameters. The distance from the common centre to their centres will be $\frac{1}{2}(r_2-r_1)$, their common radius $\frac{1}{2}(r_2+r_1)$. To find the angles at which they intersect, we have

$$\cos\theta = \frac{-(r_2-r_1)^2+\frac{1}{2}(r_2+r_1)^2}{\frac{1}{2}(r_2+r_1)^2} = \frac{\frac{1}{2}(r_2+r_1)^2-4r^2}{\frac{1}{2}(r_2+r_1)^2},$$

$$\tan^2\tfrac{1}{2}\theta = \frac{r^2}{r_1r_2} = \tan^2\frac{m\pi}{n}. \qquad (6)$$

We thus get, recalling 9],

Theorem 36.] *Let two non-intersecting circles be given, and let the line of centres meet the first in P_1Q_1 and the second in P_2Q_2; the points P_1Q_2 separating the points P_2Q_1. A necessary and sufficient condition that it should be possible to construct a finite succession of circles tangent to the given ones and successively tangent to one another is that the circles constructed on the segments (P_1Q_2) and (P_2Q_1) as diameters*

of Mathematics, vol. vii, 1878, and his brother W. W. Taylor, 'On the Ring of Circles touching two Circles', ibid. See also Lachlan, 'On Poristic Systems of Circles', ibid., vol. xvi, 1887. Our present treatment follows Vahlen, 'Ueber Steinersche Kugelketten', *Zeitschrift für Mathematik und Physik*, vol. xli, 1896. For an interesting generalization see Emch, 'An Application of Elliptic Functions', *Annals of Mathematics*, Series 2, vol. ii, 1901.

should intersect at an angle commensurable with π. The denominator of the measure of such an angle when expressed in terms of 2π and reduced to its lowest terms will give the number of circles in the succession and the numerator the number of complete circuits formed by them. If one such circuit exist, there will be an infinite number of them, one circle being perfectly arbitrary except for the types of contact with the given circles. The points of contact of successive circles in all of these circuits lie on one circle.

We may pursue this subject further. If we take as a circle of inversion any circle orthogonal to the two given ones, they are, by 24], anallagmatic therein, the line of centres becomes a circle orthogonal to the two given circles, the circles on (P_1Q_2) and (P_2Q_1) as diameters, become circles tangent to the original circles, and orthogonal to a circle orthogonal to them. We may thus state our condition in slightly more general terms by means of the angle of these last two circles. Suppose, then, that we have a ring of circles, and that two circles of the ring touch the given circles at four points of one same circle orthogonal to the original ones. By two successive inversions we may go back to the concentric case where, in our previous notations two circles of the ring have (P_1P_2) and (Q_1Q_2) as diameters. The concentric circles will be two out of a ring tangent to the circles on (P_1P_2) and (Q_1Q_2) and to one another in turn, and the circles on (P_1Q_2) and (P_2Q_1) as diameters play the same rôle with regard to both rings. If, then, m_1n_1 be the numbers for the new ring, we have

$$2\pi\frac{m_1}{n_1} = 2\pi\frac{m}{n} \text{ or else } 2\pi\frac{m_1}{n_1} = \pi - 2\pi\frac{m}{n}\cdot$$

The decision between these two possibilities requires delicate handling.* Let us first remark that, $\frac{m}{n}$ being given, these two equations give different values for $\frac{m_1}{n_1}$ except in the case where

* Vahlen, loc. cit., overlooks the necessity for making both assumptions.

$\theta = \frac{\pi}{2}\cdot$ As θ changes continuously the correct value for $\frac{m_1}{n_1}$ cannot leap from being a root of one equation to being the root of the other, except, perhaps, when θ passes through the value $\frac{\pi}{2}$. First take $r_1 = 0$,

$$m = 1, \quad n = 2, \quad \theta = \pi, \quad m_1 = 1, \quad n_1 = \infty,$$

since the circles on (P_1P_2) and (Q_1Q_2) can be simultaneously inverted into parallel lines. Here, surely,

$$\frac{m}{n} + \frac{m_1}{n_1} = \frac{1}{2},$$

and this will hold for $\theta > \frac{\pi}{2}\cdot$ On the other hand, if we take $r_1 = r_2$,

$$m = 1, \quad n = \infty.$$

To find $\frac{m_1}{n_1}$ notice that if two extremely small circles lie without one another and be inverted into concentric circles, the one becomes tiny, and $m_1 = 1$, $n_1 = 2$.

Theorem 37.] *Given two non-intersecting circles which possess the property that a ring of n circles may be constructed all tangent to them and successively tangent to one another making m complete circuits, and if two circles of the ring touch the original ones at points on one circle orthogonal to these two, then the original circles are members of a ring of n_1 circles making m_1 complete circuits, all tangent to the two of the first ring, where*

$$\frac{m}{n} + \frac{m_1}{n_1} = \frac{1}{2}\cdot \tag{7}$$

This theorem so far astonished Steiner that he called it one of the most remarkable in all geometry.*

We know that two mutually tangent circles can be inverted into parallel lines. Let us do so for two internally tangent circles c, $\bar{c}$. The circles tangent to these two lines will all have the same radius ρ'; let c_0' be that circle of the system whose centre lies on the perpendicular on the lines from the

* Collected Works, vol. i, p. 136.

centre of inversion, the circles of a system of successively tangent circles, which touch the parallel lines shall be $c_0' c_1' c_2' \ldots c_k'$, their centres $C_0' C_1' \ldots C_k'$. Inverting back we get our original circles with the system of circles $c_0 c_1 \ldots c_k$ tangent to them and to one another in succession. The centre

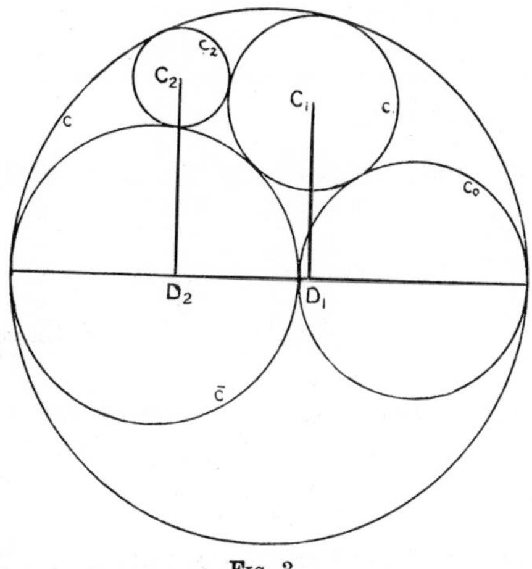

FIG. 3.

of c_n shall be C_n, the perpendicular thence to the line of centres of the original circles shall meet the latter in D_n. Since C_n and C_n' are collinear with O the centre of inversion

$$\frac{(C_n D_n)}{(OC_n)} = \frac{2n\rho'}{(OC_n')}.$$

But since O is a centre of similitude for C_n and C_n'

$$\rho' = \rho_n \frac{(OC_n)'}{(OC_n)}, \quad C_n D_n = 2n\rho_n. \tag{8}$$

Theorem 38.] *Given two circles c_0 and $\bar{c}$ externally tangent to one another and a third circle c having as diameter the sum of their collinear diameters. Then if a series of circles $c_0 c_1 \ldots c_k$ be all drawn tangent to $\bar{c}$ and c, and successively to one another, the distance from the centre of c_n to the line of centres of $\bar{c}$, c' is n times the diameter of c_n.*

This theorem is sometimes called the 'Ancient Theorem' of Pappus. Steiner deduces a number of rather dull corrollaries therefrom.

The figure bounded by the halves of $c, \bar{c}, c_0$ which lie on one side of the line of centres was first studied by Archimedes and named by him 'The Shoemaker's knife'.* Let A be the point of contact of c and $\bar{c}$, B that of c and c_0, while D is the point of contact of $\bar{c}$ and c_0. A perpendicular to AB at D shall meet c again in E. The following theorems are then easily proved.†

Theorem 39.] *The area of the knife is equal to that of the circle on (DE) as diameter.*

Theorem 40.] *The perimeter of the knife is equal to the circumference of c.*

Theorem 41.] *The point A has the same power with regard to all circles which touch c internally and DE on the same side as c_0.*

Theorem 42.] *The two circles which touch c internally and DE on opposite sides while one is externally tangent to c_0 and the other to $\bar{c}$ are equal.*

Theorem 43.] *The common tangent to the first of these and to c_0 passes through A.*

Theorem 44.] *The circle on (DE) as diameter passes through the points where c_0 and $\bar{c}$ touch a common tangent, while its centre is the intersection of this tangent with DE.*

We next pass to an invariant of two circles. Let them be $c_1 c_2$ with centres $C_1 C_2$ and radii $\rho_1 \rho_2$. The centre and radius of inversion being O and r,

$$(OC_1') = (OC_1)\frac{\rho_1'}{\rho_1} \quad (OC_2') = (OC_2)\frac{\rho_2'}{\rho_2},$$

$$(C_1'C_2')^2 = OC_1'^2 + OC_2'^2 - 2(OC_1')(OC_2')\cos\measuredangle C_1'OC_2$$

$$= (OC_1^2)\frac{\rho_1'^2}{\rho_1^2} + (OC_2^2)\frac{\rho_2'^2}{\rho_2^2} \mp \frac{\rho_1'\rho_2'}{\rho_1\rho_2}[(OC_1)^2 + (OC_2^2) - (C_1C_2)^2];$$

* Cf. Heath, *The Works of Archimedes*, Cambridge, 1897, pp. 304 ff.

† For an account of the authorship of the theorems concerning the knife, see Simon, *Ueber die Entwickelung der Elementar-Geometrie im XIXten Jahrhundert*, Leipzig, 1906, pp. 87, 88.

from these and formula (5) we find

$$\frac{(C_1'C_2')^2-(\rho_2'-\rho_1')^2}{4\rho_2'\rho_1'} = \pm\frac{(C_1C_2)^2-(\rho_2-\rho_1)^2}{4\rho_2\rho_1};$$

$$\frac{(C_1'C_2')^2-(\rho_2'+\rho')^2}{4\rho_2'\rho_1'} = \pm\frac{(C_1C_2)^2-(\rho_1+\rho_2)^2}{4\rho_2\rho_1}.$$

The numerators of the left-hand sides of these equations are the squares of the direct and transverse common tangential segments, when these exist. Suppose that we have four mutually external circles c_1, c_2, c_3, c_4 tangent to a fifth.* Either all are on one side thereof, or two on one and two on the other, or three on one and one on the other. We may invert them into four mutually external circles c_1', c_2', c_3', c_4' tangent to a line. Let them touch it at points P_1', P_2', P_3', P_4', which will be connected by the identity

$$(\overrightarrow{P_1'P_2'})(\overrightarrow{P_3'P_4'})+(\overrightarrow{P_1'P_3'})(\overrightarrow{P_4'P_2'})+(\overrightarrow{P_1'P_4'})(\overrightarrow{P_2'P_3'}) = 0.$$

If t_{xy} indicate a common tangential segment of c_x and c_y, we may write this

$$t_{12}'t_{34}' \pm t_{13}'t_{42}' \pm t_{14}'t_{23}' = 0.$$

Here t_{ij}' must indicate a direct common tangential segment if c_i' and c_j' touch the line on the same side, otherwise a transverse one. Dividing through by the square root of the product of the diameters we get a form invariant for inversion, hence dropping the primes and multiplying the diameters out again, we get Casey's condition for four circles tangent to a fifth.†

Theorem 45.] *Four mutually external circles tangent to a fifth are connected by a relation*

$$t_{12}t_{34} \pm t_{13}t_{42} \pm t_{14}t_{23} = 0. \tag{9}$$

Here all the t_{ij}'s denote common direct tangential segments, or those connecting two pairs with no common member denote direct tangents and the other four transverse, or those which lack one subscript denote direct, and those which include it transverse tangential segments.†

* In problems dealing with four circles tangent to a fifth, a straight line and a point must be treated as limiting forms of circles.

† See his greatly overrated *Sequel to Euclid*, London, 1881, p. 101. The ingenious writer makes two characteristic mistakes. He assumes that in proving the theorem he has also proved the converse. Secondly, he omits

Theorem 46.] *If a convex quadrilateral be inscribed in a circle, the sum of the products of the opposite sides is equal to the product of the diagonals.*

This is Ptolemy's famous theorem. Let us proceed to the converse of 45]. We assume that we have four mutually external circles connected by that relation. We shall call them c_1, c_2, c_3, c_4 and suppose that ρ_1 is the smallest radius. We shrink the radius of c_1 by ρ_1 and shrink by that same amount the radius of each of the given circles whose common tangential segment with c_1 is direct, but increase the radius by ρ_1 if the tangential segment be transverse. We thus get four circles c_1', c_2', c_3', c_4', whereof c_1' is a point-circle C_1' connected by

$$t_{12}'t_{34}' \pm t_{13}'t_{42}' \pm t_{14}'t_{23}' = 0.$$

These circles are still mutually external. Let us next invert with C_1' as a centre, we get three new circles c_2'', c_3'', c_4'',

$$t_{34}'' = t_{34}' \sqrt{\frac{\rho_3''\rho_4''}{\rho_3'\rho_4'}}, \quad t_{12}'' = r\sqrt{\frac{\rho_2'}{\rho_1'}},$$

$$t_{34}'' \pm t_{42}'' \pm t_{23}'' = 0.$$

Let us show that these three circles, which are also external to one another, will touch a line. Once more shrink the smallest circle until it becomes a point shrinking or increasing the radii of the other two as before. We have a point so related to two mutually external circles that the sum of its tangential segments with them is equal to a common tangential segment of theirs. If the point lie on a common tangent to the two circles such a condition will be fulfilled, and if it move off on a circle concentric with the one, the condition will be unfulfilled until it fall again on the like common tangent. Hence the point lies on a common tangent to the two circles; hence c_2'', c_3'', c_4'' touch a line, c_2', c_3', c_4' touch a circle through C_1', and c_1, c_2, c_3, c_4 touch one circle.

to require his circles to be mutually external. But in that case it is easy to find four circles tangent to a fifth whereof one surrounds the three others and has no common tangential segments with them, in the real domain.

Theorem 47.] *If there exist among the common tangential segments of four mutually external circles an equation of the type (9) with the same requirements as to direct and transverse tangents as there obtained, then these four circles are tangent to a fifth.**

Theorem 48.] *If the sum of the products of the opposite sides of a convex quadrilateral be equal to the product of the diagonals, the vertices are concyclic.*

As a second application of our formula (9) let us prove the justly celebrated theorem of Feuerbach.†

We start with a triangle with the standard notation explained on p. 21. Construct the three altitude lines, and let A_iH meet the circumscribed circle again at B_i. We have then

$$\measuredangle B_iA_jA_k = \measuredangle B_iA_iA_k = \frac{\pi}{2} - \measuredangle A_k = \measuredangle B_jA_jA_k.$$

This shows that Ha_i is mid-way between H and B_i. If we take H as a centre of similitude and a ratio $\frac{1}{2}$, the given triangle becomes that whose vertices are half-way from H to the given vertices, and the circumscribed circle is transformed into the circle through these three half-way points, and also through the feet of the altitudes. These six points are thus concyclic. Again, if we take the ΔHA_jA_k the orthocentre is A_i; the feet of the altitudes are the same points as before, the points M_j, M_k are half-way from the new orthocentre to two of the vertices. We thus get the first part of our theorem, namely, the feet of the altitudes of a triangle, the middle points of the sides, and the points half-way from the orthocentre to the vertices lie on one circle. We next construct the escribed circle c_i tangent to (A_jA_k) and to the prolonga-

* This proof is substantially taken from Lachlan, *Treatise on Pure Geometry*, London, 1893, pp. 245 ff. See also Allardice, 'Note on Four Circles Tangent to a Fifth', *Proceedings Edinburgh Mathematical Society*, vol. xix, 1901. Neither writer takes the pains to require the circles to be mutually external. It might thus happen that c_1 surrounded c_2 and the proof would break down.

† First published in 1822. The number of proofs in existence is almost transfinite, a recent writer adding nine. Sawavama, 'Nouvelles démonstrations d'un théorème relatif au cercle de neuf points', *L'Enseignement mathématique*, vol. xiii, 1911.

tions of (A_iA_j) and (A_iA_k) beyond A_j and A_k respectively. Let x be the tangential segment from A_j to this circle. The equality of the two tangential segments to this circle from A_i gives

$$a_k+x = a_j+a_i-x,$$

$$x = s-a_k, \quad a_i-x = s-a_j.$$

Let us take this as our circle c_4, while the middle points of the sides shall be the point-circles c_1, c_2, c_3,

$$t_{ij} = \tfrac{1}{2}a_k, \quad t_{ik} = \tfrac{1}{2}a_j, \quad t_{jk} = \tfrac{1}{2}a_i,$$

$$t_{i4} = \pm\tfrac{1}{2}(a_k-a_j), \quad t_{k4} = \tfrac{1}{2}(a_i+a_j), \quad t_{j4} = \tfrac{1}{2}(a_i+a_k),$$

$$t_{ij}t_{k4}-t_{ik}t_{j4}\mp t_{i4}t_{jk} = 0.$$

A similar relation will be found connecting the new circle with the inscribed circle; we thus get the theorem in its entirety.

Theorem 49.] *The middle points of the sides of a triangle, the feet of the altitudes and the points half-way from the orthocentre to the vertices lie on a circle which is tangent to the inscribed and the three escribed circles.*

This circle is, for obvious reasons, called the *nine-point*

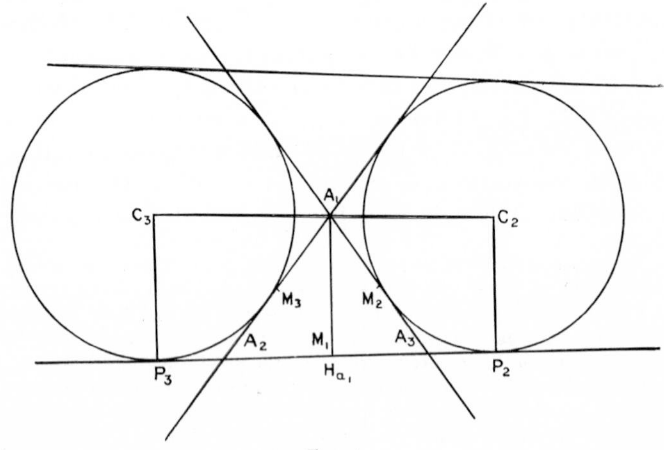

Fig. 4.

circle. Let us give another proof that it touches the inscribed and escribed circles.*

Let the circles c_2 and c_3 be escribed to the given triangle and touch the line A_2A_3 in the points P_2 and P_3 respectively. Let S be the point of concurrence of c_2c_3 (the line of centres), A_2A_3, and the fourth common tangent to c_2 and c_3. A_1 and S are thus the centres of similitude of c_2 and c_3. Moreover, if we recall the original definition of centres of similitude, we see that A_1 and S are harmonically separated by C_2 and C_3, or Ha_1 and S are harmonically separated by P_2 and P_3. The tangent at M_1 to the nine-point circle makes with M_1M_3, and so with A_1A_3, an angle equal $\measuredangle A_2$ and so is parallel to the fourth common tangent. The nine-point circle is thus the inverse of the fourth common tangent in a circle whose centre is M_1 and radius is equal to $(M_1P_2) = (M_1P_3)$. The nine-point circle must thus touch the escribed circles c_2, c_3, which are anallagmatic in this last circle. By similar means we show that it touches the inscribed circle also.

If a triangle have an obtuse angle, the orthocentre lies without it. The feet of the altitudes lie in pairs on the three circles on the sides of the given triangle as diameters. The orthocentre has the same positive power with regard to these three, so that the product of the distances from the orthocentre to each vertex and the foot of the corresponding altitude is a constant positive number.

Theorem 50.] *The circumscribed and nine-point circles of an obtuse-angled triangle are mutually inverse in a circle whose centre is the orthocentre.*

It is to be noted that this is the only circle with regard to which the given triangle is self-conjugate in the sense of modern geometry.

Feuerbach's theorem may be extended in a number of ways. The second part states that the inscribed and escribed circles of a triangle touch another circle. By inversion this

* Fontené, 'Sur le Théorème de Feuerbach', *Nouvelles Annales de Mathématiques*, Series 4, vol. viii, 1907. This proof possesses the advantage over the other of showing where the points of contact are.

will hold if we replace the triangle by a curvilinear one formed by concurrent circles. Let us try to remove the restriction that the three original circles should be concurrent. We start with three intersecting circles c_1, c_2, c_3, the intersections of $c_j c_k$ being points $A_i A_i'$. Eight circular triangles are thus formed whose angles are connected in simple ways. It is intuitively evident that a circle may be inscribed in each of these triangles. In particular let us take the triangle $A_1 A_2 A_3$ which we assume to be convex, and the three triangles A_i', A_j, A_k which we shall call *associated* with it. The four inscribed circles shall be c_0', c_1', c_2', c_3'. If we write t_{ij}' to indicate a direct common tangential segment and $\overline{t_{ij}}'$ a transverse one, we have three equations of the type

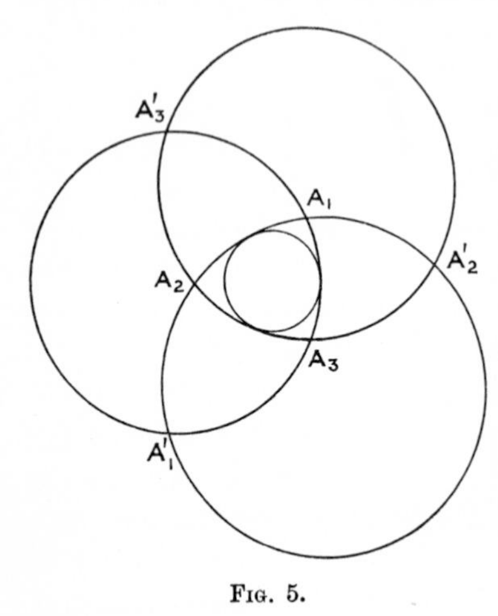

FIG. 5.

$$\overline{t_{0i}}' t_{jk}' \pm t_{0j}' \overline{t_{ki}}' \pm t_{0k}' \overline{t_{ij}}' = 0.$$

Let us determine the signs more specifically. In the arcual triangle A_1, A_2, A_3 two of our circles c_0', c_i' touch the circle of each side between the vertices, but with opposite contacts. Suppose, to fix our ideas, that in making the circuit of the triangle we meet the vertices and points of contact with the tangent circles in the following order

$$A_1 c_3' c_0' \ A_2 c_0' c_1' \ A_3 c_0' c_3' :$$

We have the following orders on our original three circles:

on c_1,	$c_3' \ A_2 c_0' c_1' \ A_3 c_2'$,
on c_2,	$c_1' \ A_3 c_0' c_2' \ A_1 c_3'$,
on c_3,	$c_2' \ A_1 c_3' c_0' \ A_2 c_1'$.

These will yield the following equations:

$$\begin{aligned} t_{02}'\overline{t_{31}}' &= \overline{t_{01}}'t_{23}' + t_{03}'\overline{t_{12}}', \\ t_{03}'\overline{t_{12}}' &= t_{01}'\overline{t_{23}}' + \overline{t_{02}}'t_{31}', \\ t_{02}'\overline{t_{31}}' &= t_{01}'\overline{t_{23}}' + \overline{t_{03}}'t_{12}'. \end{aligned}$$

Hence

$$\overline{t_{03}}'t_{12}' = \overline{t_{01}}'t_{23}' + \overline{t_{02}}'t_{31}'.$$

We thus get Hart's theorem.*

Theorem 51.] *The inscribed circle of a convex circular triangle and those of three associated triangles are touched by a circle which has contact of one sort with the first, and of the opposite sort with the other three.*

This new circle is called a *Hart circle* of the first three. It may coincide with one of the four inscribed circles. It will exist even when the given triangle is not convex; our proof is not, however, necessarily valid in that case, for the four may not lie external to one another. These delicate considerations are usually ignored in the geometrical treatment of this subject.

Let the Hart circle be called c_4. The following will give the system of contacts.

c_0'	touches	c_1, c_2, c_3, c_4 internally.		
c_1'	,,	c_2, c_3	internally c_1, c_4	externally.
c_2'	,,	c_3, c_1	,, c_2, c_4	,,
c_3'	,,	c_1, c_2	,, c_3, c_4	,,

The essential thing to notice is that c_i has an opposite sort of contact with c_i' from what it has with c_0', c_j', c_k'.

Theorem 52.] *If four circles be given whereof one is the Hart circle for a convex circular triangle formed by the other three, then each of the four is a Hart circle for the remainder.*†

* 'On the extension of Terquem's Theorem', *Quarterly Journal of Mathematics*, vol. iv, 1860. For a much simpler proof see p. 165, foot-note.

† For an elaborate treatment of this and similar theorems see an unusually badly written article by Orr, 'The Contact Relations of Certain Systems of Circles', *Transactions Cambridge Philosophical Society*, vol. xvi, 1898.

§ Circles related to a Triangle.

Suppose that two circles are so related that a triangle can be inscribed to the one and circumscribed to the other. Their radii shall be r and ρ respectively, while the distance of their centres O and O' is d. Let OO' meet the circumscribed circle

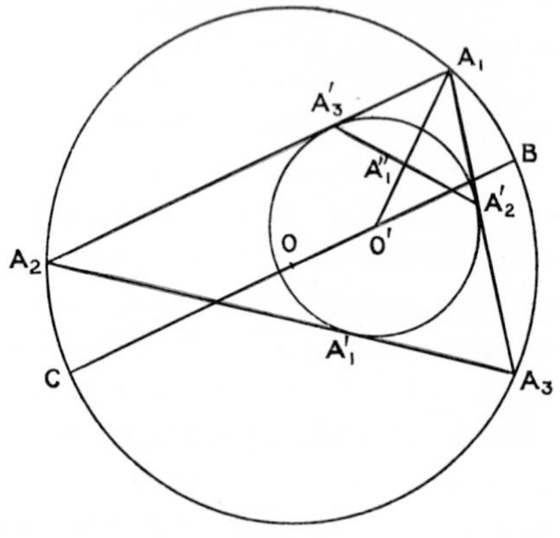

Fig. 6.

in BC. Let A_jA_k touch the inscribed circle in A_i', while $O'A_i$ meets $A_j'A_k'$ in A_i'' the middle point of $(A_j'A_k')$ and the inverse of A_i in the inscribed circle.

Theorem 53.] *If two circles be so related that a triangle inscribed in the one is circumscribed to the other, then the former is the inverse in the latter of the nine-point circle of the triangle whose vertices are the points of contact.**

The nine-point circle is circumscribed to a similar triangle of one-half the size of the original, so that its radius is one-half that of the circumscribed circle. If the inverses of B and C be B'' and C'' respectively,

$$(O'B'') = \frac{\rho^2}{(O'B)} = \frac{\rho^2}{r-d}, \quad (O'C'') = \frac{\rho^2}{r+d},$$

$$(B''C'') = \rho = \frac{\rho^2}{r+d} + \frac{\rho^2}{r-d}.$$

* The treatment of this and the four following theorems is taken direct from Casey, loc. cit., Book VI.

Theorem 54.] *The radii of the circles circumscribed and inscribed to a triangle are connected by the equation*

$$\frac{1}{r+d} + \frac{1}{r-d} = \frac{1}{\rho}, \tag{10}$$

where d is the distance of their centres.

This necessary condition is also sufficient if r be greater than ρ, for the inverse of the nine-point circle of the triangle whose vertices are the points of contact with the smaller circle of a triangle circumscribed thereto and having two vertices in the larger circle will be that larger circle which thus goes through the third vertex. Let us pursue our inquiry further and find a necessary and sufficient condition that it should be possible to inscribe a quadrilateral to one circle which is circumscribed to the other. We need two preliminary theorems.

Theorem 55.] *If a variable chord of a circle subtend a right angle at a fixed point not on the circle, the locus of the intersection of the tangents at its extremities is a circle.*

This locus is, in fact, the inverse of that of the middle points of the chord. The sum of the squares of the distances of this middle point from the fixed point and from the centre of the circle is easily seen to be constant, so that it traces a circle about the point half way between the centre of the given circle and the given point.

Suppose, now, that we have indeed a quadrilateral inscribed in one circle and circumscribed to the other. The sum of the opposite angles is π, double the angle formed by the lines connecting opposite points of contact.

Theorem 56.] *If a quadrilateral be inscribed in one circle and circumscribed to another, the lines connecting the points of contact of opposite sides are mutually perpendicular.*

Theorem 57.] *If two circles be so related that a triangle or quadrilateral may be inscribed in the one and circumscribed to the other, then an infinite number of such triangles or quadrilaterals may be found, one vertex being taken at random on the other circle.*

Let us take this random vertex on the line of centres: call it A_1; the opposite vertex A_3 will clearly be on this line also. The pairs of sides which do not meet in these vertices and are not opposite to one another are mutually perpendicular, as are the radii of the inner circle to their points of contact. If thus A_1' and A_3' be the intersections of the line of centres

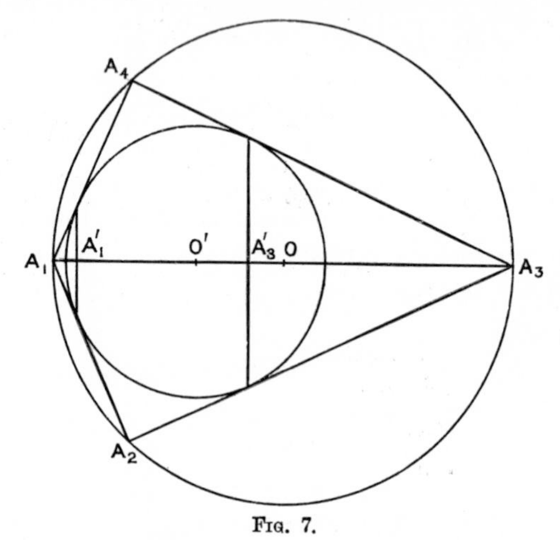

FIG. 7.

with the chords of contact to the inner circle of the tangents from A_1 and A_3, i.e. the inverses of these points

$$(OA_1')^2 + (OA_3')^2 = \rho^2,$$

$$\frac{1}{(r-d)^2} + \frac{1}{(r+d)^2} = \frac{1}{\rho^2}.$$

As before, we have no difficulty in showing that this necessary condition is also sufficient, hence

Theorem 58.] *If r and ρ be the radii of two circles, the former surrounding the latter, while d is the distance of their centres, a necessary and sufficient condition that it should be possible to construct a quadrilateral inscribed in the one and circumscribed to the other is that* *

$$\frac{1}{(r+d)^2} + \frac{1}{(r-d)^2} = \frac{1}{\rho^2}. \tag{11}$$

* There is a considerable body of literature connected with equations 10 and 11; see Simon, loc. cit., pp. 108, 109. They are originally due to Euler.

Continuing with the inscribed quadrilateral of vertices A_1, A_2, A_3, A_4, let P be any point of the circumscribed circle.

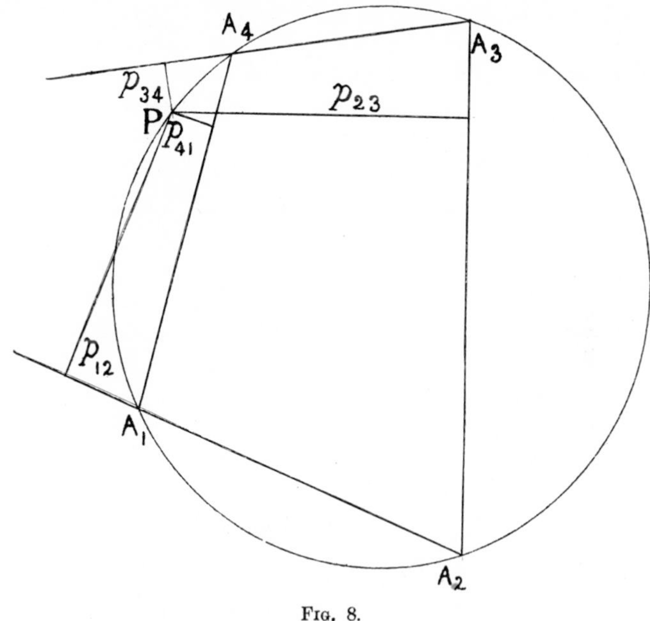

Fig. 8.

If p_{ij} indicate the distance from P to the side-line A_iA_j, we have

$$\frac{p_{12}p_{34}}{p_{23}p_{41}} = \frac{\sin \measuredangle PA_1A_2}{\sin \measuredangle PA_1A_4} \cdot \frac{\sin \measuredangle PA_3A_4}{\sin \measuredangle PA_3A_2} = 1.$$

Theorem 59.] *The product of the distances from a point on a circle to one pair of opposite side-lines of an inscribed quadrilateral is equal to the product of the distances to the other pair of side-lines, and to the product of the distances to the diagonal lines.*

If a polygon of an even number of sides be inscribed in a circle, it may be divided into one or two less sides and an inscribed quadrilateral. We thus get by mathematical induction

Theorem 60.] *If a polygon of an even number of sides be inscribed in a circle, the product of the distances of any point of the circle to the even numbered side-lines is equal to the product of its distances to the odd numbered ones.*

Theorem 61.] *If a polygon be inscribed in a circle and tangents be drawn at all of its vertices, the product of the distances of any point of the circle from these tangents is equal to the product of its distances from the side-lines.*

The circle circumscribed to a triangle is, on the whole,

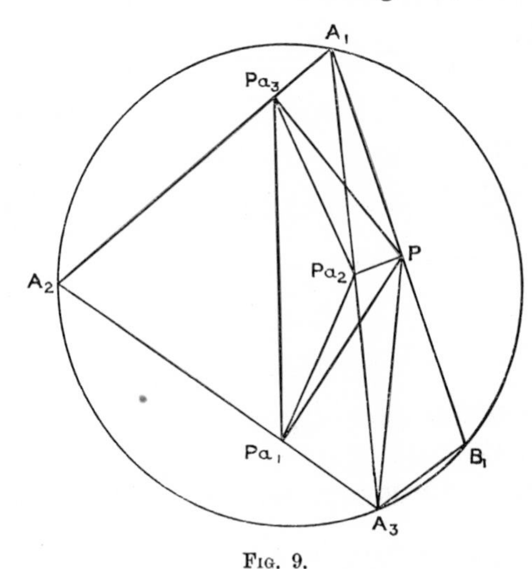

FIG. 9.

more interesting than the inscribed one. Let us take a triangle in standard notation and consider the *pedal* triangle $Pa_1 Pa_2 Pa_3$ of a point P. Let PA_1 meet the circumscribed circle again in B_1. To fix our ideas we shall take P outside the triangle, near A_3,

$$\measuredangle Pa_1 Pa_2 Pa_3 = \measuredangle Pa_1 P Pa_3 + \measuredangle Pa_2 Pa_1 P + \measuredangle Pa_2 Pa_3 P.$$

Since, however, the quadrilateral $P Pa_1 Pa_2 A_3$ is cyclic, i.e. inscriptible in a circle,

$$\begin{aligned}\measuredangle Pa_1 Pa_2 Pa_3 &= \pi - \measuredangle A_2 + \measuredangle A_3 A_1 P + \measuredangle A_1 A_3 P,\\ &= \pi - \measuredangle PA_3 B_1,\end{aligned}$$

$$\Delta Pa_1Pa_2Pa_3 = \tfrac{1}{2}(Pa_1Pa_2)(Pa_3Pa_2)\sin\measuredangle PA_3B_1,$$

$$(Pa_jPa_k) = (PA_i)\sin\measuredangle A_i;\ (PA_3)\sin\measuredangle PA_3B_1 = (PB_1)\sin\measuredangle A_2$$

$$\Delta Pa_1Pa_2Pa_3 = \tfrac{1}{2}(PA_1)(PB_1)\prod_{i=1}^{i=3}\sin\measuredangle A_i$$

$$= \pm\tfrac{1}{2}[r^2-(OP)^2]\prod_{i=1}^{i=3}\sin\measuredangle A_i.$$

Theorem 62.] *The locus of the points whose pedal triangles with regard to a given triangle have a given area is a circle concentric with the circumscribed circle.*

Theorem 63.] *The locus of the points so situated that the feet of the perpendiculars from them to the side-lines of a triangle are collinear is the circumscribed circle to the given triangle.*

This line is called the *pedal* or *Simson* line of the given point.

Let the value of $\measuredangle \overrightarrow{A_kA_iP}$ be α_i, while

$$\measuredangle \overrightarrow{A_jA_iP} = \alpha_i',$$

$$\prod_{i=1}^{i=3}\frac{\sin\alpha_i}{\sin\alpha_i'} = -1. \tag{12}$$

Conversely, if three lines be drawn through the three vertices of a triangle in such a way that this equation is satisfied, these lines will be concurrent or parallel. If, then, starting with P we take the reflection of A_iP in the bisector of $\measuredangle A_iA_jA_k$, we get three other lines concurrent in a point P' called the *isogonal conjugate* of P with regard to the given triangle, or else three parallel lines.

Theorem 64.] *Every point not on the circumscribed circle to a triangle has a single definite isogonal conjugate. The relation between the two is symmetrical.*

Let us consider the *pedal circles* of two isogonally conjugate points, i.e. the circumscribed circles of their pedal triangles.

$$\frac{\overrightarrow{(A_jPa_i')}}{\overrightarrow{(A_jPa_k')}} = \frac{\cos\alpha_j}{\cos\alpha_j'} = \frac{\overrightarrow{(A_jPa_k)}}{\overrightarrow{(A_jPa_i)}},$$

$$\overrightarrow{(A_jPa_i)}\times\overrightarrow{(A_jPa_i')} = \overrightarrow{(A_jPa_k)}\times\overrightarrow{(A_jPa_k')}.$$

The points Pa_i, Pa_i', Pa_k, Pa_k' are thus concyclic. The six points Pa_i, Pa_i' could not lie by fours on three circles, for the common chords of these circles would be the side-lines of

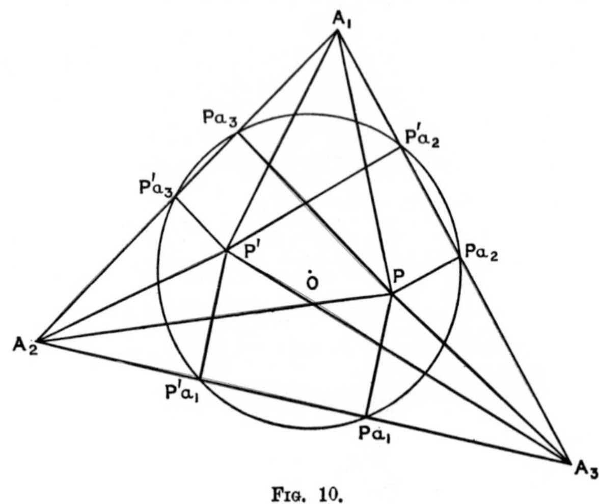

FIG. 10.

the triangle, instead of being concurrent. Hence the six points are concyclic. We thus get a generalization of the first part of Feuerbach's theorem.

Theorem 65.] *Two isogonally conjugate points have the same pedal circle.*

Theorem 66.] *If from the foot of each altitude of a triangle a perpendicular be dropped on the remaining side-lines, the six points so determined are concyclic.*

A generalization of 65] is found as follows. It is not necessary in the above proof to assume $\measuredangle PPa_j A_k = \frac{\pi}{2}$; we merely need $\measuredangle PPa_i A_j = \measuredangle P'Pa_i' A_k = \theta$.

The $\Delta A_j P Pa_i$ is thus similar to $\Delta A_j P' Pa_k'$,

$$(\overrightarrow{A_j Pa_i}) \times (\overrightarrow{A_j Pa_i'}) = (\overrightarrow{A_j Pa_k}) \times (\overrightarrow{A_j Pa'_k}).$$

Hence Pa_i, Pa_i', Pa_k, Pa_k' are concyclic, and, as before,

Theorem 67.] *If through a chosen point not on the circumscribed circle of a triangle three lines be drawn each making a fixed angle with one side-line of the triangle so oriented as to trace the whole circuit in one sense, and if through the isogonally conjugate point three others be drawn making the supplementary angles with their oriented side-lines, the six points where the lines of the two concurrent triads meet the corresponding side-lines are concyclic.**

Let us see where the pedal circle of a point P meets the nine-point circle. The intersection of the lines $Pa_j Pa_k$ and $M_j M_k$ shall be $\bar{A}_i$. We intend to show that the three lines $A_i Pa_i$ are concurrent in a point L of the nine-point circle.

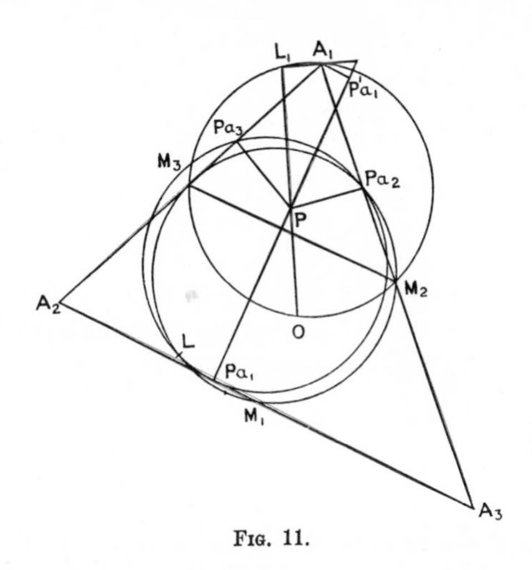

Fig. 11.

Construct the circle $A_i M_j M_k$. It will contain O which, parenthetically, is the orthocentre of the $\Delta M_1 M_2 M_3$, and is diametrically opposite to A_i. Let PO meet this circle again in L_i. The points $L_i Pa_j Pa_k$ are the vertices of three right triangles on $(A_i P)$ as common hypotenuse, and so are concyclic with A_i and P. This circle will also contain Pa_i', the reflection of Pa_i in $M_j M_k$. Moreover, the points $A_i L_i Pa_i'$

* Cf. Barrow, 'A Theorem about Isogonal Conjugates', *American Mathematical Monthly*, vol. xx, 1913, p. 25.

are collinear. For L_i lies on the circles $A_iM_jM_k$, $A_iPa_jPa_k$; hence the feet of the perpendiculars thence to the four lines A_iM_j, A_iM_k, M_jM_k, Pa_jPa_k are collinear by 63], so that L_i lies on the circle $\bar{A}_iM_kPa_k$.

$$\angle \bar{A}_iL_iPa_k = \angle A_iM_kPa_k.$$

The pentagon $A_iPPa_kPa_i'L_i$ is inscriptible, as we have just seen, and

$$-\angle Pa_kL_iPa_i' = -\angle Pa_kA_iPa_i' = \angle Pa_i'PPa_k = \angle \bar{A}_iM_kPa_k,$$

the sides being perpendicular each to each.

$$-\angle \bar{A}_iL_iPa_k = \angle Pa_k, L_iPa_i'.$$

Hence $\bar{A}_iL_iPa_i'$ are collinear. Now let the reflection of L_i in M_jM_k be L. It lies on the line $\bar{A}_iPa_i$ and also on the nine-point circle. Also

$$(\overrightarrow{A_iL})(\overrightarrow{\bar{A}_iPa_i}) = (\overrightarrow{\bar{A}_iL_i})(\overrightarrow{A_iPa_i'}) = (\overrightarrow{A_iPa_j})(\overrightarrow{\bar{A}_iPa_k}).$$

Hence L is the intersection of the nine-point and pedal circles. If P move along a fixed line through O the points L_iL remain fixed, whence *

Theorem 68.] *If a point move along a fixed line through the centre of the circumscribed circle, its pedal circle will contain a fixed point of the nine-point circle.*

The other intersection of the nine-point and pedal circles will be similarly obtained from the isogonal conjugate of P, whence

Theorem 69.] *A necessary and sufficient condition that the pedal circle of a point should touch the nine-point circle is that the point and its isogonal conjugate should be collinear with the centre of the circumscribed circle.*

We deduce Feuerbach's theorem, second part, at once from this by noticing that the centres of the inscribed and escribed circles are their own isogonal conjugate.

* This theorem and the next are due to Fontené, 'Extension du théorème de Feuerbach', *Nouvelles Annales de Mathématiques*, Series 4, vol. v, 1905. The proof here given is that of Bricard, under the initials R. B., and inserted in the next volume of the same journal.

We have already noticed that the orthocentre of a triangle is one centre of similitude for the nine-point and circumscribed circles. The other centre of similitude will be the harmonic conjugate of the orthocentre with regard to the centre of the nine-point circle and the point O. This must be the centre of gravity, since the foot of the perpendicular from there on $A_j A_k$ divides $(A_j A_k)$ in the ratio $1:2$.

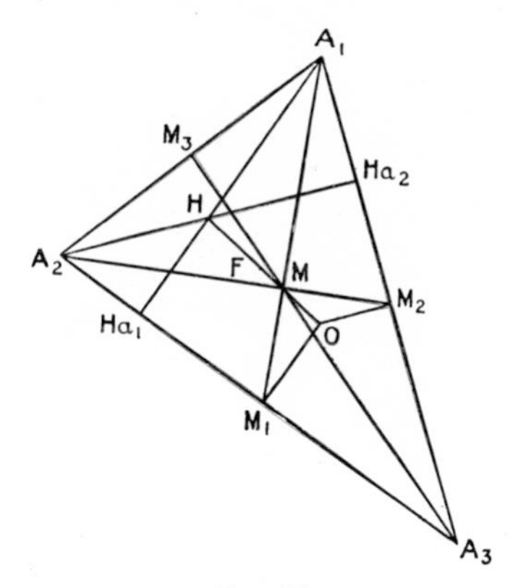

FIG. 12.

Theorem 70.] *The orthocentre and the centre of gravity are centres of similitude for the nine-point and circumscribed circles, the ratios of similitude being* $1:2$ *and* $-1:2$ *respectively.*

There is another circle much less well known than the nine-point circle but possessing a number of analogous properties.* Let the inscribed circle touch $(A_j A_k)$ in A_i' while the escribed circle corresponding to this side touches it in A_i''.

$$(A_j A_i') = s - a_j, \quad (A_k A_i') = (s - a_k), \quad (A_j A_i'') = s - a_k,$$
$$(A_k A_i'') = s - a_j.$$

The lines $A_i A_i''$ are thus concurrent in a point N.† J shall

* Spieker, 'Ein merkwürdiger Kreis um den Schwerpunkt des Perimeters des geradlinigen Dreiecks als Analogon des Kreises der neun Punkte', *Grunert's Archiv*, vol. li, 1870.

† This is Nagel's point: *Untersuchungen über die wichtigsten zum Dreiecke gehörigen Kreise*, 1836 (inaccessible to present author). It corresponds to Gergonne's point where meet lines from the vertices to the points of contact of the opposite sides with the inscribed circle.

be the centre of the inscribed circle. Applying Menelaus's theorem to $A_iA_jA_i''$, and the line A_kA_k''.

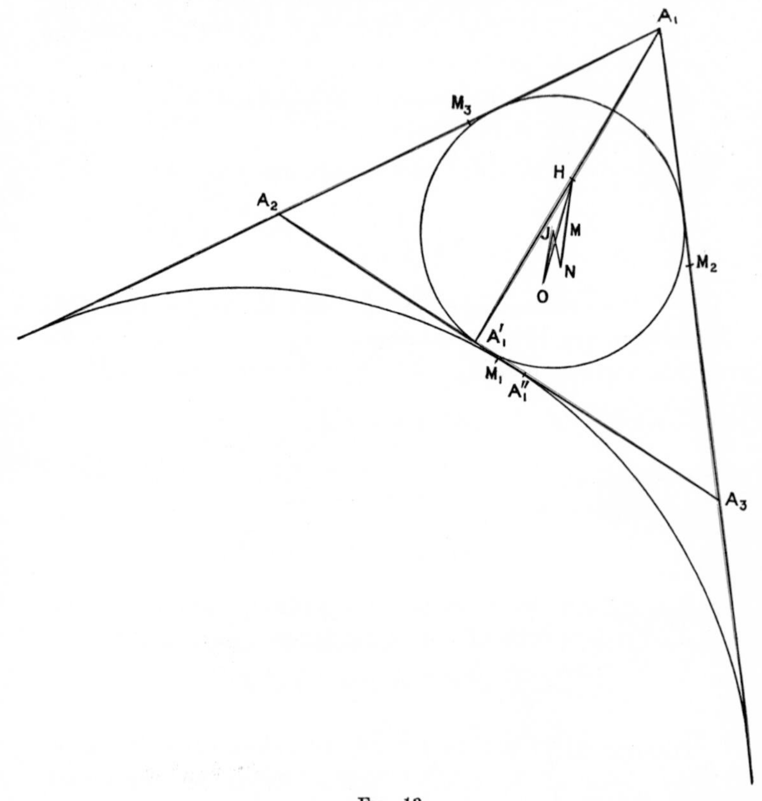

FIG. 13.

$$\frac{(NA_i'')}{(NA_i)} \cdot \frac{(A_iA_k'')}{(A_jA_k'')} \cdot \frac{(A_jA_k)}{(A_i''A_k)} = 1,$$

$$\frac{(NA_i'')}{(NA_i)} = \frac{s-a_i}{a_i},$$

$$\frac{(NA_i)}{(A_i''A_i)} = \frac{a_i}{s}.$$

We have further

$$(A_iHa_i) = \frac{2\Delta}{a_i}, \quad (JA_i') = \frac{\Delta}{s},$$

$$(A_j Ha_i) = a_k \cos \measuredangle A_j,$$

$$(Ha_i A_i'') = (s - a_k) - a_k \cos \measuredangle A_j = \frac{2s}{a_i}[\tfrac{1}{2}a_i - (s - a_j)],$$

$$(A_i' M_i) = \tfrac{1}{2}a_i - (s - a_j), \quad \frac{(Ha_i A_i'')}{(A_i' M_i)} = \frac{(A_i Ha_i)}{(JA_i')}.$$

The triangles $A_i Ha_i A_i''$ and $JA_i' M_i$ are thus similar.

$$\frac{(JM_i)}{(A_i A_i'')} = \frac{a_i}{2s} = \tfrac{1}{2}\frac{NA_i}{(A_i A_i'')}; \quad (JM_i) = \tfrac{1}{2}(NA_i).$$

Hence (JN) meets $(A_i M_i)$ in M, and is divided internally thereby in the ratio $1:2$. We see also by 70] that $OJHN$ are the vertices of a trapezoid whose diagonals meet in M,

$$(JO) = \tfrac{1}{2}(HN).$$

Now let P be the middle point of (JN). Join A_i' with J and M_i with P, and draw $A_i M_i$,

$$(JP) = \tfrac{1}{2}(JN), \quad (JM) = \tfrac{1}{3}(JN).$$

It then appears that if we take the centre of gravity as centre of similitude, a ratio of $-1:2$, the following are interchanged

$$M_i \sim A_i, \quad O \sim H, \quad J \sim N.$$

Theorem 71.] *The centre of the inscribed circle is the Nagel point of the triangle whose vertices are the middle points of the sides.*

We have further

$$\frac{(JM)}{(MP)} = \frac{(A_i M)}{(MM_i)} = \frac{2}{1}.$$

Hence $A_i J$ is parallel to $M_i P$, or $M_i P$ bisects $\measuredangle M_j M_i M_k$ so that P is the centre of the circle inscribed in the triangle $M_1 M_2 M_3$. Its radius is one-half that of the inscribed circle, and N is a centre of similitude. We shall call this the P circle, and exhibit its analogies to the nine-point circle as follows:

Nine-point circle.	*P circle.*
Circumscribed to the triangle whose vertices are the middle points of the sides.	Inscribed in the triangle whose vertices are the middle points of the sides.
Radius one-half that of circumscribed circle.	Radius one-half that of inscribed circle.
Centre of gravity and orthocentre are internal and external centres of similitude for nine-point and circumscribed circles, ratios being $-1:2$ and $1:2$ respectively.	Centre of gravity and Nagel point are internal and external centres of similitude for P circle and inscribed circle, ratios being $-1:2$ and $1:2$ respectively.
Nine-point circle passes through points half-way from orthocentre to the vertices of the triangle.	P circle touches the sides of the triangle whose vertices lie half-way between the Nagel point and the vertices of the given triangle.
Nine-point circle cuts the sides of triangle where they meet the corresponding altitudes.	P circle touches the sides of the middle point triangle where they meet the lines from the Nagel point to the corresponding vertices of the given triangle.

To prove the last statement on the right let us suppose that N_i' is the point of contact of (M_jM_k) with the P circle. Let JA_i' meet $A_i'N$ in M_i',

$$(A_i'M_i) = \tfrac{1}{2}a_i-(s-a_j) = \tfrac{1}{2}(a_k-a_j),$$

$$(A_i'A_i'') = (a_k-a_j) = 2(A_i'M_i).$$

Hence, since JM_i is parallel to A_iA_i'', J is the middle point of $(A_i'M_i')$,

$$(JM_i') = (A_i'J) = \rho = 2(PN_i'), \quad PN_i' \parallel JM_i'.$$

N_i is thus the middle point of (NM_i') and on M_jM_k,

Nine-point circle.

Meets the lines through the points mid-way from the orthocentre to the given vertices parallel to the corresponding side-lines where they meet the perpendicular bisectors of the given sides.

P circle.

Touches the sides of the triangle whose vertices are half-way from the Nagel point to the given vertices at the points where each meets the line from the centre of the inscribed circle to the middle point of the corresponding sides of the original triangle.

The last statement is at once proved by noticing that JM_i bisects (NA_i').

Theorem 72.] *The nine-point circle passes through twelve notable points, the P circle touches six notable lines at notable points. Each is obtained from a notable circle by either of two similarity transformations, the ratios being* $-1:2$ *and* $1:2$, *while the centres of similitude are notable points whereof the centre of gravity is one.*

Returning to the Nagel point we saw that

$$\frac{(NA_i'')}{(NA_i)} = \frac{s-a_i}{a_i}, \quad \frac{(NA_i)}{(A_i''A_i')} = \frac{a_i}{s}.$$

The altitude (A_iHa_i) has the length $\frac{2\rho s}{a_i}$. Hence the orthogonal projection of (A_iN) thereon has the length 2ρ. Again, if $A_1'''A_2'''A_3'''$ be the vertices of the triangle whose side-lines each pass through one of the original vertices parallel to the opposite side-line, we see that N is the centre of the inscribed circle to $\triangle A_1'''A_2'''A_3'''$. Since A_iJ passes through the middle point of the arc $\widehat{A_jA_k}$ of the circumscribed circle, $A_i'''N$ passes through the reflection of this point in A_jA_k. Call this $\mathfrak{A}_i'$; the points $A_i'''A_j$, $H\mathfrak{A}_i'A_k$ are concyclic, since the reflection of H in A_jA_k is on the circumscribed circle, and HA_i''' is a diameter since H and A_i''' are at the same distance from the diameter $\perp$ to A_jA_k $\measuredangle H\mathfrak{A}_i'N = \measuredangle H\mathfrak{A}_i'A_i''' = \frac{\pi}{2}$.

Hence

Theorem 73.] *The circle on the segment from the Nagel point to the orthocentre as diameter passes through those three points on the altitudes whose distances from the corresponding vertices are equal to the diameter of the inscribed circle, and the reflections in the side-lines of the given triangle of the middle points of the corresponding arcs of the circumscribed circle.*

This circle is known as Fuhrmann's circle.*

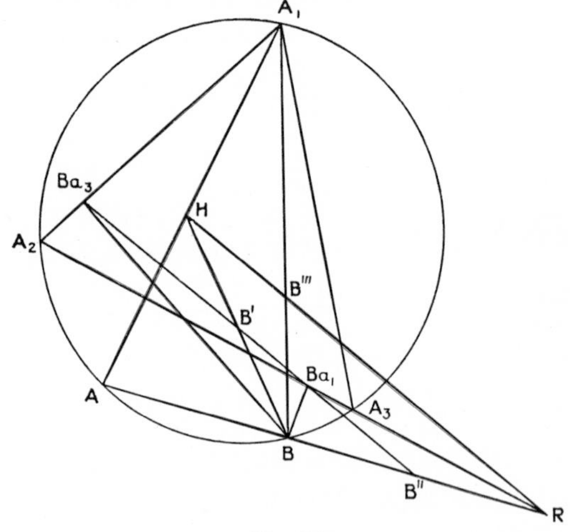

FIG. 14.

Let us continue to study the relations of a triangle to the circumscribed circle. Let A_iH meet the circle again in $\bar{A}_i$ so that $(HHa_i) = (Ha_i\bar{A}_i)$. Let B be any other point of the circumscribed circle; $B\bar{A}_i$ shall meet A_jA_k in R. Draw HR. The Simson line Ba_jBa_k of B shall meet BH in B', while it meets $B\bar{A}_i$ in B''. Let RBa_i meet HR in B'''. We see from the cyclic quadrilateral $BBa_iBa_kA_j$,

$$\angle Ba_kBa_iB = \angle Ba_kA_jB,$$

$$\angle B''Ba_iB = \angle A_i\bar{A}_iB = \angle Ba_iBB''.$$

* *Synthetische Beweise planimetrischer Sätze*, Berlin, 1890. This and the Brocard circle presently to be discussed are special cases of a more general circle discovered by Hagge, 'Der Fuhrmannsche Kreis und der Brocardsche Kreis', *Zeitschrift für mathematischen Unterricht*, vol. xxxviii, 1907.

The triangles $HR\bar{A}_i$, $Ba_iB''B$ are similar isosceles triangles.

$$\measuredangle Ba_iB''B = \measuredangle B''RBa_i + \measuredangle B''Ba_iR,$$
$$= \measuredangle HR\bar{A}_i = 2\measuredangle Ba_iRB,$$
$$\measuredangle Ba_iRB'' = \measuredangle B''Ba_iR$$
$$(BB'') = (Ba_iB'') = (B''R)$$
$$Ba_iB'' \parallel HR.$$

Theorem 74.] *The middle point of a segment bounded by a point of the circumscribed circle and the orthocentre lies on the corresponding Simson line and the nine-point circle.*

If we drop a perpendicular from A_i on the Simson line of B its lesser angle with A_iA_j will be equal to

$$\angle BBa_kBa_i = \angle BA_iA_k.$$

Theorem 75.] *The isogonal conjugate with regard to an angle of a triangle of a line through the vertex of that angle is perpendicular to the Simson line of the second intersection of the given line with the circumscribed circle.*

Let us next take a fourth point A_4 on the circumscribed circle, let H_l be the orthocentre of the $\Delta A_iA_jA_k$. The line from M_i to the middle point of (H_lA_i) bisects (H_lO), being a diameter of the nine-point circle, and $(A_iH) = 2\,(OM_i)$. Hence, in our present case, $(A_iH_j) = (A_jH_i)$, and their lines are parallel. We assume that A_i and A_j are on the same side of A_kA_l.

Theorem 77.] *If four points be taken upon a circle, the nine-point circles of the four triangles which they determine three by three are concurrent in a point common to the Simson line of each point with regard to the triangle of the others.**

Let us for the moment call this the point S.

Theorem 78.] *The perpendicular from the middle point of (A_iA_j) on A_kA_l passes through S, and the distance from S to*

* Lachlan, loc. cit., p. 69, assigns the credit of this theorem to the Cambridge Tripos of 1886. It will be found much earlier in rather a clumsy article by Greiner, 'Ueber das Kreisviereck', *Grunerts Archiv*, vol. lx, 1877. For this, and the five following without proof, see Kantor, 'Ueber das Kreisviereck und Kreisvierzeit', *Wiener Akademie, Sitzungsberichte*, vol. lxxvi, section v, 1877.

the middle point of (A_iA_j) *is equal to the distance from* O *to* A_kA_l.

Since the diagonals of a parallelogram bisect one another,

Theorem 79.] *The segments connecting the middle points of the pairs of segments* $(A_iA_j)(A_kA_l)$ *bisect one another in the middle point of* (OS).

Theorem 80.] *The four orthocentres are the vertices of a quadrilateral congruent to that with the vertices* $A_1A_2A_3A_4$ *and having the same point* S. *Each is a reflection of the other in this point.*

Theorem 81.] *The centres of the four nine-point circles are vertices of a quadrilateral similar to that with vertices* A_i, *and bearing thereto a ratio* 1 : 2. *It is inscribed in a circle of centre* S.

We see, in fact, that the distance of each nine-point centre from S is $\frac{1}{2}r$. Remembering the relations of OM_iH developed in the study of the P circle,

Theorem 82.] *The centres of gravity of the four triangles are vertices of a quadrilateral similar to that having the vertices* A, *and bearing thereto the ratio* 1 : 3.

§ 5. The Brocard Figures.

Besides the inscribed, circumscribed, nine-point, and P circles there are many others which bear simple and striking relations to the triangle. For example, let us construct three circles through the pairs of points A_iA_j tangent respectively to A_jA_k. If Ω be the intersection of two of these,

$$\measuredangle A_1\Omega A_3 = \pi - \measuredangle A_1;\ \measuredangle A_2\Omega A_1 = \pi - \measuredangle A_2;$$

hence

$$\measuredangle A_3\Omega A_2 = \pi - \measuredangle A_3.$$

It thus appears that the three are concurrent in Ω, which is called the *positive Brocard point* of the triangle. Had we constructed circles through A_iA_j tangent to A_kA_i we should have had three concurrent in the *negative Brocard point* Ω'.*

* In the study of the Brocard figures which follows we shall lean heavily on an admirable little book by Emmerich, *Die Brocardschen Gebilde*, Berlin, 1891. This gives not only proofs, but bibliography and historical notices. The

The distinguishing characteristic of these points is exhibited by the equations

$$\measuredangle \Omega A_1 A_2 = \measuredangle \Omega A_2 A_3 = \measuredangle \Omega A_3 A_1 = \omega;$$
$$\measuredangle \Omega A_2 A_1 = \measuredangle \Omega A_1 A_3 = \measuredangle \Omega A_3 A_2 = \omega'.$$

Conversely, it is easily seen that if we seek a construction

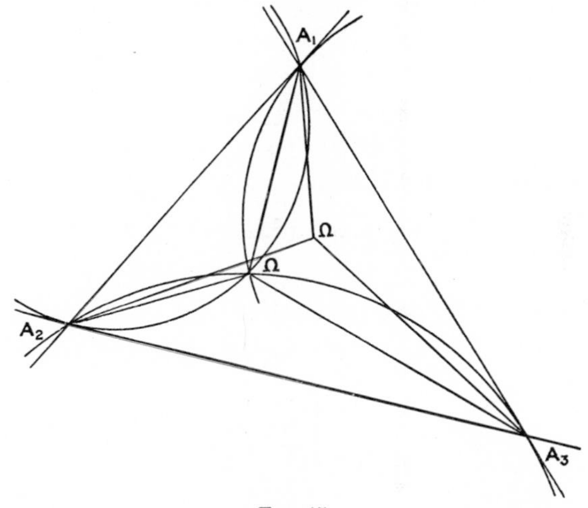

FIG. 15.

for points to satisfy these equations we shall fall back upon the Brocard points. To calculate ω

$$(\Omega A_2) : a_3 = \sin\omega : \sin\measuredangle A_2,\ (\Omega A_2) : a_1 = \sin\measuredangle(A_3 - \omega) : \sin\measuredangle A_3,$$

$$\frac{\sin\measuredangle A_3}{\sin\measuredangle A_1} = \frac{\sin\measuredangle(A_3-\omega)\sin\measuredangle A_2}{\sin\omega\sin\measuredangle A_3},$$

$$\operatorname{ctn}\omega = \frac{\sin\measuredangle A_3}{\sin\measuredangle A_1\sin\measuredangle A_2} + \operatorname{ctn}\measuredangle A_3 = \sum_{i=1}^{i=3}\operatorname{ctn}\measuredangle A_i. \quad (13)$$

The symmetry of this expression shows that $\omega = \omega'$. It is called the *Brocard angle.*

Brocardian geometry, like the study of nine-point and *P* circles, is part of the modern 'Geometry of the Triangle'. This subject has attained colossal proportions almost over night. Vigarié, 'La bibliographie de la géométrie du triangle', *Mathésis*, Series 2, vol. vi, 1896, estimates that, up to 1895, 603 articles had been written dealing therewith. The subject was only started in the seventies.

Theorem 83.] *The two Brocard points are isogonal conjugates of one another.*

$$\csc^2\omega = \csc^2 \measuredangle A_1 + \text{ctn}^2 A_2 + \text{ctn}^2 \measuredangle A_3 + 2\Sigma\, \text{ctn} \measuredangle A_i\, \text{ctn} \measuredangle A_j.$$

But
$$\text{ctn} \measuredangle A_3 = \frac{1 - \text{ctn} \measuredangle A_1\, \text{ctn} \measuredangle A_2}{\text{ctn} \measuredangle A_1 + \text{ctn} \measuredangle A_2},$$

$$\csc^2\omega = \sum_{i=1}^{i=3} \csc^2 \measuredangle A_i. \tag{14}$$

$$\sin^2\omega = \frac{\prod_{i=1}^{i=3} \sin \measuredangle A_i}{\sum_{k=1}^{k=3} \sin^2 \measuredangle A_i \sin^2 \measuredangle A_j} = \frac{4\,\Delta^2}{\sum_{k=1}^{k=3} a_i^2 a_j^2}. \tag{15}$$

$$\begin{aligned} 16\,\Delta^2 &= 16\, s\, \Pi\,(s - a_i) \\ &= 2\,\Sigma\, a_i^2 a_j^2 - \Sigma\, a_i^4, \end{aligned}$$

$$\cos^2\omega = \frac{\left[\sum_{i=1}^{i=3} a_i^2\right]^2}{\sum_{k=1}^{k=3} a_i^2 a_j^2}. \tag{16}$$

From (13)
$$\text{ctn}\,\omega = \frac{\sum_{i=1}^{i=3} \sin^2 \measuredangle A_i}{2\prod_{i=1}^{i=3} \sin \measuredangle A_i},$$

$$\text{ctn}\,\omega = \frac{\sum_{i=1}^{i=3} a_i^2}{4\,\Delta}. \tag{17}$$

$$(\Omega\Omega a_k) = (\Omega A_i) \sin\omega = a_j \frac{\sin^2\omega}{\sin A_i},$$

$$(\Omega\Omega a_k) = 2r \sin^2\omega \frac{a_j}{a_i}, \quad (\Omega'\Omega' a_{k'}) = 2r\sin^2\omega \frac{a_i}{a_j}. \tag{18}$$

$$\frac{\sin(\measuredangle A_k-\omega)}{\sin\omega}=\frac{\sin^2\measuredangle A_k}{\sin\measuredangle A_i\sin\measuredangle A_j}. \tag{19}$$

$$\frac{(A_j\Omega_i)}{a_i-(A_j\Omega_i)}=\frac{a_k^{\,2}}{a_i^{\,2}},$$

$$\frac{a_k}{(A_j\Omega_i)}=\frac{a_i}{a_k}+\frac{a_k}{a_i},$$

$$\frac{\sin(\measuredangle A_j+\omega)}{\sin\omega}=\frac{a_i}{a_k}+\frac{a_k}{a_i}>2.$$

$$\sin(\measuredangle A_j+\omega)>2\sin\omega.$$

$$\sin\omega<\tfrac{1}{2}.$$

Theorem 84.] *The Brocard angle is not greater than one-third of a right angle.*

$$\frac{(A_i\Omega)}{a_j}=\frac{\sin\omega}{\sin\measuredangle A_i}=\frac{(A_i\Omega')}{a_k}.$$

Theorem 85.] *The distances from each vertex to the two Brocard points are proportional to the two sides including that vertex.*

The three triangles into which the original one is divided by connecting the vertices with the positive Brocard point are similar to those obtained by connecting them with the negative one.

The area of $\Delta A_i\Omega A_j$ is

$$\tfrac{1}{2}(A_i\Omega)\,a_k\sin\omega=\tfrac{1}{2}\frac{a_k^{\,2}\sin(\measuredangle A_k-\omega)}{\sin\measuredangle A_j}\sin\omega=r\sin^2\omega\,\frac{a_ja_k}{a_i}.$$

Theorem 86.] *The triangles into which the given triangle is divided by connecting its vertices with the positive Brocard point are equal to those obtained by connecting them with the negative one.*

As the Brocard points are isogonal conjugates they have the same pedal circle by 65], and so by 62] are at equal distances from O. Let $A_i\Omega$ meet the circumscribed circle again at $\bar{A}_i$.

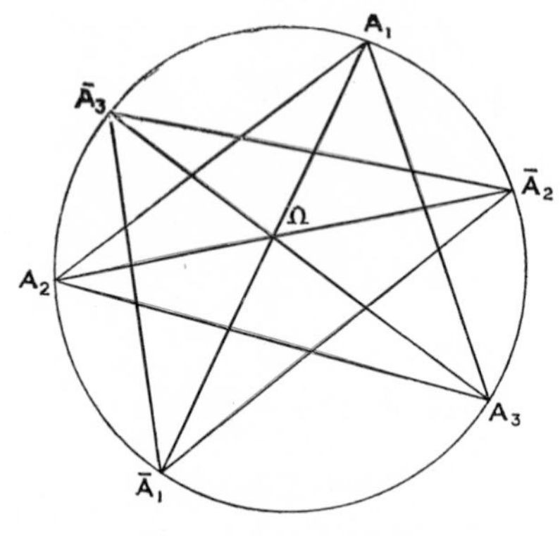

Fig. 16.

$$\measuredangle A_k = \measuredangle A_i\bar{A}_k A_k + \measuredangle A_k\bar{A}_k\bar{A}_j,$$
$$= \measuredangle \bar{A}_i A_i A_k + \measuredangle A_k A_j \bar{A}_j,$$
$$= \measuredangle A_i - \omega + \omega = \measuredangle A_i.$$

We have thus two similar triangles inscribed in the same circle, i. e.

Theorem 87.] *The points where the lines from a Brocard point to the vertices of a triangle meet the circumscribed circle again are vertices of an equal triangle.*

Since $\measuredangle \overrightarrow{A_k A_j \bar{A}_j} = \omega, \quad \measuredangle \overrightarrow{A_k O \bar{A}_j} = 2\omega.$

We may pass from $\Delta A_1A_2A_3$ to $\Delta \bar{A}_3\bar{A}_1\bar{A}_2$ by a rotation about O through an angle whose measure is 2ω. Moreover, since $\measuredangle \bar{A}_j\bar{A}_k\Omega = \omega$,

Theorem 88.] *Ω is the negative Brocard point for the $\Delta \bar{A}_1\bar{A}_2\bar{A}_3$.*

Theorem 89.] *The six triangles $A_i\Omega\bar{A}_k$, $\bar{A}_k\Omega A_j$ are similar to the given triangles.*

We have but to compare the various base angles.

$$(\Omega A_i):(A_i\bar{A}_k) = (A_iA_j):(\bar{A}_i\Omega),$$
$$(\overrightarrow{\Omega A_i}) \times (\overrightarrow{\Omega \bar{A}_i}) = -4r^2\sin^2\omega.$$

Theorem 90.] *The power of a Brocard point with regard to the circumscribed circle is minus the square of the chord determined by a central angle equal to the Brocard angle.*

$$r^2-(O\,\Omega)^2 = 4\,r^2\sin^2\omega,$$

$$O\,\Omega = r\sqrt{1-4\sin^2\omega},$$

$$= r\sqrt{\frac{\cos 3\,\omega}{\cos\omega}}. \tag{20}$$

$$(\Omega\Omega') = 2\,r\sin\omega\sqrt{\frac{\cos 3\,\omega}{\cos\omega}}. \tag{21}$$

We have here a second proof of 84].

There is another notable point of the triangle which bears the closest relation to the Brocard points. We reach it as follows. Let a transversal meet A_iA_j and A_iA_k in two such points B_k and B_j respectively that

$$\measuredangle B_kB_jA_i = \measuredangle A_j, \quad \measuredangle B_jB_kA_i = \measuredangle A_k.$$

Such a line is said to be *antiparallel* to A_jA_k.* The distances from the middle point of (B_jB_k) to (A_iA_k) and (A_iA_j) are proportional to $a_j : a_k$.

The locus of the points is thus a line, called a *symmedian*. Incidentally, the tangent to the circumscribed circle at A_i is antiparallel to A_jA_k.

The three symmedians of a triangle meet in a point called the *symmedian point*,† and indicated in our present scheme by the letter K. It is the isogonal conjugate of the centre of gravity, and its distances from the side-lines are proportional to the lengths of the corresponding sides. Three antiparallels pass through this point, and it is the centre of the three equal segments determined by each two sides on the antiparallel to the third.

Theorem 91.] *The symmedian point is the centre of a circle meeting each side of the triangle where the latter meets the two*

* This term is said to be due to Leibnitz.

† In German works this is referred to as Grebe's, and in French ones as Lemoine's point. We are not in a position to decide the question of priority, so use the usual English term.

antiparallels to the other side which pass through this symmedian point.

This circle is called *Lemoine's second circle.*

Having premised this account of K, let us draw through Ω_i a line $\parallel A_iA_j$ and let it meet A_iA_k in K'. The distances from Ω_i to A_iA_k and A_iA_j are proportional to $\sin\omega$, $\sin(\measuredangle A_j-\omega)$; K' is at the same distance from A_iA_j as is Ω_i; its distance from A_jA_k is $(\Omega_iK')\sin\measuredangle A_j$, and so bears to the distance from Ω_i to A_iA_k the ratio $a_k:a_j=\sin\measuredangle A_k, \sin\measuredangle A_j$. The ratio of the distances from K' to A_jA_k and A_iA_k is thus, by (18), $a_i:a_j$. $\quad K'=K_j$.

Theorem 92.] Ω_iK_j *is parallel to* A_iA_j.

We have already seen that

$$\frac{(\overrightarrow{A_j\Omega_i})}{(\overrightarrow{\Omega_iA_k})}=\frac{a_k^2}{a_i^2},\quad \frac{(\overrightarrow{A_kK_j})}{(\overrightarrow{K_jA_i})}=\frac{a_i^2}{a_k^2},\quad \frac{(\overrightarrow{M_kA_i})}{(\overrightarrow{A_jM_k})}=\frac{1}{1}.$$

Theorem 93.] *The line from* A_i *to the positive Brocard point, the symmedian through* A_j, *and the median through* A_k *are concurrent.*

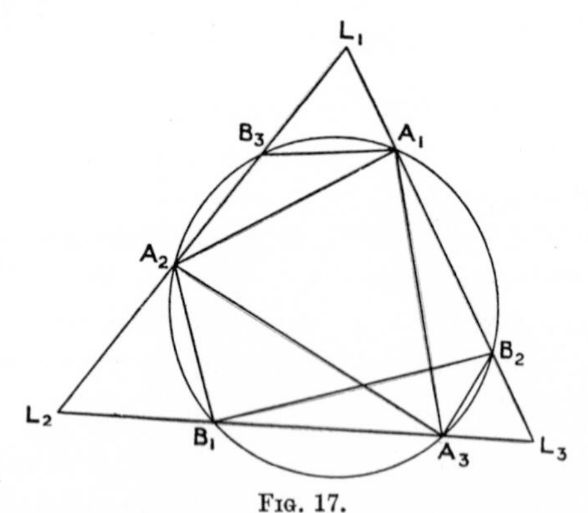

Fig. 17.

Let the point of the circumscribed circle diametrically opposite to A_i be B_i, and let A_iB_j meet A_jB_k in L_i. We proceed to prove

Theorem 94.] *The triangles* $A_1A_2A_3$, $L_3L_1L_2$ *are similar figures with the double point* Ω

$$\measuredangle B_jA_kA_i = \measuredangle A_jA_kB_i, \ \measuredangle L_k = \measuredangle A_i.$$

The quadrilateral $\Omega\, A_jL_jA_k$ is inscriptible, since

$$\measuredangle A_j\Omega A_k = \pi - \measuredangle A_k,$$

$$\measuredangle \Omega L_jL_k = \measuredangle \Omega L_jA_k = \measuredangle \Omega A_jA_k = \omega.$$

Hence Ω is the positive Brocard point for the triangle $L_3L_1L_2$. To find the ratio of similarity we have

$$(L_iL_j) : (A_jA_k) = (\Omega L_k) : (\Omega A_i),$$

$$= \sin\left(\frac{\pi}{2} - \omega\right) : \sin\omega,$$

$$= \operatorname{ctn}\omega.$$

Since A_iB_i is antiparallel to L_iL_j, we have

Theorem 95.] *The centre of the circumscribed circle is the symmedian point for* $\Delta L_3L_1L_2$.

Let us next notice that we pass from $A_1A_2A_3$ to $L_3L_1L_2$ by rotating through an angle $-\frac{\pi}{2}$ about Ω, and altering radii vectores (distances from Ω) in the ratio $\operatorname{ctn}\omega : 1$. It is evident that we might have reached a similar triangle $L_3'L_1'L_2'$ by rotating about Ω' through an angle $\frac{\pi}{2}$. This yields the important result

Theorem 96.] *The centre of the circumscribed circle and the symmedian point subtend right angles at the Brocard points.*

We have from our previous formula (20)

$$(O\Omega) = (O\Omega') = r\sqrt{1 - 4\sin^2\omega}. \tag{20}$$

$$(K\Omega) = (K\Omega') = r\tan\omega\sqrt{1 - 4\sin^2\omega}. \tag{22}$$

$$(\Omega\Omega') = 2r\sin\omega\sqrt{1 - 4\sin^2\omega}. \tag{21}$$

$$(OK) = 2r\sec\omega\sqrt{1 - 4\sin^2\omega} = 2r\sqrt{1 - 3\tan^2\omega}. \tag{23}$$

The Brocard points play an important rôle in the problem

of inscribing in a given triangle a second similar thereto. Let P_i be such a point of $(A_j A_k)$ that $\measuredangle A_j \Omega P_i = \theta$.

$$\measuredangle P_j \Omega P_k = \measuredangle A_k \Omega A_i,$$

$$(OP_j) = (\Omega\Omega a_j) \csc(\omega + \theta),$$

$$= (\Omega A_k) \frac{\sin \omega}{\sin(\omega + \theta)}.$$

The $\Delta P_3 P_1 P_2$ is thus similar to $\Delta A_1 A_2 A_3$ and has Ω as its positive Brocard point. Conversely, if the $\Delta P_3 P_1 P_2$ be similar to the given triangle, P_i lying on $A_j A_k$, then the three circles $A_i P_j P_k$ will be seen to pass through such a point that the angle subtended there by $P_j P_k$ will be $\pi - \measuredangle A_i$, and this is easily found to be the common positive Brocard point for both triangles. In like manner from the negative Brocard point and the angle $-\theta$ we get another inscribed similar triangle $P_3' P_1' P_2'$. The six points $P_i P_j'$ are concyclic by 67]. Let O_θ be the centre of this circle

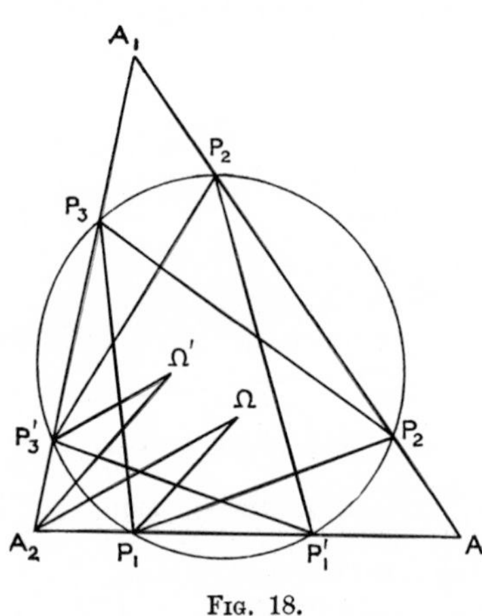

Fig. 18.

$$(\Omega P_j) : (\Omega A_k) = (\Omega O_\theta) : (\Omega O).$$

$(\Omega P_j) : (\Omega O_\theta)$ is a ratio independent of θ, and since $\measuredangle P_j \Omega O_\theta = \measuredangle A_k \Omega O$ the locus of O_θ is a straight line. This line goes through O corresponding to $\theta = O$, and through the middle point of $(\Omega\Omega')$ corresponding to $\theta = \frac{\pi}{2} - \omega$. It is therefore the line OK.

Theorem 97.] *The six points $P_i P_j'$ lie on a circle whose centre is on OK.*

Such a circle is called a *Tucker circle*.

Theorem 98.] *The line P_iP_j' is parallel to A_iA_j.*

Theorem 99.] *The line $P_i'P_j$ is antiparallel to A_iA_j.*

The proofs of these latter theorems come immediately from the definition of the Tucker circle. They also give a means for constructing a Tucker circle.

Theorem 100.] *The three segments $(P_i'P_j)$ are equal to one another.*

We see, in fact, that the lines of any two are equally inclined to one side-line of the triangle, and the segments are comprehended between parallel lines.

Theorem 101.] *The triangle formed by the three lines P_iP_j' is similar to the original triangle, the double point being a symmedian point for each.*

We see, in fact, that the sides of the two are parallel in pairs, and in the parallelogram having as three vertices A_k, P_j, P_i' a diagonal goes from A_k to a vertex of the second triangle and, being a symmedian, passes through K.

Theorem 102.] *The triangle formed by the three lines $P_i'P_j$ bears such a relation to the original triangle that lines connecting corresponding vertices are concurrent in K.*

We have but to find the ratio of the distances of a vertex of the first triangle from two sides of the second.

Theorem 103.] *The perpendiculars on the side-lines of the given triangle from the corresponding vertices of that triangle whose side-lines are $P_i'P_j$ are concurrent in the centre of the Tucker circle.*

Let us take up certain special cases of the Tucker circle obtained by giving to θ special values.

$\theta = 0$. The Tucker circle is the circumscribed circle.

$\theta = \frac{\pi}{2} - \omega$. The Tucker circle is the pedal circle of the Brocard points.

$\theta = \frac{\pi}{2}$. Here, by theorem 96], the centre of the Tucker circle is the symmedian point. Moreover, we shall have

$P_i'P_j' \parallel P_jP_k$. Hence the lines $P_i'P_j$ are concurrent in the centre of the Tucker circle. But these are also antiparallels to the side-lines of the original triangle, whence,

Theorem 104.] *The Tucker circle where* $\theta = \frac{\pi}{2}$ *is the second Lemoine circle.*

The segments which this circle cuts on the sides of the triangle will be bases of isosceles triangles whose base angles are equal to the angles of the original triangle.

Theorem 105.] *Lemoine's second circle cuts on each side of the triangle a segment proportional to the cosine of the opposite angle.*

For this reason Lemoine's second circle is sometimes called the *Cosine Circle.* The perpendicular from Ka_i on (P_jP_k') bisects the latter at a point of A_iM_i, and the symmedian point is half-way from there to Ka_i. Hence M_iK bisects (A_iHa_i).

Theorem 106.] *The lines connecting the middle points of the sides of a triangle with the middle points of the corresponding altitudes are concurrent in the symmedian point.*

$\theta = \omega$. Here P_i is equidistant from A_j and Ω, and P_i' is equidistant from A_k and Ω', $P_jP_k' \parallel A_jA_k$. Moreover, $\measuredangle O\Omega O_\theta = \omega = \frac{1}{2}\measuredangle \Omega O \Omega'$, and the centre of this circle, called *Lemoine's first circle,* is the middle point of (OK). The three lines P_iP_j' must be concurrent in the second Brocard point of $\Delta P_iP_jP_k$, or the first Brocard point of $\Delta P_i'P_j'P_k'$. This is K since $\measuredangle KO\Omega' = \omega$.

Theorem 107.] *In the case of Lemoine's first circle the segments* (P_jP_k') *are bisected at the symmedian point, and the centre of the circle is half-way from there to the centre of the circumscribed circle. The symmedian point is a Brocard point for each of the triangles.*

This circle is easily obtained by drawing through the symmedian point parallels to the side-lines of the triangle.

$$(P_iP_i') : a_i = (KKa_i) : (A_iHa_i) = (KKa_i) \times \frac{a_i}{2\Delta}.$$

But, by the fundamental symmedian property, (KKa_i) is proportional to a_i and

Theorem 108.] *The segments which Lemoine's first circle cuts on the sides of the triangle are proportional to the cubes on those sides.*

For this reason this circle is sometimes called the *triplicate ratio circle.*

There is one more Tucker circle which merits special attention; it is, however, more easily approached from another point of view.

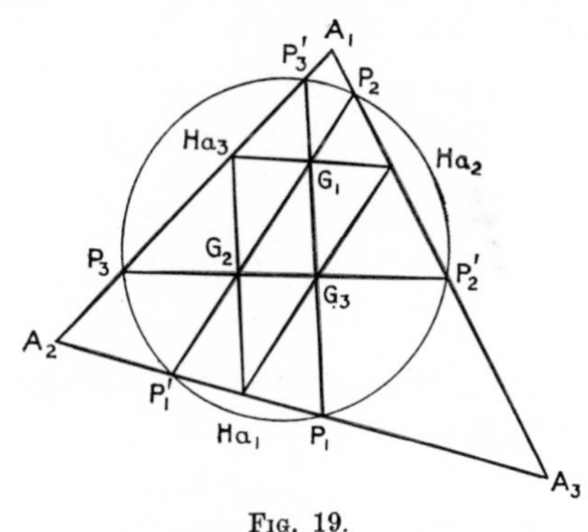

Fig. 19.

Let G_i be the middle point of Ha_jHa_k, and let G_iG_j meet A_iA_k in P_j, and A_jA_k in P_i'. It is easy to see that the length of our segment $(P_i'P_j)$ is equal to the semi-perimeter of the pedal triangle of H

$$\measuredangle P_kP_k'P_i = \measuredangle A_k$$

$$\measuredangle P_kP_i'P_i = \measuredangle A_i + \measuredangle P_kP_i'G_j = \measuredangle A_i + \measuredangle P_i'P_kG_j'$$
$$= \measuredangle A_j + \measuredangle A_i.$$

Since

$$(P_kG_j) = (Ha_kG_j) = (G_jHa_i) = (P_i'G_j),\ \measuredangle P_kG_jP_i' = \pi - 2\measuredangle A_j.$$

Hence the four points P_i, P_i', P_j, P_j' are concyclic, and so all six points P_x, P_y' lie on one circle. This circle is called *Taylor's circle.* Since the $\Delta\ G_jP_kP_i'$ is isosceles, the perpendicular bisector of (P_kP_i') bisects also $\measuredangle G_j$.

Theorem 109.] *Taylor's circle is concentric with the circle inscribed in the triangle whose vertices lie midway between the feet of the altitudes.*

Let us show that Taylor's circle is a Tucker circle. The three lines A_iG_i are concurrent in the symmedian point.

$$P_kP_i' \parallel A_iA_k \text{ since } \measuredangle P_kP_i'A_j = \pi - \measuredangle A_i - \measuredangle A_j.$$

The triangles $P_iP_jP_k$, $P_k'P_i'P_j'$ are equal by three sides, and

$$\measuredangle P_iP_jP_k = \pi - \measuredangle P_kP_i'P_i = \measuredangle A_k.$$

Hence these equal triangles are similar to the original one, and

Theorem 110.] *Taylor's circle is a Tucker circle.*

The process of finding the corresponding value of θ is a bit difficult. Let Θ_i be the foot of the perpendicular from O_θ on G_jG_k.

$$\measuredangle P_kO_\theta\Theta_i = \measuredangle P_kP_jP_j' = \pi - \theta,$$

$$(P_k\Theta_i) = \tfrac{1}{2}(P_kP_j') = \tfrac{1}{4}\Sigma(Ha_iHa_j),$$

$$= \tfrac{1}{4}\Sigma a_i \cos\measuredangle A_i = r\Pi\sin\measuredangle A_i.$$

$O_\theta\Theta_i$ is, by 109], the radius of the inscribed circle in a triangle whose sides are $\frac{1}{2}a_i\cos\measuredangle A_i$

$$(O_\theta\Theta_i) = r\Pi\cos\measuredangle A_i,$$

$$\tan\theta = -\Pi\tan\measuredangle A_i,$$

$$(G_jHa_k) = (G_jP_k) = (G_jP_i') = (G_jHa_i).$$

The circle on (Ha_iHa_k) as diameter passes through P_i', P_k.

Theorem 111.] *Taylor's circle contains the intersections of each side-line with the perpendiculars from the feet of the altitudes on the other two.**

$$(A_jP_i') = (A_jHa_k)\cos\measuredangle A_j = a_i\cos^2\measuredangle A_j,$$

$$(A_kP_i) = a_i\cos^2\measuredangle A_k,$$

$$(P_iP_i') = a_i(1-\cos^2\measuredangle A_j - \cos^2\measuredangle A_k),$$

$$= a_i(\sin^2\measuredangle A_j\sin^2\measuredangle A_k - \cos^2\measuredangle A_j\cos^2\measuredangle A_k).$$

Theorem 112.] *The segment cut by Taylor's circle on the side (A_jA_k) has the value*

$$a_i\cos\measuredangle A_i\cos(\measuredangle A_j - \measuredangle A_k).$$

* Cf. Theorem 66.

The centre of the circumscribed circle is the orthocentre of the $\Delta M_1M_2M_3$. Hence, by 74], the Simson line of Ha_i with regard to this triangle passes through the middle point of (Ha_iO). The segment (A_iHa_i) is bisected perpendicularly by M_jM_k so that the before-mentioned Simson line of Ha_i is $\parallel OA_i$. The $\Delta\, G_jG_kO_\theta$ is similar to the triangle whose vertices are Ha_kHa_j and the orthocentre of $\Delta\, A_iHa_kHa_j$, the ratio of similarity being $1:2$, while Ha_i is the centre of similitude. Hence O_θ is the middle point of the segment from Ha_i to the orthocentre of the $\Delta\, A_iHa_kHa_j$, which point lies on A_iO.

Theorem 113.] *The centre of Taylor's circle lies on the Simson line of the foot of each altitude with regard to the triangle whose vertices are the middle points of the sides of the given triangle.*

The perpendicular from M_i on Ha_jHa_k bisects (Ha_iHa_k) since $(M_iHa_j) = (M_iHa_k)$. The perpendiculars from M_i on Ha_iHa_j and Ha_iHa_k make equal angles with A_jA_k. Hence the Simson line of M_i with regard to $\Delta\, Ha_iHa_jHa_k$ is the perpendicular on A_jA_k or on P_jP_k' from the middle point of Ha_jHa_k, and so is the line G_iO_θ.

Theorem 114.] *The centre of Taylor's circle lies on the Simson line of the middle point of each side with regard to the triangle whose vertices are the feet of the altitudes.*

$$(\overrightarrow{A_iP_j}) \times (\overrightarrow{A_iP_j'}) = a_j^2 \cos^2 \measuredangle A_i \cos^2 \measuredangle A_k.$$

This last expression is equal to the square of the distance from A_i to Ha_jHa_k. But A_i is the centre of a circle escribed to the $\Delta\, Ha_iHa_jHa_k$.

Theorem 115.] *Taylor's circle cuts at right angles the circles escribed to the triangle whose vertices are the feet of the altitudes.*

Enough has now been said about the Tucker circles. Returning to the figures more nearly associated with the name of Brocard, we remember that we originally found the Brocard points by constructing circles through A_iA_j tangent to A_jA_k for Ω or to A_kA_i for Ω'. The centre of the first of these circles shall be called X_j, that of the second X_i'.

Theorem 116.] *The triangles $X_1X_2X_3$ and $X_1'X_2'X_3'$ are similar to the original triangle, the double points being the positive and negative Brocard points respectively, and the ratio of similitude being* $1:2$ *sin* ω.

Theorem 117.] *The centre of the circumscribed circle is the negative Brocard point for $\Delta X_1X_2X_3$ and the positive Brocard point for $\Delta X_1'X_2'X_3'$.*

We see, in fact, that X_i lies on the perpendicular from O on A_kA_i, while X_kX_i is the perpendicular bisector of (ΩA_k). Hence $\measuredangle\, \Omega X_iX_k = \measuredangle\, \Omega A_kA_i = \omega$.

We have already seen that

$$\measuredangle\, \Omega O \Omega' = 2\omega, \quad \measuredangle\, O\Omega K = \frac{\pi}{2}\cdot$$

Hence, if Z be the middle point of (OK),

$$(Z\Omega) = (ZK) = \tfrac{1}{2}(OK) = \frac{(\Omega K)}{2\sin\omega}\cdot$$

Theorem 118.] *The centre of the first Lemoine circle is the common symmedian point for $\Delta X_1X_2X_3$* and *$\Delta X_1'X_2'X_3'$.*

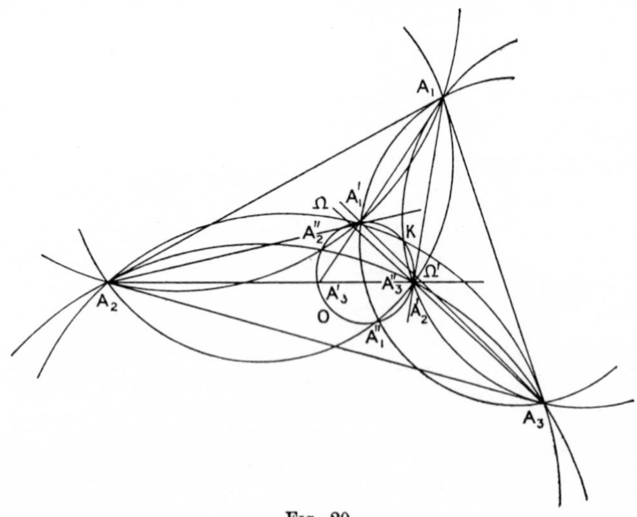

FIG. 20.

Let the circles whose centres are X_i and X_i' intersect, not only at A_i, but again at A_i''. $\Delta A_1''A_2''A_3''$ is called *Brocard's second triangle.*

$$\measuredangle A_jA_i''A_k = \measuredangle A_i + \measuredangle A_i''A_jA_i + \measuredangle A_i''A_kA_i = 2\measuredangle A_i.$$

Theorem 119.] *The points $A_jA_kOA_i''$ are concyclic.*

Theorem 120.] *A_i'' lies on the symmedian through A_i.*

We see, in fact, that the triangles $A_iA_jA_i''$, $A_kA_iA_i''$ are similar; hence the altitudes from A_i'' have the ratio $a_k:a_j$. We notice also, since $A_jA_kOA_i''$ are concyclic,

$$\angle OA_i''A_j = \angle OA_kA_j = \frac{\pi}{2} - \angle A_i.$$

But
$$\angle A_jA_i''A_i = (\pi - A_i)\angle A_j.$$

Theorem 121.] *A_i'' is the projection of O on A_iK.*

Theorem 122.] *The three points A_i'' lie on the circle on (OK) as diameter.*

We have thus, remembering 96], seven points on this important circle, which is called *Brocard's circle.* We find three more as follows. Let A_i' be the intersection of $A_j\Omega$ with $A_k\Omega'$. The $\Delta A_1'A_2'A_3'$ is called *Brocard's first triangle.*

Theorem 123.] *The three triangles $A_i'A_jA_k$ are similar isosceles triangles.*

The distance from A_i' to A_jA_k is $\frac{1}{2}a_i \tan\omega$, and this is also the distance from the symmedian point to that line by (17).

Theorem 124.] *The three lines through the points A_i' parallel to the corresponding side-lines A_jA_k are concurrent in the symmedian point.*

Since
$$\measuredangle \Omega A_i'\Omega' = 2\omega$$

Theorem 125.] *The vertices of Brocard's first triangle lie on Brocard's circle.*

Since $(A_j'A_k')$ subtends at A_k' and at K an angle $= \measuredangle A_i$

Theorem 126.] *Brocard's first triangle is similar to the given triangle.*

We get from formula (23)

Theorem 127.] *The ratio of similitude of Brocard's first triangle and the given triangle is*

$$\sqrt{1-3\tan^2\omega} : 1.$$

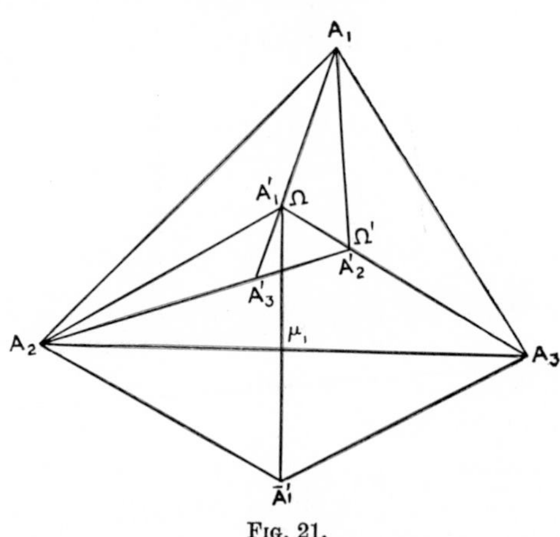

FIG. 21.

Let A_i' be the reflection of A_i' in A_jA_k, so that

$$(A_i'M_i) = (\bar{A}_i'M_i).$$

Connect $\bar{A}_i'$ with A_j and A_k, also connect A_j and A_k' with A_k and A_j'. Then $\Delta\, A_j'A_k\bar{A}_i'$ is similar to $\Delta\, A_iA_kA_j$ since

$$\measuredangle\, A_j'A_k\bar{A}_i' = \measuredangle\, A_k - \omega + \omega = \measuredangle\, A_k;$$

$$(A_kA_j') : (A_k\bar{A}_i') = a_j : a_i.$$

Hence also $A_k'A_j\bar{A}_i'$ is similar to $\Delta\, A_iA_jA_k$, and as

$$(A_j\bar{A}_i') = (A_k\bar{A}_i');\quad \Delta\, A_kA_j'\bar{A}_i' = \Delta\, \bar{A}_i'A_k'A_j;$$

$$(A_j'\bar{A}_i') = (A_k'A_j) = (A_k'A_i).$$

Similarly $(A_k'\bar{A}_i') = (A_j'A_i')$ and $A_iA_j'\bar{A}_i'A_k'$ are the vertices of a parallelogram. Hence the median from A_i' in $\Delta\, A_i'A_j'A_k'$ is the median from A_i' in $\Delta\, A_i'A_i\bar{A}_i'$. A second median of this triangle is A_iM_i. The median of $\Delta\, A_i'A_j'A_k'$ through A_i' divides (A_iM_i) in the ratio $2:1$, i.e. goes through M.

Theorem 128.] *Brocard's first triangle has the same centre of gravity as the given triangle.*

The quadrilaterals $A_iA_jA_i''A_k'$, $A_kA_iA_i''A_j'$ are equiangular and similar, so that

$$(A_k'A_i''):(A_j'A_i'') = a_k:a_j = (A_i'A_j'):(A_i'A_k')$$
$$= \sin \measuredangle A_k'A_i'A_i'' : \sin \measuredangle A_j'A_i'A_i''.$$

Hence $A_i'A_i''$ is a median of the $\Delta\, A_i'A_j'A_k'$.

Theorem 129.] *The lines connecting the corresponding vertices of Brocard's two triangles are concurrent in the common centre of gravity of the first Brocard and the given triangle.*

The triangles $A_1'A_2'A_3'$, $A_1A_2A_3$ are similar, but are easily seen to be arranged in opposite order. It is easy to see that under the similarity transformation of the plane thus defined, a line through $A_i \parallel A_j'A_k'$ will pass into one through $A_i' \parallel A_jA_k$.

Theorem 130.] *The lines through the vertices of a triangle parallel to the corresponding side-lines of Brocard's first triangle are concurrent on the circumscribed circle.*

This point of concurrence is called *Steiner's point.* That diametrically opposite is *Tarry's point.*

Theorem 131.] *The lines through the vertices of a triangle perpendicular to the corresponding side-lines of Brocard's first triangle are concurrent in Tarry's point.*

Suppose that A_iO meets the Brocard circle again in T_i. Let us find the magnitude of $\measuredangle T_iOA_i'$.

$$OA_i' \perp A_jA_k;\ \measuredangle OA_iHa_i = \measuredangle A_k - \measuredangle A_j;$$
$$\measuredangle T_iA_j'A_i' = \measuredangle A_k - \measuredangle A_j;\ \measuredangle T_iA_j'A_k' = \measuredangle A_k'.$$

Theorem 132.] *The angle between A_jA_k and $A_j'A_k'$ is equal to $\measuredangle KOA_i$.*

It appears at once from the construction of Fig. 14 that the Simson line of any point P makes with A_jA_k an angle equal to the angle formed therewith by $P\bar{A}_i$, and this is equal to $\frac{\pi}{2} - (\measuredangle A_j - \measuredangle PA_jA_k)$. The angle which A_iO makes with A_jA_k is $\measuredangle A_k + \left(\frac{\pi}{2} - \measuredangle A_j\right)$.

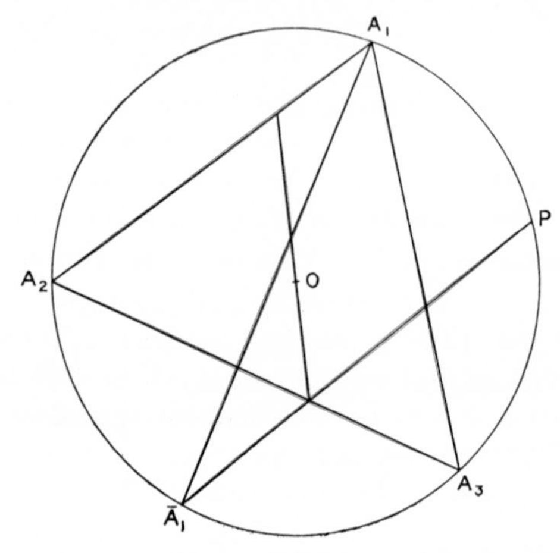

Fig. 22.

Hence the angle which the Simson line makes with OA_i is $\measuredangle A_k - \measuredangle PA_jA_k$, and this is the angle of PA_i with A_jA_k. The Simson line of Steiner's point and OK are equally inclined to OA_i. They must, thus, be parallel, or make with OA_i an angle whose algebraic sum is zero. But two lines cannot simultaneously make with the three concurrent lines pairs of angles differing only in sign.

Theorem 133.] *The Simson line of Steiner's point is parallel to the line from the centre of the circumscribed circle to the symmedian point, while the Simson line of Tarry's point is perpendicular thereto.*

Suppose that we have given the side (A_jA_k) of our original triangle, and the Brocard angle, what will be the locus of the opposite vertex? Restricting ourselves to one side of A_jA_k, we construct an arc at whose points (A_jA_k) subtends an angle equal to ω. Suppose that A_i has been found, and that

A_jA_i meets this arc again at Y_k. Draw Y_kA_k, and $A_i\Omega$ which is $\parallel Y_kA_k$ and meets A_jA_k in Ω_i.

$$(A_j\Omega_i):(\Omega_iA_k) = a_k^2 : a_j^2, \quad (\overrightarrow{A_iA_j}) \times (\overrightarrow{A_iY_k}) = -a_i^2.$$

A_i has thus a constant power with regard to a given circle; its locus is the arc of a second circle concentric therewith.

Theorem 134.] *The locus of the vertex of a triangle whose opposite side and Brocard angle are given is formed by the arcs of two circles concentric with those containing all points whereat the given side subtends the given Brocard angle.*

These circles are called *Neuberg circles* and have many interesting properties whereof we shall give but a few.* If the original triangle be given there are three pairs of Neuberg circles; let us restrict ourselves to those three whose centres lie on the same sides of the side-lines as the opposite vertices of the original triangle, and call these the Neuberg circles of the given triangle. Let the centre of the Neuberg circle corresponding to A_jA_k be N_i. Then $\measuredangle N_iA_jA_k = \frac{\pi}{2} - \omega$. The distances from N_i to A_iA_k and A_iA_j are in the ratio $\cos(\measuredangle A_k + \omega) : \cos(\measuredangle A_j + \omega)$. Now if a point lie on the perpendicular from A_i on $A_j'A_k'$, i. e. on the line from A_i to Tarry's point, the ratio of its distances from A_iA_j and A_iA_k will be, by 132],

$$\cos\measuredangle KOA_k' : \cos\measuredangle KOA_j' = \sin\measuredangle OKA_k' : \sin\measuredangle OKA_j'.$$

The sine of the angle of OK and A_iA_j, or of OK and KA_k', is, by (23),

$$\frac{(KKa_k)-(OM_i)}{(OK)} = \frac{\sin A_k \tan\omega - \cos A_k}{\sqrt{1-3\tan^2\omega}}.$$

Hence

$$\sin\measuredangle OKA_k' : \sin\measuredangle OKA_j' = \cos(\measuredangle A_k + \omega) : \cos(\measuredangle A_j + \omega).$$

Theorem 135.] *The lines connecting the vertices of a triangle with the centres of the corresponding Neuberg circles are concurrent in Tarry's point.*

* Emmerich, loc. cit., pp. 133 ff.

We see that if one angle and the Brocard angle of a triangle be given, the other angles are determined by symmetrical

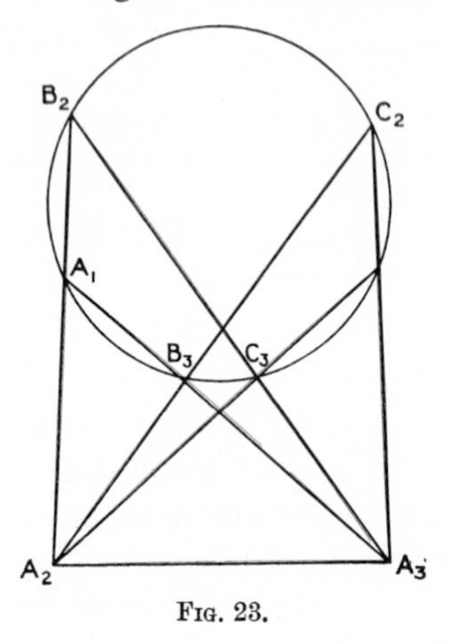

FIG. 23.

equations. Hence the various possible triangles with these data are similar.

Theorem 136.] *If A_iA_j and A_iA_k meet the corresponding Neuberg circle again in B_j and B_k respectively, then $\Delta A_jB_jA_k$ and $\Delta A_jB_kA_k$ are similar to $\Delta A_iA_jA_k$.*

Theorem 137.] *The power of A_j or A_k with regard to the corresponding Neuberg circle is a_i^2.*

If the points A_j and A_k be given, there will be ∞^1 circles with regard to which each has the power $(A_jA_k)^2$, and these will all be Neuberg circles. Let A_jB_k meet such a Neuberg circle again at C_j, while A_kB_j meet it at C_k. Then, by the preceding, A_kC_k and A_jC_j will intersect on the Neuberg circle, which gives the curious theorem:

Theorem 138.] *If a circle bear such a relation to two points that the power of each with regard to it is the square of the distance of the points, then ∞^1 re-entrant hexagons may be inscribed in the circle such that alternate side-lines pass through the one or the other given point.*

We obtain an interesting sidelight on the Brocard configuration by a study of three similar figures to which we now turn our attention.* Two figures are directly similar if corresponding distances be proportional and corresponding angles equal in magnitude and sense; when the signs of corresponding angles are opposite, the figures are inversely similar. A relation of direct similarity will be determined as soon as we know the two points $A_j'A_k'$ which correspond to two given points A_jA_k. The locus of the points whose distances from A_j and A_j' bear the ratio $(A_jA_k):(A_j'A_k)$ is a circle, and a similar circle may be found for A_k and A_k'. These circles intersect in two points which are the double points of the similarity transformations determined by the corresponding segments.

Suppose that we have three similar figures f_1, f_2, f_3. The double point of f_j and f_k shall be S_i, the three ratios of similitude $r_1:r_2:r_3$. Let D_1, D_2, D_3 be the vertices of a triangle whose sides lie along three corresponding lines. The distances from S_i to D_iD_k and D_iD_j are proportional to $r_j:r_k$. Hence, by Ceva's theorem,

Theorem 139.] *If three similar figures be given, the three lines connecting each double point to the corresponding vertex of a triangle whose side-lines correspond in the three figures are concurrent.*

Let us call this point of concurrence C. Notice that if not only the side-line but the actual sides are corresponding, it will be the symmedian point. The angles of $\Delta D_1D_2D_3$ depend merely on the transformation, as do the angles which S_iD_i make with D_iD_j and D_iD_k, since their sum and the ratio of their sines are constant. Hence the angles $\angle S_iCS_j$ are constant in size.

Theorem 140.] *The locus of the points of concurrence of lines from each double point to the corresponding vertex of a triangle whose side-lines correspond is the circle through the three double points.*

If we draw through C three lines parallel to the three lines D_iD_j they will intersect this circle again in points R_k. They will also be three corresponding lines as their angles are those

* McCleland, *A Treatise on the Geometry of the Circle*, London, 1891, ch. ix.

of any three corresponding lines, and the distances from S_i to CR_j and CR_k are in the ratio $r_k : r_j$. Also the points R_i are fixed, since $\angle R_iCS_j$ has a constant value. Conversely, if three corresponding lines be concurrent, the locus of their point of concurrence is, by 140], this circle.

Theorem 141.] *The locus of points where three corresponding lines are concurrent is the circle through the three double points; the three corresponding lines must pass through fixed points of this circle.*

These fixed points on the concurrent lines are called *invariable points.*

Theorem 142.] *The lines connecting the double points to the corresponding invariable points are concurrent.*

We see, in fact, that the invariable points are surely corresponding.

$$\begin{aligned} r_j : r_k &= (S_iR_j) : (S_iR_k) \\ &= \sin \measuredangle S_iR_kR_j : \sin \measuredangle S_iR_jR_k \\ &= \sin \measuredangle S_iR_iR_j : \sin \measuredangle S_iR_iR_k. \end{aligned}$$

Hence the three lines S_iR_i meet in a point M.

Suppose that we have P_1, P_2, P_3 three corresponding points which are collinear. The angles of $\Delta\, S_iP_jP_k$ are constant in magnitude, hence $\angle S_kP_iS_j$ has a constant value, or the locus of P_i is a circle through S_j and S_k. If S_i' be the point which corresponds to S_i in f_i, the line S_iS_i' must correspond to two other lines through S_i, namely S_iR_j and S_iR_k, so that S_i' is on S_iR_i. Again, $\angle S_kP_iP_j$ is constant, so that P_iP_j meets the P_i circle in a fixed point, namely M, and this is common to all three circles. Conversely, there are surely ∞^1 sets of corresponding collinear triads, generating three circles which correspond, and if we take P_1, P_2, P_3 three corresponding points on them $\angle S_kP_iP_j$ has a fixed value, so that P_iP_j goes through a fixed point, namely M, and P_k lies on P_iP_j.

Theorem 143.] *The loci of three collinear points in three directly similar figures are three circles each through two double points. There is one point common to all three circles, and sets of three collinear corresponding points are collinear with this.*

Theorem 144.] *If three directly similar figures be constructed on the three sides of a triangle following one another in cyclic order,*

(*a*) *The vertices of the second Brocard triangle will be the double points.*

(*b*) *The vertices of the first Brocard triangle will be the invariable points.*

(*c*) *The lines connecting corresponding vertices of the original and second Brocard triangle will be concurrent in the symmedian point of the former which lies on the Brocard circle.*

(*d*) *The lines connecting corresponding vertices of the two Brocard triangles are concurrent in the common centre of gravity of the given and first Brocard triangle.*

(*e*) *The symmedian point of every triangle formed by three corresponding segments in cyclic order will lie on the Brocard circle.*

(*f*) *If three corresponding lines be concurrent they pass through the vertices of the first Brocard triangle, and their point of concurrence is on the Brocard circle.*

(*g*) *The loci of three corresponding collinear points are the three circles through two vertices of Brocard's second triangle and the centre of gravity of the given triangle.*

The three circles mentioned in (*g*) are called *McCay circles* and deserve some further notice. The three lines $A_i'A_i''$ pass through M, which lies between A_i' and A_i''.

Theorem 145.] *The McCay circles are the reflections in the centre of gravity of the given triangle of the inverses of the sides of the first Brocard triangle in a fixed circle whose centre is that centre of gravity.*

Theorem 146.] *The McCay circles intersect at angles equal to those of the given triangle.*

As M is the centre of gravity of the first Brocard triangle, it is the middle point of three segments each on a line parallel to one side-line of this triangle and terminated by the other two.

Theorem 147.] *The centre of gravity of the given triangle is the middle point of the three segments which each two McCay circles cut on the tangent to the third at that point.*

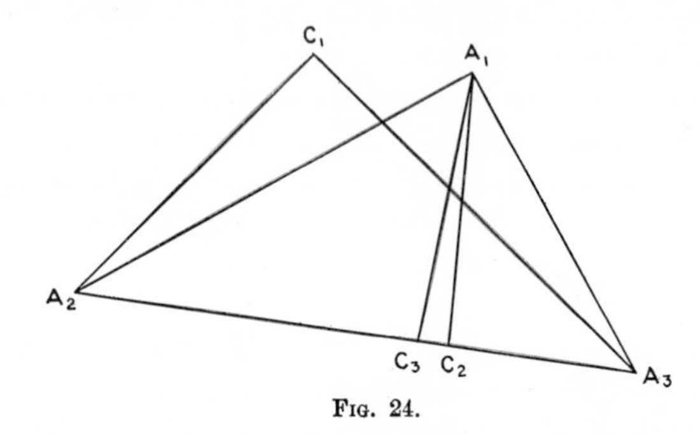

FIG. 24.

Starting with our original triangle, we may construct three others similar to it as follows:

C_1 shall be such a point on the same side of A_2A_3 as A_1 that

$$\measuredangle C_1A_2A_3 = \measuredangle A_3, \quad \measuredangle C_1A_3A_2 = \measuredangle A_2.$$

C_2 shall be such a point on A_2A_3 that

$$\measuredangle C_2A_1A_3 = \measuredangle A_2.$$

C_3 shall be such a point on A_2A_3 that

$$\measuredangle C_3A_1A_2 = \measuredangle A_3.$$

The centres of gravity of our three triangles $C_1A_2A_3$, $A_1C_2A_3$, $A_1A_2C_3$ lie on the line through $M \parallel A_2A_3$ and are corresponding points. The centre of gravity of $\Delta\, C_1A_2A_3$ is thus on the McCay circle through $A_2''A_3''$ and is the reflection of M in the perpendicular bisector of (A_2A_3).

Theorem 148.] *The centre of each McCay circle lies on the perpendicular bisector of the corresponding side of the original triangle.*

We shall show in the next chapter that M_i is a centre of similitude for the corresponding McCay and Neuberg circles. The geometric proof seems to be, however, decidedly intricate.*

* McCleland, loc. cit., pp. 213 ff.

§ 6. Concurrent Circles and Concyclic Points.

We have so far had certain examples of circles through a number of notable points: the nine-point circle passed through twelve, the Brocard circle through ten. We shall next proceed to find, by induction, circles which contain notable points *ad libitum*. Suppose that on each side-line A_jA_k of our typical triangle we take a random point B_i. If P be the intersection of the circles $A_1B_2B_3$, $A_2B_3B_1$,

$$\angle B_2PB_3 = \angle A_3A_1A_2, \quad \angle B_3PB_1 = \angle A_1A_2A_3;$$

hence $$\angle B_1PB_2 = \angle A_2A_3A_1,$$

and our three circles are concurrent.

Theorem 149.] *If a point be marked on each side-line of a triangle, the three circles each through a vertex and the adjacent marked points are concurrent.**

The number of corollaries which flow from this truly admirable theorem is almost transfinite. Suppose that P lies within the triangle, the most important case,

$$\measuredangle A_iPA_j = \measuredangle A_k + \measuredangle A_kA_iP + \measuredangle A_kA_jP,$$
$$\measuredangle A_kA_iP = \measuredangle B_jB_kP,$$
$$\measuredangle A_iPA_j = \measuredangle A_k + \measuredangle B_k.$$

A similar result is easily found when P is not within. It appears also that if the angles of the $\Delta\, B_1B_2B_3$ be known the point P is also known.

Theorem 150.] *If a triangle with known angles have its vertices anywhere on specified side-lines of a given triangle, the three circles each through one vertex of the fixed triangle and two adjacent ones of the variable triangle are concurrent in a fixed point.*

The most interesting case is where the two triangles are similar. If $\measuredangle A_i = \measuredangle B_i$ we may take for B_i the point M_i.

* The earliest proof of this theorem known to the author is that of Miquel, 'Théorèmes de géométrie', *Liouville's Journal*, vol. iii, 1838.

If $\measuredangle A_i = \measuredangle B_k$ we take B_k infinitely near to A_i on A_iA_j, and similarly if $\measuredangle A_k = B_i$.

Theorem 151.] *If three points be so taken on the side-lines of a triangle that they are vertices of a triangle similar to the given one, then the three circles each through a vertex of the given triangle, and the two adjacent vertices of the new triangle are concurrent either in the centre of the circumscribed circle of the given triangle or in one of the Brocard points.**

Let the reader prove:

Theorem 152.] *The only case where the lines A_iB_i are concurrent in the point P is where they are the altitude lines of the triangle.*

We easily find from 67]:

Theorem 153.] *If the intersections of a circle with the side-lines of a triangle be divided into two groups of three, each group containing one point on each side-line, then the point of concurrence of the three circles each through one vertex and the adjacent points of the first group, and that of circles through each vertex and the adjacent points of the second group, are isogonal conjugates.*†

It is immediately evident by inversion that our fundamental theorem 149] holds equally well when the side-lines of the triangle are replaced by concurrent circles. It may then be reworded as follows:

Theorem 154.] *If four points on a circle or line be taken in sequence and if each successive pair be connected by a circle, the remaining intersections of successive pairs of circles are concyclic or collinear.*

Still another form for the theorem is as follows:

Theorem 155.] *If four circles be arranged in sequence, each two successive circles intersecting, and a circle pass*

* McCleland, loc. cit., ch. iii, takes this as the basis of the whole Brocard theory.

† This excellent theorem is due to Barrow, loc. cit., p. 252.

*through one point of each such pair of intersections, then the remaining intersections lie on another circle or a line.**

Let us give another proof of this theorem depending on different considerations. If a triangle be formed by the arcs of three circles c_1, c_2, c_3, and if $\measuredangle \overrightarrow{c_1c_2}$ mean the oriented angle of the half-tangents to two circles at a vertex of the triangle, those halves being taken which correspond to the positive orientation of the circle, then, if the three circles be concurrent, we have

$$\measuredangle \overrightarrow{c_1c_2} + \measuredangle \overrightarrow{c_2c_3} + \measuredangle \overrightarrow{c_3c_1} = 0.$$

Conversely, if this equation holds, it is easy to see that the circles are concurrent. Suppose now that we have a sequence of four circles c_1, c_2, c_3, c_4, and that one intersection of each two successive lines lies on c,

$$\measuredangle \overrightarrow{cc_1} + \measuredangle \overrightarrow{c_1c_2} + \measuredangle \overrightarrow{c_2c} = \measuredangle \overrightarrow{cc_2} + \measuredangle \overrightarrow{c_2c_3} + \measuredangle \overrightarrow{c_3c}$$
$$= \measuredangle \overrightarrow{cc_3} + \measuredangle \overrightarrow{c_3c_4} + \measuredangle \overrightarrow{c_4c} = \measuredangle \overrightarrow{cc_4} + \measuredangle \overrightarrow{c_4c_1} + \measuredangle \overrightarrow{c_1c} = 0.$$
$$\measuredangle \overrightarrow{c_1c_2} + \measuredangle \overrightarrow{c_2c_3} + \measuredangle \overrightarrow{c_3c_4} + \measuredangle \overrightarrow{c_4c_1} = 0.$$

Conversely, when this equation holds, the circle through three properly chosen intersections passes through the fourth. But when we move from one intersection on c_1 and c_2 to the other we have merely to reverse the sign of $\measuredangle \overrightarrow{c_1c_2}$; the theorem is thus proved.

Let us next suppose that we have given not three lines but four, no two being parallel nor any three concurrent. Let each line be used to determine the marked points on the other three; we thus get

Theorem 156.] *If four lines be given, whereof no two are parallel nor any three concurrent, the circumscribing circles of the triangles which they form three by three are concurrent.*

Let us call this the *Miquel point* of the four lines. If we invert with this as centre we get a second figure entirely analogous to the given one, but the present circles become

* Miquel, 'Mémoire de géométrie', *Liouville's Journal*, vol. ix, 1844, p. 23.

lines and the present lines become circles. The feet of the perpendiculars on the four new lines from the Miquel point are on the Simson line of the four new triangles for this point; the four reflections of the Miquel point in the four new lines are also collinear; hence, inverting back and remembering 34],

Theorem 157.] *The centres of the circles which circumscribe the triangle formed by four lines lie on a circle through the Miquel point.**

The following theorem is interesting in this connexion, though the proof is based upon different considerations which we leave to the reader.

Theorem 158.] *The centres of the circles which touch sets of three out of four given lines, whereof no three are concurrent or parallel, lie by fours on four circles.*

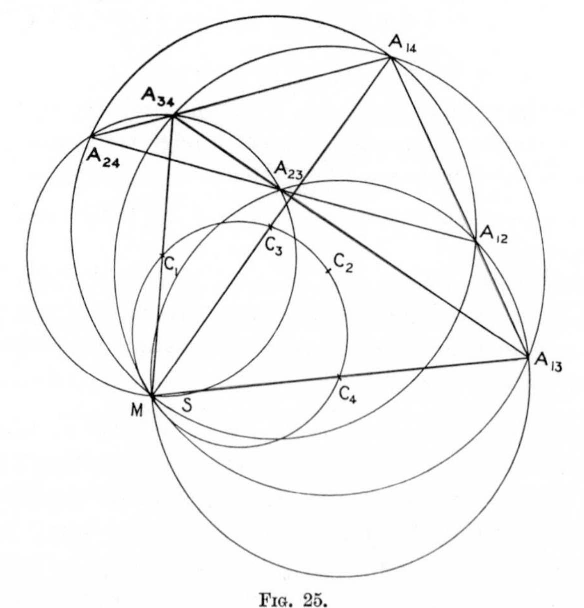

Fig. 25.

* Cf. Steiner, *Collected Works*, vol. i, p. 223.

The circle in theorem 157] seems to contain five notable points; we may easily find five others thereon. Let the original lines be l_1, l_2, l_3, l_4. The lines l_i and l_j shall intersect in A_{ij}, while the circle about the triangle formed by $l_i l_j l_k$ shall be c_l, its centre C_l, the Miquel point M.

We shall temporarily use $|\angle XYZ|$ for the positive value of $\angle XYZ$.

$$|\angle A_{jl} M C_i| = \frac{\pi}{2} - |\angle A_{jl} A_{jk} M|,$$

$$|\angle A_{jl} M C_k| = \frac{\pi}{2} - |\angle A_{jl} A_{ij} M|,$$

$$|\angle C_i M C_k| = |\angle A_{ik}|.$$

Let $A_{kl} C_i$ meet $A_{il} C_k$ in S,

$$\begin{aligned} |\angle C_i S C_k| &= |\angle A_{jl} A_{kl} C_i| - |\angle A_{jl} A_{il} C_k| \\ &= \frac{\pi}{2} - |\angle A_{jk}| - \left(\frac{\pi}{2} - |\angle A_{ij}|\right) \\ &= |\angle A_{ik}|. \end{aligned}$$

Theorem 159.] *Given four lines in a plane, no two parallel and no three concurrent. The lines connecting each vertex of a triangle formed by three of the lines with centre of the circle circumscribed to the triangle formed by the two lines meeting in this vertex and the fourth line, are concurrent on the circle through the centres of the four circumscribing circles.*

Suppose that five lines are given l_1, l_2, l_3, l_4, l_5. Omitting each in turn, we have five Miquel points. Consider the circles circumscribing the triangles with l_5 as a common side-line. Successive circles intersect on l_5; hence, by 154], their other intersections, which are Miquel points, are concyclic.

Theorem 160.] *If five lines be given, no two parallel and no three concurrent, the five Miquel points which they determine four by four are concyclic or collinear.*

Theorem 161.] *If a pentagon be given, and five triangles be constructed each having as vertices two adjacent vertices of the pentagon and the intersection of the remaining side-lines* *through them, then, if circles be circumscribed to these five triangles, the remaining intersection of pairs of successive circles are concyclic or collinear.**

Let us tabulate the results so far attained.

One line may be associated with a circle of infinite radius, the line itself.

Two lines may be associated with their point of intersection.

Three lines may be associated with the circle circumscribed to their triangle.

Four lines may be associated with their Miquel point.

Five lines may be associated with a circle or line through the five Miquel points which they determine four by four.

We are thus led by analogy to announce the following theorem:

Theorem 162.] *Given n lines in a plane, no two parallel and no three concurrent. If n be odd there is associated therewith a circle, and if n be even a point. The circle will contain the n points associated with the n sets of lines obtained by neglecting each of the given lines in turn; the point will lie on each of the n-circles obtained by neglecting each of the lines in turn.*†

It is to be understood for the purposes of this theorem that a line is considered as a special form of circle. Let us begin with the case where n is even. We take the three sets of lines $(l_2 l_3 \dots l_n)$, $(l_1 l_3 \dots l_n)$, $(l_1 l_2 l_4 \dots l_n)$. The associated circles shall be c_1, c_2, c_3, and, in general, the circle associated with the system obtained by omitting the line l_i shall be c_i. If lines $l_i l_j$ be omitted, the point associated with the others shall be P_{ij}, and so on.

* It is to this theorem alone that the name of Miquel is usually attached.

† This theorem is due to Clifford, 'A Synthetic Proof of Miquel's Theorem', *Messenger of Mathematics*, vol. v, 1870. Independently given by Fuortes, 'Ricerche geometriche', *Battaglini's Journal*, vol. xvi, 1878, and Kantor, 'Ueber den Zusammenhang von *n* Geraden in der Ebene', *Wiener Berichte*, vol. lxxvi, section v, 1877. Recently given without demonstration and in incorrect form by Hagge, 'Ueber Umkreise und Transversalen des vollständigen *n*-seits', *Zeitschrift für mathematischen Unterricht*, vol. xxxvi, 1905.

Circle c_{124} contains $P_{12}, P_{14}, P_{24}, P_{1234}$.
Circle c_{234} contains $P_{23}, P_{24}, P_{34}, P_{1234}$.
Circle c_{134} contains $P_{13}, P_{14}, P_{34}, P_{1234}$.

We have thus exactly the figure of three concurrent circles corresponding to **149**] generalized by inversion.

Points P_{12}, P_{13}, P_{14} lie on c_1.
Points P_{12}, P_{23}, P_{24} lie on c_2.
Points P_{13}, P_{23}, P_{34} lie on c_3.

Hence these three circles are concurrent, and as they are any three of our system the theorem is proved for n even, provided it holds for $n-1$. We now imagine that we have an odd number n of lines; let us show that any four of the points P_1, P_2, P_3, P_4 are concyclic or collinear.

The circles c_{12} and c_{23} meet in P_2 and P_{123}.
The circles c_{23} and c_{34} meet in P_3 and P_{234}.
The circles c_{34} and c_{41} meet in P_4 and P_{341}.
The circles c_{41} and c_{12} meet in P_1 and P_{412}.

But the four points $P_{123}, P_{234}, P_{341}, P_{412}$ are on the circle c_{1234}; hence the four points P_1, P_2, P_3, P_4 are concyclic or collinear, and so all of our points are on a circle or line.

Let us try another method of generalizing **149**]. We start with four lines l_1, l_2, l_3, l_4, and on each line l_i mark a point P_i. If these four be concyclic or collinear, then, by **154**], the six circles each of which passes through the intersection of two lines and the marked points thereon will pass by threes through four points on one circle or line. Suppose, next, that we have five lines l_1, l_2, l_3, l_4, l_5, and five concyclic marked points thereon. The point marked on l_i shall be P_i. The point obtained by omitting l_i and l_j shall be P_{ij}. The circle obtained from what immediately precedes by neglecting l_i shall be c_i; that which contains $P_iP_jP_{kl}$ shall be c_{ij}.

c_1 contains $P_{12}, P_{13}, P_{14}, P_{15}$.
c_2 contains $P_{12}, P_{23}, P_{24}, P_{25}$.
c_3 contains $P_{13}, P_{23}, P_{34}, P_{35}$.
c_{25} contains P_{13}, P_{14}, P_{34}.
c_{35} contains P_{12}, P_{24}, P_{14}.
c_{15} contains P_{23}, P_{24}, P_{34}.

But the circles c_{15}, c_{25}, c_{35} are concurrent in P_5; hence the circles c_1, c_2, c_3 are concurrent, and so all five are. The extension to n is as before, and, as before, we define a line as a special case of a circle.

Theorem 163.] *Let n concyclic or collinear points be marked on n lines whereof no two are parallel and no three concurrent. If n be even there is associated therewith a circle, and if n be odd a point; the circle will contain the n points associated with the n sets of $n-1$ lines obtained by neglecting each of the given lines in turn; the point will lie on each of the circles obtained by neglecting each of the lines in turn.**

In these two generalizations there is a distinction between n even and n odd. In the remarkable one which follows this disappears. Four coplanar lines are given, no two parallel and no three concurrent. Each line is associated with the circle circumscribing the triangle formed by the other three. The centres of these four circles are concyclic, and the circles themselves pass through the Miquel point.

Theorem 164.]‡ *Given n lines whereof no two are parallel and no three concurrent. Each set of $n-1$ will be associated with a circle in such a way that all n circles pass through a point, and their centres lie on a circle which is associated with the n given lines.*†

We shall assume that the theorem has been proven for $n-1$ lines. We use the previous notation for the circle associated with certain lines, its centre being indicated conformably, while the point associated with certain lines shall be indicated by the letter M with suitable subscript. We shall also assume that

$$\angle C_{ik} C_k C_{jk} = \angle C_{ik} M_k C_{jk} = \angle l_j l_i,$$

this equation being certainly true in the case $n = 4$, if C_{ij}

* Due to Grace, 'Circles, Spheres, and Linear Complexes', *Cambridge Philosophical Transactions*, vol. xvi, 1898.

† Pesci, 'Dei cercoli circonscritti ai triangoli formati di n rette in un piano', *Periodico di Matematica*, vol. v, 1891. The case $n = 5$ was given by Kantor, 'Ueber das vollständige Vierseit', *Wiener Berichte*, lxxviii, section 2, 1878.

‡ The credit for discovering this is, apparently, due to de Longchamps. See his 'Note de géométrie', *Nouvelles correspondances de mathématiques*, vol. iii, 1887, pp. 306 ff.

indicate the intersection of $l_k l_l$. Suppose that c_1 and c_2 intersect in M,

c_1 contains $C_{12}, C_{1i}, C_{1j}, M$.

c_2 contains $C_{12}, C_{2i}, C_{2j}, M$.

$$\angle C_{1i}MC_{12} = \angle C_{1i}C_{1l}C_{12} = \angle l_2 l_i;$$
$$\angle C_{12}MC_{2i} = \angle C_{12}C_{2m}C_{2i} = l_i l_1.$$
$$\angle C_{1i}MC_{2i} = \angle l_2 l_1.$$

But $$\angle C_{1i}C_{ji}C_{2i} = \angle l_2 l_1.$$

Hence M lies on c_i and the circles are concurrent. Again,

c_1 and c_i meet in C_{1i} and M.

c_i and c_2 meet in C_{2i} and M.

$$\angle C_1 C_i C_2 = \angle C_{1i}MC_{2i} = \angle l_2 l_1.$$

Hence all points C_i are concyclic, and the theorem is proved.

The following corollary is rather curious.

Theorem 165.] *If n be greater than four, M will not lie on c.*

Suppose, on the contrary, $\angle C_1 C_2 M = \angle C_1 C_3 M$.

$(C_{12}M)$ is the common chord of c_1 and c_2,

$$\angle C_1 C_2 M = \angle C_{12}C_{23}M,$$
$$\angle C_1 C_3 M = \angle C_{13}C_{23}M.$$

Hence $$\angle C_{12}C_{23}M = \angle C_{13}C_{23}M,$$
$$\angle C_{13}C_{23}C_{12} = 0.$$

But $C_{12}C_{23} \perp C_{123}M_2$ and $C_{23}C_{13} \perp C_{123}M_3$,

$$\angle M_2 C_{123} M_3 = 0.$$

This, however, is impossible since these three points lie on C_{23}. Hence, if C_{123} exist or $n > 4$, the point M cannot be concyclic with all C_i's.

In the theorem last given we associated n lines with a circle and a point, the circle being the locus of the centres of n others. In the theorem before we associated n circles or points with n lines and n concyclic points. Here is another form of association akin to both.

Theorem 166.]† *Given n points on a fixed circle $n \geqq 4$. We may associate with them a point and a circle in the following manner:*

(*a*) *The point is the centre of the circle.*

(*b*) *The radius of the circle is one-half that of the fixed circle.*

(*c*) *The point lies on the n circles each associated with $n-1$ points obtained by omitting each of the given points in turn.*

(*d*) *The circle contains the centres of these circles.**

Let us call the radius of the fixed circle 2, for convenience. When $n = 2$ we shall associate with two points the point midway between them. When $n = 3$ we associate the nine-point circle whose radius is here unity. When $n = 4$ we have, by 77], four nine-point circles passing through a common point. Their centres lie, therefore, on a circle of radius 1 about that point as centre. The theorem thus holds when $n = 4$. To prove it in the case where $n = 5$ we proceed, exactly as in the case of 162], to prove that the circles are concurrent. P_{ij} will be the centre of the nine-point circle of the $\Delta\, P_k P_l P_m$. c_{ijk} will be the circle through the centres of the three nine-point circles associated with P_l and P_m, i. e. the locus of points at a unit distance from the middle point of $(P_l P_m)$. P_{ijkl} will be the point midway between P_m and the centre of the fixed circle, which is at a unit's distance from the middle point of each of the chords $(P_i P_m)$, $(P_j P_m)$, $(P_k P_m)$, $(P_l P_m)$, and so on all three circles c_{jkl}, c_{kil}, c_{ijl}. We may thus repeat our previous reasoning word for word; the five circles c_1, c_2, c_3, c_4, c_5 are concurrent, and as all have a unit radius their centres lie on a unit circle about the point of concurrence as centre. For $n > 5$ we proceed in exactly the same way.

Here is a second proof of the foregoing that has the advantage of being easily extended to the analogous case in three dimensions, while our first proof cannot be so enlarged. Take $n = 4$, the centre of gravity of the four points will be

* See the Author's 'Circles Associated with Concyclic Points', *Annals of Mathematics*, Series 2, vol. xii, 1910.

† This is contained implicitly in an elaborate theorem due to Burgess, 'Theorems connected with Simson's Line', *Proceedings Edinburgh Math. Soc.*, vol. xxiv, 1906, p. 126.

the point of concurrence of the segments, each bounded by one given point and the centre of gravity of the other three, and will divide these segments in the ratio 1 : 3. The centres of gravity of the four triangles will thus lie on a circle whose radius is one-third the radius of the given circle; hence, by what precedes 72], the centres of the four nine-point circles lie on a circle of half the radius of the circumscribed circle, and whose distance from the fixed centre is $\frac{4}{2}$ the distance to the centre of gravity. The theorem thus holds for $n=4$. Assume that it is true for $n-1$ points, and that the centre of their circle is collinear with the centre of the fixed circle, and the centre of gravity of the $n-1$ points, but the distances from the centre of the fixed circle to these points is in the ratio $\frac{n-1}{2}$. If n points be given, we have n centres of gravity of groups of $n-1$ points. These lie on a circle whose radius bears to that of the fixed circle the ratio $1 : n-1$. Hence the n points lie on a circle whose radius is one-half that of the fixed circle, and the n associated circles pass through a fixed point at the proper distance from the fixed centre.

§ 7. Coaxal Circles.

We have defined the power of a point with regard to a circle as the product of its oriented distances to any two points of the circle collinear with it. When the point is outside the circle this is the square of the length of the tangential segment. The sum of the power and the square of the radius is seen to be the square of the distance from the point to the centre. We see, thus, that if a point move along a line perpendicular to the line of centres of two non-concentric circles, the difference of its powers with regard to the two is constant.

Theorem 167.] *The locus of points having like powers with regard to two non-concentric circles is a line perpendicular to the line of centres.*

This line is called the *radical axis.* It is the common secant when the circles intersect, the common tangent when they touch.

Theorem 168.] *The radical axes formed by the pairs of three given circles whereof no two are concentric, are concurrent or parallel.*

The point of concurrence, when it exists, is called the *radical centre* of the three. It is the only point having equal powers with regard to all three, and when these powers are all positive it is the centre of a circle whose radius is the square root of this power, and which cuts the three given circles at right angles.

Let us calculate the difference of the powers of a point with regard to two given circles. When the circles are concentric, it is the difference of the squares of the radii. Suppose them non-concentric. Their centres shall be CC', their radii $r\,r'$, while the distance of their centres shall be d. Let F be the intersection of the radical axis with the line of centres

$$(\overrightarrow{CF})^2-(\overrightarrow{C'F})^2 = r^2-r'^2$$

$$(\overrightarrow{CF})-(\overrightarrow{C'F}) = \pm d$$

$$(CF) = \frac{r^2-r'^2+d^2}{2\,d}. \tag{24}$$

Now let P be any point, H the foot of the perpendicular from there on the line of centres, its powers with regard to the two circles p and p'. We easily find

$$p-p'= 2\,(\overrightarrow{FH})\,d. \tag{25}$$

Theorem 169.] *The numerical value of the difference of the powers of a point with regard to two non-concentric circles is twice the product of its distance from the radical axes multiplied by the distance between the centres.*

If a point be taken upon the circle of similitude of two circles, outside of both, and a tangent be drawn thence to each circle, the two not separated by the centres, it will be found at once from 26] that the chords which the circles determine on the line connecting the points of contact are equal, so that the power of each point of contact with regard to the other circle is the same; the converse will also hold, hence

Theorem 170.] *If two points be taken on two unequal and non-concentric circles in such a way that each has the same power with regard to the other circle, and the tangents at these points are not separated by the centres of the circles, then the intersection of these tangents is on the circle of similitude of the two.*

Let the distance from C to a point of CC' be x, this point being the centre of a circle of radius ρ. If

$$\frac{r^2-r'^2+d^2}{2\,d}=\frac{r^2-\rho^2+x^2}{2\,x},$$

$$x^2-\frac{(r^2-r'^2+d^2)}{d}\,x+r^2=\rho^2;$$

then each two of our three circles have the same radical axis. Let us put $\rho=0$, and consider the discriminant of the resulting quadratic in x; we assume $r^2 \geqq r'^2$.

$$\frac{(r^2-r'^2+d^2)^2}{d^2} \gtreqless 4\,r^2.$$

Assume, first, $\qquad (r-d)^2-r'^2 > 0.$

The two original circles did not intersect; there are two real values of x for which $\rho=0$, i.e. two points which may be looked upon as limiting circles of radius zero. These shall be called the *limiting points* of the system of circles. The power of a point with regard to a point circle shall, naturally, be defined as the square of its distance from the point to which the circle has shrunk. Any point outside the segment of these limiting points may be taken as the centre of a circle having with either of the original circles the same radical axis as they have with one another.

Suppose, secondly, $\quad (r-d)^2-r'^2=0.$

Here the original circles are tangent to one another. Their point of contact is the single limiting point of the system, every other point of the line of centres in the centre of a circle touching the two at the limiting point.

Lastly, let $\qquad (r-d)^2-r'^2 < 0.$

Here the given circles intersect in real points. Any point on their line may be taken as the centre of a circle through their intersections; the least possible radius for such a circle will be one-half the distance between these two common points, and there are no limiting points in the system.

If a system of circles be so related that each two have the same radical axis, they are said to be *coaxal circles*. Circles through two common points or touching the same line at the same point are examples of such systems.

A system of circles through two points will cut interesting ranges on any line through either point.* Let two such points be A and B, and let two lines through B meet the various circles in the ranges $P_1P_2 \ldots P_n$ and $Q_1Q_2 \ldots Q_n$ respectively.

Since $\triangle AP_jP_k$ and $\triangle AQ_jQ_k$ are similar,

$$(P_iP_j):(Q_iQ_j) = (AP_j):(AQ_j) = (P_jP_k):(Q_jQ_k),$$
$$(P_iP_j):(P_jP_k):(P_kP_l) = (Q_iQ_j):(Q_jQ_k):(Q_kQ_l).$$

Theorem 171.] *A system of circles through two points cut such ranges on any two lines through one of the points that corresponding distances are proportional, and, conversely, if two ranges be given on intersecting lines in such a way that corresponding distances are proportional and the point of intersection does not correspond to itself, then the lines connecting corresponding points in the two ranges are concurrent in a point common to all circles containing a pair of corresponding points and the point of intersection of the two lines.*

Since the Simson line of A is the same for all triangles BP_jQ_j,

Theorem 172.] *If a system of circles through two points cut ranges on two lines through one of these points, then the feet of the perpendiculars from the other point on all lines connecting corresponding pairs of points of the two ranges are collinear.*

Theorem 173.] *If two circles cut a third either orthogonally or in two pairs of diametrically opposite points of the latter,*

* Casey, loc. cit., ch. v.

then the centre of the third circle is on the radical axis of the two, and every point of the radical axis not between the intersections of the circles, when such exist, is the centre of a circle cut by both at right angles, while every point between such intersections is the centre of a circle cut by both in pairs of diametrically opposite points.

We see, in fact, that if a point have the same positive power with regard to two circles it is the centre of a circle cutting both orthogonally, while if it have the same negative power with regard to both it is the centre of one cut by both in pairs of diametrically opposite points, the radius being in the first case the square root of the power, and in the second the square root of the negative of the power.

Theorem 174.] *If two circles intersect two others orthogonally, then every circle coaxal with (orthogonal to) one pair is orthogonal to (coaxal with) the other. The radical axis of each system is the line of centres of the other.*

We see that the plane is thus covered with a double net-work of circles in such a way that every point not on either radical axis is the intersection of two circles, one of each net-work, and these circles cut orthogonally. Remembering that the limiting points of a coaxal system are point circles of the system,

Theorem 175.] *If two circles intersect, the coaxal system of circles cutting them orthogonally will have their points of intersection as limiting points; if two non-concentric circles do not intersect, the limiting points of their coaxal system are common to all circles orthogonal to them.*

Theorem 176.] *If a system of circles be tangent to one another at any point, they are orthogonal to a second system tangent at this point.*

Two coaxal systems of mutually orthogonal circles are said to be *conjugate.*

Theorem 177.] *The limiting points of a coaxal system of circles are mutually inverse with regard to every circle of the system.*

Theorem 178.] *The inverse of a coaxal system is either a coaxal or concentric system, or a pencil of concurrent or parallel lines.*

Theorem 179.] *If three non-coaxal circles be given, no two concentric or intersecting, the three pairs of limiting points which they determine two by two are concyclic or collinear.*

Theorem 180.] *Two mutually inverse circles are coaxal with the circle of inversion.*

Theorem 181.] *If two points A and C lie on a circle orthogonal to all circles through B and D, then B and D lie on a circle orthogonal to all circles through A and C.*

Two such pairs of points are said to be *orthocyclic*.

It is perfectly clear that the circles of a coaxal system with two common points may be inverted into a system of concurrent lines. A system with no common point, being the orthogonal trajectories of a system with two such points, may be inverted into a concentric system. A system all tangent at one point may be inverted into parallel lines. The following theorem has already been suggested.

Theorem 182.] *If the radical centre of three circles lie without one, and, hence, without all of them, it is the centre of a circle cutting all three orthogonally, and they may be inverted into three circles with collinear centres; if it lie within one, and, hence, within all three, it is the centre of a circle cut by all three in pairs of diametrically opposite points.*

Suppose that we have a triangle with our usual notation, a point P not on any side-line, and let P_jP_k meet A_jA_k in B_i. Applying Desargues' triangle theorem to $\triangle A_1A_2A_3$ and $\triangle P_1P_2P_3$, since the three lines A_iP_i are concurrent, the points B_i, B_j, B_k are collinear. If, in particular, P be the orthocentre, we see H_jH_k is anti-parallel to A_jA_k, and the points A_j, A_k, H_j, H_k are concyclic, i.e. B_i has the same power with regard to the nine-point and circumscribed circles.

Theorem 183.] *The radical axis of the nine-point and circumscribed circles contains the intersections of corresponding side-lines of the given triangle and the pedal triangle of the orthocentre.*

Theorem 184.] *The orthocentre of a triangle is the radical centre of any three circles each of which has a diameter whose extremities are a vertex and a point of the opposite side-line, but no two passing through the same vertex.*

We see, in fact, that, since the orthocentre is a centre of similitude for the circumscribed and nine-point circles, the product of its distances from each vertex and the foot of the corresponding altitude is constant. Suppose next that we have a complete quadrilateral.* The orthocentres of the triangles formed by the given side-lines three by three will, apparently, all be radical centres for the three circles whose diameters are the three diagonals of the complete quadrilateral. This apparent contradiction leads to the Gauss-Bodenmiller theorem.

Theorem 185.] *The three circles on the diagonals of a complete quadrilateral as diameters are coaxal.*†

We get from 74] and 156]

Theorem 186.] *The radical axis of the three circles on the diagonals of a complete quadrilateral as diameters contains the orthocentres of the four triangles determined three by three by the side-lines of the quadrilateral, and is parallel to the Simson line of the Miquel point, but twice as far from this point as is the Simson line.*

Theorem 187.] *Two pairs of circles c_1c_3 and c_2c_4 are each coaxal with a given circle; then if c_1 intersect c_2 and c_3 intersect c_4, the four points so determined are concyclic.*‡

Suppose that we have three segments each bounded by one vertex of a triangle and a point of the opposite side-line, and

* Cf. McCleland, loc. cit., p. 189. Not a little of the remainder of the present chapter is taken from this source.

† Bodenmiller, *Analytische Sphärik*, Cologne, 1830, p. 138.

‡ Chasles, loc. cit., p. 540.

all having in common a point S. Let the perpendiculars on the lines of these segments from the orthocentre meet the circles having the segments as diameters in the three pairs of points $B_i B_i'$. Let us first show that these six points are concyclic. We see, in fact, since H is the radical centre of the three circles with diameters $(A_i S_i)$, $(A_j S_j)$, $(A_k S_k)$,

$$(\overrightarrow{HB_1}) \times (\overrightarrow{HB_1'}) = (\overrightarrow{HB_2}) \times (\overrightarrow{HB_2'}) = (\overrightarrow{HB_3}) \times (\overrightarrow{HB_3'}).$$

On the other hand, the perpendicular bisectors of the segments $(B_i B_i')$ pass through S, hence S is the centre of a circle through all six. We next notice that H has the same power with regard to the three circles on diameters $(A_i S_i)$ as with regard to those with diameters $(A_j A_k)$. If we take M as a centre of similitude, and a ratio $-1:2$, we change $A_1 A_2 A_3$ into $M_1 M_2 M_3$. Let T be transformed into S.

Let $C_i C_i'$ be the points where the line through $H \perp TA_i$ meets the circle whose diameter is $(A_j A_k)$. Once more

$$(\overrightarrow{HC_i}) \times (\overrightarrow{HC_i'}) = (\overrightarrow{HC_j}) \times (\overrightarrow{HC_j'}).$$

The perpendicular bisectors of the segments $(C_i C_i')$ pass through the points M_i and are parallel to the lines $A_i T$, and correspond to them in the similarity transformation. They are thus concurrent in S. Lastly, since H has like powers with regard to all six circles,

Theorem 188.] *If S be any point, and T the point which bears to the original triangle the same relation that S bears to the middle point triangle $M_1 M_2 M_3$, then the intersections of the circles on segments $(A_i S_i)$ as diameters with the perpendiculars from the orthocentre on these segments, and of the circles on the segments $(A_j A_k)$ as diameters with the corresponding perpendiculars from the orthocentre on the lines TA_i, lie on a circle with centre S.**

Let us return to our theorem 169] from which flow a great wealth of interesting results. If a circle of radius ρ have contact of a specified kind with two others, the difference of the powers of its centre with regard to these two will be $2\rho(r \pm r')$.

* Hagge, 'Ein merkwürdiger Kreis des Dreiecks', *Zeitschrift für mathematischen Unterricht*, vol. xxxix, 1908.

Theorem 189.] *If a variable circle have contact of a fixed type with each of two given non-concentric circles, or have just the reverse contact with each, then its radius will bear a fixed ratio to the distance of its centre from the radical axis.*

If we call the distance to the radical axis δ, the fixed ratio e, and take a line parallel to the radical axis at a distance $\frac{r}{e}$ therefrom and on the proper side thereof, the distance of the centre of the variable circle therefrom is $\frac{1}{e}$ times its distance from the centre of the circle of radius r. We see, also, that the sum or difference of the distances from the variable centre to the fixed centres is constant. We thus reach the fundamental theorem for central conics.

Theorem 190.] *If a point so move that the sum or difference of its distances from two fixed points is constant, its distance from either fixed point bears a constant ratio to its distance from a corresponding fixed line perpendicular to that which connects the fixed points.*

The power of a point with regard to any circle of a coaxal system is by 169] twice the product of its distance from the radical axis multiplied by the distance from the centre of the circle to that of the circle of the coaxal system through the given point. The point is supposed, of course, not to be on the radical axis.

Theorem 191.] *If there be any points whose powers with regard to two non-concentric circles bear a fixed given ratio different from unity, they all lie on one circle coaxal with the two given ones.*

The necessity of the proviso that such points should exist is apparent when we reflect that if, for instance, the circles were very small and far apart there could be no point corresponding to such a ratio as -1.

Theorem 192.] *If a variable chord of a circle subtend a right angle at a fixed point, the foot of the perpendicular from the fixed point on the line of the chord and the point of inter-*

sections of the tangents at its extremities trace coaxal or concentric circles.

We see, in fact, that if we treat the given point as a circle of radius zero, the foot of the perpendicular on the line of the chord and the middle point of that chord trace the same circle, since the power of each with regard to the given circle is the negative of the square of its distance from the fixed point. We have then but to apply 180].

Suppose that a variable line meets one circle in points S_1T_1, and makes therewith an angle α_1, while its intersections and angle with a second circle are S_2T_2 and α_2. If we find a point where a tangent at S_1 or T_1 meets one at S_2 or T_2, we see that the tangential segments thence to the two circles bear the ratio $\sin\alpha_2 : \sin\alpha_1$.

Theorem 193.] *If a pair of tangents be drawn to each of two circles, the points of contact being collinear, then the intersections of the tangents to one circle with the tangents to the other will lie on a circle concentric or coaxal with the given circles, or on their radical axis.*

Theorem 194.] *If a variable line move in such a way that the segments cut thereon by two fixed circles have a constant ratio, then the locus of the intersections of the tangents to the first circle where it meets this line, with the corresponding tangents to the second, is a circle concentric or coaxal with the given circles or their radical axis.*

Suppose that we have a quadrilateral inscribed in a circle. If a transversal be so drawn that it makes an isosceles triangle with one pair of side-lines, or with the diagonal lines, then it will do so with the remaining side-lines or diagonal ones. Let us also momentarily extend our concept of the isosceles triangle so as to say that a line perpendicular to two parallel lines makes an isosceles triangle with them.

Theorem 195.] *If a quadrilateral be inscribed in a circle, then we may, in an infinite number of ways, find three other circles concentric or coaxal with the original circle and with one another such that each is tangent to a pair of opposite side-lines or diagonal lines of the quadrilateral.*

Theorem 196.] *If a variable triangle be inscribed in a circle, and if two of its side-lines continually touch two circles concentric or coaxal with the given circle, the same is true of the third side-line.*

A rigorous proof of this is not difficult, but a little delicate. Let us take two positions of our triangle $A_1A_2A_3$, $A_1'A_2'A_3'$. Suppose that A_1A_2 and $A_1'A_2'$ touch a certain circle, while A_1A_1', A_2A_2' touch another concentric or coaxal with this and with the given circle. In the same way A_1A_1', A_3A_3' will touch a third circle of the coaxal system. Now it is conceivable that the circle touched by A_1A_1', A_2A_2' should be different from that touched by A_1A_1', A_3A_3', for two circles of the coaxal system might well both touch A_1A_1'. If, however, we can show in a particular case that one of the circles of the coaxal system tangent to A_1A_1' is extraneous to the discussion, we shall know that in general both circles will not appear. The particular case is when A_1' is infinitely near A_1, the circle tangent to A_1A_1' is the original circle, the other coaxal circle tangent to this is distinct from this and not connected with the discussion. Hence A_1A_1', A_2A_2', A_3A_3' all touch one circle of the coaxal system, and A_2A_3, $A_2'A_3'$ also touch one of these circles.

Theorem 197.] *Poncelet's theorem. If a polygon of any number of sides be inscribed in a circle, and all of its side-lines but one each touch a fixed circle of a system concentric or coaxal with the given one, then the same is true of the remaining side-line.**

Theorem 198.] *The problem of inscribing a polygon of any given number of sides in a given circle so that its side-lines shall also touch a second given circle has an infinite number of solutions if it have any at all.*

There are certain special cases coming under our theorem 191] which deserve particular notice. If the fixed ratio be unity we do not get a circle, but the radical axis. Let us rather look at the case where the ratio is -1. The locus is in this

* For the most full discussion of this and allied questions see Weill, 'Sur les polygones inscrits et circonscrits à la fois à deux cercles', *Liouville's Journal*, Series 3, vol. iv, 1878; also Bützberger, loc. cit., pp. 7–32.

case called the *radical circle* of the two original ones.* It will actually exist if the circles intersect, or if, not intersecting, they lie on the same side of their radical axis, or if they be concentric. The centre will be half-way between their centres. We leave the verification of these statements to the reader.

Theorem 199.] *Two intersecting or tangent circles, or two non-intersecting circles which are concentric or else lie on the same side of their radical axis, have a radical circle which is the circle coaxal with them whose centre is mid-way between their centres.*

The slight modification needed in the case of concentric circles is easily made.

Theorem 200.] *If three circles be given whereof no two are concentric, the radical circle of each pair is identical with that of the radical circles determined by the circles of the pair severally with the third circle.*

The truth of this theorem is, of course, contingent on the existence of all the radical circles in question. We see, moreover, that the radical axes of both pairs of circles are parallel, for one is orthogonal to a side-line of a triangle while the other is perpendicular to the line connecting the middle points of the other two sides. Moreover, the radical centre of three original circles is easily seen to be the radical centre of the radical circles which they determine two by two. The theorem is thus proved.

If a circle be cut by a second orthogonally, while it is cut by a third in a pair of diametrically opposite points, its centre has powers with regard to the other two circles which differ only in sign.

Theorem 201.] *The radical circle of two given circles is the locus of the centres of circles cut by the one orthogonally, and by the other in diametrically opposite points.*

* Cf. Duran-Loriga, 'Ueber Radicalkreise', *Grunerts Archiv*, Series 2, vol. xv, 1896.

Theorem 202.] *The pedal circle of two isogonally conjugate points is the radical circle of any pair of circles whose centres are these points, and each of which cuts orthogonally the three circles whose diameters are the segments cut by the other on the side-lines of the triangle.**

Theorem 203.] *The nine-point circle of an obtuse-angled triangle is the radical circle of the circumscribed circle and a circle of anti-similitude of this and the pedal circle of the orthocentre.*

Theorem 204.] *The circles on two sides of a triangle as diameters have the circle whose diameter is the included median as their radical circle.*

Besides the radical circle there is one other circle of the coaxal family that is interesting. We see at once from 27]

Theorem 205.] *The circle of similitude of two given circles is coaxal with them.*

Consider, now, three circles with non-collinear centres. The three circles of similitude which then determine two by two cut orthogonally the common orthogonal circle of the original three, when that exists, and the circle through the centres of the original three by 13]. This latter will be the radical circle of the common orthogonal circle and the circle cutting each of the original three in a pair of diametrically opposite points.

Theorem 206.] *If three circles have non-collinear centres, and their radical centre lies outside of them, then the circle through their three centres is the radical circle of their common orthogonal circle and that circle, when it exists, which cuts each of the three in a pair of diametrically opposite points. The coaxal system conjugate to that determined by these new circles contains the circles of similitude determined two by two by the given circles.*

* Roberts, 'On the Analogues of the Nine-point Circle in the Space of Three Dimensions', *Proceedings London Mathematical Society*, vol. xix, 1878.

Theorem 207.] *If a circle move so that each of two given points has a constant power with regard to it, it will trace a coaxal system.*

The line connecting the points is a radical axis for any two positions of the circle.

Theorem 208.] *If a circle so move that it cuts two others in diametrically opposite points, or cuts one in diametrically opposite points and the other orthogonally, it will generate a coaxal system.*

Theorem 209.] *If three mutually external circles be given, their centres being non-collinear, three other circles may be found each cutting two of them orthogonally and the third in diametrically opposite points, and three each cutting two in diametrically opposite points and the third orthogonally.**

Theorem 210.] *Given two non-concentric circles. If there be a circle coaxal with them whose centre is the reflexion of the centre of the first circle in that of the second, then this third circle will cut in diametrically opposite points all circles orthogonal to the first circle whose centres lie on the second.*

Theorem 211.] *Given two non-concentric circles. If there be a circle coaxal with them whose centre is the reflexion of the centre of the first circle in that of the second, then this circle will cut orthogonally all circles cut by the first in diametrically opposite points, and having their centres on the second.*

Let us start with two fixed circles. These may be inverted into concentric circles or into two lines. We thus get

Theorem 212.] *If a variable circle cut two fixed circles at given angles, it will cut every circle coaxal with them either at a fixed angle or at the supplementary angle.*

It is clear that there would be advantage in sharpening our idea of the angle of two circles in such a way as to remove the ambiguity in this statement. We do so as follows. Let each circle be looked upon as generated by a point

* For the three theorems which follow see Affolter, 'Zur Geometrie des Kreises und der Kugel', *Grunerts Archiv*, vol. lvii, 1875.

tracing its circumference in a positive or counter-clockwise sense. At a point of intersection draw the half-tangents which lie on the same sides of their respective diameters as near by-points of the circle traced subsequent to the point of contact. At each intersection these half-tangents will make the same angle, except for algebraic sign, and this shall be defined as the angle of the two circles.* Analytically it is the angle θ, where

$$\cos\theta = \frac{r^2 + r'^2 - d^2}{2rr'}\,. \tag{24}$$

Theorem 213.] *If a variable circle cut two fixed circles at given angles, it will cut at a given angle every circle concentric or coaxal with them.*

Suppose that we have a circle cutting three given circles at chosen angles. If we simplify the figure by inversion, we see that there will be a second circle cutting them at the same angles or cutting all three at just the supplementary angles; the two are mutually inverse in the common orthogonal circle of the first three, when this circle exists. The problem of finding a circle cutting three given circles at assigned angles or at exactly the supplementary angles has thus, usually, more than one solution. The circles will be orthogonal to three circles each coaxal with two of the given circles. These three new circles must be coaxal, as otherwise they would have but one common orthogonal circle. The circles sought will thus belong to a coaxal system, and touch six given circles, but every circle of the system touching one of the six will touch the other five.

Theorem 214.] *The problem of constructing a circle to cut three non-concentric and non-coaxal circles at preassigned angles or at just the supplements of these angles has, at most, two solutions. The construction may be effected by the aid of ruler and compass.*

We shall postpone to a subsequent chapter the explanation of the actual construction to be employed. For the moment

* The angle, so defined, will be transformed into its supplement by inversion if one circle surrounds the centre of inversion, and the other does not.

let us consider the problem of constructing a circle to meet certain given circles at equal angles. We easily see by inverting two circles into concentric circles or into two lines that a circle cutting them at equal angles will be orthogonal to one particular circle of anti-similitude, when such exists, and, conversely, every circle orthogonal to this circle of anti-similitude will cut them at equal angles, while a circle orthogonal to the other circle of anti-similitude will cut them at supplementary angles. To be more specific, we see that if two circles intersect, both circles, of anti-similitude, exist; the circles which cut them at equal angles are orthogonal to the external circle of anti-similitude, i. e. to that whose centre is the external centre of similitude; a circle cutting them at supplementary angles will be orthogonal to the internal circle of anti-similitude. If two circles lie outside one another, there is no internal circle of anti-similitude, and circles cutting them at equal angles are orthogonal to the external circle of anti-similitude, or to the radical axis when the radii are equal. When one circle surrounds the other there is no external circle of anti-similitude, and the internal one is orthogonal to those circles which cut the two at supplementary angles.

Theorem 215.] *If a circle cut two others at equal angles it is orthogonal to their external circle of anti-similitude when this circle exists, and every such circle cuts them at equal angles if at all. If a circle cut two others at supplementary angles it will be orthogonal to their internal circle of anti-similitude when such a circle exists, and every circle orthogonal to an internal circle of anti-similitude will cut the given circles at supplementary angles if it cut them at all.*

If a circle cut two others and be orthogonal to a circle of anti-similitude, it is anallagmatic with regard to the inversion in that latter circle (which interchanges the original circles). The intersections with the original circles are thus collinear in pairs with the centre of this circle of anti-similitude. If the circle of anti-similitude do not exist, and be not replaced by the radical axis, the given circles are interchanged by the product of a reflexion in the centre of similitude, and an inversion in a circle with this as centre, and every circle invariant for such a transformation will cut the original circles in equal or supplementary angles. Conversely, if a circle cut two others at equal or supplementary angles, yet be not orthogonal to a circle of anti-similitude or radical axis, it is easily seen to be carried into itself by such a transformation.

Theorem 216.] *If a circle intersect two other non-concentric circles of unequal radius at equal angles, the points of intersection are collinear two by two with the external centre of similitude; if it intersect two others at supplementary angles, the points of intersection are collinear in pairs with the internal centre of similitude.*

Theorem 217.] *If each of two non-concentric circles cut two other non-concentric ones at one same angle, then the radical axis of each pair passes through the external centre of similitude of the other pair or is parallel to their line of centres when the circles of the second pair have equal radii. If each of two non-concentric circles make supplementary angles with each of two other non-concentric circles, and each circle of the second pair make supplementary angles with each of the first, then the internal centre of similitude of each pair lies on the radical axis of the other.*

The radical axis of two circles will replace the external circle of similitude when, and only when, they have equal radii, whence

Theorem 218.] *If a centre of inversion be taken on a circle of anti-similitude, the inverses of two given circles will have equal radii.*

Suppose next that we have three circles. A line connecting two external centres of similitude will pass through the third, when the latter exists; a line connecting two internal centres of similitude will pass through an external centre.

Theorem 219.] *If three circles be given with non-collinear centres, the circles cutting them at equal angles form a coaxal or concentric system, as do those which cut one at angles supplementary to those cut on the other three. The locus of the centres is the perpendicular from the radical centre of the original three on a line containing three of the centres of similitude which they determine two by two.*

Theorem 220.] *If four circles be given, no three having collinear centres, there is at most one circle cutting all at equal angles, four cutting one in angles supplementary to the angles cut in the other three, and three cutting one pair in angles supplementary to those cut in the other pair.*

Theorem 221.] *A necessary and sufficient condition that it should be possible to invert three circles simultaneously into three circles with equal radii is that a circle of anti-similitude of one pair should intersect such a circle of another pair in a point outside all three given circles.*

It is a parlous undertaking to suggest possible lines of further advance in the subject of plane geometry. On the one hand, the subject has shown itself inexhaustibly fertile, new discoveries have come in such numbers at times when a superficial observer would have felt sure that the last word had been said, that it would be highly unwise to assert that with a little patience one might not strike oil by working in any portion of the subject. On the other hand, the existing literature is so vast that there is a large antecedent probability that any new seeming result may have been discovered decades if not centuries before.

It seems likely that there are other simple criteria for various systems of tangent circles like Casey's condition for four circles tangent to a fifth, Vahlen's criterion for poristic systems, or the Euler conditions that there may be a triangle or quadrilateral inscribed in one circle which is circumscribed to the other. There seem to be limitless possibilities for finding circles through notable points or tangent to notable lines. There must be other circles analogous to the P circle. It seems likely also that there are other special cases of Tucker circles which are worthy of attention. Moreover, it may be possible to generalize the Tucker systems in interesting ways as suggested by 67]. It seems likely that there are other chains of concurrent circles and concyclic points besides those noticed in theorems 162–6. The Brocard figures seem to offer an inexhaustible store of theorems. It is quite likely also that in coaxal systems of circles there may be other interesting circles besides the special ones which we have discussed. For instance, the following theorem came to our notice too late to be inserted in its proper place.

Theorem 222.] *If a transversal through the centre of the circumscribed circle meet the side-lines of a triangle in the points* B_1, B_2, B_3, *the circles on* (A_iB_i) *as diameters are concurrent on the circumscribed and nine-point circles.**

The concurrence on the nine-point circle comes from 68], that on the circumscribed circle comes from 184] and the remark immediately following.

* Thébault, 'Sur quelques théorèmes de géométrie élémentaire', *Nouvelles Annales de Math.*, Series 4, vol. x, 1910.

CHAPTER II

THE CIRCLE IN CARTESIAN PLANE GEOMETRY

§ 1. The Circle studied by means of Trilinear Coordinates.

ALL figures studied in the present chapter are supposed to exist in one plane which has been rendered a perfect continuum by the adjunction of the line at infinity. The complex domain is included as well as the real. We call this the *cartesian plane.* The assemblage of all points in such a plane may be put into one to one correspondence with that of all triads of homogeneous coordinate values, not all simultaneously zero.

In studying circles in the cartesian plane, three types of coordinates may properly be used. We start with the least fruitful, trilinear coordinates.* Let us take a fundamental triangle whose side-lines have the equations

$$\cos \alpha_i x + \sin \alpha_i y - \pi_i = 0, \quad i = 1, 2, 3. \tag{1}$$

We take as the trilinear coordinates of any finite point, whose cartesian rectangular coordinates are (x, y), the three quantities

$$p_i \equiv -(\cos \alpha_i x + \sin \alpha_i y) + \pi_i. \tag{2}$$

We assume that the triangle surrounds the cartesian origin, so that each coordinate of the origin is positive. Every other point within the triangle will also have three positive

* Cf. Whitworth, *Trilinear Coordinates*, London, 1866; Casey, *Analytic Geometry of the Point, Line, Circle, and Conic Sections*, Dublin, 1893, a brilliant but untrustworthy book. Also Clebsch-Lindemann, *Vorlesungen über Geometrie*, second ed., Leipzig, 1906, vol. i, part 2, pp. 312 ff.

coordinates. The coordinates of any finite point will be connected by the fundamental identity

$$\sum_{i=1}^{i=3} a_i p_i \equiv 2\Delta. \tag{3}$$

If the left-hand side of this equation be equated to zero, we have the equation of the line at infinity, whose points may be put into one to one correspondence with sets of coordinate values satisfying such an equation. If the radius of the circumscribed circle be r, our fundamental identity may also be written

$$\sum_{i=1}^{i=3} \sin \measuredangle A_i p_i \equiv \frac{2\Delta}{r}. \tag{4}$$

Let us begin by finding the equation of this circumscribed circle. Since it is a conic circumscribing the triangle of reference, it must come under the form

$$\sum_{i=1}^{i=3} \lambda_i p_j p_k = 0.$$

We determine the coefficients by noticing that the tangent at a vertex will have an equation

$$\sin \measuredangle A_j p_k + \sin \measuredangle A_k p_j.$$

Hence the equation of our circle may be put in any one of the three forms

$$\sum_{i=1}^{i=3} \sin \measuredangle A_i p_j p_k = 0, \quad \sum_{i=1}^{i=3} a_i p_j p_k = 0, \quad \sum_{i=1}^{i=3} \frac{a_i}{p_i} = 0. \tag{5}$$

We may proceed in similar fashion to find the equation of the inscribed circle. In line coordinates it must have an equation of the form

$$\sum_{i=1}^{i=3} \mu_i u_j u_k = 0.$$

The point of contact of a side-line will have the equation

$$\cos^2 \tfrac{1}{2} \measuredangle A_k u_j + \cos^2 \tfrac{1}{2} \measuredangle A_j u_k = 0.$$

Hence the equation of the inscribed circle will be

$$\sum_{i=1}^{i=3} \cos^2 \tfrac{1}{2} \measuredangle A_i u_j u_k = 0. \tag{6}$$

$$\sum_{i=1}^{i=3} \cos^4 \tfrac{1}{2} \measuredangle A_i p_i^2 - 2\sum_{i=1}^{i=3} \cos^2 \tfrac{1}{2} \measuredangle A_j \cos^2 \tfrac{1}{2} \measuredangle A_k p_j p_k = 0.$$

This last equation is factorable, giving the reduced form for the equation

$$\cos \tfrac{1}{2} \measuredangle A_i \sqrt{p_i} \pm \cos \tfrac{1}{2} \measuredangle A_j \sqrt{p_j} \pm \cos \tfrac{1}{2} \measuredangle A_k \sqrt{p_k} = 0. \tag{7}$$

The escribed circle corresponding to the side a_i will be likewise

$$i \cos \tfrac{1}{2} \measuredangle A_i \sqrt{p_i} \pm \sin \tfrac{1}{2} \measuredangle A_j \sqrt{p_j} \pm \sin \tfrac{1}{2} \measuredangle A_k \sqrt{p_k} = 0.$$

The equations of circles circumscribed or inscribed to a polygon of n sides may be found in like shape.* Suppose, in fact, that the sides of an inscribed polygon have the lengths $a_1 \ldots a_n$; the perpendiculars on these side-lines from any point of the circle shall be $p_1 \ldots p_n$. Taking this point as centre of inversion, we transform our circle into a straight line. Let p' be the distance from the centre of inversion to this line. Then

$$p_i = p' \frac{a_i}{(\overrightarrow{A_i' A_{i+1}'})}, \qquad \sum_{i=1}^{i=n} (\overrightarrow{A_i' A_{i+1}'}) = 0.$$

With regard to signs, we may take all of these segments except the extreme one as positive, while the latter, which comes from that side of the original polygon (supposed to be convex) which shuts the given point from the other sides, is negative. On the other hand, the number p_i corresponding to this side is negative, while the other numbers p_j are positive. Hence, for every point of our circle,

$$\sum_{i=1}^{i=n} \frac{a_i}{p_i} = 0.$$

The equation of the circle will be a factor of this.

* See a highly ingenious article by Casey, 'On the Equations of Circles', *Transactions Royal Irish Academy*, vol. xxvi, 1878.

To find the equation of an inscribed circle we see that if B_i be the point of contact of the side $(A_i A_{i+1})$, the circle must have an equation of the type

$$p_{i-1} p_i = \sigma p_i'^2,$$

where p_i' is the distance to $(B_{i-1} B_i)$. We see also that $\sigma = 1$, since our circle passes through the middle point of the circle inscribed in the triangle $B_{i-1} A_i B_i$. But, by the formula for the circumscribed circle,

$$\sum_{i=1}^{i=n} \frac{(B_{i-1} B_i)}{p_i'} = 0, \quad (B_{i-1} B_i) = 2\rho \cos \tfrac{1}{2} \measuredangle A_i.$$

Hence for our inscribed circle

$$\sum_{i=1}^{i=n} \frac{\cos \frac{1}{2} \measuredangle A_i}{\sqrt{p_{i-1} p_i}} = 0.$$

When the polygon has an even number of sides, the equation of the circumscribed circle may be put into much simpler form by means of I. 60], namely

$$p_1 p_3 p_5 \cdots p_{2n-1} + p_2 p_4 p_6 \cdots p_{2n} = 0.$$

In the case of the quadrilateral this gives

$$p_1 p_3 + p_2 p_4 = 0.$$

When this is reduced to rectangular cartesian form the coefficient of $x^2 + y^2$ is

$$\tfrac{1}{2} [\cos(\alpha_1 - \alpha_3) + \cos(\alpha_2 - \alpha_4)].$$

Since properly oriented opposite angles of our quadrilateral are equal,

$$\cos(\alpha_1 - \alpha_2) = -\cos(\alpha_4 - \alpha_3),$$

$$\cos(\alpha_1 - \alpha_4) = -\cos(\alpha_2 - \alpha_3).$$

Now, suppose that this same quadrilateral is circumscribed to another circle of radius ρ, the distance of the centres being d and the radius of the circumscribed circle r as before. Taking the centre of the inscribed circle as origin, the cartesian coordinates of the centre of the circumscribed circle will be

$$x_0 = \frac{-\rho \sum\limits_{i=1}^{i=4} \cos \alpha_i}{\cos(\alpha_1 - \alpha_3) + \cos(\alpha_2 - \alpha_4)},$$

$$y_0 = \frac{-\sum\limits_{i=1}^{i=4} \sin \alpha_i}{\cos(\alpha_1 - \alpha_3) + \cos(\alpha_2 - \alpha_4)}.$$

The power of the centre of the inscribed circle with regard to the circumscribed may be obtained by substituting $(\rho\rho\rho)$ in the equation of the latter when the coefficient of x^2+y^2 has been divided out, and found to be

$$-(r^2-d^2) = \frac{4\rho^2}{\cos(\alpha_1 - \alpha_3) + \cos(\alpha_2 - \alpha_4)}.$$

On the other hand, if we first find d we get

$$r^2+d^2 = \frac{2\rho^2\left[\left(\sum\limits_{i=1}^{i=4} \cos \alpha_i\right)^2 + \left(\sum\limits_{i=1}^{i=4} \sin \alpha_i\right)^2\right] - 4\rho^2[\cos(\alpha_1 - \alpha_3) + \cos(\alpha_2 - \alpha_4)]}{[\cos(\alpha_1 - \alpha_3) + \cos(\alpha_2 - \alpha_4)]^2}$$

$$\frac{2(r^2+d^2)}{(r^2-d^2)^2} = \frac{1}{\rho^2}\left[\frac{\left(\sum\limits_{i=1}^{i=4} \cos \alpha_i\right)^2 + \left(\sum\limits_{i=1}^{i=4} \sin \alpha_i\right)^2 - 2[\cos(\alpha_1 - \alpha_3) + \cos(\alpha_2 - \alpha_4)]}{4}\right]$$

Multiplying out on the right, and remembering the identities recently found,

$$\frac{2(r^2+d^2)}{(r^2-d^2)^2} = \frac{1}{\rho^2}, \quad \frac{1}{(r+d)^2} + \frac{1}{(r-d)^2} = \frac{1}{\rho^2}.$$

This is our old formula I (11).

A circle concentric with that circumscribed to our triangle will have an equation

$$\sum_{i=1}^{i=3} \sin \measuredangle A_i p_j p_j = \text{const.}$$

The left-hand side of this equation is the double area of the pedal triangle of the point (p) which proves I. 62]. Let us inquire under what circumstances the general equation of the second degree

$$\sum_{\substack{i=1\\j=1}}^{\substack{i=3\\j=3}} u_{ij} p_i p_j = 0, \qquad a_{ij} = a_{ji}$$

will represent a circle. It is necessary and sufficient that it should be possible to rewrite this

$$\lambda \sum_{i=1}^{j=1} \sin \measuredangle A_i p_j p_k + \mu \sum_{i=1}^{i=3} u_i p_i \sum_{i=1}^{i=3} \sin \measuredangle A_i p_i = 0.$$

We have three equations

$$u_i = \sigma \frac{a_{ii}}{\sin \measuredangle A_i},$$

and three others

$$\sin \measuredangle A_i + \rho \left[\frac{a_{jj}}{\sin \measuredangle A_j} \sin \measuredangle A_k + \frac{a_{kk}}{\sin \measuredangle A_k} \sin \measuredangle A_j \right] = \rho\,(a_{jk} + a_{kj}).$$

Eliminating ρ

$$a_{jj} \sin^2 \measuredangle A_k + a_{kk} \sin^2 \measuredangle A_j - 2 a_{jk} \sin^2 \measuredangle A_j \sin \measuredangle A_k = \text{const.}$$

In the special case where $a_{jk} = 0,\ j \neq k$,

$$\sigma a_{ii} = \sin^2 \measuredangle A_i [\sin^2 \measuredangle A_i - \sin^2 \measuredangle A_j - \sin^2 \measuredangle A_k]$$
$$= \sin 2 \measuredangle A_i \prod_{j=1}^{j=3} \sin \measuredangle A_i.$$

We thus find the equation of the only circle with regard to which the triangle is self-conjugate,

$$\sum_{i=1}^{i=3} \sin 2 \measuredangle A_i p_i^2 = 0. \tag{8}$$

Let us find the coefficient of $x^2 + y^2$ when the circle $\Sigma a_{ij} p_i p_j = 0$ is changed to its cartesian form. The coefficient of x^2 is $\Sigma a_{ij} \cos \alpha_i \cos \alpha_j$, that of y^2 is $\Sigma a_{ij} \sin \alpha_i \sin \alpha_j$. As these two are equal we may replace both by half their sum, namely

$$\tfrac{1}{2} [a_{11}{}^2 + a_{22}{}^2 + a_{33}{}^2 - \sum_i a_{jk} \cos \measuredangle A_i].$$

For the circumscribed circle this becomes

$$-\frac{1}{4}\sum_{i=1}^{i=3}\sin 2\measuredangle A_i = -\prod_{i=1}^{i=3}\sin\measuredangle A_i.$$

This will also be the coefficient of x^2+y^2 in the cartesian equation corresponding to the trilinear form

$$\sum_{i=1}^{i=3}\sin\measuredangle A_i p_j p_k + \sum_{i=1}^{i=3} u_i p_i \sum_{i=1}^{i=3}\sin\measuredangle A_i p_i = 0,$$

since the second factor of the last term is a constant. If the coordinates of a point be substituted in the equation of a circle and the result be divided by the coefficient of x^2+y^2, we shall get the power of the point with regard to the circle. Thus, if we take $(\rho\rho\rho)$ the coordinates of the centre of the inscribed circle, and substitute in the equation of the circumscribed circle,

$$d^2-r^2 = -\frac{\rho^2\sum_{i=1}^{i=3}\sin\measuredangle A_i}{\prod_{i=1}^{i=3}\sin\measuredangle A_i} = -2r\rho,$$

$$\frac{1}{r+d}+\frac{1}{r-d}=\frac{1}{\rho}.$$

This is our previous formula I (10).

It is geometrically evident that the centre of our circle (8) is the orthocentre of the triangle, for the polar of any point with regard to a circle is perpendicular to the line from that point to the centre. We know from I. 50] that this is a circle of antisimilitude for the circumscribed and nine-point circles, so that the equation of the latter will be of the type

$$\sum_{i=1}^{i=3}\sin 2\measuredangle A_i p_i^2 - \lambda\sum_{i=1}^{i=3}\sin\measuredangle A_i p_j p_k = 0.$$

We find λ by requiring this circle to pass through the foot of one altitude.

The nine-point circle has thus the equation

$$\sum_{i=1}^{i=3} \sin 2\measuredangle A_i p_i^2 - 2\sum_{i=1}^{i=3} \sin \measuredangle A_i p_j p_k = 0. \qquad (9)$$

We next notice the identity *

$$\sum_{i=1}^{i=3} \sin 2\measuredangle A_i p_i^2 + 2\sum_{i=1}^{i=3} \sin \measuredangle A_i p_j p_k$$
$$\equiv 2\sum_{i=1}^{i=3} \sin \measuredangle A_i p_i \sum_{i=1}^{i=3} \cos \measuredangle A_i p_i.$$

Hence the equation of the nine-point circle may be written

$$2\sum_{i=1}^{i=3} \sin \measuredangle A_i p_j p_k - \sum_{i=1}^{i=3} \sin \measuredangle A_i p_i \sum_{i=1}^{i=3} \cos \measuredangle A_i p_i = 0. \quad (10)$$

The equation of the inscribed circle was seen to be

$$\sum_{i=1}^{i=3} \cos^4 \tfrac{1}{2}\measuredangle A_i p_i^2 - 2\sum_{i=1}^{i=3} \cos^2 \tfrac{1}{2}\measuredangle A_j \cos^2 \tfrac{1}{2}\measuredangle A_k p_j p_k = 0, \quad (7)$$

and this may also be written

$$\sum_{i=1}^{i=3} \frac{\cos^4 \frac{1}{2}\measuredangle A_i}{\sin \measuredangle A_i} p_i \sum_{i=1}^{i=3} \sin \measuredangle A_i p_i$$
$$-4\prod_{i=1}^{i=3} \frac{\cos^2 \frac{1}{2}\measuredangle A_i}{\sin \measuredangle A_i} \sum_{i=1}^{i=3} \sin \measuredangle A_i p_j p_k = 0.$$

The radical axis of the nine-point and inscribed circle will be

$$\prod_{i=1}^{i=3} \sin \measuredangle A_i \sum_{i=1}^{i=3} \frac{\cos^4 \frac{1}{2}\measuredangle A_i}{\sin \measuredangle A_i} p_i$$
$$-2\prod_{i=1}^{i=3} \cos^2 \tfrac{1}{2}\measuredangle A_i \sum_{i=1}^{i=3} \cos \measuredangle A_i p_i = 0,$$

* Whitworth, loc. cit., pp. 294 ff.

$$\sum_{i=1}^{i=3} \cos\tfrac{1}{2}\measuredangle A_i \sin\tfrac{1}{2}(\measuredangle A_i - \measuredangle A_j)\sin\tfrac{1}{2}(\measuredangle A_i - \measuredangle A_k)\, p_i = 0,$$

$$\sum_{i=1}^{i=3} \frac{\cos\tfrac{1}{2}\measuredangle A_i}{\sin\tfrac{1}{2}(\measuredangle A_j - \measuredangle A_k)}\, p_i = 0.$$

The coordinates of this line are seen to satisfy our equation (6), so that the nine-point circle touches the inscribed one. In like manner we may prove that it touches the escribed circles.

At this point let us make a short digression into the geometry of conic sections.* We start with the familiar theorem that, if the side-lines of two triangles touch a conic, their vertices lie on another conic. If the first conic be a parabola and one triangle be formed by the tangents through the focus and the line at infinity, we see that a circle circumscribed to a triangle circumscribed to a parabola will pass through the focus. The Miquel point of four lines is thus the focus of the parabola which touches them. Let this be the point F, and let us find the polar reciprocal of our figure with regard to another circle of centre C. The polar of the parabola will be a conic through C. A triangle with vertices A_1, A_2, A_3 will be inscribed in this conic, and another conic with focus at C will touch the side-lines of the triangle. This last conic, regardless of the positions of A_1, A_2, A_3 on the conic through them, will always touch a fixed line, the polar of F with regard to the circle of reciprocation. The foot of the perpendicular from C on any line is the inverse of the pole of this line with regard to the circle whose centre is C; the pedal circle of C with regard to $\triangle A_1, A_2, A_3$ is the inverse of the circle through F and will pass through the inverse of F, a fixed point.

Theorem 1.] *The pedal circle of a chosen finite point of a conic with regard to all triangles inscribed in this conic passes through a fixed point.*

If the conic be a rectangular hyperbola, we see, by taking the special case where the vertices are the extremities of the

* Mannheim, 'Solutions de questions 1798 et 1803', *Nouvelles Annales de Mathématiques*, Series 4, vol. ii, 1902.

asymptotes and either vertex, that this fixed point will be the centre.*

Let us return to the geometry of the circle. Every conic through the vertices and orthocentre of a triangle is a rectangular hyperbola, for the involution determined by such conics on the line at infinity has three pairs of mutually perpendicular directions. The locus of the centres of these conics is a conic, namely, the nine-point circle. We thus get †

Theorem 2.] *If four finite points be given, whereof no three are collinear, which are not the vertices and orthocentre of a triangle, the four nine-point circles which they determine three by three, and the pedal circle of each with regard to the triangle of the other three, are concurrent.*

This theorem enables us at once to deduce Fonténé's extension of Feuerbach's theorem which we had before in I. 68]. For if a point move along a line through the centre of the circumscribed circle, its isogonal conjugate, whose coordinates are proportional to the reciprocals of its own, will trace a conic through the vertices and orthocentre of the triangle, i.e. a rectangular hyperbola, and the pedal circle of the moving point and its isogonal conjugate will continually pass through the centre of this hyperbola.

For the sake of reference it may be worth while to give the trilinear coordinates of the various notable points of the triangle which appear in connexion with the Brocard figures. We get from formulae (13) to (23) of Ch. I,

Point O, $p_1 : p_2 : p_3 = \cos \measuredangle A_1 : \cos \measuredangle A_2 : \cos \measuredangle A_3.$

Point K, $p_1 : p_2 : p_3 = \sin \measuredangle A_1 : \sin \measuredangle A_2 : \sin \measuredangle A_3$
$= a_1 : a_2 : a_3.$

Point Ω, $p_1 : p_2 : p_3 = \dfrac{\sin \measuredangle A_3}{\sin \measuredangle A_2} : \dfrac{\sin \measuredangle A_1}{\sin \measuredangle A_3} : \dfrac{\sin \measuredangle A_2}{\sin \measuredangle A_1}$
$= a_1 a_3^2 : a_2 a_1^2 : a_3 a_2^2.$ (11)

* See a remark by 'G.', *Nouvelles Annales de Mathématiques*, Series 4, vol. v, 1905.

† Fonténé, ibid., p. 504, speaks of this as a well-known theorem.

Point Ω', $p_1 : p_2 : p_3 = \dfrac{\sin \measuredangle A_2}{\sin \measuredangle A_3} : \dfrac{\sin \measuredangle A_3}{\sin \measuredangle A_1} : \sin \dfrac{\measuredangle A_1}{\sin \measuredangle A_1}$
$= a_1 a_2^2 : a_2 a_3^2 : a_3 a_1^2.$

Point A_i', $p_i : p_j : p_k = \sin \omega : \sin (\measuredangle A_k - \omega) : \sin (\measuredangle A_j - \omega)$
$= a_1 a_2 a_3 : a_k^3 : a_j^3.$

The vertices of the second Brocard triangle are not quite so easy to determine.

The equation of any circle through A_j and A_k will be of the form

$$\sum_{i=1}^{i=3} a_i p_j p_k + l p_k \sum_{i=1}^{i=3} a_i p_i = 0.$$

We wish this to touch $A_i A_k$ at A_i. Putting $p_j = 0$,

$$a_j p_k p_i + l p_k (a_i p_i + a_k p_k) = 0.$$

There will be two roots $p_k = 0$ if $l = -\dfrac{a_j}{a_i}$. The equation of our circle is, then,

$$(a_i^2 - a_j^2) p_j p_k + a_i a_k p_i p_j - a_j a_k p_k^2 = 0.$$

To find where this meets the line from A to K we put

$$\rho p_j = a_j, \quad \rho p_k = a_k.$$

We thus get our desired coordinates.

Point A_i'', $p_i : p_j : p_k = a_j^2 + a_k^2 - a_i^2 : a_i a_j : a_i a_k.$ (12)

Let us find the equation of a Tucker circle. If P_i be such a point of $A_j A_k$ that

$$\measuredangle A_j \Omega P_i = \theta,$$

$$(A_j P_i) = (\Omega A_j) \frac{\sin \theta}{\sin (\omega + \theta)} = \frac{a_k \sin \omega \sin \theta}{\sin (\omega + \theta) \sin \measuredangle A_j'}.$$

The distance from P_i to $A_i A_j$ is, thus,

$$\frac{a_k \sin \omega \sin \theta}{\sin (\omega + \theta)}.$$

The equation of the line $P_i P_j'$ is, by I. 98],

$$p_k \sin (\omega + \theta) - \frac{a_k \sin \omega \sin \theta}{2 \Delta} \sum_{i=1}^{i=3} a_i p_i = 0.$$

The cubic curve

$$\left[\prod_{i=1}^{i=3}\left(2\Delta\sin(\omega+\theta)\,p_i - a_i\sin\omega\sin\theta\sum_{j=1}^{j=3}a_j p_j\right)\right] - \lambda p_1 p_2 p_3 = 0$$

will contain the six points $P_i P_j'$. If we can so choose λ that this equation contains $\sum_{i=1}^{i=3} a_i p_i$ as a factor, the other factor will give the Tucker circle for the angle θ. We have but to take $\lambda = 8\Delta^3\sin^3(\omega+\theta)$. The Tucker circle is

$$\frac{1}{\sum_{i=1}^{i=3} a_i p_i}\left[\prod_{i=1}^{i=3}(2\Delta\sin(\omega+\theta)\,p_i - a_i\sin\omega\sin\theta\sum_{j=1}^{j=3}a_j p_j)\right.$$

$$\left. - 8\Delta^3\sin^3(\omega+\theta)\,p_1 p_2 p_3\right] = 0. \qquad (13)$$

In the special case of the first Lemoine circle $\sin\theta = \sin\omega$,

$$16\cos^2\omega\,\Delta\sum_{i=1}^{i=3} a_i p_j p_k - 4\,\Delta\sin\omega\cos\omega\sum_{i=1}^{i=3} a_i p_i\sum_{i=1}^{i=3} a_j a_k p_i$$

$$+ a_1 a_2 a_3\sin^2\omega\left(\sum_{i=1}^{i=3} a_i p_i\right)^2 = 0.$$

The Brocard circle is concentric with this, and so has an equation of the type

$$16\cos^2\omega\,\Delta^2\sum_{i=1}^{i=3} a_i p_j p_k - 4\,\Delta\sin\omega\cos\omega\sum_{i=1}^{i=3} a_i p_i\sum_{i=1}^{i=3} a_j a_k p_i$$

$$+ (a_1 a_2 a_3\sin^2\omega - \lambda)\left(\sum_{i=1}^{i=3} a_i p_i\right)^2 = 0.$$

This must pass through the symmedian point whose coordinates are proportional to the sides of the triangle. Remembering that $\text{ctn}\,\omega = \dfrac{\Sigma a_i^2}{4\Delta}$, $\lambda = a_1 a_2 a_3\sin^2\omega$, we get the following different forms for the equation of the Brocard circle:

$$4\,\Delta \operatorname{ctn}\omega \sum_{i=1}^{i=3} a_i p_j p_k - \sum_{i=1}^{i=3} a_j a_k p_i \sum_{i=1}^{i=3} a_i p_i = 0.$$

$$\sum_{i=1}^{i=3} a_i^2 \sum_{i=1}^{i=3} a_i p_j p_k - \sum_{i=1}^{i=3} a_j a_k p_i \sum_{i=1}^{i=3} a_i p_i = 0. \qquad (14)$$

$$\sum_{i=1}^{i=3} \sin^2 \measuredangle A_i \sum_{i=1}^{i=3} \sin \measuredangle A_i p_j p_k$$
$$- \sum_{i=1}^{i=3} \sin \measuredangle A_j \sin \measuredangle A_k p_i \sum_{i=1}^{i=3} \sin \measuredangle A_i p_i = 0.$$

It will be found by direct substitution that the circle with this equation does effectively pass through our ten points. Radical axis of Brocard and circumscribed circle

$$\sum_{i=1}^{i=3} \frac{p_i}{a_i} = 0.$$

The area of the pedal triangle of a point (p) is

$$\frac{\Delta}{a_1 a_2 a_3} \sum_{i=1}^{i=3} a_i p_j p_k.$$

The sum of the squares of the lengths of its sides will be

$$2 \sum_{i=1}^{=3} p_i^2 + 2 \sum_{i=1}^{i=3} p_j p_k \cos \measuredangle A_i$$
$$= 2 \sum_{i=1}^{i=3} p_i^2 + \frac{1}{a_1 a_2 a_3} \sum_{i=1}^{i=3} a_i (a_j^2 + a_k^2 - a_i^2) p_j p_k.$$

The cotangent of the Brocard angle of this triangle will be

$$\operatorname{ctn}\omega' = \frac{2\, a_1 a_2 a_3 \sum_{i=1}^{i=3} p_i^2 + \sum_{i=1}^{i=3} a_i (a_j^2 + a_k^2 - a_i^2) p_j p_k}{4\,\Delta \sum_{i=1}^{i=3} a_i p_j p_k}.$$

Writing the equation of the Brocard circle

$$a_1 a_2 a_3 \sum_{i=1}^{i=3} p_i^2 = \sum_{i=1}^{i=3} a_i^3 p_j p_k, \tag{15}$$

$$\operatorname{ctn} \omega' = \frac{\sum_{i=1}^{i=3} a_i^2}{4\Delta} = \operatorname{ctn} \omega,$$

we thus reach an interesting theorem due to Schoute.*

Theorem 3.] *The locus of points whose pedal triangles have the same Brocard angle as the given triangle is the Brocard circle.*

Theorem 4.] *The locus of points whose pedal triangles have a given Brocard angle is a circle coaxal with the circumscribed and Brocard circles.*

Let us find the equation of the Neuberg circle corresponding to $(A_j A_k)$.

$$\sum_{i=1}^{i=3} a_i p_j p_k + \sum_{i=1}^{i=3} u_i p_i \sum_{i=1}^{i=3} a_i p_i = 0.$$

As this is to contain A_i, while A_j and A_k are to have like powers with regard to it,

$$u_i = 0, \qquad u_j = \lambda a_j, \qquad u_k = \lambda a_k.$$

$$\sum_{i=1}^{i=3} a_i p_j p_k + \lambda (a_j p_j + a_k p_k) \sum_{i=1}^{i=3} a_i p_i = 0.$$

The coordinates of A_j are $0, \dfrac{2\Delta}{a_j}\ 0$. Its power with regard to the Neuberg circle is, by I. 137], a_i^2, and the coefficient of x^2+y^2 in the corresponding cartesian equation is

$$-a_i \sin \measuredangle A_j \sin \measuredangle A_k.$$

$$\frac{-\lambda 4 \Delta^2}{a_j^2} \times \frac{a_j^2}{a_i \sin \measuredangle A_j \sin \measuredangle A_k} = a_i^2, \quad \lambda = -\frac{a_i}{a_j a_k}.$$

* 'Over een nauwer verband tusschen hoek en cirkel van Brocard', *Amsterdam Transactions*, Series 3, vol. iii, 1887.

The equation of the Neuberg circle is

$$a_j a_k \sum_{i=1}^{i=3} a_i p_j p_k - a_i (a_j p_j + a_k p_k) \sum_{i=1}^{i=3} a_i p_i = 0. \tag{16}$$

We turn to the closely related **McCay** circle. The radical axis of the **McCay** and Brocard circles is $A_j''A_k''$ whose equation is

$$(a_i^4 - a_j^4 - a_k^4 + a_j^2 a_k^2)\, p_i + a_i a_j\, (2a_k^2 - a_i^2 - a_j^2)\, p_j$$
$$+ a_i a_k\, (2a_j^2 - a_i^2 - a_k^2) = 0.$$

The **McCay** circle will thus have an equation of the type

$$\sum_{i=1}^{i=3} a_i^2 \sum_{i=1}^{i=3} a_i p_j p_k - \sum_{i=1}^{i=3} a_i p_i \{[a_j a_k + \lambda (a_i^4 - a_j^4 - a_k^4 + a_j^2 a_k^2)]\, p_i$$
$$+ [a_k a_i + \lambda a_i a_j\, (2a_k^2 - a_i^2 - a_j^2)]\, p_j$$
$$+ [a_i a_j + \lambda a_i a_k\, (2a_j^2 - a_k^2 - a_i^2)]\, p_k\} = 0.$$

Moreover, by I. 148] A_j and A_k have like powers with regard to the **McCay** circle,

$$\frac{a_i a_k + \lambda a_i a_j\, (2a_k^2 - a_i^2 - a_j^2)}{a_j} = \frac{a_i a_j + \lambda a_i a_k\, (2a_j^2 - a_i^2 - a_k^2)}{a_k}$$

$$\lambda = -\frac{1}{3a_j a_k};$$

$$3a_j a_k \sum_{i=1}^{i=3} a_i^2 \sum_{i=1}^{i=3} a_i p_j p_k - \sum_{i=1}^{i=3} a_i p_i \{[(a_j^2 + a_k^2)^2 - a_i^4]\, p_i$$
$$+ a_i \sum_{i=1}^{i=3} a_i^2\, [a_j p_j + a_k p_k]\} = 0. \tag{17}$$

Substituting $p_i = \dfrac{\rho}{a_i}$, the coefficient of ρ^2 will be

$$3\left(\sum_{i=1}^{i=3} a_i^2\right)^2 - 3\left[\frac{(a_j^2 + a_k^2)^2 - a_i^4}{a_i} + 2a_i \sum_{i=1}^{i=3} a_i^2\right]$$
$$= \frac{3}{a_i}\left[\left(\sum_{i=1}^{i=3} a_i^2\right)^2 - \left(\sum_{i=1}^{i=3} a_i^2\right)^2\right] = 0.$$

This shows that the centre of gravity lies on the McCay circle. M_i has the coordinates

$$p_i = 0, \quad p_j = \frac{a_i a_k}{4r}, \quad p_k = \frac{a_i a_j}{4r}.$$

Its power with regard to the Neuberg circle is

$$\frac{-1}{a_1 a_2 a_3 \sin \measuredangle A_j \sin \measuredangle A_k} \left| - 3 a_i a_j^2 a_k^2 \times \frac{a_i^2}{16 r^2} \right| = \frac{3 a_i^2}{4}.$$

Its power with regard to the McCay circle is

$$\frac{a_i^2}{12}.$$

The ratio of these is $1:9$, A_i is three times as far from M_i as is M, hence the second intersection of $A_i M$ with the Neuberg circle is three times as far from M as is the second intersection with the McCay circle.

Theorem 5.] *The middle point of a side of a triangle is a centre of similitude for the corresponding McCay and Neuberg circles, the ratio of similitude being $\frac{1}{3}$.*

This justifies a remark made after I. 148]. Remembering the original definition of McCay circles, we have

Theorem 6.] *The McCay circle corresponding to a particular side of a given triangle is the locus of the centre of gravity of a triangle having the given side and Brocard angle, its vertex also lying on a specified side of the given side-line.*

§ 2. Fundamental Relations, Special Tetracyclic Coordinates.

It is clear that the trilinear coordinates which we have so far used are not adapted to dealing with the circle in any broad way, and, in fact, are of use only in studying those properties of a circle which are related to a particular triangle. Let us now turn to homogeneous rectangular cartesian coordinates $x:y:t$, and define, once for all, as a *circle* in the cartesian plane every locus which corresponds to an equation of the type

$$x_0 i(x^2+y^2+t^2)+x_1(x^2+y^2-t^2)+x_2(2xt)+x_3(2yt) = 0. \quad (18)$$

The quantities (x) shall be called the *coordinates* of the circle; they are homogeneous, and subjected only to the restriction that all may not vanish at once. We distinguish the following types of circles:

$$\sum_{i=0}^{i=3} y_i z_i \equiv (yz).$$

(*a*) *Proper circles* $(xx) \neq 0, \quad ix_0+x_1 \neq 0.$

(*b*) *Non-linear null circles* $(xx) = 0, \quad ix_0+x_1 \neq 0.$

These consist in pairs of finite lines through the circular points at infinity.

(*c*) *Non-isotropic line circles* $(xx) \neq 0, \quad ix_0+x_1 = 0.$

These consist in a non-isotropic line and the line at infinity.

(*d*) *Linear null circles* $(xx) = 0, \quad ix_0+x_1 = 0.$

These consist in an isotropic line and the line at infinity, or the line at infinity counted twice.

The four multipliers of x_0, x_1, x_2, x_3 in (18) shall be called the *special tetracyclic coordinates* of the point (x, y, t), or rather, any four quantities not all zero which are proportional to them. The reason for this curious designation will appear later. The relation between our homogeneous cartesian coordinates and our special tetracyclic ones may be written

$$\begin{gathered} y_0 : y_1 : y_2 : y_3 \equiv i(x^2+y^2+t^2) : (x^2+y^2-t^2) : 2xt : 2yt, \\ x : y : t \equiv y_2 : y_3 : -(iy_0+y_1). \end{gathered} \quad (19)$$

Every finite point has thus a definite set of special tetracyclic coordinates (y) for which

$$(yy) = 0, \quad iy_0+y_1 \neq 0.$$

Conversely, every set of homogeneous values which satisfy these relations will correspond to a single definite finite point. Returning to our circle (x), which we assume to be not a line circle, we have for the radius

$$r = \frac{\sqrt{(xx)}}{ix_0+x_1}. \quad (20)$$

This expression is, of course, double valued as it stands Where the circle is real we assume that such a sign has been attached to the radical that $r \geqq 0$. The concept of a circle with a negative radius will be treated most fully in a subsequent chapter. Let the reader show that the special tetracyclic coordinates of a point are nothing more nor less than the coordinates of that null circle whose centre the point is. The special tetracyclic coordinates of the centre of (x) are

$$\begin{aligned} \rho x_0' &= x_0 - \frac{i\,(xx)}{2\,(ix_0+x_1)}, \\ \rho x_1' &= x_1 - \frac{i\,(xx)}{2\,(ix_0+x_1)}, \\ \rho x_2' &= x_2, \\ \rho x_3' &= x_3. \end{aligned} \tag{21}$$

The coordinates of the circle concentric with (x) and orthogonal thereto are

$$\begin{aligned} \sigma \overline{x_0} &= x_0 - \frac{i\,(xx)}{(ix_0+x_1)}, \\ \sigma \overline{x_1} &= x_1 - \frac{i\,(xx)}{(ix_0+x_1)}, \\ \sigma \overline{x_2} &= x_2, \\ \sigma \overline{x_3} &= x_3. \end{aligned} \tag{22}$$

The power of the finite point (y) with regard to the proper circle (x) will be

$$\frac{-2\,(xy)}{(iy_0+y_1)\,(ix_0+x_1)}. \tag{23}$$

This formula holds even when (x) is null, if it be not a line circle, and gives the square of the distance of the finite points (x) and (y). If the power of a finite point with regard to a proper circle be divided by the radius, the quotient is

$$\frac{-2\,(xy)}{\sqrt{(xx)}\,(iy_0+y_1)}. \tag{24}$$

This expression has a meaning when the circle is a non-isotropic line circle. In fact we see that if a point remain fixed while the radius of a certain circle increase indefinitely, the ratio of power to radius will approach as a limit double the distance from the point to that line which is the limit of the circle. If we extend the phrase *ratio of power to radius* to include this limiting case, it is easy to see that this ratio for the circle $x_i = 1$, $x_j = 0$ will be

$$\frac{-2y_i}{iy_0+y_1}.$$

The special tetracyclic coordinates of a point are thus proportional to the ratio of power to radius with regard to four mutually orthogonal circles, namely, the y axis, the x axis, the unit circle around the origin as centre, and the concentric circle the square of whose radius is -1. It is this aspect of our coordinates which we shall subsequently generalize. If two circles be given which are not null, their angle θ will be given by

$$\cos\theta = \frac{(xy)}{\sqrt{(xx)}\sqrt{(yy)}}. \tag{25}$$

In the case of real circles the radicals in the denominator should be so taken as to make the radius of each positive. The formula is then

$$\cos\theta = \frac{r_1^2+r_2^2-d^2}{2r_1r_2}. \tag{26}$$

The condition for orthogonal intersection is

$$(xy) = 0. \tag{27}$$

For internal or external contact we shall have

$$\frac{(xy)^2}{(xx)(yy)} = 1. \tag{28}$$

Before proceeding further, let us look at our tetracyclic coordinates from still another point of view.* The homogeneous coordinates (x) may be taken to represent a point in a three-dimensional space S, which we shall assume has an

* One of the earliest writers to look upon circles as corresponding to points in a three-dimensional space seems to have been Mehmke, 'Geometrie der Kreise in einer Ebene', *Zeitschrift für Mathematik und Physik*, vol. xxiv, 1879. He does not, however, make use of the idea of elliptic measurement. The reader not familiar with non-Euclidean geometry will find this measurement fully treated in all books on the subject, e.g. the Author's *Elements of Non-Euclidean Geometry*, Oxford, 1909.

elliptic type of measurement, the equation of the absolute quadric being

$$(xx) = 0.$$

Our formula (25) for the cosine of the angle of two circles (x) and (y) will give exactly the cosine of the distance of two points in our non-Euclidean space. The totality of circles whose coordinates are linearly dependent on those of two will give the pencil of circles through the intersections of two given circles. When the given circles are proper, this will be a coaxal system as defined in the last chapter. We shall extend the term *coaxal system* to include the pencil in every case. Our correspondence may thus be written:

Plane π.	Space s.
Circle.	Point.
Null circle.	Point of Absolute.
Angle of two not null circles.	Distance of two points not on Absolute.
Mutually orthogonal circles.	Points conjugate with regard to Absolute.
Coaxal system of circles.	Line.
Pencil of tangent circles.	Line tangent to Absolute.
Circles mutually inverse in proper circle, or reflexions of one another in a non-isotropic line.	Points collinear with a given point and equidistant therefrom.
Circle of anti-similitude of two circles.	Centre of gravity of two points.
Inversion, or reflexion in a line.	Reflexion in a point.

As an example of the sort of theorems that correspond in the two domains, we take the following:

Plane π.	Space S.
The circles of anti-similitude of three non-coaxal circles are coaxal in threes.	*The centres of gravity of pairs formed by three given points are collinear by threes.**

* See the Author's *Non-Euclidean Geometry*, cit., p. 102.

We may establish our correspondence of circles in π with points in S by a direct geometric process without recurrence to non-Euclidean notions. Starting with our typical circle (18), the cone with the vertex $(0, 0, 1, i)$ through that circle will have the equation

$$x_0 i\,[x^2+y^2+t^2-z^2-2\,itz]+x_1\,[x^2+y^2+z^2-t^2+2\,itz]$$
$$+x_2\,(2\,xt-2\,ixz)+x_3\,(2\,yt-2\,iyz) = 0.$$

This may be written

$$(ix_0+x_1)\,(x^2+y^2+z^2+t^2)$$
$$-2\,i\,(z+it)\,(x_2 x_3+x_3 y+x_0 z+x_1 t) = 0.$$

This cone will thus cut the sphere

$$x^2+y^2+z^2+t^2 = 0$$

in a circle whose plane is

$$x_2 x+x_3 y+x_0 z+x_1 t = 0.$$

The coordinates of the pole of this plane with regard to the sphere in question will be

$$x:y:z:t = x_2:x_3:x_0:x_1.$$

The coordinates of a circle in the cartesian plane may be interpreted as the coordinates of a point in space whose polar plane with regard to a fundamental sphere cuts that sphere in a circle whose stereographic projection is the given circle.

Let us in this connexion give the formulae for inversion. Suppose that we have a point (y) and a circle of inversion (x). Since every circle through (y) and (y') is orthogonal to (x), and these relations are expressed by linear equations of like type, the coordinates of (y') must be linearly dependent on those of (x) and (y).

$$y_i' = \lambda y_i+\mu x_i,$$
$$(y'y') = (yy) = 0,$$
$$\rho y_i' = (xx)\,y_i-2\,(xy)\,x_i. \tag{29}$$

We may go further. Suppose in this equation (y) is any circle. Then if (t) lie on (y) we shall find that its inverse lies on (y'). Our formula will thus give the inverse of any chosen circle. We next turn to the non-homogeneous cartesian coordinates, taking for our circle of inversion

$$x^2+y^2=1;$$

the inverse of (x, y) will be

$$x'=\frac{x}{x^2+y^2},\quad y'=\frac{y}{x^2+y^2},$$

$$dx'^2+dy'^2=\frac{dx^2+dy^2}{(x^2+y^2)^2},$$

$$\frac{dx'\delta x'+dy'\delta y'}{\sqrt{dx'^2+dy'^2}\sqrt{\delta x'^2+\delta y'^2}}\equiv\pm\frac{dx\delta x+dy\delta y}{\sqrt{dx^2+dy^2}\sqrt{\delta x^2+\delta y^2}}.$$

This last equation shows that the angle between two curves is equal or supplementary to that of their inverses.

§ 3. The Identity of Darboux and Frobenius.*

It is now time to take up an important identity connecting the coordinates of any ten circles, which plays a fundamental rôle in much of our theory. Let us suppose that we have two groups of five circles each, $(x)\,(y)\,(z)\,(s)\,(t)$ and $(x')\,(y')\,(z')\,(s')\,(t')$. Multiplying together the two determinants $|\,x\;y\;z\;s\;t\;0\,|$ and $|\,x'\,y'\,z'\,s'\,t'\,0\,|$ we get the fundamental identity

$$\begin{vmatrix} (xx') & (xy') & (xz') & (xs') & (xt') \\ (yx') & (yy') & (yz') & (ys') & (yt') \\ (zx') & (zy') & (zz') & (zs') & (zt') \\ (sx') & (sy') & (sz') & (ss') & (st') \\ (tx') & (ty') & (tz') & (ts') & (tt') \end{vmatrix} \equiv 0. \tag{30}$$

As a first special case, let (x'), (y'), (z'), (s') be four finite points, no three collinear, nor are all four concyclic. (x) shall

* It is rather a delicate question to know to whom one should give the credit for the identity which forms the subject of the present section. It was first given in a particular form by Darboux, 'Groupes de points, de cercles et de sphères', *Annales de l'École Normale*, Series 2, vol. i, 1872. Frobenius thereupon announced that he had long been familiar with it, and proceeded to publish his results, 'Anwendungen der Determinantentheorie auf die Geometrie des Masses', *Crelle's Journal*, vol. lxxix, 1875. Another elaborate discussion is in an important article, by Lachlan, 'On Systems of Circles and Spheres', *Philosophical Transactions of the Royal Society*, vol. clxxvii, 1886.

be the circle circumscribed to the triangle whose vertices are (y'), (z'), (s'), and so for (y), (z), and (s): (t) and (t') shall be the line at infinity.

$$0 = \begin{vmatrix} (xx') & 0 & 0 & 0 & (ix_0+x_1) \\ 0 & (yy') & 0 & 0 & (iy_0+y_1) \\ 0 & 0 & (zz') & 0 & (iz_0+z_1) \\ 0 & 0 & 0 & (ss') & (is_0+s_1) \\ (ix_0'+x_1') & (iy_0'+y_1') & (iz_0'+z_1') & (is_0'+s_1') & 0 \end{vmatrix}$$

$$= \begin{vmatrix} \frac{-2\,(xx')}{(ix_0+x_1)(ix_0'+x_1')} & 0 & 0 & 0 & 1 \\ 0 & \frac{-2\,(yy')}{(iy_0+y_1)(iy_0'+y_1')} & 0 & 0 & 1 \\ 0 & 0 & \frac{-2\,(zz')}{(iz_0+z_1)(iz_0'+z_1')} & 0 & 1 \\ 0 & 0 & 0 & \frac{-2\,(ss')}{(is_0+s_1)(is_0'+s_1')} & 1 \\ 1 & 1 & 1 & 1 & 0 \end{vmatrix}.$$

If p_1 be the power of (x') with regard to the circle (x), and so for p_2, p_3, p_4,

$$\frac{1}{p_1} + \frac{1}{p_2} + \frac{1}{p_3} + \frac{1}{p_4} = 0.$$

Theorem 7.] *If four finite points be given of which no three are collinear nor do all four lie on one circle, then the sum of the reciprocals of the powers of each point with regard to the circle passing through the other three is zero.*

If none of our circles be null or isotropic line circles, we may divide the various rows and columns in the left side of (30) by expressions of the type $\sqrt{(xx)}$. If, then, we indicate the angle formed by the circles (x) and (x') by $\measuredangle\,(xx')$,

$$\begin{vmatrix} \cos\measuredangle xx' & \cos\measuredangle xy' & \cos\measuredangle xz' & \cos\measuredangle xs' & \cos\measuredangle xt' \\ \cos\measuredangle yx' & \cos\measuredangle yy' & \cos\measuredangle yz' & \cos\measuredangle ys' & \cos\measuredangle yt' \\ \cos\measuredangle zx' & \cos\measuredangle zy' & \cos\measuredangle zz' & \cos\measuredangle zs' & \cos\measuredangle zt' \\ \cos\measuredangle sx' & \cos\measuredangle sy' & \cos\measuredangle sz' & \cos\measuredangle ss' & \cos\measuredangle st' \\ \cos\measuredangle tx' & \cos\measuredangle ty' & \cos\measuredangle tz' & \cos\measuredangle ts' & \cos\measuredangle tt' \end{vmatrix} = 0. \quad (31)$$

On the other hand, suppose that (t) and (t') are both the line at infinity, so that the last row and column are divided by $it_0 + t_1$ and $it_0' + t_1'$, we have

$$\begin{vmatrix} \cos\measuredangle xx' & \cos\measuredangle xy' & \cos\measuredangle xz' & \cos\measuredangle xs' & \dfrac{1}{r_x} \\ \cos\measuredangle yx' & \cos\measuredangle yy' & \cos\measuredangle yz' & \cos\measuredangle ys' & \dfrac{1}{r_y} \\ \cos\measuredangle zx' & \cos\measuredangle zy' & \cos\measuredangle zz' & \cos\measuredangle zs' & \dfrac{1}{r_z} \\ \cos\measuredangle sx' & \cos\measuredangle sy' & \cos\measuredangle sz' & \cos\measuredangle ss' & \dfrac{1}{r_s} \\ \dfrac{1}{r_{x'}} & \dfrac{1}{r_{y'}} & \dfrac{1}{r_{x'}} & \dfrac{1}{r_{s'}} & 0 \end{vmatrix} = 0. \qquad (32)$$

r_x, r_y, r_z, r_s are the radii of the first four circles and $r_{x'}, r_{y'}, r_{z'}, r_{s'}$ those of the second four. Again, suppose that our circles are non-linear null circles. We have, for any two groups of five finite points,

$$\begin{vmatrix} d_{xx'}^2 & d_{xy'}^2 & d_{xz'}^2 & d_{xs'}^2 & d_{xt'}^2 \\ d_{yx'}^2 & d_{yy'}^2 & d_{yz'}^2 & d_{ys'}^2 & d_{yt'}^2 \\ d_{zx'}^2 & d_{zy'}^2 & d_{zz'}^2 & d_{zs'}^2 & d_{zt'}^2 \\ d_{sx'}^2 & d_{sy'}^2 & d_{sz'}^2 & d_{ss'}^2 & d_{st'}^2 \\ d_{tx'}^2 & d_{ty'}^2 & d_{tz'}^2 & d_{ts'}^2 & d_{tt'}^2 \end{vmatrix} = 0. \qquad (33)$$

Here $d_{xx'}$ means the distance from the point (x) to the point (x'). If the second set of five proper circles or non-isotropic lines be identical with the first, we have

$$\begin{vmatrix} 1 & \cos\measuredangle xy & \cos\measuredangle xz & \cos\measuredangle xs & \cos\measuredangle xt \\ \cos\measuredangle yx & 1 & \cos\measuredangle yz & \cos\measuredangle ys & \cos\measuredangle yt \\ \cos\measuredangle zx & \cos\measuredangle zy & 1 & \cos\measuredangle zs & \cos\measuredangle zt \\ \cos\measuredangle sx & \cos\measuredangle sy & \cos\measuredangle sz & 1 & \cos\measuredangle st \\ \cos\measuredangle tx & \cos\measuredangle ty & \cos\measuredangle tz & \cos\measuredangle ts & 1 \end{vmatrix} = 0. \qquad (34)$$

We get similarly from (32)

$$\begin{vmatrix} 1 & \cos\measuredangle xy & \cos\measuredangle xz & \cos\measuredangle xs & \frac{1}{r_x} \\ \cos\measuredangle yx & 1 & \cos\measuredangle yz & \cos\measuredangle ys & \frac{1}{r_y} \\ \cos\measuredangle zx & \cos\measuredangle zy & 1 & \cos\measuredangle zs & \frac{1}{r_z} \\ \cos\measuredangle sx & \cos\measuredangle sy & \cos\measuredangle sz & 1 & \frac{1}{r_s} \\ \frac{1}{r_x} & \frac{1}{r_y} & \frac{1}{r_z} & \frac{1}{r_s} & 0 \end{vmatrix} = 0. \quad (35)$$

If each set of five be made up of four finite points and the line at infinity, we get Euler's identical relation for any four (finite) points in the plane,

$$\begin{vmatrix} 0 & d_{xy}^2 & d_{xz}^2 & d_{xs}^2 & 1 \\ d_{yx}^2 & 0 & d_{yz}^2 & d_{ys}^2 & 1 \\ d_{zx}^2 & d_{zy}^2 & 0 & d_{zs}^2 & 1 \\ d_{sx}^2 & d_{sy}^2 & d_{sz}^2 & 0 & 1 \\ 1 & 1 & 1 & 1 & 0 \end{vmatrix} = 0. \quad (36)$$

If we take four finite concyclic points, and the circle through them,

$$\begin{vmatrix} 0 & d_{xy}^2 & d_{xz}^2 & d_{xs}^2 \\ d_{yx}^2 & 0 & d_{yz}^2 & d_{ys}^2 \\ d_{zx}^2 & d_{zy}^2 & 0 & d_{zs}^2 \\ d_{sx}^2 & d_{sy}^2 & d_{sz}^2 & 0 \end{vmatrix} = 0.$$

$$(d_{xy}d_{zs}+d_{xz}d_{ys}+d_{xs}d_{yz})\,(d_{xy}d_{zs}+d_{xz}d_{ys}-d_{xs}d_{yz})$$
$$(d_{xy}d_{zs}-d_{xz}d_{ys}+d_{xs}d_{yz})\,(-d_{xy}d_{zs}+d_{xz}d_{ys}+d_{xs}d_{yz})=0. \quad (37)$$

This last equation gives Ptolemy's theorem for a quadrilateral inscribed in a circle.

If three circles have the coordinates (y), (z), (s), their equations are

$$(xy) = (xz) = (xs) = 0.$$

The coordinates of their common orthogonal circle will be

$$\rho x_i = \frac{\partial}{\partial t_i}\,|\, tyzs \,|. \quad (38)$$

A necessary and sufficient condition that this should be null is

$$\begin{vmatrix} (yy) & (yz) & (ys) \\ (zy) & (zz) & (zs) \\ (sy) & (sz) & (ss) \end{vmatrix} = 0. \tag{39}$$

When all of our given circles are proper this may be written

$$\begin{vmatrix} 1 & \cos \measuredangle xy & \cos \measuredangle xz \\ \cos \measuredangle yx & 1 & \cos \measuredangle yz \\ \cos \measuredangle zx & \cos \measuredangle zy & 1 \end{vmatrix} = 0. \tag{40}$$

A necessary and sufficient condition that four circles (y), (z), (s), (t) should be orthogonal to a fifth is

$$| yzst | = \sqrt{\begin{vmatrix} (yy) & (yz) & (ys) & (yt) \\ (zy) & (zz) & (zs) & (zt) \\ (sy) & (sz) & (ss) & (st) \\ (ty) & (tz) & (ts) & (tt) \end{vmatrix}} = 0. \tag{41}$$

When none of them are null we may write

$$\begin{vmatrix} 1 & \cos \measuredangle yz & \cos \measuredangle ys & \cos \measuredangle yt \\ \cos \measuredangle zy & 1 & \cos \measuredangle zs & \cos \measuredangle zt \\ \cos \measuredangle sy & \cos \measuredangle sz & 1 & \cos \measuredangle st \\ \cos \measuredangle ty & \cos \measuredangle tz & \cos \measuredangle ts & 1 \end{vmatrix} = 0. \tag{42}$$

On the other hand, if we have four proper circles, (x), (y), (z), (s), each two of which are orthogonal, we get from (35)

$$\frac{1}{r_x^{\,2}} + \frac{1}{r_y^{\,2}} + \frac{1}{r_z^{\,2}} + \frac{1}{r_s^{\,2}} = 0. \tag{43}$$

Theorem 8.] *The sum of the squares of the reciprocals of the radii of four mutually orthogonal proper circles is zero.*

We defined as the *special tetracyclic coordinates* of a point numbers proportional to the ratio of power to radius with regard to four mutually orthogonal circles which were not null; extending the meaning of this ratio to the cases where some of the circles were non-isotropic lines. Suppose, now, that we have any four mutually orthogonal circles not null and we take the ratio of power to radius with regard to each, interpreting this ratio as before for line circles. If the four ratios be proportional to s_1, s_2, s_3, s_4, we have

$$\begin{vmatrix} 1 & 0 & 0 & 0 & -s_1 \\ 0 & 1 & 0 & 0 & -s_2 \\ 0 & 0 & 1 & 0 & -s_3 \\ 0 & 0 & 0 & 1 & -s_4 \\ -s_1 & -s_2 & -s_2 & -s_4 & 0 \end{vmatrix} = 0.$$

$$s_1^2 + s_2^2 + s_3^2 + s_4^2 \equiv 0. \tag{44}$$

If, then, we define these ratios as the *general tetracyclic coordinates of a point,** we see that they are linear in the special tetracyclic coordinates, and connected by the same quadratic identity; the sum of the squares vanishes.

Theorem 9.] *The passage from one set of tetracyclic coordinates to another is effected by a quaternary orthogonal substitution.*†

The sum of the squares of the four variables will be a relative invariant for all such substitutions, as will be the polar of this form, hence the expression for the angle of two not null or isotropic circles will be invariant, and we have in the general tetracyclic coordinates for two circles (x) and (y)

$$\cos\theta = \frac{(xy)}{\sqrt{(xx)}\,\sqrt{(yy)}}. \tag{25}$$

The determination of the signs of the radicals in the denominator can only be effected by a further knowledge of the relation of the present coordinate system to the original one. It is to be noted also that our formula (29) for the inverse of a point or circle will hold equally well here.

* Strictly speaking, perhaps, the term *general* should be extended to the case of any four circles where the simple identity would be replaced by a more complicated quadratic relation. The restriction to the orthogonal case is highly useful in the case of tetracyclic coordinates, and sanctioned by custom.

† The term *orthogonal substitution* is sometimes restricted to the case where the square of the determinant is unity. We do not impose this restriction, and merely require the invariance of the sum of the squares of the variables.

Theorem 10.] *The equation of a circle will be linear in the general system of tetracyclic coordinates, and the expression for the cosine of the angle of two not null or isotropic circles will be invariable in form.*

If two proper circles cut two others orthogonally, the radical axis of one pair is the line of centres of the other.

Theorem 11.] *If four mutually orthogonal proper circles be given, their vertices are the vertices and orthocentre of a triangle.*

Let $(y), (z), (s)$ be three proper circles, $(y')\,(z')\,(s')$ the vertices of an arcual triangle determined by them. Let (x) be the circle circumscribed to this triangle, (t) the common orthogonal circle to $(y)\,(z)\,(s)$, while (y'') is orthogonal to $(z)\,(s)\,(t)$. Taking the two groups of circles

$$(y)\,(z)\,(s)\,(x)\,(t), \quad (y')\,(z')\,(s')\,(y'')\,(t),$$

$$\begin{vmatrix} (yy') & 0 & 0 & (yy'') & 0 \\ 0 & (zz') & 0 & 0 & 0 \\ 0 & 0 & (ss') & 0 & 0 \\ 0 & 0 & 0 & (xy'') & (xt) \\ (ty') & (tz') & (tt') & 0 & (tt) \end{vmatrix}$$

$$(yy')\,(xy'')\,(tt) + (yy'')\,(xt)\,(ty') = 0.$$

$$\frac{(xy'')}{\sqrt{(xx)}\,\sqrt{(y''y'')}} = -\frac{(xt)}{\sqrt{(xx)}\,\sqrt{(tt)}}\left[\frac{(yy'')\,(ty')}{(yy')\,\sqrt{(tt)}\,\sqrt{(y''y'')}}\right].$$

Now, since $(y')\,(y'')\,(t)$ are orthogonal to (z) and (s),

$$t_i = py_i' + qy_i'',$$

$$0 = p(yy') + q(yy''), \quad (ty') = q(y'y''),$$

$$(tt) = p(ty') = 2pq\,(y'y'') + q^2\,(y''y'') = 2p\,(ty') + q^2\,(y''y''),$$

$$(tt) = p\,(ty') = -q^2\,(y''y''),$$

$$\frac{(yy'')\,(ty')}{(yy')\,\sqrt{(tt)}\,\sqrt{(y''y'')}} = \pm i; \quad \frac{(xy'')}{\sqrt{(xx)}\,\sqrt{(y''y'')}} = \pm i\,\frac{(xt)}{\sqrt{(xx)}\,\sqrt{(tt)}}.$$

The right-hand side of this equation is unaltered when we permute the letters $(y), (z), (s)$.

We thus get an interesting theorem due to Study:*

Theorem 12.] *The circles circumscribed to the arcual triangles formed by three non-concurrent proper circles cut at equal or supplementary angles the three circles each orthogonal to two of the given circles and to the common orthogonal circle.*

If four proper circles touch one another externally,

$$\begin{vmatrix} 1 & -1 & -1 & -1 & \frac{1}{r_x} \\ -1 & 1 & -1 & -1 & \frac{1}{r_y} \\ -1 & -1 & 1 & -1 & \frac{1}{r_z} \\ -1 & -1 & -1 & 1 & \frac{1}{r_s} \\ \frac{1}{r_x} & \frac{1}{r_y} & \frac{1}{r_z} & \frac{1}{r_s} & 0 \end{vmatrix} = 0.$$

$$\sum_x \frac{1}{{r_x}^2} = \sum_{x,\,y} \frac{1}{r_x r_y}. \tag{45}$$

Each term on the right appears twice. This formula is due to Steiner.†

If a circle (s) be externally tangent to (x), (y), (z), three not null circles, while (t) is a point thereon, and if s_1, s_2, s_3 be the ratio of power to radius for (t) with regard to (x), (y), (z),

$$\begin{vmatrix} 1 & \cos\measuredangle xy & \cos\measuredangle xz & -1 & s_1 \\ \cos\measuredangle yx & 1 & \cos\measuredangle yz & -1 & s_2 \\ \cos\measuredangle zx & \cos\measuredangle zy & 1 & -1 & s_3 \\ -1 & -1 & -1 & 1 & 0 \\ s_1 & s_2 & s_3 & 0 & 0 \end{vmatrix} = 0.$$

$$\sin\tfrac{1}{2}\measuredangle yz\,\sqrt{s_1} \pm \sin\tfrac{1}{2}\measuredangle zx\,\sqrt{s_2} \pm \sin\tfrac{1}{2}\measuredangle xy\,\sqrt{s_3} = 0.$$

We see that formula (7) for the inscribed circle is a special

* See a condensed but important article, 'Das Apollonische Problem', *Mathematische Annalen*, vol. xlix, 1897. Owing to a mistake in sign, the present theorem is there given too great an extension.

† 'Einige geometrische Sätze', *Crelle's Journal*, vol. i, 1826, p. 274.

case of this. Suppose, more generally, that (x) cuts (y), (z), (s) at angles α_1, α_2, α_3. We get from (33)

$$\begin{vmatrix} 1 & \cos\measuredangle yz & \cos\measuredangle ys & \cos\alpha_1 & \dfrac{1}{r_y} \\ \cos\measuredangle zy & 1 & \cos\measuredangle zs & \cos\alpha_2 & \dfrac{1}{r_z} \\ \cos\measuredangle sy & \cos\measuredangle sz & 1 & \cos\alpha_3 & \dfrac{1}{r_s} \\ \cos\alpha_1 & \cos\alpha_2 & \cos\alpha_3 & 1 & \dfrac{1}{r} \\ \dfrac{1}{r_y} & \dfrac{1}{r_z} & \dfrac{1}{r_s} & \dfrac{1}{r} & 0 \end{vmatrix} = 0. \qquad (46)$$

The condition * that there should be a real circle cutting the three real circles at these three real angles is that the discriminant of this quadratic equation in $\frac{1}{r}$ should be greater than zero. This condition is easily transformed by means of the familiar determinant identity

$$\frac{\partial\Delta}{\partial a_{ii}}\frac{\partial\Delta}{\partial a_{jj}} - \frac{\partial\Delta}{\partial a_{ij}}\frac{\partial\Delta}{\partial a_{ji}} \equiv \Delta\frac{\partial^2\Delta}{\partial a_{ii}\,\partial a_{jj}}, \qquad (47)$$

thus giving

$$\begin{vmatrix} 1 & \cos\measuredangle yz & \cos\measuredangle ys & \cos\alpha_1 \\ \cos\measuredangle zy & 1 & \cos\measuredangle zs & \cos\alpha_2 \\ \cos\measuredangle sy & \cos\measuredangle sz & 1 & \cos\alpha_3 \\ \cos\alpha_1 & \cos\alpha_2 & \cos\alpha_3 & 1 \end{vmatrix}$$

$$\times \begin{vmatrix} 1 & \cos\measuredangle yz & \cos\measuredangle ys & \dfrac{1}{r_y} \\ \cos\measuredangle zy & 1 & \cos\measuredangle zs & \dfrac{1}{r_z} \\ \cos\measuredangle sy & \cos\measuredangle sz & 1 & \dfrac{1}{r_s} \\ \dfrac{1}{r_y} & \dfrac{1}{r_z} & \dfrac{1}{r_s} & 0 \end{vmatrix} \geqq 0.$$

* This holds only in the case where the given circles have non-collinear centres. When the centres are collinear, the second factor of the left-hand side of the equation at the bottom of the page, which is stated, p. 144, to be negative, is zero. Conversely, suppose that three real circles are cut at

The second of these factors may be written

$$\begin{vmatrix} \frac{y_0}{\sqrt{(yy)}} & \frac{y_1}{\sqrt{(yy)}} & \frac{y_2}{\sqrt{(yy)}} & \frac{y_3}{\sqrt{(yy)}} \\ \frac{z_0}{\sqrt{(zz)}} & \frac{z_1}{\sqrt{(zz)}} & \frac{z_2}{\sqrt{(zz)}} & \frac{z_3}{\sqrt{(zz)}} \\ \frac{s_0}{\sqrt{(ss)}} & \frac{s_1}{\sqrt{(ss)}} & \frac{s_2}{\sqrt{(ss)}} & \frac{s_3}{\sqrt{(ss)}} \\ i & 1 & 0 & 0 \end{vmatrix}^2$$

The three original circles being real and proper, we see from (18) that this is essentially negative. Our criterion for a real circle is thus

$$\begin{vmatrix} 1 & \cos \measuredangle yz & \cos \measuredangle ys & \cos \alpha_1 \\ \cos \measuredangle zy & 1 & \cos \measuredangle zs & \cos \alpha_2 \\ \cos \measuredangle sy & \cos \measuredangle sz & 1 & \cos \alpha_3 \\ \cos \alpha_1 & \cos \alpha_2 & \cos \alpha_3 & 1 \end{vmatrix} \leqq 0. \qquad (48)$$

One or more roots of our equation (46) may be negative. Arithmetically speaking, the sum of the roots of the equation is the sum or difference of the reciprocals of the radii satisfying the given conditions. We thus get

$$\frac{1}{r} \pm \frac{1}{\bar{r}} = A_1 \cos \alpha_1 + A_2 \cos \alpha_2 + A_3 \cos \alpha_3.$$

Replacing two angles by their supplements, and keeping the other one fixed,

$$\frac{1}{r'} \pm \frac{1}{\bar{r}'} = A_1 \cos \alpha_1 - A_2 \cos \alpha_2 - A_3 \cos \alpha_3.$$

Permuting the three angles, we get two other similar equations. Adding,

$$\frac{1}{r} \pm \frac{1}{r'} \pm \frac{1}{r''} \pm \frac{1}{r'''} = \frac{1}{\bar{r}} \pm \frac{1}{\bar{r}'} \pm \frac{1}{\bar{r}''} \pm \frac{1}{\bar{r}'''}. \qquad (49)$$

preassigned real angles by a circle with real radius and imaginary centre. They will meet the circle with the same real radius and conjugate imaginary centre at the same real angles, and so have collinear centres.

To find a circle which meets four others at one same angle ϕ.

$$\begin{vmatrix} 1 & \cos \measuredangle yz & \cos \measuredangle ys & \cos \measuredangle yt & \cos\phi \\ \cos \measuredangle zy & 1 & \cos \measuredangle zs & \cos \measuredangle st & \cos\phi \\ \cos \measuredangle sy & \cos \measuredangle sz & 1 & \cos \measuredangle st & \cos\phi \\ \cos \measuredangle ty & \cos \measuredangle tz & \cos \measuredangle ts & 1 & \cos\phi \\ \cos\phi & \cos\phi & \cos\phi & \cos\phi & 1 \end{vmatrix} = 0.$$

$$\frac{\begin{vmatrix} 1 & \cos \measuredangle yz & \cos \measuredangle ys & \cos \measuredangle yt \\ \cos \measuredangle zy & 1 & \cos \measuredangle zs & \cos \measuredangle zt \\ \cos \measuredangle sy & \cos \measuredangle sz & 1 & \cos \measuredangle st \\ \cos \measuredangle ty & \cos \measuredangle tz & \cos \measuredangle ts & 1 \end{vmatrix}}{\begin{vmatrix} 1 & \cos \measuredangle yz & \cos \measuredangle ys & \cos \measuredangle yt & 1 \\ \cos \measuredangle zy & 1 & \cos \measuredangle zs & \cos \measuredangle zt & 1 \\ \cos \measuredangle sy & \cos \measuredangle sz & 1 & \cos \measuredangle st & 1 \\ \cos \measuredangle ty & \cos \measuredangle tz & \cos \measuredangle ts & 1 & 1 \\ 1 & 1 & 1 & 1 & 0 \end{vmatrix}} = \cos^2\phi. \qquad (50)$$

This equation will be unaltered if we change ϕ into $\pi-\phi$. If we do not specify which of the two angles ϕ and $\pi-\phi$ we wish, the equation for ϕ becomes,

$$\begin{vmatrix} 1 & \cos \measuredangle yz & \cos \measuredangle ys & \cos \measuredangle yt & \epsilon_1 \\ \cos \measuredangle zy & 1 & \cos \measuredangle zs & \cos \measuredangle zt & \epsilon_2 \\ \cos \measuredangle sy & \cos \measuredangle sz & 1 & \cos \measuredangle st & \epsilon_3 \\ \cos \measuredangle ty & \cos \measuredangle tz & \cos \measuredangle ts & 1 & \epsilon_4 \\ \epsilon_1 & \epsilon_2 & \epsilon_3 & \epsilon_4 & \sec^2\phi \end{vmatrix} = 0.$$

$$\epsilon_1{}^2 = \epsilon_2{}^2 = \epsilon_3{}^2 = \epsilon_4{}^2 = 1.$$

There are usually eight distinct circles which touch three given circles. It is easy to distinguish the cases where the number is less, but we confine ourselves to the general case. The angle which one of our eight circles makes with the common orthogonal circle to the three is given only through the square of its cosine in (34), and we see that when the common orthogonal circle is not null, the eight circles are in pairs inverse therein. Such a pair of tangent circles are said to be *coupled.* When the orthogonal circle is a non-isotropic line, inversion is, of course, replaced by reflection. The radical axis of the couple and the orthogonal circle will, by I. 217], pass through a centre of similitude of each two of the three. Let the given circles be (y) (z) (s), (t) the circle sought, (x) a point thereon,

$$\begin{vmatrix} (yy) & (yz) & (ys) & (yt) & (yx) \\ (zy) & (zz) & (zs) & (zt) & (zx) \\ (sy) & (sz) & (ss) & (st) & (sx) \\ (ty) & (tz) & (ts) & (tt) & 0 \\ (xy) & (xz) & (xs) & 0 & (0) \end{vmatrix} = 0.$$

Multiplying through by $|(yy)\,(zz)\,(ss)|$ and remembering (47),

$$\begin{vmatrix} (yy) & (yz) & (ys) & (yx) \\ (zy) & (zz) & (zs) & (zx) \\ (sy) & (sz) & (ss) & (sx) \\ (ty) & (tz) & (ts) & 0 \end{vmatrix} \pm |yzst| \times |yzsx| = 0. \qquad (51)$$

But $\qquad (ty) = \epsilon_1 \sqrt{(tt)}\,\sqrt{(yy)}\,; \; \epsilon_1^2 = 1$, &c.

Let $\quad \measuredangle zs = \measuredangle A_1; \; \measuredangle sy = \measuredangle A_2; \; \measuredangle yz = \measuredangle A_3.$

$$\begin{vmatrix} 1 & \cos\measuredangle A_3 & \cos\measuredangle A_2 & \dfrac{(yx)}{\sqrt{(yy)}} \\ \cos\measuredangle A_3 & 1 & \cos\measuredangle A_1 & \dfrac{(zx)}{\sqrt{(zz)}} \\ \cos\measuredangle A_2 & \cos\measuredangle A_1 & 1 & \dfrac{(sx)}{\sqrt{(ss)}} \\ \epsilon_1 & \epsilon_2 & \epsilon_3 & 0 \end{vmatrix}$$

$$\pm\sqrt{\begin{vmatrix} 1 & \cos\measuredangle A_3 & \cos\measuredangle A_2 & \epsilon_1 \\ \cos\measuredangle A_3 & 1 & \cos\measuredangle A_1 & \epsilon_2 \\ \cos\measuredangle A_2 & \cos\measuredangle A_1 & 1 & \epsilon_3 \\ \epsilon_1 & \epsilon_2 & \epsilon_3 & 1 \end{vmatrix}}\;\frac{|yzsx|}{\sqrt{(yy)}\,\sqrt{(zz)}\,\sqrt{(ss)}} = 0. \qquad (52)$$

$$\epsilon_1^2 = \epsilon_2^2 = \epsilon_3^2 = 1.$$

This is the equation of a circle touching our given three; the radicals in the denominator of the second part have known signs. The problem of constructing a circle tangent to two circles and orthogonal to a third has clearly four solutions, for the coordinates of the circle sought are limited by one linear and two quadratic equations. We thus get

Theorem 13.] *Any two couples of circles tangent to three given circles are tangent to a fourth circle also.*

It is easy to see by examining the case, where the common orthogonal circle is a line, that no two of the four circles of this theorem can fall together unless two circles of a couple become null. But then our three original circles would be coaxal, and the whole theorem goes to pieces. The three original circles and the fourth found by this theorem are said to form a *Hart system of the second sort.** The discussion of the Hart systems of the first sort is much more difficult, but, in compensation, reveals a number of most interesting theorems. To this we now turn our attention. We start with two circles (y') and (z'). Let (y) have external contact with (y') and internal contact with (z'); (z) has external contact with (z') and internal contact with (y'). Let (s) have internal contact with (y') and (z'), while (x) has external contact with both. From $|\,x\,s\,y\,z\,y'\,0\,|^2 = 0$,

$$\begin{vmatrix} 1 & \cos\measuredangle xs & \cos\measuredangle xy & \cos\measuredangle xz & -1 \\ \cos\measuredangle sx & 1 & \cos\measuredangle sy & \cos\measuredangle sz & 1 \\ \cos\measuredangle yx & \cos\measuredangle ys & 1 & \cos\measuredangle yz & -1 \\ \cos\measuredangle zx & \cos\measuredangle zy & \cos\measuredangle zs & 1 & 1 \\ -1 & 1 & -1 & 1 & 1 \end{vmatrix} = 0;$$

$$\begin{vmatrix} 0 & \cos^2\frac{\measuredangle xs}{2} & \sin^2\frac{\measuredangle xy}{2} & \cos^2\frac{\measuredangle xz}{2} \\ \cos^2\frac{\measuredangle xs}{2} & 0 & \cos^2\frac{\measuredangle ys}{2} & \sin^2\frac{\measuredangle zs}{2} \\ \sin^2\frac{\measuredangle xy}{2} & \cos^2\frac{\measuredangle ys}{2} & 0 & \cos^2\frac{\measuredangle yz}{2} \\ \cos^2\frac{\measuredangle zx}{2} & \sin^2\frac{\measuredangle zs}{2} & \cos^2\frac{\measuredangle zy}{2} & 0 \end{vmatrix} = 0.$$

$$\cos\frac{\measuredangle xs}{2}\cos\frac{\measuredangle yz}{2} \pm \sin\frac{\measuredangle xy}{2}\sin\frac{\measuredangle zs}{2} \pm \cos\frac{\measuredangle xz}{2}\cos\frac{\measuredangle ys}{2} = 0.$$

* Study, loc. cit., p. 537.

If we replace (y') by (z') and interchange (y) and (z),

$$\cos\frac{\measuredangle xs}{2}\cos\frac{\measuredangle yz}{2}\pm\sin\frac{\measuredangle xz}{2}\sin\frac{\measuredangle ys}{2}\pm\cos\frac{\measuredangle xy}{2}\cos\frac{\measuredangle zs}{2}=0.$$

$$\measuredangle xy\pm\measuredangle xz=\pm(\measuredangle sy\pm\measuredangle sz).$$

The left side of this equation is independent of (s). If, then, we drop the terminology of speaking of internal or external contact, which is meaningless in the complex domain, and refer to the circles which are tangent to two given circles as belonging to the one or the other system, according to the circle of similitude to which they are orthogonal, we have

Theorem 14.] *If two circles of one system be taken tangent to two fixed circles, neither of which is null, the sum or difference of their angles with all tangent circles of the other system is constant.*

Let us now sharpen our concept of angle as we did for the second proof of I. 155]. Let us measure the oriented angle of two circles at a point by measuring the angle at that point from the half-tangent to the first, which starts there and is oriented in the positive sense of rotation (for a real circle) to the similarly oriented tangent to the second. The angles which two circles make at their two intersections will thus differ in sign. By choosing the proper intersection for each two successive circles above, we may write the congruence

$$\measuredangle\overrightarrow{xy}-\measuredangle\overrightarrow{ys}+\measuredangle\overrightarrow{sz}+\measuredangle\overrightarrow{zx}\equiv 0,\ \text{mod. } 2\pi. \tag{53}$$

If three circles (a), (b), and (c) be concurrent, we have

$$\measuredangle\overrightarrow{ab}+\measuredangle\overrightarrow{bc}+\measuredangle\overrightarrow{ca}\equiv 0,\ \text{mod. } 2\pi.$$

Conversely, if this equation hold, since the cosine of the negative of an angle is the cosine of the angle, we may deduce

$$\frac{(ab)}{\sqrt{(aa)}\ \sqrt{(bb)}}=\frac{(bc)\,(ca)}{\sqrt{(bb)}\ \sqrt{(aa)}\,(cc)}\pm\frac{\sqrt{(bb)\,(cc)-(bc)^2}\ \sqrt{(cc)\,(aa)-(ca)^2}}{\sqrt{bb}\ \sqrt{aa}\,(cc)}.$$

$$\begin{vmatrix} (aa) & (ab) & (ac) \\ (ba) & (bb) & (bc) \\ (ca) & (cb) & (cc) \end{vmatrix} = 0.$$

The last equation shows that the common orthogonal circle of the three given circles is null, they are concurrent, or have their centres on an isotropic, and each two have but one finite intersection.

We now return to equation (53), and explicitly exclude the possibility that two of the circles should have their centres on an isotropic. This equation distinguishes two sets of four points, each point being the intersection of two successive circles of the sequence. Let a circle (t) pass through the intersection of (x) and (y), that of (y) and (s), and that of (s) and (z) in one set. We have

$$(\measuredangle \overrightarrow{ty} - \measuredangle \overrightarrow{tx}) - (\measuredangle \overrightarrow{ts} - \measuredangle \overrightarrow{ty}) + (\measuredangle \overrightarrow{tz} - \measuredangle \overrightarrow{ts}) + \measuredangle \overrightarrow{zx} \equiv 0, \text{ mod. } 2\pi.$$

Now the two expressions for $\measuredangle \overrightarrow{ty}$ are equal with opposite sines since they are taken at the two intersections of (t) and (y), and the same will hold for the two expressions $\measuredangle \overrightarrow{ts}$.

$$\measuredangle \overrightarrow{xt} + \measuredangle \overrightarrow{tz} + \measuredangle \overrightarrow{zx} = 0.$$

Theorem 15.] *If four proper circles be given tangent to two fixed proper circles, two belonging to the first system and two to the second, but no two having their centres on an isotropic, the intersections of the two circles of the first system with the two of the second lie by fours on two circles.**

Suppose, conversely, that (y) (z) (s) are given, tangent to (y') and (z'), where (y) (z) belong to one system and (s) to the other. If P be an arbitrary point on (y), we may find two points (Q) on (z) where it meets the circles through P, and through an intersection of (y) (s) and one of (z) (s), which give properly chosen signs to $\measuredangle \overrightarrow{ys}$ and $\measuredangle \overrightarrow{sz}$. These points are the

* See a carelessly written paper by Orr, 'On the Contact Relations of Certain Systems of Circles', *Transactions Cambridge Philosophical Society*, vol. xvi, 1895. Theorem 16 is from the same source.

two intersections of (z) with the two circles through P tangent to (y') and (z') and belonging to the same system as (s).

Theorem 16.] *If three proper circles be given tangent to two fixed proper circles, two belonging to one system and one to the other, yet no two having their centres on an isotropic, and if a point be taken on each of the first two concyclic with a properly chosen intersection of each of the two with the third, then these two points lie on a circle tangent to the fixed circles and belonging to the same system as the third.*

Let us next assume that (53) holds, that $(y)\,(z)\,(s)$ have the same contacts with $(y')\,(z')$ as before, and that (x) is tangent to (y'). The intersections of (x) and (s) with (y) and (z) lie on two circles (t). But, by (16), such pairs of points lie on circles touching both (y') and (z'). Hence (x) touches (z') also. We are thus led once more to the Hart system of the first sort developed in Ch. I. We start with $(y)\,(z)\,(s)$, and suppose that circles $(y')\,(z')\,(s')$ are all tangent to them, circles given by the same letter having external contact, while those given by different letters have internal contact. We then take (x') having internal contact with $(y)\,(z)\,(s)$, and, lastly, (x) having external contact with $(y')\,(z')\,(s')$. Since (y) and (z) have unlike contacts with (y') and (z'), while (x) has like contacts with both, and (s) has also, (53) holds. But (y) has like contacts with (x') and (s'), (z) has like contacts with them also, (s) has unlike contacts with them, and (x) touches (s') externally. Hence (x) touches (x') internally, and we have indeed the Hart system. In the complex domain, of course, the words external and internal contact lose their geometric significance, and depend merely on the sign of a complex radical. Our Hart system may be arranged in three sequences:

$$(x)\,(y)\,(s)\,(z),\quad (x)\,(y)\,(z)\,(s),\quad (x)\,(s)\,(y)\,(z).$$

Each sequence gives rise to two circles of the type above, thus leading to two beautiful propositions due to Larmor.*

* Cf. Lachlan, 'On the Properties of some Circles connected with a Triangle formed by Circular Arcs', *Proceedings London Math. Soc.*, vol. xxi, 1890, p. 267. Also Study, loc. cit., p. 521.

Theorem 17.] *The intersections of the circles of a Hart system of the first sort fall into two groups of six points each; each system is the total intersection of three circles.*

Theorem 18.] *The circles circumscribing the arcual triangles formed by three non-concurrent proper circles are two Hart systems of the first sort, mutually inverse in the common orthogonal circle of the given circles.*

These two theorems may also be established in the following manner, which is of interest in itself. Let us start with a fundamental proper circle c. Each finite point P, except the centre of c, and its inverse P' with regard to c, may be associated with the circle coaxal with the null circles (P), (P') and orthogonal to c. Conversely, c and any circle orthogonal thereto but not concentric will determine a pencil or coaxal system whose limiting points are inverse in c. When the circles are concentric we take the centre as one limiting point, and treat every straight line as though it were a circle through the other limiting point.*

We next notice that two circles mutually inverse in c, if looked upon as point loci, will be transformed into two other such circles, considered as envelopes and vice versa, and that tangency of circles is an invariant property. A Hart system will go into a Hart system. We start with a Hart system of the first sort, and take c_1, c_2, c_3 as three circles of the complementary Hart system, c being the common orthogonal circle. The original Hart system, and its inverse in c, looked upon as envelopes, will go into the eight circles circumscribed to the arcual triangles of c_1, c_2, c_3, and these eight will be seen to form two Hart systems. Clearly there is nothing special about the circles c_1, c_2, c_3, so that 18] is proved. To prove 17] we have but to show that there is nothing special about the type of Hart systems formed by the circles circumscribing eight arcual triangles. But this is evident when we remember that we may choose three circles, so that three of the surrounding

* This transformation is due to Lachlan, 'On Coaxal Systems of Circles', *Quarterly Journal of Mathematics*, vol. xxvi, 1892. If we take the corresponding transformation on a sphere, and take for c the circle at infinity, we have the correspondence of a great circle to its two poles.

Hart circles shall intersect at any three chosen angles not congruent to zero, modulo π. But we may pass from any Hart system where three circles meet at specified angles to any other where the same angles appear by means of inversions and transformations of central similitude; hence any Hart system may be so transformed into one surrounding eight arcual triangles, and 17] is proved.

The Hart systems of the second sort are simpler; their properties are intuitively evident when we replace the circle of inversion by a straight line.

Theorem 19.] *The relation of a Hart system of the second sort to the four circles tangent to them is reciprocal; the common orthogonal circle of one system is a circle of anti-similitude of each pair of the other.**

Theorem 20.] *The intersection of a system of Hart circles of the second sort fall into two groups. The pairs of intersections of couples of circles lie on the common orthogonal circle of the complementary system, the remaining eight lie by fours on two circles orthogonal to this orthogonal circle.*

Theorem 21.] *If of the twelve intersections of four circles six are the total intersections of three other circles, then the four belong to a Hart system.*

§ 4. Analytical Systems of Circles.

We have now given a sufficient number of examples of our fundamental Frobenius identity (30); let us pass on and consider systems of many circles. The theorems concerned with concyclic points and concurrent circles which we took up in the last chapter are, for the most part, better handled by geometric means than by analytic ones. This rule, like all others, however, has exceptions. For instance, take I. 149]. The three circles each through a vertex of a triangle and two marked points of the adjacent side-lines will constitute, with the adjunction of these side-lines, a system of three cubic curves through eight common points. Such cubics have always

* For this theorem and the two following see Study, loc. cit., p. 525.

a ninth point in common, hence the circles are concurrent. Let us next repeat Clifford's own proof of I. 162].*

A curve of class $n+1$ is required to have the line at infinity as a multiple tangent of order n and to touch $2n+1$ given finite lines, no three of which are concurrent, and no two parallel. The number of linear conditions imposed on the coefficients is

$$2n+1+\frac{n(n+1)}{2}=\frac{(n+1)(n+4)}{2}-1.$$

If these conditions were not independent, we could have ∞^1 curves touching the line at infinity n times and $2(n+1)$ common finite tangents. Two such curves would have $(n+1)^2+1$ common tangents, which is absurd. The conditions are independent, and we have a one-parameter family of curves; all are linearly dependent on two of their number. From each circular point at infinity we may draw one more tangent to each curve, and these two tangents will clearly generate projective pencils; the locus of their intersections, the finite focus, is thus a circle. Among our curves are $n+1$, which degenerate and consist in the infinite point of one of our finite tangents, and a curve of class n, touching $2n$ given lines. We thus get $2n+1$ curves of class n, each touching $2n$ of our given lines and having their foci on a circle. If another line were added there would be one focus associated with $2n+2$ lines lying on as many circles each through $2n+1$ foci, and so on.

The analytic discussion of I. 155] will bring to light a new theorem not easily reached geometrically. We started with four points on a circle, and arranged them in order. Through each two successive points we passed a circle, and showed that the remaining intersections of successive circles were concyclic. Now the four points may be arranged in three different cyclic orders, so that they are connected in pairs by six circles, and three new circles are produced. The points being $P_1, P_2,$

* loc. cit. For a proof by an ingenious analysis apparently invented *ad hoc*, see Morley, 'On the Metric Geometry of the Plane n line', *Transactions American Math. Soc.*, vol. i, 1900. A proof is also given of Pesci's theorem, I. 164].

P_3, P_4 on the circle c, the points P_i, P_j shall be connected by the circle c_{ij}. The circles c_{ij} and c_{jk} will intersect again in P_{ijk}.

Let $P_{123}, P_{234}, P_{341}, P_{412}$ lie on c_1,
let $P_{132}, P_{314}, P_{142}, P_{423}$ lie on c_2,
let $P_{124}, P_{243}, P_{431}, P_{312}$ lie on c_3. $P_{ijk} = P_{kji}$.

The sextic $c_{13}\,c_{24}\,c_1$ contains every point common to $c_{12}\,c_{34}$ and $c_{41}\,c_{23}$, and has a triple point at each circular point at infinity; hence, by Nöther's fundamental theorem, we have an identity

$$c_{13}c_{24}c_1 \equiv \phi c_{12}c_{34} + \psi c_{41}c_{23}.$$

The curves ϕ and ψ are circles, since they are curves of the second order passing through the circular points at infinity, and they contain the remaining points P_{ij}. Hence they are the circles c_2, c_3.

$$c_{13}c_{24}c_1 \equiv c_{34}c_{12}c_2 + c_{41}c_{23}c_3.$$

But this shows that c_1, c_2, c_3 are coaxal.

Theorem 22.] *If four points on a circle be arranged in three cyclic orders, each two points be joined by a circle, and each cyclic order be associated with that circle which contains the remaining intersections of successive circles joining pairs of points in the given cyclic order, then will the three associated circles be coaxal.**

The advantages of the analytic as compared with the synthetic method are nowhere more apparent than when we come to study coaxal systems of circles. We shall extend that term to include every system through the intersections of two, i.e. every system linearly dependent on two circles. If (x) be the coordinates of a point on a circle coaxal with two proper circles (y) and (z), we have an equation,

$$\lambda\,(yx) + \mu\,(zx) = 0, \quad \frac{(yx)}{iy_0 + y_1} = \kappa\,\frac{(zx)}{iz_0 + z_1}.$$

This proves immediately the important theorem, I. 191].

* See the Author's *Circles Associated*, &c., cit.

The equations of the most interesting circles coaxal with (y) and (z) are written immediately.

Radical axis

$$(iz_0+z_1)(yx)-(iy_0+y_1)(zx) = 0. \tag{54}$$

Radical circle

$$(iz_0+z_1)(yx)+(iy_0+y_1)(zx) = 0. \tag{55}$$

Circle of similitude

$$\frac{(iy_0+y_1)(yx)}{(yy)}-\frac{(iz_0+z_1)(zx)}{(zz)} = 0. \tag{56}$$

Circles of antisimilitude

$$\sqrt{(zz)}\,(yx)\pm\sqrt{(yy)}\,(zx) = 0. \tag{57}$$

We easily see that the two circles represented by these equations are mutually orthogonal, and bisect the angles made by the circles (y) and (z) when these are not null or isotropic. The limiting points of the coaxal system will have the coordinates

$$\rho x_i = (yy)\,z_i-\left[(yz)\pm\sqrt{(yz)^2-(yy)(zz)}\right]y_i. \tag{58}$$

If r and r' be the radii of (y) and (z), while their angle is θ, the limiting points are

$$\rho x_i = \sqrt{(yy)}\,z_i-e^{\pm i\theta}\sqrt{(zz)}\,y_i.$$

The fact that if two circles be orthogonal to two others, every circle coaxal with (orthogonal to) one pair is orthogonal to (coaxal with) the other appears at once, for if

$$(yy) = (yz') = (zy') = (zz') = 0,$$

then

$$\sum_{i=0}^{i=3}(\lambda y_i+\mu z_i)(\lambda' y_i'+\mu' z_i') = 0.$$

Such conjugate coaxal systems will appear in three dimensions as pairs of lines conjugate with regard to the absolute quadric. The circle coaxal with (y) and (z) which is orthogonal to (s) will be

$$(ys)(z)-(yz)(s).$$

Theorem 23.] *If three circles be given, the three circles each coaxal with two and orthogonal to the third are coaxal.*

Theorem 24.] *If three circles be given, the three circles each coaxal with two of them and orthogonal to a fourth circle are coaxal.*

The concurrence of the radical axes appears as a limiting form of this. Let the reader devise an analytic proof of I. 201], namely, the radical circle is the locus of the centres of circles cut by one circle orthogonally, and by the other in diametrically opposite points.

A system of circles whose coordinates are proportional to analytic functions of one variable yet not bearing to one another constant ratios shall be called a *series of circles.* A coaxal system is the simplest type of series, and the only one lacking a curved envelope. If the circles be orthogonal to a fixed not null or isotropic circle, the envelope is anallagmatic with regard to this fixed circle. This was proved geometrically in what followed I. 15]; the easiest analytic proof is found by taking the fixed circle as fundamental for a tetracyclic coordinate system; the corresponding coordinate will be lacking in the generating circles and in the envelope.

Plane π.	Space S.
Anallagmatic envelope.	Plane curve.

In general the circles of a series will touch their envelope in pairs of distinct points. In special cases there will be but one point of contact. It is tolerably clear geometrically that this occurs when we have the circles osculating a given curve, and then only. Let us give an analytic demonstration. Let the variable circle be

$$y_i = y_i(t),$$

then, if adjacent circles tend to touch one another,

$$(yy)(y'y')-(yy')^2 = 0, \quad y_i' \equiv \frac{dy_i}{dt},$$

$$(yy)(y'y'') = (yy')(yy'').$$

The point of contact will have the coordinates

$$\rho x_i = (yy)\, y_i' - (yy')\, y_i.$$

But from this

$$(yx) = (yx') = (yx'') = 0,$$

which shows that the circle osculates the envelope. Conversely, take three adjacent points of the envelope

$$(x),\ (x)+(x')\,dt,\ (x)+2\,(x')\,dt+(x'')\,dt^2.$$

We have identically

$$(xx)=(xx')=0.$$

The osculating circle will have the equation

$$(yX)\equiv|\,xx'x''X\,|=0.$$

The adjacent osculating circle is

$$|\,xx'x''X\,|+|\,xx'x'''X\,|\,dt\equiv(yX)+(y'X)\,dt=0.$$

The condition of contact for (y) and $(y)+(y')\,dt$ gives

$$(xx'')\,(x'x')\,(xx'')\,(xx''')\,(x'x')\,(xx''')-[(xx''')\,(x'x')\,(xx'')]^2=0.$$

Theorem 25.] *A necessary and sufficient condition that the circles of a series should touch their envelope but once each is that they should be the osculating circles thereof.*

Plane π.	Space S.
Series of osculating circles.	Curve of length zero.

Next to the linear or coaxal series, the simplest are those whose coordinates are quadratic functions of the variable. Such will correspond to a conic in S, and we shall call it a *conic series.* We exclude the case where the series is reducible.

Theorem 26.] *If a circle move so that it is orthogonal to a fixed circle not null or isotropic, and the sum or difference of its angles with two fixed circles be constant, it traces a conic series.**

We see, in fact, that in S we have the intersection of a plane with a quadric of revolution. If we accept that the properties of confocal quadrics (which are nearly the same in non-Euclidean as in Euclidean space), in particular the relations of their focal conics, we have, from the known relations of three such conics,

* For the proofs of the theorems about non-Euclidean conics and quadrics on which our present circle theorems are based see the Author's *Non-Euclidean Geometry*, cit., ch. xii and xiii.

Theorem 27.] *The general conic series contains four distinct null circles. If such a series be given, there are associated therewith two other general conic series. The sum or difference of the angles which all circles of one series make with any two of another series depends merely on the choice of the latter.*

We shall prove this theorem in a later chapter without the use of non-Euclidean geometry.

Theorem 28.] *The radical axes of the circles of a conic series and a fixed circle will envelop a conic; the radical centres of these circles and two fixed circles generate a trinodal quartic.*

Theorem 29.] *The locus of a circle orthogonal to a fixed circle, and to corresponding circles in two projective pencils, neither of which includes the fixed circle, is a conic series.*

Since a central non-Euclidean conic has three axes of symmetry.

Theorem 30.] *A conic series which includes four distinct null or isotropic circles is anallagmatic in three mutually orthogonal circles, all orthogonal to that circle which is orthogonal to all circles of the series.*

From the focus and directrix property of central conics.

Theorem 31.] *If a circle move so that it is orthogonal to a fixed not null or isotropic circle, and the sine of its angle with one circle orthogonal thereto bears a constant ratio to the cosine of its angle with another circle also orthogonal thereto, it generates a conic series.*

Since the coordinates of a circle of a conic series are quadratic functions of an auxiliary variable, the same is true of the cartesian coordinates of their centres.

Theorem 32.] *The locus of the centres of the circles of a conic series is a conic.*

To find the envelope of the circles of a conic series, we put

$$y_i \equiv a_i r^2 + 2\, b_i rs + c_i s^2. \tag{59}$$

We then eliminate r and s between $\left(x \frac{\partial y}{\partial r}\right) = 0$ and $\left(x \frac{\partial y}{\partial r}\right) = 0$, and replace the x_i's by their cartesian values.

Theorem 33.] *The envelope of the circles of a conic series is, in general, a curve of the fourth order with a double point at each circular point at infinity.*

As we shall study this curve in some detail in a subsequent chapter, we shall say no more about it now.

We pass next to the general cubic series. We shall define this as an algebraic series whose members are not all orthogonal to one circle, but whereof three are orthogonal to an arbitrary circle. In three dimensions we have a non-planar curve which is algebraic and of the third order, and there is only one such type of curve (under the general projective group).

Theorem 34.] *The common orthogonal circles to corresponding triads in three projective pencils of circles whereof no two have a common member will generate a general cubic series, and every general cubic series may be so generated in ∞ ways.**

Theorem 35.] *The coordinates of the circles of a general cubic series are homogeneous functions of the third order of two variables.*

Theorem 36.] *The locus of the centres of the circles of a general cubic series is a rational curve of the third order.*

Since the osculating planes of a space cubic generate a developable envelope whose properties are dual to those of the curve.

Theorem 37.] *The common orthogonal circles to sets of three successive circles of a general cubic series generate another such series. The relation between the two series is reciprocal.*

Theorem 38.] *The envelope of the radical axes of successive circles of a general cubic series is the locus of the centres of the circles of the reciprocal series.*

A theorem analogous to this is clearly true of any series not orthogonal to one circle. The general cubic series is

* For a general purely geometrical account of this series see Timerding, 'Ueber eine Kugelschar', *Crelle's Journal*, vol. cxxi, 1899. Also Tauberth, *Die Abbildung des ebenen Kreissystemes auf den Raum*, Dissertation, Jena, 1885.

distinguished by the fact that it is not the same type as the reciprocal series.

Theorem 39.] *The envelope of the circles of a general cubic series is a curve of the eighth order with a quadruple point at each circular point at infinity.*

Theorem 40.] *The tangents to the loci of the centres of the circles of two reciprocal general cubic series can be put into such one to one correspondence that corresponding lines are mutually orthogonal. The asymptotes to one curve will correspond to the inflexional tangents to the other.*

A two-parameter family of circles, that is, a system whose coordinates are proportional to analytic functions of two independent variables, not having ratios all functions of one variable, shall be defined as a *congruence* of circles. Such a system, when algebraic, is best represented by means of an equation

$$f(x_0 x_1 x_2 x_3) = 0.$$

Remembering that in non-Euclidean space, as in Euclidean, every surface not a developable circumscribed to the Absolute, is covered by a double network of curves of zero length, isotropic curves, we have

Theorem 41.] *Every congruence of circles may be either generated on a one-parameter family of pencils of tangent circles, or, in two ways, by the osculating circles of a one-parameter family of curves.*

If (x) be a circle of the congruence, the circle $\left(\frac{\partial f}{\partial x}\right)$ shall be called its *correlative circle.* It is orthogonal to (x), and, to the first degree of approximation, to all infinitely near circles of the congruence. If we take two adjacent circles of our congruence, the pencil which they determine is not, in general, orthogonal to the pencil determined by their correlative circles. If we take the pencils determined by $(x)\,(x+dx)$ and $(x)\,(x+\delta x)$, then, if the first be orthogonal to the pencil determined by the correlatives of the second two circles, the second pencil is orthogonal to that determined by the correlatives of the first two circles. Two such pencils are said to be *pseudo-conjugate*; they correspond to conjugate directions on the surface in S which corresponds to our congruence.* Since the only surface where the asymptotic lines fall together is a developable, we have

Theorem 42.] *A congruence of circles is either determined by a one-parameter family of coaxal systems each determined by successive circles of a series, or else each coaxal system determined by a circle of the congruence in general position and an adjacent circle is pseudo-conjugate to another such coaxal system. Each circle will belong to two coaxal systems pseudo-conjugate to themselves which cannot coincide for every circle of the congruence.*

We mean by a circle in *general position* one whose correlative exists. Since there are two sorts of ruled surfaces in space,

Theorem 43.] *Congruences of circles generated by one-parameter families of coaxal systems are of two sorts. In the first case the coaxal systems are determined by adjacent circles of a series, in the second case they are not so determined. In the first case all circles of a coaxal system have the same correlative circle, in the second case no two have the same.*

If we define as the *order* of an algebraic congruence the number of its members in an arbitrary coaxal system, we see that this is equal to the order of the equation of the congruence. A congruence of the first order is the system of circles orthogonal to one circle.

The most interesting congruences of circles are the *quadric* ones. We shall define such a congruence as the totality of circles satisfying an equation

$$\sum_{i,j=0}^{i,j=3} a_{ij}\, x_i x_j = 0. \tag{60}$$

We may classify these in various ways. The broadest classification is under the fifteen-parameter group of all linear

* We call these coaxal systems 'pseudo-conjugate', as 'conjugate' coaxal systems have already been otherwise defined, p. 99.

transformations of our circle coordinates. Here we have the following types:*

$$\text{I. } |a_{ij}| \neq 0.$$

$$\text{II. } |a_{ij}| = 0, \quad \frac{\partial |a_{ij}|}{\partial a_{kl}} \not\equiv 0.$$

$$\text{III. } \frac{\partial |a_{ij}|}{\partial a_{kl}} \equiv 0, \quad \frac{\partial^2 |a_{ij}|}{\partial a_{pq}\, \partial a_{rj}} \not\equiv 0.$$

$$\text{IV. } \frac{\partial^2 |a_{ij}|}{\partial a_{pq}\, \partial a_{rj}} \equiv 0.$$

We shall call I the *general quadratic congruence.*

Theorem 44.] *The general quadratic congruence contains two families of coaxal systems; each circle belongs to one coaxal system of each family, each two systems of different families share a circle, but not two of the same family have any common circle. The congruence may be generated in* $2\infty^2$ *ways by the coaxal systems, determined by corresponding members in two given projective coaxal systems which have no common circle. The lines of centres of the coaxal circles of the two families envelop one same conic.*

To prove this last part of the theorem, the line of centres of a coaxal system in π will correspond in S to the point where the polar in the Absolute of the line corresponding to the coaxal system meets that plane which represents the totality of straight lines. The totality of lines of centres will be represented by the intersection of this plane with the polar in the Absolute of the quadric representing the series.

Theorem 45.] *The assemblage of all circles meeting a given not null or isotropic circle at a given angle or its supplement is a quadric congruence.*

Theorem 46.] *The correlative of a general quadric congruence is a second such congruence.*

* The best discussion of these congruences is in a pleasantly written paper by Loria, 'Remarques sur la géométrie analytique des cercles du plan', *Quarterly Journal of Mathematics*, vol. xxii, 1886.

Theorem 47.] *If two correlative quadric congruences be given, the coaxal systems of one will correspond to those of the other. All circles of one coaxal system cut all those of the correlative system at right angles.*

Theorem 48.] *The locus of the centres of the null circles of a quadric congruence includes the locus of the points common to coaxal systems of the correlative congruence.*

Two quadric congruences which have the same null and isotropic circles shall be called *homothetic*; if their correlatives have the same null circles they shall be called *confocal.*

Plane π.	Space S.
Homothetic quadric congruences.	Homothetic quadric surfaces.
Confocal quadric congruences.	Confocal quadric surfaces.

Theorem 49.] *There are ∞^1 general quadric congruences confocal with a given general congruence; an arbitrary circle will belong to three of these.*

The system of congruences confocal with (60) will be

$$\begin{vmatrix} A_{00}+\lambda & A_{11} & A_{02} & A_{03} & x_0 \\ A_{10} & A_{11}+\lambda & A_{12} & A_{13} & x_1 \\ A_{20} & A_{21} & A_{22}+\lambda & A_{23} & x_2 \\ A_{30} & A_{31} & A_{32} & A_{33}+\lambda & x_3 \\ x_0 & x_1 & x_2 & x_3 & 0 \end{vmatrix} = 0. \qquad (61)$$

Theorem 50.] *In a homothetic system of quadric congruences there will, in general, be four congruences of type II. The correlative to each of these will be a conic series of circles which envelop the locus of the centres of the null circles of the given homothetic congruences.*

The meaning of the words *in general* as here used will appear more fully in Ch. IV.

Theorem 51.] *The assemblage of all circles the sum or difference of whose angles with two given not null or isotropic circles is constant is a quadric congruence.*

In non-Euclidean space there are two types of parallel lines. The first are *Lobachevski parallels* and intersect on the Absolute, the second are *Clifford parallels* and intersect the same two generators of one set of the Absolute. Let us reserve the name *parallel* for the first kind and use *paratactic* for the second.

Plane π.	Space S.
Coaxal systems with common limiting point.	Parallel lines.
Coaxal systems whose null circles are orthogonal in pairs.	Paratactic lines.

Theorem 52.] *If a coaxal system of circles be given with two distinct null circles, an arbitrary not null circle will belong to two coaxal systems each sharing one limiting point with this coaxal system, and to two whose limiting points are in pairs at null distances from those of the given system.*

In special cases the coaxal system may be concentric and have no limiting points; the reader can easily find for himself the slight modification here needed.

Paratactic lines are at a constant distance from one another, and have an infinite number of common non-Euclidean perpendiculars. These generate a quadric, whose generators of each set are paratactic.*

Theorem 53.] *If two coaxal systems have their limiting points in pairs at null distances from one another, but no point is at a null distance from all four, nor do they lie on one isotropic line, then their circles may be so paired that corresponding circles make a constant angle with one another, the least angle which any proper circle of one system makes with one of the other. The coaxal systems determined by such sets of circles will generate a quadric congruence. Two coaxal systems of the same family in this congruence will have their limiting points in pairs at null distances.*

* See the Author's *Non-Euclidean Geometry*, cit., pp. 114, 129, 130.

It is clear that our quadric congruence of type II will correspond to a cone in S, and, as we have seen, its correlative is a conic series.

Theorem 54.] *A quadric congruence of type II may be generated in* ∞^2 *ways by coaxal systems determined by one fixed circle and the circles of a conic series.*

Theorem 55.] *A quadric congruence of type III is reducible, and consists in the totality of circles orthogonal to either of two distinct circles. A congruence of type IV consists in the circles orthogonal to a given circle all counted twice.*

It is clear that although the subject-matter of the present chapter does not offer such a wide field for further study as did that of Ch. I, yet there is room for further advance. It is probable that there is little to be gained by a further study of the circle in trilinear coordinates. On the other hand, there is no knowing how much more may be obtained by a further study of the Frobenius identity. The subject of Hart circles and the circles inscribed or circumscribed to arcual triangles seems almost illimitable. It seems likely that the Frobenius identity should yield a simpler proof of the existence of the Hart circle than any yet found, and this would be a real gain. There is also room for much new material connected with the interpretation of non-Euclidean three-dimensional space in the geometry of the circle.*

* An extended account of how the geometry of the circle may be used to interpret non-Euclidean geometry will be found in Weber und Wellstein, 'Encyklopädie der Elementar-Mathematik', Second Edition, vol. iv, Leipzig, 1907.

CHAPTER III

FAMOUS PROBLEMS IN CONSTRUCTION

THERE has been one conspicuous lack in all the work that we have done so far in the geometry of the circle; we have paid next to no attention to any problems in construction. This omission, let us hasten to say, has been intentional, as it is much easier to attack such problems satisfactorily if both algebraic and geometric methods are available. No one would ever have found by the aid of pure geometry alone that it was impossible to square the circle. The time has now come when certain problems in construction must be seriously faced. It is clear that the number of such problems is illimitable; we shall restrict ourselves to a very few which have become famous in the history of the subject.

In discussing problems of geometrical construction one has frequently to face the question, 'Which of the various solutions is the simplest?' Such a query cannot be answered categorically. What is a simple solution? Is it one that involves very little drawing, or one that is based on elementary theorems, or one that can be explained and proved in a few words? These desiderata seem to vary almost independently of one another; there must be a great measure of arbitrariness in any criterion of simplicity.

The best known and least undesirable tests for the simplicity of a geometrical construction are those originally devised by Émile Lemoine.* Three distinct operations are recognized for the compass, two for the ungraded ruler:

(1) To place one point of the compass in a given position.

(2) To place one point of the compass on a given line.

* His various writings on this subject are summed up in his *Géométrographie*, Paris, 1902. For convenience we shall refer to this work by page number.

(3) To draw a circle.

(4) To place one edge of the ruler on a given point.

(5) To draw a line.

The sum of the number of times that all of these operations are performed is called the *simplicity* of the construction, the sum of the number of times that the first, second, and fourth are performed is called its *exactitude*. Lemoine recognizes that these names are ill chosen, and suggests that the word simplicity might better be replaced by 'measure of complication', but neither he nor his followers have seen fit to adopt this improvement in terminology. Moreover, as tests they are of the roughest. As the area of a parallelogram is equal to the product of its altitudes divided by the sine of the angle formed by intersecting sides, the exactitude of the operation of drawing a line through the intersection of two others will vary directly with the sine of their angle. It is not, however, our present business to devise tests of geometrical simplicity, but to apply certain recognized tests to concrete problems. We shall start with the most famous of all, the problem of Apollonius, *To construct a circle tangent to three given circles.**

Let us begin by examining how many real solutions can be found for the problem.† The answer to this is intuitively evident in any particular case by examining the figure. It is more sportsmanlike, however, to use II (48), which we rewrite for the case of contact,

$$\begin{vmatrix} 1 & \cos\measuredangle yz & \cos\measuredangle ys & \epsilon_1 \\ \cos\measuredangle zy & 1 & \cos\measuredangle zs & \epsilon_2 \\ \cos\measuredangle sy & \cos\measuredangle sz & 1 & \epsilon_3 \\ \epsilon_1 & \epsilon_2 & \epsilon_3 & 1 \end{vmatrix} \leqq 0, \quad \epsilon_1{}^2 = \epsilon_2{}^2 = \epsilon_3{}^2 = 1.$$

Remembering that $2rr'\cos\theta = r^2 + r'^2 - d^2$,

if (y) and (z) lie outside one another,

$$\cos\measuredangle yz < -1, \ \sin^2\tfrac{1}{2}\measuredangle yz > 0, \ \cos^2\tfrac{1}{2}\measuredangle yz < 0.$$

* Simon, loc. cit., pp. 98 ff., mentions some seventy works dealing with this problem which appeared in the nineteenth century.

† See the note on P. 143.

If they intersect in real points,

$$-1 < \cos \measuredangle yz < 1, \ \sin^2 \tfrac{1}{2} \measuredangle yz > 0, \ \cos^2 \tfrac{1}{2} \measuredangle yz > 0.$$

If one include the other,

$$\cos \measuredangle yz > 1, \ \sin^2 \tfrac{1}{2} \measuredangle yz < 0, \ \cos^2 \tfrac{1}{2} \measuredangle yz > 0.$$

(A) A circle having like contact with all three,

$$\sin^2 \tfrac{1}{2} \measuredangle zs \sin^2 \tfrac{1}{2} \measuredangle sy \sin^2 \tfrac{1}{2} \measuredangle yz \geq 0, \quad \epsilon_1 = \epsilon_2 = \epsilon_3.$$

The construction of two circles satisfying the given conditions is real unless one circle separate the other two, or unless two intersect and the third surrounds or lies within the one but not the other.

(B) A circle having with (y) a contact opposite to that with (z) and (s),

$$\sin^2 \tfrac{1}{2} \measuredangle zs \cos^2 \tfrac{1}{2} \measuredangle sy \cos^2 \tfrac{1}{2} \measuredangle yz \geqq 0, \quad -\epsilon_1 = \epsilon_2 = \epsilon_3.$$

The construction is possible unless two circles are separated by the third, or (z) and (s) intersect, while (y) lies within one but not within the other, or surrounds the one but not the other.

The first method which we shall employ for the solution of the problem is that ascribed to Apollonius himself.*

Problem 1.] To construct a circle which shall pass through two given points and touch a given circle.

It is clear that to obtain a real solution we must have two points not separated by the circle. We see also that the common secants of the given circle and all circles through the two points will be concurrent on the line through these two points.—We therefore make the following construction. *Draw a convenient circle through the two points, find where the radical axis meets the line through the two points, and draw tangents thence to the given circle. A circle through the given points and either point of contact will satisfy the given conditions, and there are no other circles which do so.*

Let us apply Lemoine's criteria. To construct a circle through two given points involves drawing two circles with

* Killing-Hovestadt, *Handbuch des mathematischen Unterrichts*, Leipzig, 1910, p. 414. A very clear and easy discussion of the method will be found in Cranz, *Das apollonische Berührungsproblem*, Stuttgart, 1891.

the same compass opening and the given points as centres, and a third circle with the same radius and a given centre S. 6, E. 3.

We next connect the two intersections of two circles by a line, S. 3, E. 2. Then draw tangents to a given circle from an exterior point, S. 18, E. 12 (p. 33; the usual construction has S. 19). Then construct two circles through two common points, one through each of two given points, S. 23, E. 14. We have for the total construction

Simplicity 38, Exactitude 25.

Problem 2.] To construct a circle through a given point tangent to two given circles.

Let us, to be specific, take a point P external to both circles and imagine them external to one another. *A shall be the external centre of similitude. We find Q on AP so that* $(\overrightarrow{AP})(\overrightarrow{AQ})$ *is the square of the radius of the circle of antisimilitude corresponding to A. Then a circle through P and Q tangent to one of the given circles is tangent to the other also.*

We first construct the common tangents to two mutually external circles, S. 35, E. 22. (These are Lemoine's numbers, p. 43; the usual construction runs much higher.) Starting with one centre of similitude, let R and R' be corresponding points of contact on the same tangent which are mutually inverse in the circle of antisimilitude. We must find Q on AP so that $(\overrightarrow{AP}) \times (\overrightarrow{AQ}) = (\overrightarrow{AR}) \times (\overrightarrow{AR'})$. To accomplish this we draw AP and PR, S. 6, E. 4, and through R' a line making a given angle with AR, S. 11, E. 7. There is another point Q found in similar fashion from the other centre of similitude. These operations give S. 34, E. 22. We then must solve problem 1] twice in succession. Our total numbers are

Simplicity 145, Exactitude 94.

As an alternative we offer the closely allied solution. *Take any convenient circle about the given point as centre for a circle of inversion, and find the inverses of the given circles. Find the common tangents to these circles and invert back.*

We see that this construction is simpler than the last, in the sense that it is described in fewer words. To construct our circle of inversion, which we shall imagine cuts the given circles in real points, we have S. 1, E. 0. We next find the inverses of two given points, one on each circle, S. 19, E. 12 (p. 54). To find the inverses of our given circles we must find the inverse of a point on each and construct two circles each through three points; each of these latter constructions involves S. 15, E. 9. We next construct the common tangents to two circles, S. 35, E. 22. Assuming that these intersect the circle of inversion, the construction of their inverses will amount merely to drawing a circle through three points four times, one point being the same in each case; this will require S. 54, E. 43. For our total construction,

Simplicity 139, Exactitude 95.

Problem of Apollonius. To construct a circle tangent to three given circles.* We assume for the sake of definiteness that they lie outside of one another, so that there are effectively eight real solutions. *Let* C_1 *be the centre of the circle of smallest radius* r_1. *Construct a circle or circles through* C_1 *having external contact with the two circles concentric with the other two given circles, but whose radii are less than the radii of these by the quantity* r_1. *Two of the required circles are concentric with those last constructed, but their radii are* r_1 *greater. To construct circles tangent externally to some of our circles and internally to others we must shrink some radii by* r_1, *and increase others by like amount; on the other hand, we shall not in any one case need more than two out of the four circles through a given point tangent to two given circles.* The processes of finding direct and transverse common tangents to two circles have nothing in common except the drawing of the line of centres, hence the construction of one pair of common tangents involves S. 19, E. 12. The construction of two of the four circles through a point touching two circles will require S. 97, E. 60. This operation will have to be

* An elaborate geometrographic discussion of various solutions of this problem will be found in Bodenstedt, 'Das Berührungsproblem des Apollonius', *Zeitschrift für mathematischen Unterricht*, vol. xxxvii, 1906.

performed four times. To shrink or swell a radius by a given amount will involve S. 10, E. 8, and this operation must be performed twice on two of the given circles, and once on each of eight constructed. We have, all told,

Simplicity 508, Exactitude 336.

It is certain that these numbers can be very greatly reduced by ingenuity in construction; they are sufficiently exact to show, however, that the problem is not of the simplest.

As a second solution of the Apollonian problem we give the neatest and most famous of all, that of Gergonne.* We saw in I. 217] that if two circles intersect two others at equal or supplementary angles the radical axis of each pair passes through a centre of similitude of the other. When the given circles are mutually external there will exist a pair of circles which have either a preassigned type of contact with each, or else exactly the opposite type of contact with each. The radical axes of the circles sought will be the lines which contain triads of centres of similitude for pairs of the given circles. On the other hand, a centre of similitude of a pair of solutions (which have each the same or exactly opposite contacts with each of the three given ones) will lie on the radical axis of each two given circles, i.e. be their radical centre. The line connecting the points of contact of a pair of circles sought with one given circle will go through this radical centre, and through the pole with regard to this chosen circle of the corresponding line containing three centres of similitude, for the pole of this line will lie on the radical axis of the pair. We thus get Gergonne's construction. *Find the poles with regard to the given circles of the lines containing triads of their centres of similitude two by two. The lines connecting the corresponding poles with the radical centre of the three circles will meet these circles in the points of contact with one pair of the circles sought.*

Let us examine this geometrographically. The determination of the radical centre of non-intersecting circles involves

* 'Recherche du cercle qui en touche trois autres dans un plan', *Annales de Mathématiques*, vol. vii, 1817. Inaccessible to the Author.

(p. 57) S. 26, E. 16. The construction of the three tetrads of common tangents calls for S. 105, E. 66. The determination of the lines containing triads of centres of similitude, S. 12, E. 8. Determining their four poles of each (p. 55), S. 60, E. 36. Twenty-four points of contact, S. 36, E. 24. Construction of the eight circles through given points, S. 120, E. 96. Totals,

Simplicity 479, Exactitude 318.

As an example of how much the manual labour of geometry may be shortened by using constructions which are difficult to remember, and ingenious rather than obvious, let us mention that, apparently, these numbers can be reduced to

Simplicity 199, Exactitude 129.*

It is geometrically evident that Gergonne's construction fails when the centres of the three circles are collinear. Here, however, we may employ a very simple method. All circles tangent externally to c_1 and c_2 will cut the radical axis at a fixed angle by I. 212], the angle which this axis makes with a direct common tangent, or the angle which either circle makes with the corresponding polar of the external centre of similitude. The polar and radical axis are corresponding lines in a transformation of central similitude between c_1 and the circle sought, the centre of similarity being the point of contact. The radical axis of c_1 and c_2 being a_2, while the polar is l_2, and c is the centre of the circle sought,

$$\frac{\overrightarrow{(c_1 l_2)}}{\overrightarrow{(c a_2)}} = -\frac{r_1}{r}, \quad \frac{\overrightarrow{(c_1 l_3)}}{\overrightarrow{(c a_3)}} = -\frac{r_1}{r}, \quad r = \frac{r_1 \overrightarrow{(a_2 a_3)}}{\overrightarrow{(l_2 l_3)}}.$$

The value of r is thus easily found, and so the circle sought.† Gergonne's construction is also at fault when the radii of two given circles reduce to zero. The solution by other means is, however, extremely easy in this case, as we have already seen.

Another problem closely allied to that of Apollonius is

* Reusch, *Planimetrische Konstruktionen in geometrographischer Ausführung*, Leipzig, 1904, p. 84. Gerard, *Scientia*, vol. vi (inaccessible to the Author), is said to give a construction of S. 152; Lemoine, *Géométrographie*, cit. p. 62, gives one of S. 154.

† Cranz, loc. cit., p. 157.

the problem of Steiner, to construct a circle meeting three given circles at given angles.* The easiest way here is to throw the problem back on the preceding one. We have already seen in I. 212] that all circles which make given angles with two given circles will make constant angles with every circle coaxal with them, and this may also be easily shown analytically. If, therefore, we assume that the three circles lie outside one another, we have

To construct a circle cutting three given circles c_1, c_2, c_3 at the angles θ_1, θ_2, θ_3 respectively. Let P_i and P_j be two convenient points on the circles c_i and c_j respectively. Through them draw lines which make with the radii angles $\frac{\pi}{2}-\theta_i$ and $\frac{\pi}{2}-\theta_j$ *respectively, and on these lines take Q_i and Q_j so that $(P_iQ_i)=(P_jQ_j)$. Find the intersections of the circles with centres C_i and C_j, and radii (C_iQ_i) (C_jQ_j), and with one of these points as centre and radii equal to (P_iQ_i) construct a circle c_k'. This will intersect c_i and c_j in the angles θ_i and θ_j respectively. Construct c_k'' coaxal with c_i and c_j and tangent to c_k'. The circles required will touch c_i'', c_j'', c_k''.*

It is to be noted that whenever the problem can be solved at all we shall get the solution by this method. Let us see how much has been added geometrographically to our original problem. One circle c_k' will involve S. 36, E. 23 (p. 22). Three such circles will cost but S. 63, E. 39. We have supposed that both θ_i and $\frac{\pi}{2}-\theta_i$ were known, i.e. constructed. It would be easy to find c_k'' if we supposed c_i and c_j intersected, but for the purposes of our present problem it is better to suppose them external to one another. We draw the radical axis of c_i and c_k', S. 3, E. 2, and the radical axis of c_i and c_j, which costs, if cleverly done (p. 56), S. 16, E. 10. The radical centre of c_i, c_j, c_k' is thus found, and from here we draw tangents to c_k' which will (p. 33) involve S. 18, E. 12. We next must draw a circle coaxal with c_ic_j, and passing through a given point of contact. We know a line c_ic_j through the

* 'Einige geometrische Betrachtungen', *Crelle's Journal*, vol. i, 1826, p. 162.

centre of such a circle, and one of its points. We find the centre then as the intersection of $c_i c_j$ with the diametral line of c_k' through the point of contact. The total labour on c_k'' has been S. 46, E. 30. Multiplying by 3, and adding to the price of c_k', we have finally

Simplicity 201, Exactitude 129.

Here again it is certain that great reductions could be effected by sufficient geometrographic ingenuity.

We pass now to another problem of an analogous sort. *To construct a circle cutting four given circles at equal or supplementary angles.* We may determine the number of real solutions from II (50). The circles sought are orthogonal to a circle of antisimilitude of each pair of the given circles. Among such circles of antisimilitude we may always find three which are not coaxal. The problem then resolves itself into that of finding the common orthogonal circle of three given circles. Instead of supposing that the given circles are mutually external, let us this time assume that each two intersect. We first draw tangents to two circles at an intersection (p. 32), S. 18, E. 12. Draw the bisectors of the angles of the tangents, S. 12, E. 10. Since the two circles of antisimilitude of intersecting circles are mutually orthogonal, the tangents to one intersect in the centre of the other. Hence the construction of two such circles will involve in addition S. 6, E. 4. Three such pairs of circles must be constructed. The construction on the common orthogonal circles of three given circles involves (p. 57) S. 44, E. 28, if done in the most improved fashion. Hence we may construct the eight solutions of our problem for the small cost of

Simplicity 460, Exactitude 302.

Our next problem has also to deal with contact of circles, and is nearly as well known as the others; the celebrated and often-discussed problem of Malfatti. *To construct three circles, each of which shall touch the other two, and two sides of a given triangle.** One reason for the popularity of the problem is

* *Memorie di matematica e di fisica della Società Italiana delle Scienze*, vol. x, Modena, 1803. Inaccessible to the Author. Simon, loc. cit., pp. 147ff., gives some forty titles bearing thereon.

that Steiner * left the classical solution without proof as an example of the power of his methods. His solution is as follows:

Let the vertices of the triangle be A_1, A_2, A_3. Let I be the centre of the inscribed circle. Inscribe a circle in each of the triangles IA_jA_k. The circles inscribed in ΔIA_iA_j and ΔIA_jA_k have IA_j as one transverse common tangent. Construct D_jE_j, the other such common tangent. The circles required are inscribed in the quadrilaterals whose side-lines are A_iA_j, A_iA_k, D_jE_j, D_kE_k.

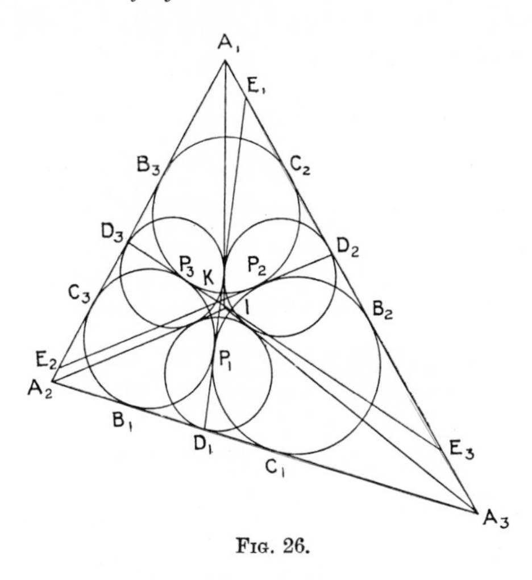

FIG. 26.

The simplest proof, beyond a peradventure, is that of Hart.† Suppose, first, that the figure has been drawn. The two circles which touch (A_jA_k) shall touch one another in P_i. Their common tangent thereat shall meet A_jA_k in D_i. The radical centre of our three circles, the point of concurrence of the tangents D_iP_i, shall be K (not supposed here to be the symmedian point). The points of contact on A_jA_k shall be B_iC_i, the former being supposed to be the nearer to A_j. Each of the lines P_iD_i meets two sides of the triangle. Suppose,

* *Einige geometrische, etc.*

† 'Geometrical investigation of Steiner's Solution of Malfatti's Problem', *Quarterly Journal of Mathematics*, vol. i, 1856.

to be specific, that P_1D_1 and P_3D_3 both meet A_1A_3 in E_1 and E_3 respectively.

$$(E_1D_2)-(E_3D_2) \equiv (E_1B_2)-(E_3C_2) = (E_1P_1)-(E_3P_3)$$
$$= (E_1K)-(E_3K).$$

It thus appears that D_2 is the point of contact of A_1A_3, with the circle inscribed in the $\Delta\, E_1KE_3$. The reasoning would hold equally well if E_1, or E_3, or both, were not between A_1 and A_3. We shall therefore inscribe circles in the three triangles with side-lines KE_k, KE_i, A_kA_i, the points of contact being D_j with (A_kA_i), F_i with (E_iK), and G_k with (E_kK). We next notice that

$$\begin{aligned}(A_1D_2)-(A_1D_3) &= (C_2D_2)-(B_3D_3)\\ &= (P_3G_3)-(P_2F_2)\\ &= (P_1F_1)-(P_1G_1) = (F_1G_1).\end{aligned}$$

Hence the other transverse common tangent goes through A_1, and a similar phenomenon holds for A_2 and A_3.

$$\begin{aligned}(D_2F_2) &= (D_2P_2)+(P_2F_2)\\ &= (D_2C_2)+(D_3B_3)\\ &= (P_3G_3)+(D_3P_3) = (D_3G_3).\end{aligned}$$

The circles $D_2F_1G_3$ and $D_3F_2G_1$ cut equal segments on (D_2D_3), and so, by I. 170], A_1 is on their circle of similitude, and, by I. 28] converse, the other transverse common tangent will bisect the $\measuredangle A_1$. If there be a solution of Malfatti's problem this will be it. Conversely, if a very small circle be drawn tangent to two sides of the triangle, the two circles each touching this little circle and two other sides will surely intersect. But if the little circle swell up, always touching the two sides till it become the inscribed circle, the other two circles are eventually separated by it. Hence, for some intermediate value of the little circle, the three will touch. Hart's solution is thus complete.

It has been objected to Hart's proof that it makes use of theorems which probably Steiner did not know, but were invented *ad hoc* by Hart.* The criticism seems to us trivial, and certainly not of sufficient importance to justify the great pains bestowed by subsequent writers to devise less simple proofs of the construction. There is a suspicion which naturally arises that, if the first discoverer of a proof had been of Steiner's own nationality, less trouble would have been given to disparaging his work.

Let us find what geometrographic numbers should be attached to Steiner's construction. We first bisect the angles of a triangle (p. 27), S. 21, E. 12. Inscribe circles in three adjacent triangles (p. 27), S. 80, E. 46. If we take two of these circles, we have already one common transverse tangent. To draw the other, we find the intersection of this tangent with the line of centres, sweep out an arc with a radius equal to the given tangential segment to one circle, and thus find the point of contact for the other common transverse tangent. This tangent will involve S. 9, E. 6. We draw three such common tangents, then inscribe circles in three given triangles, which can be done at a cost of S. 63, E. 36, since some bisectors are already known. The totals will be

Simplicity 191, Exactitude 112.

Let us give another solution of the problem, which depends on finding the point of contact of the circles.† The lengths of the sides of the triangle shall be, as usual, a_1, a_2, a_3, the distance from A_i to the points of contact of the circle which touches $(A_iA_j)(A_iA_k)$ shall be x_i. We also write, by definition,

$$\sum_{i=1}^{i=3} a_i = 2s, \quad \sqrt{\frac{s-a_i}{s}} = b_i, \quad \sqrt{\frac{a_i}{s}} = c_i. \tag{1}$$

The radii of our three circles shall be r_1, r_2, r_3. The

* See Schröter in *Crelle's Journal*, vol. lxxvii, 1874, p. 232. As a matter of fact some of the theorems objected to were discovered by Plücker long before Hart's time, though after Steiner's.

† These formulae were first found by Schellbach, *Sammlung und Auflösung mathematischer Aufgaben*, Berlin, 1863, pp. 100 ff. The form here given is from Mertens, 'Die Malfattische Aufgabe für das geradlinige Dreieck', *Zeitschrift für Mathematik und Physik*, vol. xxi, 1886.

distance from A_i to the centre of the inscribed circle shall be d_i. We have the following additional relations:

$$\rho = s\, b_1\, b_2\, b_3. \tag{2}$$

$$\rho = d_i \sin \tfrac{1}{2} \measuredangle A_i. \tag{3}$$

$$\rho = d_i \cdot \frac{b_j b_k}{c_j c_k}. \tag{4}$$

$$d_i = s b_i\, c_j\, c_k. \tag{5}$$

The side $(A_i\, A_j)$ is made up of distances from A_i and A_j to two points of contact, and a common tangential segment

$$x_i + x_j + \sqrt{(r_i + r_j)^2 - (r_i - r_j)^2} = a_k.$$

The radical reduces to the simple value $2\sqrt{r_i\, r_j}$. We have also

$$r_i = x_i \tan \tfrac{1}{2} \measuredangle A_i,\ r_j = x_j \tan \tfrac{1}{2} \measuredangle A_j,\ 2\sqrt{r_i\, r_j} = 2\sqrt{x_i}\sqrt{x_j} b_k.$$

$$x_i + x_j + 2 b_k \sqrt{x_i}\sqrt{x_j} = a_k. \tag{6}$$

$$\sqrt{x_i} + b_k\sqrt{x_j} = c_k\sqrt{s - x_j},$$

$$\sqrt{x_j} + b_k\sqrt{x_i} = c_k\sqrt{s - x_i}.$$

Multiplying these together, and subtracting (6) multiplied by b_k,

$$(1 - b_k^2)\sqrt{x_i}\sqrt{x_j} = c_k^2\sqrt{s - x_i}\sqrt{s - x_j} - a_k\, b_k,$$

$$\sqrt{x_i}\sqrt{x_j} - \sqrt{s - x_i}\sqrt{s - x_j} = -s b_k, \tag{7}$$

$$\sqrt{x_i} = -b_k\sqrt{x_j} + c_k\sqrt{s - x_j},$$

$$\sqrt{x_j} = -b_k\sqrt{x_i} + c_k\sqrt{s - x_i},$$

$$\sqrt{x_i}\sqrt{x_j} = b_k^2\sqrt{x_i}\sqrt{x_j} + c_k^2\sqrt{s - x_i}\sqrt{s - x_j} - b_k c_k\left[\sqrt{x_i}\sqrt{s - x_j} + \sqrt{x_j}\sqrt{s - x_i}\right],$$

$$\sqrt{x_i}\sqrt{s - x_j} + \sqrt{x_j}\sqrt{s - x_i} = s c_k. \tag{8}$$

From (7) and (8) we get, permuting the subscripts,

$$(\sqrt{x_i} + i\sqrt{s - x_i})(\sqrt{x_j} + i\sqrt{s - x_j}) = s(-b_k + i c_k), \tag{9}$$

$$(x_i + i\sqrt{s - x_i})^2 = \frac{s(-b_j + i c_j)(-b_k + i c_k)}{(-b_i + i c_i)}.$$

But $$b_i^2+c_i^2=1.$$

$$\sqrt{x_i}+i\sqrt{s-x_i}=\sqrt{s(-b_i-ic_i)(-b_j+ic_j)(-b_k+ic_k)} \quad (10)$$

$$=\pm\sqrt{s}\left[\sqrt{\frac{1-b_i}{2}}-i\sqrt{\frac{1+b_i}{2}}\right]$$

$$\left[\sqrt{\frac{1-b_j}{2}}+i\sqrt{\frac{1+b_j}{2}}\right]\left[\sqrt{\frac{1-b_k}{2}}+i\sqrt{\frac{1+b_k}{2}}\right]$$

$$=u_i+iv_i,$$

$$x_i=u_i^2, \quad u_i^2+v_i^2=s,$$

$$(u_i+iv_i)^2=2u_i^2-s+2u_iv_ii,$$

$$=s(-b_i-ic_i)(-b_j+ic_j)(-b_k+ic_k),$$

$$x_i=\tfrac{1}{2}[s-sb_1b_2b_3+sb_ic_jc_k-sb_jc_kc_i-sb_kc_ic_j],$$

$$x_i=\tfrac{1}{2}[s-\rho+d_i-d_j-d_k]. \quad (11)$$

These simple equations give us another construction which is geometrographically simpler than that of Steiner. Determination of d_1, d_2, d_3, S. 27, E. 15. Determination of $2s$, S. 6, E. 4, that of $s-\rho$, S. 11, E. 6. Combining the quantities d_1, d_2, d_3 with these, the total determination of $2x_1, 2x_2, 2x_3$ involves S. 69, E. 41. We next bisect three collinear segments $2x_i$ with one common extremity, which will cost S. 17, E. 10. To find a point of contact after x_i is known requires S. 4, E. 3. We pick one point of contact for each circle, erect a perpendicular to the corresponding side-line, and, finding where it meets the corresponding bisector (p. 24), already drawn, construct circle. These will involve S. 33, E. 21, so that we have for our total construction

Simplicity 131, Exactitude 81.*

Let us now try to generalize the problem. We first replace *side* by *side-line*. The problem then reads

To construct three circles each of which shall touch the other two and two out of three given lines which form a triangle.

* These numbers also can be wonderfully reduced. Hagge, 'Zur Konstruktion der Malfattischen Kreise', *Zeitschrift für mathematischen Unterricht*, vol. xxxix, 1908, p. 588, gives S. 66, E. 42.

We begin by seeking the number of solutions. How many real solutions are possible? There will surely be no fewer solutions in the general case than in the special one, where the lines determine an equilateral triangle. To count the solutions here let us first notice that the two side-lines at any vertex form four angular openings, which we shall refer to as *inside*, *vertical*, and the two *adjacent*. We notice also that if two circles touch one another, and also the same line at different points, their contact must be external, and they lie on the same side of the line. These facts premised, it is easy to show that we have the following real solutions; the proofs come by simple considerations of continuity.

Circles in three inside openings . .	8 ways.
Circles in two inside and one vertical opening	3 ways.
Circles in one inside and two adjacent openings	15 ways.
Circles in two adjacent and one vertical opening	6 ways.

Malfatti's problem so generalized must usually have thirty-two real solutions: how shall we find them analytically?*

When we pass from the narrower to this wider form for the problem, the quantities a_i must be allowed to take either positive or negative values, the quantities a_i, s_j, $s-a_i$ will be permuted among one another. More specifically, as reversing the signs of all three quantities a_1, a_2, a_3 may be looked upon as leaving everything unaltered, we see that the quantities b_i, b_j, b_k may take the following sets of values:

$$\left(\sqrt{\frac{s-a_i}{s}}\sqrt{\frac{s-a_j}{s}}\sqrt{\frac{s-a_k}{s}}\right)\left(\sqrt{\frac{s}{s-a_i}}\sqrt{\frac{s-a_k}{s-a_i}}\sqrt{\frac{s-a_j}{s-a_i}}\right)$$

$$\left(\sqrt{\frac{s-a_k}{s-a_j}}\sqrt{\frac{s}{s-a_j}}\sqrt{\frac{s-a_i}{s-a_j}}\right)\left(\sqrt{\frac{s-a_j}{s-a_k}}\sqrt{\frac{s-a_i}{s-a_k}}\sqrt{\frac{s}{s-a_k}}\right).$$

The product, multiplied by the common denominator within

* Taken with some alteration from Pampuch, 'Die 32 Lösungen des Malfattischen Problems', *Grunerts Archiv*, Series 3, vol. viii, 1904.

the radicals, will be the radius of an inscribed or escribed circle. We now write

$$4\,(s-a_k)\,x_i x_j = (a_k - x_i - x_j)^2 s,$$
$$a_i^2 = A_i,\ a_j^2 = A_j,\ a_k^2 = A_k, \tag{12}$$

assuming that A_i, A_j, A_k are known values. These equations have sixty-four solutions, which include the thirty-two real solutions of the problem in hand and thirty-two others obtained by altering the signs of all the a_i's and x_i's, which gives nothing new geometrically. These equations will thus contain nothing extraneous if we impose the restriction $a_1 a_2 a_3 > 0$. They give the thirty-two real solutions of the problem and nothing besides. The quantities a_1, a_2, a_3 are capable of taking four sets of values. We pick out one set, and write the equation

$$x_i + x_j + 2\beta_k \sqrt{x_i}\sqrt{x_j} = d_k,$$
$$\alpha_i^2 = A_i,\ \alpha_j^2 = A_j,\ \alpha_k^2 = A_k,$$
$$\sigma = \sum_{i=1}^{i=3} \alpha_i,\quad \beta_i^2 = \frac{\sigma - \alpha_i}{\sigma}, \tag{13}$$
$$\gamma_i^2 = \frac{\alpha_i}{\sigma},\quad \delta_i = \sigma\beta_i\gamma_j\gamma_k,\quad \sigma\beta_1\beta_2\beta_3 = \lambda,$$
$$x_i + i\sqrt{\sigma - x_i} = \pm\sqrt{\sigma}\Big(\sqrt{\frac{1-\beta_i}{2}} - i\sqrt{\frac{1+\beta_i}{2}}\Big)$$
$$\Big(\sqrt{\frac{1-\beta_j}{2}} + i\sqrt{\frac{1+\beta_j}{2}}\Big)\Big(\sqrt{\frac{1-\beta_k}{2}} + i\sqrt{\frac{1+\beta_k}{2}}\Big),$$
$$x_i = [\sigma - \lambda + \delta_i - \delta_j - \delta_k].$$

Of the quantities here involved σ is single valued, β_i and γ_i double valued in (α). To be specific let us assume that $\beta_1\beta_2\beta_3 > 0$. Then, since $\beta_1\beta_2\beta_3\sigma = \lambda$ is the radius of an inscribed or escribed circle, we shall have

$$x_i = \tfrac{1}{2}\,[\sigma - \lambda \pm \delta_i \pm \delta_j \pm \delta_k]. \tag{14}$$

The quantities here involved are all single-valued functions of α_1, α_2, α_3 and the radius of the inscribed or escribed circle while δ_1, δ_2, δ_3 are the distances from the centre of that circle to the vertices of the triangle.

The analytical expression of the distance from each vertex to the points of contact of the corresponding circle in the thirty-two cases of the extended Malfatti problem is of the same type.

It would be tedious to determine which value of α_i and which sign for each δ_i should be used in every case. On the other hand, let us notice that Hart's proof may easily be extended to every case, so that

Steiner's construction may be extended to all thirty-two cases of the extended Malfatti problem, the triangles abutting at the centre of the inscribed circle being replaced in twenty-four cases by those abutting at the centre of an escribed circle. The triangles being chosen, we can associate with each, either its inscribed circle, or the escribed one which actually touches the side which it shares with the original triangle.

There is a further extension of Malfatti's problem due even to Steiner himself. *To construct three circles, each of which shall touch two out of three given non-coaxal but intersecting circles, and also the other two circles sought.**

When the three given circles are concurrent, we get the construction at once by inversion. Steiner's own construction for the general case is as follows:

Find a circle of antisimilitude of each pair of the given circles. Inscribe circles in the arcual triangles each determined by one given circle and two circles of antisimilitude. The remaining circles orthogonal to the common orthogonal circle of the original three, each touching a pair of the constructed ones, and belonging to the same system as the common tangent circle of antisimilitude will, in pairs, touch the circles sought.

The proof of this is given by Hart immediately after his proof of the simpler case. The reasoning is as follows. Hart's proof for the Steiner construction holds just as well on the surface of a sphere as in a plane, provided that straight lines be replaced by great circles, and that I. 28] and 170] be extended to the sphere, which can be done as follows. If two

* *Einige geometrische*, &c., loc. cit., p. 180.

small circles cut equal arcs on a great circle, we find, by the formulae for a right spherical triangle, that if a circular triangle be formed by this great circle and tangents to the small ones at a pair of points of intersection that do not separate the other pair, then the sines of the legs of this triangle are proportional to the tangents of the radii of the small circles, i.e. the two small circles will subtend equal angles at the opposite vertex of the triangle. On the sphere then, as in the plane, the second transverse common tangent of the circles $D_2F_1G_3$, $D_3F_2G_1$ will bisect the $\measuredangle A_1$. This established, the previous proof holds word for word. We next see that any three circles of the plane which are not concurrent may be carried by a real or imaginary stereographic projection into three great circles. We have but to take the sphere whose equator circle is concentric with but orthogonal to the common orthogonal circle of the three. This transformation is conformal and carries great circles bisecting the angles of given great circles into circles of antisimilitude in the plane. The number of solutions is seen to be sixty-four.

The most systematic attempt ever made to reduce to a uniform method the solution of all problems involving the construction of circles subject to given conditions was made by Fiedler,* and we must now give some account of his method.

In the preceding chapter we showed how the circles of a plane may be represented by the points of a three-dimensional space. A more direct method of accomplishing the same end, when none but proper circles are involved, is as follows. At the centre of each proper circle in the plane, erect a perpendicular on a specified side of the plane, which we shall call *above*, equal in length to the radius of the circle. The extremity of this perpendicular shall be taken to represent the circle. Conversely, if any point (in the finite domain) be given above the plane or upon it, the circle whose centre is the foot of the perpendicular from the point to the plane, and whose radius is the length of this perpendicular, will be

* *Cyklographie*, Leipzig, 1882. See also Müller, 'Beiträge zur Zyklographie', *Jahresbericht der deutschen Mathematikervereinigung*, vol. xiv, 1905.

the circle which is represented by the point. There is thus a one to one correspondence between the proper circles of the plane and the finite points above; the points of the plane will represent the null circles whereof they are centres.

The circles of one system tangent to two intersecting lines will be represented by the points of two half-lines above the plane, intersecting in the intersection of the lines, and making, with the plane, angles whose cotangents are equal to the cosecant of the corresponding half-angle of the given lines. Conversely, the points of every half-line above the plane will be represented by circles tangent to two intersecting or parallel lines which will be real if the angle which the half-line makes with the plane be $\leqq \frac{\pi}{4}$. The reflection of the opposite half-line in the same plane will represent the remaining circles tangent to the two lines and belonging to the same system.

The points of a half-plane above the given plane, and of the reflection in that plane of the opposite half-plane, will represent the circles intersecting at a fixed angle the line common to the two half-planes and the given plane. The cosine of this angle will be the cotangent of the angle between the half-plane and the given plane. Conversely, every such system of circles will be represented by a half-plane and the reflection of its opposite.

We next observe that every line in space, not parallel to our plane or lying therein, may be represented by its intersection with the plane, and by the intersection therewith of a parallel to the given line through a fixed point above the plane. The line connecting the two points will be the intersection of our given plane with the plane through the given line and the fixed point. The circles tangent externally to a given circle will be represented by the portion above the given plane of a cone of revolution through the given circle, with its vertex at a radius distance below. The circles which touch the given circle internally will be represented by the reflection in that plane of the remainder of the same cone. The word *cone* is here used in its widest sense to indicate a conical surface of two nappes.

Let us make two specific applications of these methods.

Problem 1.] *To construct a circle having contact of a pre-assigned sort with each of two given intersecting lines, and with a given proper circle.**

Analysis. The given circle shall be c with its centre C. The vertex of the corresponding cone, which we shall assume below the plane, shall be V. The lines shall be l and l' meeting in P at an angle θ. Their bisector orthogonal to the circle sought shall be b. We wish to find the intersection of the cone with a line through P whose projection on the plane shall be b, and making with the plane an angle whose cotangent is $\cos\frac{\theta}{2}\cdot$ A plane through this line and V will meet the given plane in the line from P to the intersection with a parallel to the given line passing through V, and will meet the circle c in the points of contact desired.

Construction. *Through C draw a line parallel to b, and take thereon points whose distances from C are $2\cos\frac{\theta}{2}\cdot$ Connect these points with P. These lines will intersect the given circle in the points of contact desired.*

Problem 2] of Apollonius. *To construct a circle tangent externally to three mutually external circles.*†

Analysis. Let the cones of revolution be constructed as before, the vertices being $V_1V_2V_3$. Each two of these have a common conic in the plane at infinity, hence they intersect also in a finite conic. We wish to find the intersections of two of these conics, as one intersection will represent the circle desired. Let A_1 be the intersection of the given plane with V_2, V_3, assuming no two circles are of equal radius; it is the external centre of similitude of c_2, c_3. The plane through the line $A_1V_2V_3$, tangent to c_2, c_3, will touch the finite conic, but at infinity, since it is there that the finite and infinite conic intersect. Hence a plane through V_3, parallel to

* Fiedler, loc. cit., p. 30.

† Ibid., p. 161.

the plane of the finite conic, will intersect the given plane in the polar of A_1 with regard to c_3. The plane of the finite conic will meet the given plane in the radical axis of c_2 and c_3. The line common to the planes of the three finite conics may be represented, if V_3 be the fixed point without the plane, by the radical centre and the pole with regard to c_3 of the line containing the three external centres of similitude.

Construction. *Find the poles with regard to each circle of the line containing the external centres of similitude. The lines connecting these poles with the radical centre will meet the circles in the points of contact sought.*

It is certainly striking that Fiedler's method should lead us back to the Gergonne construction.

The work which we have done in problems of construction not unnaturally raises the old question of what constructions are possible and what ones are not with the means allowed in elementary geometry, namely, the ungraded ruler and the compass. Various suggestions have also been made for substituting other instruments for these. Steiner employed the ruler and one circle completely drawn. Others have studied the constructions possible with the ruler and compass of a single opening, the two-edge ruler, and even the constructions possible with the aid of paper folding.* The most interesting attempt of this sort from our present point of view is that originally made by Mascheroni,† to see what constructions are possible with the aid of the compass alone. Mascheroni's original procedure may be greatly shortened by the aid of inversion.

What are the constructions possible with ruler and compass? To connect two points by a straight line, and to describe a circle of given radius about a given point. Clearly no compass alone will enable us to perform the first of these. At the same time the primary uses which we make of these constructions are to determine certain points, and, so con-

* For an excellent account of all these attempts, as well as the constructions that follow, see Enriques, *Questioni riguardanti la geometria elementare*, Bologna, 1900.

† *La geometria del compasso*, Pavia, 1797.

sidered, the fundamental problems are three in number: (1) To find the intersections of two circles given by radius and centre. (2) To find the intersection of a line given by two points with a circle given by radius and centre. (3) To find the intersections of two lines, each given by two points. The primary object of the geometry of the compass is to show that all three of these problems may be solved by the aid of that instrument alone. About the first nothing need be said; the last two may be thrown back upon the first by means of inversion. It is only needful to show, therefore, that with the aid of the compass alone we can find the inverse of a given point with regard to a given circle, and can find the centre of a circle through three given points.

Problem 1.] *To construct the multiples of a given segment* (AO).

Construct a circle with centre O and radius (OA). Inscribe a regular hexagon with one vertex at A. The opposite vertex B will determine a segment (AB) whose middle point is O.

Problem 2.] *To construct a fourth proportional to three given lengths* m, n, p.

Take a convenient centre O and construct concentric circles with radii m, n. Let A and B be two points of the first separated by a distance p. If $2m < p$ we replace our circles by concentric ones of radius km, kn, both $> \frac{p}{2}$, and proceed as before. With A and B as centres and the same radius, construct circles intersecting the other circle in two pairs of points. We then take the points A' and B', one belonging to each pair. $(A'B')$ is the length sought.

We may, in fact, take such a radius at A and B that the radii (OA), (OB'), (OB), (OA') follow around in order. Then, since $\Delta\, AOA' = \Delta\, BOB'$ by three sides,

$$\measuredangle AOB = \measuredangle A'OB'.$$

Hence the isosceles triangles AOB and $A'OB'$ are similar.

$$(OA) : (OA') = (AB) : (A'B').$$

Problem 3.] *To construct a circle through three non-collinear points.*

The points shall be A, B, C. With B as centre and (BA) as radius, and with C as centre and (CA), construct circles meeting again in A', the reflection of A in BC. If O be the centre of the circle sought, the triangles BAA', OAC are clearly similar. Hence the radius sought is a fourth proportional to (AA'), (AB), (AC). The radius being found, the centre is found at once, and so the circle.

Problem 4.] *To construct the inverse of a given point with regard to a given circle.*

The given point shall be P, the centre of the given circle O, and its radius r. Suppose, first, that $(OP) > \frac{r}{2}\cdot$ Take P as centre, and radius (PO), and construct a circle cutting the given circle in A and B. With A and B as centres construct circles intersecting in O and P'. Then P' is the point sought. We see, in fact, that by symmetry P' is on the line OP. Moreover, $\Delta\, OPA$ and $\Delta\, OAP'$ are similar.

$$(\overrightarrow{OP}) \times (\overrightarrow{OP'}) = (OA) \times (AP') = r^2.$$

When $(OP) < \frac{r}{2}$ let us find $(OM) = k(OP) > \frac{r}{2}\cdot$

Then, if
$$(\overrightarrow{OM}) \times (\overrightarrow{OM'}) = r^2,$$
$$k\,(\overrightarrow{OP})\,(\overrightarrow{OM'}) = r^2,$$
$$(OP') = k\,(OM').$$

We can now find the intersection of a line and a circle or of two lines by finding those of their inverses, and our fundamental problems are solved. It is surely a remarkable fact that with the single instrument we can find any individual point which normally we reach only with the aid of both.

CHAPTER IV

THE TETRACYCLIC PLANE

§ 1. Fundamental Theorems and Definitions.

ANY set of objects which can be put into one to one correspondence with the sets of essentially distinct values of four homogeneous coordinates $x_0 : x_1 : x_2 : x_3$, not all simultaneously zero, but connected by the relation

$$x_0^2 + x_1^2 + x_2^2 + x_3^2 \equiv (xx) = 0, \tag{1}$$

shall be called *points*; their assemblage shall be called a *tetracyclic plane*. The assemblage of all points (x) whose coordinates satisfy a linear equation

$$(yx) = 0,$$

where the values of (y) are not all simultaneously zero, shall be called a *circle*, to which the points (x) are said to *belong*, or be *upon*. The coefficients (y) are called the *coordinates* of the circle. If they satisfy the identity (1) the point (y) is called the *vertex* of the circle, which is then said to be *null*.

If (y) and (z) be two not null circles, the number θ defined by

$$\cos\theta \equiv \frac{(yz)}{\sqrt{(yy)}\,\sqrt{(zz)}} \tag{2}$$

is called their *angle*. If one possible value for the angle be $\frac{\pi}{2}$, the circles are said to be mutually *perpendicular* or *orthogonal*, or to cut at *right angles*. If one possible value be *O* or π the circles are said to be *tangent*. The conditions for orthogonality and tangency are, respectively,

$$(yz) = 0. \tag{3}$$

$$(yy)\,(zz) - (yz)^2 = 0. \tag{4}$$

If (y) and (z) be two mutually orthogonal null circles, i.e. two null circles whose coordinates satisfy equations (3), (4), every circle of the system

$$x_i = \lambda y_i + \mu z_i$$

is null and orthogonal to every other. The locus of the vertices of the circles (x) shall be called an *isotropic*. Through each point in the tetracyclic plane will pass two distinct isotropics which together constitute the null circle having the given point as vertex.

The coordinates of each point in the tetracyclic plane may be parametrically represented by means of the isotropics through it as follows. Let i be supposed to be a well-defined value of $\sqrt{-1}$, a given irrational adjoined to the number system. We may write

$$\begin{aligned} &ix_0 : x_1 : x_2 : ix_3 \\ &\qquad = (\lambda_1\mu_1 + \lambda_2\mu_2) : (\lambda_1\mu_1 - \lambda_2\mu_2) : (\lambda_1\mu_2 + \lambda_2\mu_1) : (\lambda_1\mu_2 - \lambda_2\mu_1), \\ &\lambda_1 : \lambda_2 = (ix_0 + x_1) : (x_2 - ix_3) = (\lambda_2 + ix_3) : (ix_0 - x_1), \\ &\mu_1 : \mu_2 = (ix_0 + x_1) : (x_2 + ix_3) = (\lambda_2 - ix_3) : (ix_0 - x_1). \end{aligned} \tag{5}$$

If (x) and (x') be two points, we shall have

$$\rho\,(xx') = (\lambda_1\lambda_2' - \lambda_2\lambda_1')\,(\mu_1\mu_2' - \mu_2\mu_1').$$

It thus appears that if we keep either $\lambda_1 : \lambda_2$ or $\mu_1 : \mu_2$ fixed we have the points of an isotropic.

The system of all circles through the intersections of two given circles, i.e. that of all circles whose coordinates are linearly dependent on those of two given circles, shall be called a *coaxal system*.

Two points are said to be mutually *inverse* in a circle (which is supposed not to be null) when every circle through them is orthogonal to the given circle. The vertices of the null circles orthogonal to those null circles whose vertices are the given points must lie on the circle of inversion; hence the coordinates of the circle of inversion are linearly dependent on those of the given points. If the points be (x) and (x'), while (y) is the circle of inversion,

$$\rho x_i' \equiv (yy)\,x_i - 2\,(xy)\,y_i. \tag{6}$$

The transformation from (x) to (x'), being linear, carries a circle into a circle, and we see also that equation (6) may be interpreted as giving the relation between any two mutually inverse circles. They are coaxal with the circle of inversion, and make equal or supplementary angles therewith. We shall also speak of (y) as a circle of *antisimilitude* for (x) and (x').

The definitions so far given have been apparently arbitrary. Let us see whether there be any sets of familiar objects which obey all the rules prescribed for the points of a tetracyclic plane. Obviously a Euclidian sphere is a perfect example of such a plane, and the definitions of angles, inversions, &c., for the tetracyclic plane are entirely in consonance with what we should have on the surface of such a sphere. Again, the relation between the tetracyclic coordinates of all finite points of the cartesian plane is the same as that for all points of the tetracyclic plane. We rewrite the equations:

$$x : y : t \equiv x_2 : x_3 : -(ix_0 + x_1), \tag{7}$$

$$x_0 : x_1 : x_2 : x_3 \equiv i(x^2 + y^2 + t^2) : (x^2 + y^2 - t^2) : 2xt : 2yt.$$

Every finite point of the cartesian plane $(t \neq 0)$ will be represented by a definite point of the tetracyclic plane for which $ix_0 + x_1 \neq 0$, and conversely. If, however, we make the cartesian plane a perfect continuum by adjoining the line at infinity, the correspondence ceases to be unique, for all infinite cartesian points other than the circular ones will correspond to the same point of the tetracyclic plane. We may extend the finite cartesian plane to a tetracyclic plane by first omitting the line at infinity, then extending the plane to be a perfect continuum as follows:*

The set of coordinates $x_0 : x_1 : x_2 : x_3 = i : 1 : 0 : 0$ shall be said to represent the *point at infinity*. Every other set of coordinates (y) satisfying the equations

$$iy_0 + y_1 = (yy) = 0$$

* Conf. Beck, 'Ein Gegenstück zur projektiven Geometrie', *Grunerts Archiv*, Series 3, vol. xviii, 1911, and Bôcher, 'The Infinite Regions of various Geometries', *Bulletin American Math. Soc.*, vol. xx, 1914.

shall be taken to represent a minimal line

$$y_2 x + y_3 y + \tfrac{1}{2}(iy_0 - y_1)\, t = 0.$$

The point at infinity and the totality of such minimal lines shall be called *improper points*. By adjoining them to the finite domain, the cartesian plane becomes once more a perfect continuum, and obeys all the laws for a tetracyclic plane. The definitions of circle, angle, inversion, &c., given in Ch. II for the cartesian plane, and here for the tetracyclic one, are entirely compatible. Care must be taken not to confuse *minimal lines*, looked upon as improper tetracyclic points, with *isotropics* which are point loci. If we take as our tetracyclic plane the cartesian plane rendered a perfect continuum in this fashion, the following expressions are synonymous:

Circle orthogonal to point at infinity.	Line.
Inversion in such a circle.	Reflection in line.
Null circle whose vertex is point at infinity.	Totality of minimal lines.
Null circle containing point at infinity.	Points of a minimal line and minimal lines parallel thereto.
Improper points of a circle not through the infinite point.	Asymptotes of a circle.

We shall mean by the *cartesian equivalent* of a tetracyclic figure the following: If the tetracyclic plane be taken as a Euclidean sphere, we take the stereographic projection of this sphere. If the tetracyclic plane be built on the cartesian one in the present fashion, we replace the coordinates of each proper tetracyclic point by their cartesian equivalents from (7), then render the plane a perfect continuum by the adjunction of the line at infinity. In either case, if we mean by the *degree* of an algebraic curve of the tetracyclic plane the number of its intersections with a circle, we see

The cartesian equivalent of an algebraic curve of order n with a multiple point of order k at infinity, is an algebraic

curve of order $n-k$ with a combined multiplicity of order $n-2k$ at the circular points at infinity.

It is worth while to look also at cross ratios in the tetracyclic plane. We start with the circle $x_3 = 0$. Let $(y_0 y_1 y_2 y_3)$ and $(y_0 y_1 y_2 - y_3)$ be any two points mutually inverse therein. Let (β) be any circle through these points, cutting the fundamental circle again in (α) and (γ),

$$\begin{vmatrix} y_0 & y_1 & y_2 \\ \alpha_0 & \alpha_1 & \alpha_2 \\ \gamma_0 & \gamma_1 & \gamma_2 \end{vmatrix}^2 = 2(\alpha y)(\gamma y) - (\alpha\gamma) y_3^2 = 0.$$

Our circle $x_3 = 0$ may be represented parametrically by the equation

$$x_i \equiv t^2 \alpha_i + t\beta_i + \gamma_i, \qquad (\beta\beta) + 2(\alpha\gamma) = 0, \qquad i = 0, 1, 2. \tag{8}$$

The circles through the points $(y_0 y_1 y_2 y_3)$ $(y_0 y_1 y_2 - y_3)$ and the points with parameter values t_1 and t_2 will be

$$(1)\ \ z_i = t_1^2 \begin{vmatrix} \alpha_j & \alpha_k \\ y_j & y_k \end{vmatrix} + t_1 \begin{vmatrix} \beta_j & \beta_k \\ y_j & y_k \end{vmatrix} + \begin{vmatrix} \gamma_j & \gamma_k \\ y_j & y_k \end{vmatrix};$$

$$(2)\ \ z_i = t_2^2 \begin{vmatrix} \alpha_j & \alpha_k \\ y_j & y_k \end{vmatrix} + t_2 \begin{vmatrix} \beta_j & \beta_k \\ y_j & y_k \end{vmatrix} + \begin{vmatrix} \gamma_j & \gamma_k \\ y_j & y_k \end{vmatrix}.$$

For the cosine of the angle between them we have

$$\cos\theta_{12} = \frac{t_1^2 t_2^2 (\alpha y)^2 + 3(t_1^2 + t_2^2)(\alpha y)(\gamma y) - 4 t_1 t_2 (\alpha y)(\gamma y) + (\gamma y)^2}{[t_1^2(\alpha y) + (\gamma y)]\,[t_2^2(\alpha y) + (\gamma y)]},$$

$$\sin^2\frac{\theta_{12}}{2} = \frac{-(\alpha y)(\gamma y)(t_1 - t_2)^2}{[t_1^2(\alpha y) + (\gamma y)]\,[t_2^2(\alpha y) + (\gamma y)]}.$$

Giving the parameter t four sets of values t_1, t_2, t_3, t_4, we have

$$\frac{\sin\dfrac{\theta_{12}}{2}\sin\dfrac{\theta_{34}}{2}}{\sin\dfrac{\theta_{14}}{2}\sin\dfrac{\theta_{32}}{2}} = \frac{(t_1 - t_2)(t_3 - t_4)}{(t_1 - t_4)(t_3 - t_2)}. \tag{9}$$

This expression is independent of (y) and is defined as a *cross ratio* of the four points of the fundamental circle

$x_3 = 0$. Had we taken any other pair of points besides (α) and (γ) to use in the parametric equation (8), we should have replaced t by $\frac{pt+q}{rt+s}$; the right-hand side of the equation would have been unaltered. Moreover, every not null circle can be expressed parametrically in this form, and our two expressions (9) for a cross ratio will be the same for all such circles. If this cross ratio have the value -1, the points t_1 and t_3 are said to separate the points t_2 and t_4 *harmonically*. The relation between the two pairs is reciprocal. If we take the harmonic pairs of parameter values $0, \infty, t, -t$, we see that the circle (β) is orthogonal to every circle through the last two points. Our four points are thus both concyclic and orthocyclic if we extend the definition of p. 100 to the tetracyclic plane, and, in fact, we find that

A necessary and sufficient condition for harmonic separation is that the concyclic points should also be orthocyclic.

If we pass a circle through the fundamental circle meeting it orthogonally at the points $t_1 t_3$, while another orthogonal circle meets it at the points $t_2 t_4$, and if θ be the angle of these circles, we easily find

$$-\tan^2\frac{\theta}{2} = \frac{(t_1-t_2)(t_3-t_4)}{(t_1-t_4)(t_3-t_2)}.$$

For harmonic sets, our new circles will be orthogonal to one another.

We must next consider the cross ratios of four points of an isotropic. If (α') and (γ') be two points of one isotropic, then every point thereof will have coordinates of the form

$$x_i' = t'\alpha_i' + \gamma_i'.$$

We shall define the right-hand side of (9) as a cross ratio of the four points corresponding to the parameter values t_1, t_2, t_3, t_4. Harmonic separation shall be as before. We find the geometric meaning of the cross ratio of four points of an isotropic as follows. What point of the circle (8) will lie on an isotropic with a given point of the isotropic (10)? Writing the condition of orthogonality for the corresponding null circles, we have an equation quadratic in t and linear in t'. This must be reducible, if looked upon as an equation in t, for one point sought is the intersection of the circle and isotropic. The other root is a fractional linear function of t', and, since a linear transformation leaves cross ratios invariant, we see that the *cross ratio of four points of an isotropic may be defined as that of the points where the other four isotropics through them meet any not null circle.*

We next take up the question of problems of construction in the tetracyclic plane.* What constructions shall be allowed here? The point at infinity shall play no special rôle, and we shall require our constructions to be invariant for inversion. We next remark that there are two different ways in which we may suppose that a circle is known. We may know all of its points, or all in a domain called *real*. Or, secondly, we may know how to find the inverse of any known point. In the first case we say that the circle is *known by points*, in the second that it is *known by inversion.*

Suppose that we took for our tetracyclic plane the real domain of a real sphere and represented each circle known by points by the pole of its plane. Points of the sphere collinear with this pole would be mutually inverse in the circle. On the other hand, if we took an interior point of the sphere it would be the pole of a self-conjugate imaginary circle of the sphere whose points are not in the domain; at the same time we know the circle by inversion, for we can join any point of the sphere with the interior pole and find where the line meets the sphere again. Moreover, in three dimensions, we assume that we can connect two points by a line, three points by a plane, and find the intersection of lines, planes, and sphere when such intersections exist. This leads us naturally to the following postulates for constructions in the tetracyclic plane.

Postulate 1.] *If three points be known, all points of their circle are known.*

* The whole question of tetracyclic constructions is elaborately discussed by Study, loc. cit.

Postulate 2.] *If two circles be known by points, their intersections, if in the known domain, are also known.**

If we consider as known the whole tetracyclic plane, then the intersections are always known. On the other hand, we might limit ourselves to such a domain as one where x_0 was proportional to a pure imaginary value, while the other coordinates were proportional to real values, in which case it is not certain that intersections will be real. We make, therefore, the further assumption

Postulate 3.] *If two not null circles known by points have one common known point, they have a second such point unless they be tangent to one another.*

Theorem 1.] *A circle is completely known by inversion if two pairs of inverse points be known.*

Suppose, in fact, that we have two pairs of inverse points QQ', RR', to find the inverse of any point P we have but to construct the other intersection of the circles PQQ', PRR'. The construction is of the first degree.

Problem 1.] *Given two circles by inversion, to find by points a circle through a given point orthogonal to them.*

We have but to find the two inverses of the point, then apply postulate 1].

Problem 2.] *Given a circle by points, to determine it by inversion.*

Take four points thereon. They may be divided into two pairs in three different ways. The product of the three inversions, each of which interchanges the members of two pairs, will be the inversion sought. The proof comes by easy analysis, which we leave to the reader.

Problem 3.] *Given two circles by inversion, to construct by points the circle coaxal with them passing through a given point.*

* Study, loc. cit., p. 53, makes a different assumption. He is not interested in separating real from imaginary, and so assumes that if two circles be mutually orthogonal, and one be known by points, their intersections are known. It will easily follow from this that if a circle be known by inversion it is also known by points.

We construct by points two circles orthogonal to these two. Then we find these same two by inversion, and then the circle through the chosen point orthogonal to them.

Problem 4.] *To construct by inversion the circle orthogonal to three non-coaxal circles given by inversion.*

Through any point pass three circles, each coaxal with two of the given circles. These three are concurrent again in the point inverse to the given one in the circle sought.

Problem 5.] *To pass a circle through two points of a circle given by points which shall be orthogonal thereto.*

We pass any circle through these points, find the inversions in both, then find by points, and so, by inversion, two circles orthogonal to the given ones. Lastly, find by points the circle orthogonal to these last two circles, and to the original one.

Problem 6.] *Given the points A, B, and C, to find the harmonic conjugate of B with regard to A and C.*

We assume that these points are not on one isotropic. We take two other pairs of points $A'C'$ and $A''C''$, both concyclic with AC. We next find the inversion which interchanges A with C, A' with C', and that which interchanges A with C, A'' with C''. We pass a circle through B coaxal with these last two circles of inversion. It will meet the circle ABC again in the point required.

Definition. Two ranges of points on the same or different circles shall be said to be *projective* if their members are in one to one correspondence, and corresponding cross ratios are equal. We have at once

Theorem 2.] *Two ranges of points on the same circle harmonically separated by two fixed points of the circle are projective.*

Two such ranges are said to form an *involution*.

Theorem 3.] *The circles through a fixed point and through the pairs of an involution will be a coaxal system.*

We see, in fact, that all of these circles will pass through the inverse of the given point in the circle orthogonal to the given circle, and passing through the two points which separate the pairs of the involution. These separating points shall be called the *double points of the involution.* We shall also extend the meaning of the word *involution* to validate the converse theorem, i.e.

Definition. The pairs of points where a fixed circle intersects the circles of a coaxal system not including this circle form an *involution.* We see that if any circle of the coaxal system (in the given domain) touch the given circle, the point of contact, which is said to be *double* for the involution, lies on the circle orthogonal to the given circle and to those of the coaxal system, and that an inversion in this circle interchanges the pairs of the involution, i.e. they are harmonically separated by the double points. The coordinates of these double points may always be found, even though the corresponding points may not be in that domain which for the purposes of our construction we define as real; hence the two definitions of involution amount to the same thing if we include the limiting case where the double points fall together.

Problem 7.] *Given two pairs of an involution, to find the double points if they exist in the given domain.*

The solution comes at once from what immediately precedes, and from problem 4].

Theorem 4.] *If two projective ranges have three self-corresponding points, every point is self-corresponding.*

The proof of this is immediate from the definition, and from the fact that a point is uniquely known as soon as we know a cross ratio determined thereby with three given points. Equally evident is

Theorem 5.] *The projective transformation between two ranges is completely determined by the fate of three points.*

The analytic formula for a projective transformation is found immediately if we write our circle in the paramethic form (8) and then make the transformation

$$t' = \frac{\alpha t + \beta}{\gamma t + \delta}, \qquad \alpha\delta - \beta\gamma \neq 0.$$

The whole theory of projective ranges may be at once deduced from this familiar analytic form by simple methods known to every student of geometry. Nevertheless, we shall continue to follow a geometric development more closely akin to the fundamental methods of the tetracyclic plane. We next have

Theorem 6.] *Four points of a circle correspond projectively to the points obtained by interchanging them two by two.*

We see, in fact, that this may be done by an inversion.*

Theorem 7.] *If, in a projective transformation of a circle into itself, MAA_1 correspond to MA_1A_2, and M and A_1 be separated harmonically by A and A_2, then M is the only self-corresponding point.*

Consider the involution with double points M and A_1. The given projective ranges are carried hereby into those determined by MA_2A_1C, MA_1AC_1, and were it possible for C_1 to be identical with C we might find an involution to carry $CMAA_1$ into MCA_1A. Hence $MA_2A_1C, \barwedge MA_1AC, \barwedge CAA_1M$, if we use the symbol $\barwedge$ for projective, and CM, AA_2, A_1A_1 are pairs of an involution. But MM, AA_2, A_1A_1 are pairs of an involution, and two involutions cannot share two pairs.

Theorem 8.] *Given two projective ranges of points. They have either a single self-corresponding point which may be found by a linear construction, or the problem of finding their self-corresponding points is the problem of finding the double points of an involution.*

If the projective ranges form an involution, nothing need be said. If not, let AA_1 in the first correspond to A_1A_2 in the second. Let H_1 be the harmonic conjugate of A_1 with regard to A and A_2. If H_1 be self-corresponding, it is the only such point, by the last theorem. If not, suppose that H_1 in the first corresponds to H_2 in the second. Suppose that there is a pair of points MN which are double for the

* For the next three theorems see Von Staudt, *Beiträge zur Geometrie der Lage*, Nuremberg, 1858, pp. 144-6, or Wiener, 'Verwandtschaften als Folgen zweier Spiegelungen', *Leipziger Berichte*, vol. xliii, 1891, pp. 651 ff.

involution A_1H_1, A_2H_2. There is also an involution with double points A_1H_1, and in this MN, AA_2, A_1A_1 are three pairs, so that

$$MNAA_1, \barwedge NMA_2A_1, \barwedge MNA_1A_2.$$

Similarly $$MNAH_1, \barwedge MNA_1H_2.$$

Hence $$MNAA_1H_1, \barwedge MNA_1A_2H_2,$$

and M and N are the self-corresponding members of our two projective ranges. The reasoning is reversible, so that the theorem is proved.

Problem 8.] *Given two pairs of an involution, to find the mate of any point.*

This comes by a simple construction which we leave to the reader.

Problem 9.] *Given* $ABC, \barwedge A'B'C'$, *to find the mate of any chosen point.*

This transformation is the product of two involutions, $ABC, \barwedge B'A'C_1$ and $B'A'C_1, \barwedge A'B'C'$. Incidentally, we have proved that in a projective transformation the mates of three members may be chosen at random.

Problem 10.] *To construct the circles of antisimilitude of two given circles.*

We mean, of course, the circles which invert the given circles into one another. Any circle orthogonal to both our circles is anallagmatic with regard to every such circle of antisimilitude, and it will intersect the circle of antisimilitude, if at all, in a pair of double points of the involution determined by the intersections with the given circles. The problem thus reduces to that of finding the double points of an involution, or the intersections of two circles given by points.

Problem 11.] *Of Apollonius. To construct a circle tangent to three given circles.*

We begin by finding their circles of antisimilitude two by two. These, when they intersect in our domain, will pass by threes through at most eight points, inverse in pairs in the circle orthogonal to the three given circles. Through each such pair of points pass a circle orthogonal to each of the given circles; when the points of contact of the circles sought exist in our domain they will be found in this way.* The proof consists in noticing that this is exactly the construction for finding the circles which touch three great circles of a sphere.

Before proceeding to discuss further loci in the tetracyclic plane, let us look once more at the parametric representation already touched upon.† We begin with a slight change of notation, writing

$$\dot{x}_0 \equiv ix_0, \quad \dot{x}_1 \equiv x_1, \quad \dot{x}_2 \equiv x_2, \quad \dot{x}_3 \equiv x_3.$$

A point shall be said to be *real* if the homogeneous coordinates $(\dot{x})$ be proportional to real values. The real domain of a sphere will serve as the best example of a real tetracyclic domain, the identical relation being

$$-\dot{x}_0{}^2+\dot{x}_1{}^2+\dot{x}_2{}^2+\dot{x}_3{}^2 \equiv 0,$$

$$\lambda_1 : \lambda_2 = (\dot{x}_0+\dot{x}_1) : (\dot{x}_2 - i\dot{x}_3) \equiv (\dot{x}_2+i\dot{x}_3) : (\dot{x}_0-\dot{x}_1), \tag{10}$$

$$\mu_1 : \mu_2 = (\dot{x}_0+\dot{x}_1) : (\dot{x}_2 + i\dot{x}_3) \equiv (\dot{x}_2-i\dot{x}_3) : (\dot{x}_0-\dot{x}_1).$$

For a real point the isotropic parameters $\lambda_1 : \lambda_2$ and $\mu_1 : \mu_2$ must take conjugate imaginary values. We therefore write

$$\lambda_1 : \lambda_2 = \xi_1 : \xi_2, \quad \mu_1 : \mu_2 = \bar{\xi}_1 : \bar{\xi}_2.$$

A real circle, or a self-conjugate imaginary one, will be given by an equation bilinear in (ξ) and $(\bar{\xi})$, which is unaltered by interchanging conjugate imaginary values, i.e. by a Hermite form

$$a\xi_1\bar{\xi}_1+\beta\xi_1\bar{\xi}_2+\bar{\beta}\bar{\xi}_1\xi_2+c\xi_2\bar{\xi}_2 = 0.$$

Here a and c are supposed to be real, β and $\bar{\beta}$ conjugate imaginaries. This may be written in a satisfactory abbreviated form by the aid of the Clebsch-Aronhold symbolic notation

$$a_\xi \bar{a}_{\bar{\xi}} = 0. \tag{11}$$

* Cf. Plücker, 'Analytisch-geometrische Aphorismen', *Crelle's Journal*, vol. x, 1833.

† See Kasner, 'The Invariant Theory of the Inversion Group', *Transactions American Mathematical Society*, vol. i, 1900.

If ρ and σ be fixed complex multipliers, while l and m are real variables, the assemblage of points

$$\xi_i = l\rho\eta_i + m\sigma\zeta_i ; \quad \bar{\xi}_i = l\bar{\rho}\bar{\eta}_i + m\bar{\sigma}\bar{\zeta}_i \tag{12}$$

are said to form a *chain*. These equations are equivalent to requiring t to be real in our equations (8). The cross ratio of any four points of a chain is real, and, conversely, every set of points on a circle, such that the cross ratios of any four are real, will belong to one chain.* If, in these equations, we allow $\frac{l}{m}$ to take all values, real and imaginary, we have a parametric representation of a circle connecting the points (η) and (ζ); by changing the constant multipliers ρ and σ, we get every circle through these two points in this fashion.

Let us write the relation between the binary and quaternary coordinates once more:

$$\begin{aligned} \dot{x}_0 &\equiv (\xi_1\bar{\xi}_1 + \xi_2\bar{\xi}_2), \\ \dot{x}_1 &\equiv (\xi_1\bar{\xi}_1 - \xi_2\bar{\xi}_2), \\ \dot{x}_2 &\equiv (\xi_1\bar{\xi}_2 + \xi_2\bar{\xi}_1), \\ \dot{x}_3 &\equiv (\xi_1\bar{\xi}_2 - \xi_2\bar{\xi}_1). \end{aligned} \tag{13}$$

If thus

$$(\dot{u}\,\dot{x}) \equiv a_\xi \bar{a}_{\bar{\xi}} = 0,$$

$$a_1\bar{a}_1 = \dot{u}_0 + \dot{u}_1, \; a_1\bar{a}_2 = \dot{u}_2 - i\dot{u}_3, \; \bar{a}_1 a_2 = \dot{u}_2 + i\dot{u}_3, \; a_2\bar{a}_2 = \dot{u}_0 - \dot{u}_1,$$

$$(-\dot{u}_0\dot{v}_0 + \dot{u}_1\dot{v}_1 + \dot{u}_2\dot{v}_2 + \dot{u}_3\dot{v}_3) = -\tfrac{1}{2} \,|\, a\, b \,| \cdot |\, \bar{a}\, \bar{b} \,|.$$

The cosine of the angle of the circles (u) and (v) will take the simple form

$$\cos\theta = \frac{|\, a\, b \,| \cdot |\, \bar{a}\bar{b} \,|}{\sqrt{|\, a\, a' \,| \cdot |\, \bar{a}\,\bar{a}' \,|}\; \sqrt{|\, b\, b' \,| \cdot |\, \bar{b}\,\bar{b}' \,|}}. \tag{14}$$

Here a and a' are equivalent symbols, as are b and b'. The condition for orthogonal intersection of two circles will be

$$|\, a\, b \,| \cdot |\, \bar{a}\,\bar{b} \,| = 0. \tag{15}$$

To find the inverse of the point (η) in the circle (11) we

* The corresponding concept in projective geometry is due to Von Staudt, loc. cit., p. 137.

have merely to require that every circle through them shall be orthogonal to this circle:

$$\lambda\zeta_1 = a_2\bar{a}_{\bar{\eta}},$$
$$\lambda\zeta_2 = -a_1\bar{a}_{\bar{\eta}}. \qquad (16)$$

§ 2. Cyclics.

The only loci which we have so far discussed in the tetracyclic plane are circles (or chains). Let us now take up others of a more complicated sort.

Definition. *The locus of the vertices of the null circles of a quadric circle congruence shall be called a cyclic.* We mean by a *quadric circle congruence* of the tetracyclic plane exactly what was meant by that term in the case of the cartesian plane. Every cyclic will have two equations of the type

$$\sum_{i,j=0}^{i,j=3} a_{ij}x_ix_j = 0, \quad (xx) = 0, \quad a_{ij} = a_{ji}. \qquad (17)$$

The first of these equations has ten different coefficients. As, however, the cyclic is unaltered if we replace that equation by

$$\sum_{i,j=0}^{i,j=3} a_{ij}x_ix_j + \lambda(xx) = 0.$$

Theorem 9.] *Eight points in general position will determine a single cyclic, but all cyclics through seven points have an eighth common point also.*

The problem of classifying cyclics under the inversion group, that is, under the group of quaternary orthogonal substitutions, is the problem of classifying the intersections of two quadric surfaces in three-dimensional projective space, of which one surely has a non-vanishing discriminant. The modern way to do this is by means of Weierstrassian elementary divisors applied to the two quadratic forms. We may take this problem as solved, merely interpreting the known results in the language of our tetracyclic plane.*

[1 1 1 1]	General cyclic.
[(1 1) 1 1]	Two circles, not tangent and neither null.
[(1 1) (1 1)]	Two isotropics of each set.
[(1 1 1) 1]	Not null circle counted twice.
[(1 1 1 1)]	No locus.
[2 1 1]	Nodal cyclic.
[(2 1) 1]	Mutually tangent not null circles.
[2 (1 1)]	Null and not null circle, not mutually orthogonal.
[(2 1 1)]	Null circle counted twice.
[2 2]	Cubic cyclic and isotropic, not tangent to it.
[(2 2)]	Two mutually orthogonal null circles.
[3 1]	Cuspidal cyclic.
[(3 1)]	Null and not null circle, mutually orthogonal.
[4]	Cubic cyclic and isotropic tangent thereto.

In what follows, unless otherwise stated, we shall confine ourselves to the first type, the *general* cyclic. A number of facts can be at once stated about this curve by considering the Cayleyan characteristics of the elliptic space curve of the fourth order, and re-interpreting them in our present terminology.†

Tetracyclic plane π.	Projective space S.
General cyclic.	Elliptic quartic space curve.
Twelve osculating circles orthogonal to given circle.	Class of developable 12.

* Cf. Bromwich, *Quadratic Forms, and their Classification by means of Invariant Factors*, Cambridge, 1906, especially pp. 46, 47. Also Kasner, loc. cit.

† Cf. Salmon, *Geometry of Three Dimensions*, Fourth ed., Dublin, 1882, p. 312.

Eight circles of arbitrary coaxal system orthogonal to the curve.	Order of developable 8.
Sixteen circles have four-point contact.	Sixteen planes have stationary contact.
Sixteen circles orthogonal to an arbitrary circle belong to two pencils of mutually tangent circles orthogonal to the curve.	Sixteen points on two tangents lie in an arbitrary plane.
Eight circles orthogonal to an arbitrary circle have double contact.	Eight planes having double contact pass through an arbitrary point.

A simple construction for a cyclic is suggested by the foregoing. There is a theorem ascribed to Chasles whereby a line meeting two skew-lines and touching a quadric will have its points of contact with the latter on a quartic. The easiest proof would seem to consist in writing the condition that the line from a point of the quadric to meet two skew-lines shall touch the quadric. If we take this quadric to correspond to our tetracyclic identity, we have

Theorem 10.] *If two coaxal systems be given, with no common circle, the locus of the points where a circle of one system touches one of the other will be a cyclic, general or special.*

Let us simplify the equation of our general cyclic. This is immediately accomplished if we remember that in the case of the elementary divisors [1 1 1 1] the two quadratic forms may be simultaneously carried by a linear transformation into two forms involving only the squared terms; in other words, keeping the identity for tetracyclic coordinates invariant, we may write the equation of the general cyclic in the form

$$(ax^2) \equiv \sum_{i=0}^{i=3} a_i x_i^2 = 0, \quad a_0 a_1 a_2 a_3 \neq 0, \quad a_i \neq a_j. \tag{18}$$

This last equation is unaltered if we change the sign of any one of the x_i's, hence

Theorem 11.] *The general cyclic is anallagmatic with regard to four mutually orthogonal circles. It may be generated in four ways by the circles of a conic series.*

The circles with regard to which the cyclic is anallagmatic shall be called the *fundamental circles.* To prove the last part of the theorem, take the circle

$$y_i = \lambda x_i + a_i x_i.$$

If this be orthogonal to one of our fundamental circles,

$$\lambda = -a_i.$$

It is, moreover, tangent to our cyclic at the point (x). Substituting for (x) in the equations (18),

$$y_i = \frac{y_j^2}{a_j - a_i} + \frac{y_k^2}{a_k - a_i} + \frac{y_l^2}{a_l - a_i} = 0. \tag{19}$$

The last part of our theorem is thus proved. Let (z) and (s) be inverse in the circle (y):

$$y_i = 0, \quad y_j = s_i z_j - s_j z_i, \quad y_k = s_i z_k - s_k z_i, \quad y_l = s_i z_l - s_l z_i,$$

$$\frac{(s_i z_j - s_j z_i)^2}{a_j - a_i} + \frac{(s_i z_k - s_k z_i)^2}{a_k - a_i} + \frac{(s_i z_l - s_l z_i)^2}{a_l - a_i} = 0.$$

Theorem 12.] *The locus of the inverse of a fixed point with regard to the generating circles of one system of a cyclic is a nodal or cuspidal cyclic, whose double point is at the fixed point.*

If we take the equivalent cartesian figure, the fixed point being the point at infinity,

Theorem 13.] *The general cartesian cyclic is a curve of the fourth order with a node at each circular point at infinity; and, conversely, every such curve is a cyclic. It may be generated in four ways as the envelope of a circle moving orthogonally to a fixed fundamental circle, while its centre traces a central conic. The four fundamental circles are mutually orthogonal, and each meets the corresponding deferent in four of the sixteen foci of the cyclic.*

We mean by a *focus* of any curve the vertex of a null circle having double contact therewith, this definition holding equally in the cartesian and the tetracyclic plane. It appears also that the inverse of a focus will be a focus.

Theorem 14.] *The general cyclic has sixteen foci lying by fours on the fundamental circles.*

If we invert in either the tetracyclic or cartesian plane the inverse of a general cyclic will, in the first case, always be a general cyclic; in the second case it will usually be such a cyclic except for special positions of the centre of inversion which we need not particularize. The foci will be inverted into foci also. Now in the case of the cartesian cyclic the foci are the intersections of isotropics, not tangent to the curve at the circular points at infinity. At each circular point there will be two tangents to the curve, and these intersect in pairs in four points called the *double foci*, which are not invariant for inversion but have a certain importance. Let the centre of a generating circle pass through a point of contact of a tangent to the deferent from a focus of the latter, i.e. a tangent from one of the circular points at infinity. The centres of two successive generating circles will lie on this line; hence the circles will touch this line and one another at a circular point at infinity, or

Theorem 15.] *The four deferent conics of the general cartesian cyclic are confocal, their foci being the double foci of the cyclic.*

The cyclic is completely determined by one fundamental circle and the corresponding deferent. The radical axis of successive generating circles is the perpendicular from the centre of the fundamental circle on the corresponding tangent to the deferent. The cyclic will cut the fundamental circle at points of contact of common tangents to circle and deferent. These four tangents form a complete quadrilateral. Let us take a pair of opposite vertices of this quadrilateral and construct two circles, with these points as centres, orthogonal to the fundamental circle, i.e. cutting it at pairs of points of contact with the tangents mentioned. These circles have double contact with the cyclic and, so, are generating circles of a second family. The common orthogonal circle to these two and to the given fundamental circle will be a second fundamental circle; the conic confocal with the given deferent and passing through the chosen pair of vertices of the complete quadrilateral of common tangents is the deferent corresponding to the second fundamental circle. We are thus enabled to pass from one generation to another.*

The locus of the centres of gravity of the intersections of a cartesian algebraic plane curve with a set of parallel lines is a line, the line-polar of the infinite point common to the parallels. In the case of a cartesian cyclic, this line will meet the line at infinity in the harmonic conjugate of the point common to the parallels with regard to the circular points at infinity, i.e. this line-polar will be perpendicular to the direction of the parallels. The line-polars corresponding to two such systems of parallels meet in a finite point O, whose first polar meets the like at infinity four times, i.e. includes the line at infinity. Hence O lies on the line-polar of every infinite point. If a point on the line at infinity approach one of the circular points as a limiting position, its conic polar with regard to a general cartesian cyclic will approach as a limit the two tangents to the cyclic at that circular point, and its line-polar will approach the line from that circular point to O. This line will be harmonically separated from the line at infinity by the two tangents to the cyclic at that circular point We thus reach an interesting theorem due to Humbert.†

Theorem 16.] *The locus of the centres of gravity of the intersections of a general cartesian cyclic with a set of parallel lines not passing through a circular point at infinity is the perpendicular on these lines from the common centre of the four deferents.*

* Darboux, *Sur une classe remarquable de courbes et de surfaces*, Paris, 1873, p. 35.

† 'Sur les surfaces cyclides', *Journal de l'École Polytechnique*, vol. lv, 1885, p. 127 ff.

If we take the common centre of the deferents as origin, the rectangular cartesian equation of the general cyclic will be

$$(x^2+y^2)^2+f^2(xy) = 0. \tag{20}$$

Here f^2 is a quadrate polynomial from which we may remove the term in xy if we choose the axes of the focal conics as axes of coordinates.

Let us return to the tetracyclic cyclic from which we have strayed. To find the coordinates of a focus, we have

$$x_i \equiv x_j^2+x_k^2+x_l^2 = \frac{x_j^2}{a_j-a_i}+\frac{x_k^2}{a_k-a_i}+\frac{x_l^2}{a_l-a_i} = 0,$$

$$x_i : x_j : x_k : x_l = 0 : \sqrt{(a_j-a_i)(a_k-a_l)} : \sqrt{(a_k-a_i)(a_l-a_j)}$$
$$: \sqrt{(a_l-a_i)(a_j-a_k)}. \tag{21}$$

Let us find the cross ratio of these four, which will clearly be an invariant of the curve. In particular, if we take the foci on $x_0 = 0$ and seek the corresponding values of $\lambda_1 : \lambda_2$ from (5), we get

$$\lambda_1 : \lambda_2 = \sqrt{(a_1-a_0)(a_2-a_3)} : [\pm \sqrt{(a_2-a_0)(a_3-a_1)}$$
$$\mp i\sqrt{(a_3-a_0)(a_1-a_2)}].$$

The cross ratio of four points of a circle will be that determined by four isotropics through them, as we have seen from the definition of the latter. Hence we have, as a cross ratio for four foci,

$$\frac{(a_0-a_2)(a_1-a_3)}{(a_0-a_3)(a_1-a_2)}. \tag{22}$$

The six possible cross ratios are obtained by permuting the four letters a_i. Hence we have the same sets of cross ratios on all four fundamental circles.

Let us next seek the points of contact of the cyclic with isotropics tangent thereto. If such a point be (y), a tangent circle there will be $y_i+\mu a_i y_i$. This will be null if $\mu^2(a^2y^2) = 0$. Hence the points sought are the intersections of the cyclic with a second general cyclic whose equation is

$$(a^2y^2) = 0.$$

The coordinates of the points of contact will be

$$\rho y_i^2 = \frac{\partial}{\partial s_i} \mid 1\, a\, a^2\, s \mid. \tag{23}$$

These eight points will also lie on the cyclic

$$2(a^2y^2) - \left[\sum_{i=0}^{i=3} a_i\right](ay^2) = 0, \tag{24}$$

which bears to the given cyclic a curious relation. Substituting the isotropic parameters from (5) in (18) and (24) we get

$$[(a_1-a_0)\,\mu_1{}^2 + (a_2-a_3)\,\mu_2{}^2]\,\lambda_1{}^2 + 2\,[(-a_0-a_1+a_2+a_3)\,\mu_1\mu_2]\,\lambda_1\lambda_2 \\ + [(a_2-a_3)\,\mu_1{}^2 + (a_3-a_0)\,\mu_2{}^2]\,\lambda_2{}^2 = 0.$$

$$[(a_1-a_0)\,\mu_1{}^2 - (a_2-a_3)\,\mu_2{}^2]\,\lambda_1{}^2 + 2\,\frac{(a_2-a_3)^2-(a_1-a_0)^2}{(a_2+a_3)-(a_1+a_0)}\,\mu_1\mu_2\,\lambda_1\lambda_2 \\ + [-(a_2-a_3)\,\mu_1{}^2 + (a_1-a_0)\,\mu_2{}^2]\,\lambda_2{}^2 = 0.$$

Keeping either parameter fixed, we may look on these as quadratic equations in the other parameter, and it will be found that the simultaneous invariant vanishes identically; hence

Theorem 17.] *A general cyclic has a covariant cyclic so related that every isotropic of either set intersects the two in pairs of harmonically separated points. The relation of the two cyclics is mutual, and at every intersection each curve is tangent to one of the isotropics through that point.**

This covariant is simply expressed in our symbolic notation.† Let our cyclic be

$$a_\xi{}^2\bar{a}_{\bar{\xi}}{}^2 = 0.$$

The covariant is

$$\mid a\, a' \mid \cdot \mid \bar{a}\, \bar{a}' \mid a_\xi\, a_\xi{}'\, \bar{a}_{\bar{\xi}}\, \bar{a}_{\bar{\xi}}{}' = 0.$$

* This excellent theorem was discovered by the Author's former pupil, Mr. Lloyd Dixon, but never published.

† Kasner, loc. cit., pp. 480 ff., gives a list of concomitants with their geometric properties. Those which follow are from this source.

We see, in fact, that if $(\bar{\xi})$ be fixed the roots of these two quadratics in (ξ) divide one another harmonically. Or we may reason otherwise. The circle

$$a_\xi a_{\xi'} \bar{a}_{\bar{\xi}} \bar{a}_{\bar{\xi}'} = 0$$

is called the *polar* circle of (ξ'). It contains the harmonic conjugate of (ξ') with regard to the intersections of the isotropics through that point with that cyclic. The covariant cyclic is the locus of points whose polar circles are null. If we add to our equation (18) such a multiple of (yy) that

$$\sum_{i=0}^{i=3} a_i = 0,$$

the equation of the covariant becomes simply

$$(a^2x^2) = 0. \tag{25}$$

The polar circle of (y) will be

$$\sum_{i=0}^{i=3} a_i y_i x_i = 0. \tag{26}$$

Another covariant circle is the *autopolar* circle

$$\sum_{i=0}^{i=3} \frac{y_i x_i}{a_i} = 0. \tag{27}$$

This is orthogonal to the polar circle of every point on the null circle whose vertex is (y). The locus of points lying on their own autopolar circles will be another covariant cyclic:

$$\left(\frac{1}{a} x^2\right) = 0. \qquad (28$$

The circles of different generations of a cyclic are connected by an interesting relation which we shall now develop. Let (y) be a circle of one generation:

$$y_i^2 = \frac{y_j^2}{a_j - a_i} + \frac{y_k^2}{a_k - a_i} + \frac{y_l^2}{a_l - a_i} = 0,$$

$$y_j{}^2 = \frac{a_i - a_j}{a_k - a_i} y_k{}^2 + \frac{a_i - a_j}{a_l - a_i} y_l{}^2,$$

$$(yy) = \frac{a_k - a_j}{a_k - a_i} y_k{}^2 + \frac{a_l - a_j}{a_l - a_i} y_l{}^2.$$

In the same way, if (z) and (z') be two circles of another generation,

$$(zz) = \frac{(a_k - a_i)}{(a_k - a_j)} z_k{}^2 + \frac{(a_l - a_i)}{(a_l - a_j)} z_l{}^2,$$

$$(z'z') = \frac{(a_k - a_i)}{(a_k - a_j)} z_k'^2 + \frac{(a_l - a_i)}{(a_l - a_j)} z_l'^2.$$

$$\cos \measuredangle yz = \frac{y_k z_k + y_l z_l}{\sqrt{\frac{(a_k - a_j)}{(a_k - a_i)} y_k{}^2 + \frac{(a_l - a_j)}{(a_l - a_i)} y_l{}^2}\ \sqrt{\frac{(a_k - a_i)}{(a_k - a_j)} z_k{}^2 + \frac{(a_l - a_i)}{(a_l - a_j)} z_l{}^2}},$$

$$\sin \measuredangle yz = \pm \frac{\sqrt{\frac{(a_k - a_j)(a_l - a_i)}{(a_k - a_i)(a_l - a_j)}}\, y_k z_l - \sqrt{\frac{(a_k - a_i)(a_l - a_j)}{(a_k - a_j)(a_l - a_i)}}\, y_l z_k}{\sqrt{\frac{(a_k - a_j)}{(a_k - a_i)} y_k{}^2 + \frac{(a_l - a_j)}{(a_l - a_i)} y_l{}^2}\ \sqrt{\frac{(a_k - a_i)}{(a_k - a_j)} z_k{}^2 + \frac{(a_l - a_i)}{(a_l - a_j)} z_l{}^2}},$$

$$\cos \measuredangle yz \cos \measuredangle yz' \mp \sin \measuredangle yz \sin \measuredangle yz'$$

$$= \frac{y_k{}^2\left[z_k z_k' + \frac{(a_k - a_j)(a_l - a_i)}{(a_k - a_i)(a_l - a_j)} z_l z_l'\right] + y_l{}^2\left[\frac{(a_l - a_j)(a_k - a_i)}{(a_l - a_i)(a_k - a_j)} z_k z_k' + z_l z_l'\right]}{\left(\frac{a_k - a_j}{a_k - a_i} y_k{}^2 + \frac{a_l - a_j}{a_l - a_i} y_l{}^2\right)\sqrt{\frac{a_k - a_i}{a_k - a_j} z_k{}^2 + \frac{a_l - a_i}{a_l - a_j} z_l{}^2}\ \sqrt{\frac{a_k - a_i}{a_k - a_j} z_k'^2 + \frac{a_l - a_i}{a_l - a_j} z_l'^2}},$$

$$= \frac{\frac{a_k - a_i}{a_k - a_j} z_k z_k' + \frac{a_l - a_i}{a_l - a_j} z_l z_l'}{\sqrt{\frac{a_k - a_i}{a_k - a_j} z_k{}^2 + \frac{a_l - a_i}{a_l - a_j} z_l{}^2}\ \sqrt{\frac{a_k - a_i}{a_k - a_j} z_k'^2 + \frac{a_l - a_i}{a_l - a_j} z_l'^2}}.$$

This expression is independent of (y), thus giving an admirable theorem.*

* Jessop, 'A Property of Bicircular Quartics', *Quarterly Journal of Math.*, vol. xxiii, 1889.

Theorem 18.] *The sum or difference of the angles which all circles of one generation of a general cyclic make with two fixed circles of another is constant.*

This theorem enables us to give an invariant geometric definition of the cyclic.

Theorem 19.] *The envelope of circles orthogonal to a fixed circle the sum of whose angles with two other fixed circles is constant is a cyclic.*

This is essentially, II. 26], proved without the aid of non-Euclidean geometry. If we pass to the limiting case where the cyclic becomes a pair of circles we reach another proof of II. 14].

The generating circles tangent to our cyclic at the point (x) will be

$$y_i = 0, \quad y_j = (a_j - a_i)\,x_j, \quad y_k = (a_k - a_i)\,x_k, \quad y_l = (a_l - a_i)\,x_l.$$

Permuting the indices, and taking the cross ratio of the four, we get

$$\frac{(a_i - a_j)\,(a_k - a_l)}{(a_i - a_l)\,(a_k - a_j)}.$$

Theorem 20.] *The cross ratios of four generating circles tangent at the same point are those of the four foci on any fundamental circle.*

This theorem can be easily generalized. Passing over to the cartesian plane, let P and Q be any two points of a general cyclic. Inverting, with P as centre of inversion, we get an elliptic cubic curve. The cross ratio of the four tangents to this curve from the inverse of Q, is independent of the position of the latter on the curve, by Salmon's theorem ; hence

Theorem 21.] *The cross ratio of four circles through two points of a cyclic tangent to the curve at other points is equal to that of four concyclic foci.*

We find the coordinates of the osculating circle at (x) as follows. We write

$$y_i = \lambda x_i + a_i x_i.$$

Since

$$(yd^2x) = 0, \quad (dxdx) = -(xd^2x), \quad \sum_{i=0}^{i=3} a_i dx_i^2 = -\sum_{i=0}^{i=3} a_i x_i d^2 x_i,$$

$$\lambda(dxdx) + (adx^2) = 0.$$

Let us assume $(udx) = 0.$

$$(xdx) = \sum_{i=0}^{i=3} a_i x_i dx_i = (udx) = 0,$$

$$\rho dx_i = \begin{vmatrix} x_j & x_k & x_l \\ a_j x_j & a_k x_k & a_l x_l \\ u_j & u_k & u_l \end{vmatrix},$$

$$\rho \frac{dx_i}{\sqrt{a_j a_k a_l}} = \begin{vmatrix} \sqrt{a_j} x_j & \sqrt{a_k} x_k & \sqrt{a_l} x_l \\ \frac{x_j}{\sqrt{a_j}} & \frac{x_k}{\sqrt{a_k}} & \frac{x_l}{\sqrt{a_l}} \\ \frac{u_j}{\sqrt{a_j}} & \frac{u_k}{\sqrt{a_k}} & \frac{u_l}{\sqrt{a_l}} \end{vmatrix},$$

$$\lambda(a^2x^2) + a_0 a_1 a_2 a_3 \left(\frac{1}{a} x^2\right) = 0,$$

$$y_i = a_0 a_1 a_2 a_3 \left(\frac{1}{a} x^2\right) x_i - (a^2 x^2) a_i x_i. \qquad (29)$$

Theorem 22.] *Twelve osculating circles to a general cyclic are orthogonal to an arbitrary circle.*

Theorem 23.] *The evolute of the general cartesian cyclic is of the twelfth order.*

We may get the class of a cartesian cyclic, and also of its envelope, from one same formula. The circle tangent at (x), which is orthogonal to (s), will be

$$y_i = \left[\sum_{j=0}^{j=3} a_j x_j s_j\right] x_i - (xs) a_i x_i.$$

This will be orthogonal to (t) also if

$$\sum_{j=0}^{j=3} a_j x_j s_j(xt) - \sum_{j=0}^{j=3} a_j x_j t_j(xs) = 0.$$

Adjoining the equation (18) we see

Theorem 24.] *Eight circles of an arbitrary coaxal system will touch a general cyclic.*

Theorem 25.] *The class of the general cartesian cyclic is eight.*

This agrees with Plücker's equations. If the coaxal system be a concentric one, we see

Theorem 26.] *The class of the evolute of a general cartesian cyclic is eight.*

A circle in the cartesian plane is an adjoint curve to the general cyclic. We may thus apply Nöther's fundamental theorem and the residue theorem.

Theorem 27.] *If a circle meets a cyclic in $ABCD$, while a second meets it in ABC_1D_1, and a third meets it in A_1B_1CD, then $A_1B_1C_1D_1$ are concyclic.*

When the cyclic has a node we may invert into a conic. The theorem is easily proved for a conic; hence it is true of the universal cyclic.

Numerous simple and easy corollaries follow from this theorem.*

Problem 12.] *To construct a tangent circle at a given point of a cyclic passing through another given point.*

Let the given points be A_1B_1. Suppose that the pair of points A_2B_2 is coresidual to the pair A_1B_1 on the cyclic, that is, both are concyclic with the same pair of the cyclic. Let A_1B_1' be concyclic with A_2B_2, and on the cyclic, i.e. residual to A_2B_2; then A_1B_1 and A_1B_1' are residual, or the circle through $A_1B_1B_1'$ is tangent at A_1.

* Cf. Saltel, 'Théorèmes sur les cycliques planes', *Bulletin de la Société mathématique de France* vol. iii, 1874, pp. 96 ff.

Problem 13]. *To construct the osculating circle at a given point.*

Let A_1 be the point. Construct a tangent circle there, and let $A_2 B_2$ be the residual pair. Let $A_3 B_3$ be residual to $A_2 B_2$, and let A_1' be residual to $A_1 A_3 B_3$. Then the circle tangent at A_1 and passing through A_1', which can be constructed by the last problem, is the circle required.

Theorem 28.] *The locus of pairs of points concyclic with each of three given pairs of points, no two of which are concyclic, is a cyclic.*

We may, in fact, pass a cyclic through the six given points and through one pair of the locus. The residuation with regard to this cyclic will give pairs of points concyclic with the given pairs.

Problem 14.] *To construct a cyclic through eight given points.*

Let the points be $A_1, B_1, A, B, C, D, E, F$. Omitting the point F, we have ∞^1 cyclics with one other common point L, by 9]. We find this point as follows. Let the circles $A_1 B_1 C$ and ABC meet again in C'; let the circles $A_1 B_1 D$ and ABD meet again in D'; let $CC'E$ and $DD'E$ meet again in E'. The cyclic determined by pairs of points concyclic with $A_1 B_1$, AB, EE' will contain all of our given points but F. A second such cyclic may be found by interchanging the rôles of D and E. Now take an arbitrary circle c through $A_1 B_1$. The pencil of cyclics through A_1, B_1, A, B, C, D, E meets c in pairs of points. The circles through such pairs, and through a fixed point V, will be a one-parameter family linearly dependent on two of its members, i.e. a coaxal system. Two circles of the system may be determined from the two cyclics just found, and so the other fixed point V_1 of the coaxal system. Replacing E by F we find V_2, which plays the rôle formerly played by V_1. Let the circle $V V_1 V_2$ meet c in $A_2 B_2$. Then $A_2 B_2$ are two points of the cyclic sought. We may find two such on every circle through $A_1 B_1$; the construction is thus complete. We may also, with the aid of the two preceding theorems, find tangent and osculating circles to the cyclic given by eight points.

Suppose that we have in the cartesian plane two sets of four circles. By taking one circle from each set we have, in all, sixteen pairs of circles. Let one intersection of each of fifteen pairs lie on a given cyclic; one intersection of the sixteenth pair will lie thereon also. We see, in fact, that a linear combination of the product of the equations of the first four circles and of the product of the equations of the last four will be a curve of the eighth order, with each circular point at infinity as a quadruple point. These curves have sixteen infinite and fifteen finite fixed intersections with our cyclic. If there were one variable intersection the cyclic would be a rational curve, which it is not. Hence our cyclic contains sixteen intersections of pairs of circles. It is to be noted that the other sixteen lie on another cyclic, for a curve of the family containing a seventeenth point of the given cyclic would degenerate into that and another cyclic. We may restate our theorem in better form.

Theorem 29]. *If three circles meet a general cyclic in* $A_1A_2A_3A_4$, $B_1B_2B_3B_4$, $C_1C_2C_3C_4$ *respectively, and if each of the four points* D_i *be residual to the corresponding triad* $A_iB_iC_i$, *then* $D_1D_2D_3D_4$ *are concyclic.**

The limiting cases of this theorem are more interesting than the general one.

Theorem 30.] *The osculating circles at four concyclic points of a cyclic meet the curve again in four concyclic points.*

If the first three circles have four-point contact.

Theorem 31.] *A circle which meets a cyclic in three points where the osculating circles have four-point contact meets the curve again in such a point.*

There are, as we have seen, sixteen of these points. Let us look a little more closely at their position. To begin with, a circle with four-point contact is a generating circle, so that our points lie by fours on four fundamental circles. If we take two of our points on one fundamental circle they are mutually

* Lachlan, 'On a Theorem relating to Bicircular Quartics', *Proceedings London Math. Soc.*, vol. xxi, 1891, pp. 276 ff. Cf. Schröter, *Grundzüge einer reingeometrischen Theorie der Raumkurven vierter Ordnung*, Leipzig, 1890.

inverse in one of the other fundamental circles. Another pair of points of four-point contact not on either fundamental circle so far mentioned, but mutually inverse, can be found in four ways. We may thus find two pairs of points of four-point contact mutually inverse in one of the fundamental circles in forty-eight ways. Lastly, we may find four points of four-point contact, one on each fundamental circle, in sixty-four ways. There are thus 116 circles, each of which meets the cyclic in four points of four-point contact. The coordinates of these points are easily found by taking the intersections of the curve with each fundamental circle:

$$x_i=0,\ x_j=\pm\sqrt{a_k-a_l},\ x_k=\pm\sqrt{a_l-a_j},\ x_l=\pm\sqrt{a_j-a_k}. \quad (30)$$

We have already seen that twelve osculating circles are orthogonal to a given circle. If the given circle be null, its vertex on the curve, three of these will be accounted for by the osculating circle at that point.

Theorem 32.] *Let the osculating circles at the points* $A_1, A_2, \ldots, A_9$ *meet the general cyclic again at* A, *those at* $B_1, B_2, \ldots, B_9$ *meet it at* B, *those at* $C_1, C_2, \ldots, C_9$ *meet it at* C, *and those at* $D_1, D_2, \ldots, D_9$ *meet it at* D; *then, if the points* A, B, C, D *be concyclic, the points* A_i, B_j, C_k, D_l *lie on* 729 *circles.*

Theorem 33.] *If the osculating circles at* $A_1, A_2, \ldots, A_9$ *meet the general cyclic again at* A, *while those at* $A, B_1, \ldots, B_8$ *meet it at* B, *then the points* A_i, A_j, A_k *lie by threes on eighty-four circles, each of which contains a point* B_l.

Theorem 34.] *Let* A, B, C, D *be four concyclic points of a general cyclic. Let the four generating circles which touch at* A *touch the curve again respectively at* A_1, A_2, A_3, A_4, *and so for* B, C, D. *Then the points* A_i, B_j, C_k, D_l *lie by fours on sixty-four circles.*

The theorems of intersection and residuation for the general cyclic are best handled by the parametric representation of the curve with the aid of elliptic functions. This, of course, is essentially a familiar process, being one of the classical examples of the application of elliptic functions to geometry.* We first replace our equations (18) by

$$(a_0-a_3)\,x_0^2+(a_1-a_3)\,x_1^2+(a_2-a_3)\,x_2^2=0,$$

$$(a_0-a_2)\,x_0^2+(a_1-a_2)\,x_1^2+(a_3-a_2)\,x_3^2=0.$$

Let us then write

$$x=\sqrt{\frac{a_1-a_3}{a_3-a_0}}\,\frac{x_1}{x_0},\quad y=\sqrt{\frac{a_2-a_3}{a_3-a_0}}\,\frac{x_2}{x_0},\quad z=\sqrt{\frac{a_3-a_2}{a_2-a_0}}\,\frac{x_3}{x_0}.$$

$$x^2+y^2-1=0,$$

$$k^2x^2+z^2-1=0. \tag{31}$$

$$k^2=\frac{(a_1-a_2)\,(a_0-a_3)}{(a_1-a_3)\,(a_0-a_2)}.$$

It is to be noted that k^2 is one of our fundamental cross ratios. These equations are equivalent to

$$x=sn\,u,$$

$$y=cn\,u, \tag{32}$$

$$z=dn\,u.$$

The right-hand sides of these equations are the Legendrian elliptic functions of periods $4k$, $4ik'$. There will be a one to one correspondence between the points of the cyclic and the values of u in a period parallelogram of sides $4k$, $4ik'$. Four points u_1, u_2, u_3, u_4 will be concyclic if †

$$u_1+u_2+u_3+u_4\equiv 0 \pmod{4k,\ 4ik'}. \tag{33}$$

To prove 27], let

$$u_1+u_2+u_3+u_4\equiv 0 \pmod{4k,\ 4ik'},$$

$$u_1+u_2+v_3+v_4\equiv 0 \pmod{4k,\ 4ik'},$$

$$v_1+v_2+u_3+u_4\equiv 0 \pmod{4k,\ 4ik'}.$$

Then $$v_1+v_2+v_3+v_4\equiv 0 \pmod{4k,\ 4ik'}.$$

* Cf. e. g. Appell et Lacour, *Principes de la théorie des fonctions elliptiques*, Paris, 1897, ch. v.

† Ibid., p. 163.

To prove 29]. If

$$u_1+u_2+u_3+u_4 \equiv 0 \pmod{4k,\ 4ik'},$$
$$v_1+v_2+v_3+v_4 \equiv 0 \pmod{4k,\ 4ik'},$$
$$w_1+u_2+u_3+w_4 \equiv 0 \pmod{4k,\ 4ik'},$$
$$u_1+v_1+w_1+w_1 \equiv 0 \pmod{4k,\ 4ik'},$$
$$u_2+v_2+w_2+w_2 \equiv 0 \pmod{4k,\ 4ik'},$$
$$u_3+v_3+w_3+w_3 \equiv 0 \pmod{4k,\ 4ik'},$$
$$u_4+v_4+w_4+w_4 \equiv 0 \pmod{4k,\ 4ik'}.$$

Then
$$w_1+w_2+w_3+w_4 \equiv 0 \pmod{4k,\ 4ik'}.$$

Let us next take up 34] in detail. If a point $A = u_1$ be chosen, the other points of contact of generating circles tangent at A are

$$-u_1, \quad -u_1+2k, \quad -u_1+2ik', \quad -u_1+2k+2ik'.$$

(*a*) If the generating circles tangent at AA_i, BB_j, CC_k belong to the same generation, that tangent at DD_l belongs to the same generation also, for the circles A, B, C, D and A_i, B_j, C_k, D_l are interchanged by inverting in the corresponding fundamental circle.

(*b*) Let the circles AA_i and BB_j belong to one generation, while CC_k belongs to a second: we may write

$$A = w_1, \quad A_i = -w_1,$$
$$B = v_1, \quad B_j = -v_1,$$
$$C = w_1, \quad C_k = -w_1+2k.$$

Then $D = -(u_1+v_1+w_1)$, $D_l = (u_1+v_1+w_1)-2k$.

This shows that CC_j and DD_k belong to one generation.

(*c*) If AA_i, BB_j, and CC_k belong to different generations, then DD_l must belong to the fourth generation, as otherwise we should be in conflict with (*b*).*

* Lachlan, *Bicircular Quartics*, cit., seems rather afraid of 34], as he says, p. 278: 'But it would seem in the above reasoning that the three bitangent circles at ABC need not necessarily belong to the same system.' Of course they need not!

The osculating circle at v will meet the curve again at $u = -3v$.

The point u lies on the osculating circles at the nine points,

$$v = -\frac{u}{3} + \frac{4mk}{3} + \frac{4nik'}{3}, \quad m = 0, 1, 2, \quad n = 0, 1, 2.$$

The points where the osculating circles have four-point contact are

$$u = mk + nik', \quad m = 0, 1, 2, 3, \quad n = 0, 1, 2, 3.$$

Besides the study of individual cyclics there is not a little of importance in the study of systems of cyclics. The most interesting systems are the confocal ones. We shall define these as the loci of the vertices of the null circles of a system of confocal quadric congruences, these latter being defined exactly as in the cartesian case. Analytically, we replace our cyclic (18) by

$$\sum_{i=0}^{i=3} \frac{a_i}{1-\lambda a_i} x_i^2 = (xx) = 0, \tag{34}$$

where λ takes all possible values. The expression for the coordinates of the foci in (21) will be unaltered, so that confocal cyclics have the same foci. We shall presently see that the converse is not always the case. If we look upon (x) as fixed in (34), we have a quadratic equation in λ, for the coefficient of λ^3 will vanish in virtue of our fundamental identity. There are thus two confocal general cyclics through each point in general position. If these correspond to the parameter values λ and λ', we have for two tangent circles to the two curves at

$$y_i = lx_i + \frac{a_i x_i}{1-\lambda a_i}, \quad y_i' = l'x_i + \frac{a_i x_i}{1-\lambda' a_i}.$$

Since (x) lies on both cyclics, we have

$$\sum_{i=0}^{i=3} \frac{a_i}{1-\lambda a_i} x_i^2 = \sum_{i=0}^{i=3} \frac{a_i}{1-\lambda' a_i} x_i^2 = 0.$$

Subtracting one equation from the other we get

$$\sum_{i=0}^{i=3}\left(\frac{a_i}{1-\lambda a_i}\right)\left(\frac{a_i}{1-\lambda' a_i}\right)x_0^2 = 0.$$

This yields, however,

$$(yy') = 0.$$

Theorem 35.] *Through each point in general position in the tetracyclic plane will pass two confocal general cyclics of a given system, and these two intersect orthogonally at that point.*

We mean by a point in *general position* one where the roots of the quadratic in λ are distinct, i.e. a point not on the isotropics, which are the envelope of the system.

Confocal cyclics in the tetracyclic plane will correspond to confocal ones in the cartesian plane. There are some advantages in studying the latter rather than the former, as we shall now show. We begin with the differential equation

$$dw = \frac{du}{\sqrt{(1-u^2)(1-k^2u^2)}} = \frac{dv}{\sqrt{(1-v^2)(1-k^2v^2)}}. \qquad (35)$$

These lead to the solution

$$u = sn\, w,$$

$$v = sn(w-\alpha),$$

where α is the constant of integration. Eliminating w, we get *

$$au^2v^2+bu^2v+cuv^2+du^2+ev^2+fu+gv+h = 0.$$

If we give u and v the following values,

$$u = x+iy, \quad v = x-iy,$$

we see that we have the general cartesian cyclic. By varying α we get a one-parameter family of cyclics, and these have the same foci. We see, in fact, that in (35)

$$du = 0 \text{ if } u = \pm 1, \quad u = \pm\frac{1}{k}.$$

* Darboux, *Sur une classe*, cit., p. 76; Appell et Lacour, loc. cit., p. 129.

Hence ± 1 and $\pm \frac{1}{k}$ are the values of the isotropic parameters corresponding to the tangent isotropics of the two systems. The fact that we have the same quadratic expression on both sides corresponds to the fact, proved at once by inversion, that the two triads of tangent isotropics of the two systems have the same cross ratios. Suppose, conversely, that we have a general cyclic. The tangent isotropics of the two systems have the same cross ratios, and, by a linear fractional transformation of u and v, we may make these tangents correspond to the parameter values $\pm 1 \pm \frac{1}{k}$. Then the one-parameter family of cyclics given by (35) will include the given cyclic. If $u\ v$ be known, the two values of $\frac{dv}{du}$ obtained from (35) differ only in sign, i.e. the curves intersect at right angles.

Let us consider what will be the effect on (35) if we subject v to such a linear transformation,

$$v' = \frac{\alpha v + \beta}{\gamma v + \delta},$$

that the denominator on the right is covariant. There are four conceivable types of such transformation, when the tangent isotropics are all different.

(a) The tangent isotropics are interchanged in pairs. This will be done by the involution whose double members separate the interchanged pairs harmonically, i.e. the double members are a pair of roots of the sextic covariant of the quartic form. The sextic covariant has the property that each pair (not each two) of its roots separates harmonically two pairs of the roots of the quartic. If we take for the roots of the sextic

$$0, \quad \infty, \quad 1, \quad -1, \quad i, \quad -i,$$

the three involutory transformations of the quartic into itself are

$$v' = -v, \quad v' = \frac{1}{v_1}, \quad v' = -\frac{1}{v};$$

these will change the right-hand side of (35) at most in sign.

(*b*) The roots of the quartic are permuted cyclically. Here we reduce the sextic to the previous form. The quartic will involve only even terms, and not lack the term in v^4. The transformation will carry the sextic into itself, and it is easy to see that it will leave one pair of roots of the sextic in place. Hence we shall easily find that it is of the form *

$$\pm v, \quad \pm\frac{1}{v}, \quad \pm i\frac{v+1}{v-1}, \quad \pm i\frac{v-1}{v+1}, \quad \pm\frac{v+i}{v-i}, \quad \pm\frac{v-i}{v+i}.$$

These will all leave the whole right-hand side of (35) invariant, except for sign.

(*c*) One root of the quartic is left in place, the other two permuted. The roots here have at most two cross ratios, instead of the usual number of six, i.e. they must be equiharmonic. Under these circumstances we may rewrite (35)

$$\frac{du}{\sqrt{u^3+1}} = \frac{dv}{\sqrt{v^3+1}}; \tag{36}$$

the transformations to be effected on the right are

$$v' = \omega v, \quad v' = \omega^2 v, \quad \omega^3 = 1.$$

The right-hand side of (36) will be multiplied by $\pm\omega$ or $\pm\omega^2$, and we get two new confocal systems with the same foci. The angle of the curves $\frac{dv}{du}$, $\frac{\omega\, dv}{du}$ is, by Laguerre's projective definition, $\frac{1}{2i}$ times the logarithm of the cross ratio of the four quantities $0, \infty, \frac{dv}{du}, \frac{\omega\, dv}{du}$.

$$e^{2i\theta} = \omega,$$

$$\theta = \frac{\pi}{3}.$$

The six curves through any point make equal angles with one another.

(*d*) One pair of roots remain in place, the others are inter-

* These are the transformations of the tetrahedral group. Cf. Weber, *Lehrbuch der Algebra*, second ed., Braunschweig, 1899, vol. ii, p. 274.

changed. Here the roots must form a harmonic set. If we take two of them as 0 and ∞, we may replace (35) by

$$\frac{du}{\sqrt{u(u^2-1)}} = \frac{dv}{\sqrt{v(v^2-1)}}. \tag{37}$$

The change of v into $-v$ will multiply the right-hand side by $\pm i$.

$$e^{2i\theta} = i, \quad \theta = \frac{\pi}{4}.$$

Theorem 36.] *There is but one set of cyclics having the same foci as a given general cyclic, namely, those confocal with it, except where the tangent isotropics form a harmonic or an equiharmonic set. In the harmonic case four cyclics pass through each point in general position, making successively angles of* $\frac{\pi}{4}$; *in the harmonic case six cyclics will pass through a general point, making successively angles of* $\frac{\pi}{6}$.*

The tetracyclic plane does not seem to offer such a promising field for further study as some other parts of our subject. The subject of problems of construction might certainly be carried further. Something might be done to line up our analytic work with the large amount of literature dealing with the geometry on a sphere in general, and the study of spheroconics in particular. There are doubtless also numerous theorems of interest concerning special types of cyclics still to be discovered. Some of these, such as the lemniscate, have been already extensively treated. It is never safe to say that any branch of mathematics has been explored to the end; merely, in this case, the outlook is less promising than in some others.

* The equiharmonic case seems to have been discovered by Roberts, 'On foci, and confocal Plane Curves', *Quarterly Journal of Mathematics*, vol. xxxv, 1904. It is not clear how he was first led to his results: had he made an exhaustive study of the transformations of the quartic into itself he could not have overlooked the simpler harmonic case.

CHAPTER V

THE SPHERE IN ELEMENTARY GEOMETRY

§ 1. Miscellaneous Elementary Theorems.

THE elementary geometry of the sphere is closely allied to that of the circle, or, rather, to certain portions of the latter. Theorems about the circle, which are largely descriptive in character, carry over easily into three dimensions. On the other hand, the sphere has no simple property corresponding to the invariance of the angle inscribed in a given circular arc. For this reason we fail to find in the case of a sphere many theorems corresponding to the most beautiful metrical ones associated with the circle.

The likeness between circles and spheres extends beyond individual theorems to general methods of proof. Often the procedure which is applicable in one case may be directly transferred to the other. Furthermore, a goodly number of theorems, where all the spheres involved have collinear centres, may be obtained from the corresponding circle theorems by rotation about an axis. For this reason it will be possible in the present chapter to omit the proofs of a considerable proportion of the theorems, leaving to the reader the task of referring back to the corresponding cases in Ch. I. To facilitate such reference we shall follow much the same order as there prevailed.

All figures considered in the present chapter shall be supposed to exist in the finite real domain of Euclidean space, the domain of elementary solid geometry. Points, lines, and angles have the same meaning as before. Let us mean by a *plane* the surface generated by lines meeting in distinct points any two sides of a given triangle. The portion of a plane on one side of one of its lines shall be called a *half-plane.* If two non-coplanar half-planes be bounded by the same line, the region which includes all segments whose extremities lie in these two half-planes shall be called their *interior dihedral angle*, or, more shortly, their *dihedral angle.* The remainder of space shall be their *exterior dihedral angle.* Four non-coplanar points determine four triangles, and the figure bounded by them is called a *tetrahedron*, the triangles are its *faces*, their planes its *face-planes*, the sides of the triangle are the *edges* of the tetrahedron, their lines its *edge-lines.* The meanings of such words as *vertex, face angle, dihedral angle, trihedral angle* of a tetrahedron are immediately evident. A line through a vertex perpendicular to the opposite face-plane is called an *altitude line*, the portion between the vertex and the intersection with the plane is the *altitude*, the extremities of the altitude are the vertex and its *foot.*

The locus of points at a given distance from a given point shall be called a *sphere. Centre, radius, diameter, diametral line* have meanings conformable to those used for a circle. Spheres of equal radius shall be called equal.

Let a sphere be given with centre O and radius r. Let P and P' be two such points collinear with O that

$$(\overrightarrow{OP}) \times (\overrightarrow{OP'}) = r^2. \tag{1}$$

Each is said to be the *inverse* of the other in the given sphere. The sphere is called the *sphere of inversion*, its centre and radius the *centre* and *radius of inversion.*

Theorem 1.] *Every point except the centre of inversion has a single definite inverse with regard to a given sphere.*

Theorem 2.] *The sphere of inversion is the locus of points which are their own inverses. Points outside the sphere will invert into points within, points within, other than the centre, will invert into points without.*

Theorem 3.] *Mutually inverse points are harmonically separated by the intersections of their line with the sphere of inversion.*

Theorem 4.] *If A, B, C, D be four points, and A', B', C', D' their inverses,*

$$\frac{(AB)(CD)}{(AD)(CB)} = \frac{(A'B')(C'D')}{(A'D')(C'B')}.$$

Theorem 5.] *The angle at which two curves intersect is equal in absolute value to that made by their inverses.*

This is easily proved when we remember that two trihedral angles are symmetrical if two face angles and the included dihedral angle are equal to the corresponding parts in the other, but arranged in opposite senses.

Theorem 6.] *The angle at which two surfaces intersect is equal to that made by their inverses.*

Any locus which is its own inverse shall be said to be *anallagmatic.*

Theorem 7.] *An anallagmatic curve or surface cuts the sphere of inversion orthogonally at every intersection which is a simple point of the curve or surface.*

Theorem 8.] *A plane through the centre of inversion is anallagmatic.*

Theorem 9.] *A sphere through any pair of inverse points is anallagmatic; every sphere cutting the sphere of inversion orthogonally is of this sort.*

Theorem 10.] *The inverse of a plane not passing through the centre of inversion is a sphere passing through that point, and vice versa.*

Theorem 11.] *The inverse of a sphere not passing through the centre of inversion is a sphere of the same sort.*

Theorem 12.] *The inverse of a circle not passing through the centre of inversion is a like circle, the inverse of a circle passing through the centre of inversion is a line not passing through that point, and vice versa.*

Theorem 13.] *A circle passing through a pair of inverse points, or a circle or line orthogonal to the sphere of inversion, is anallagmatic.*

Theorem 14.] *If two figures be mutually inverse with regard to a sphere, their inverses with regard to a second sphere, whose centre is not on the first, are mutually inverse in the inverse of the first sphere with regard to the second. When the centre of inversion is on the first sphere, the inverses are the reflections of one another in the plane into which the first sphere is transformed.*

Theorem 15.] *If two figures be mutually inverse with regard to two spheres, they are mutually inverse with regard to the inverse of one sphere in the other, or are reflections of one another in the plane which is the inverse of one sphere in the other.*

Theorem 16.] *If an anallagmatic surface do not contain an anallagmatic series of circles, it is the envelope of a two-parameter family of anallagmatic spheres whose centres move on a fixed surface called the deferent, and, conversely, the envelope of every such system of spheres, if a surface, will be an anallagmatic one. The line connecting corresponding points on the anallagmatic surface is the perpendicular from the centre of inversion on the corresponding tangent plane to the deferent.*

Theorem 17.] *An anallagmatic surface which contains a one-parameter family of anallagmatic circles, which are lines of curvature, is the envelope of a one-parameter family of anallagmatic spheres, and vice versa.*

These last two theorems belong more properly in the domain of differential geometry, the last one arising from the well-known fact that every evolute of a circle is a point.

Theorem 18.] *If two spheres be mutually inverse, the centre of inversion is a centre of similitude for them, the ratio of similitude being in absolute value that of their radii.*

It is immediately evident that two non-concentric spheres

of unequal radii have two centres of similitude, the ratio being positive in one case, negative in the other.

Theorem 19.] *Any two spheres of unequal radius are mutually inverse with regard to one real sphere. When they intersect in a real circle they are mutually inverse with regard to a second such sphere. When they do not intersect or touch and are not concentric, there is another sphere of such a nature that the two are interchanged by an inversion in this sphere followed by a reflection in its centre.*

A sphere with regard to which two spheres are mutually inverse is called a sphere of *antisimilitude* for them; its centre is one of their centres of similitude. We shall also define as the *power* of a point with regard to a sphere its power with regard to any circle of the sphere coplanar with it.

Theorem 20.] *The locus of points whose powers with regard to two unequal and non-concentric spheres are proportional to the squares of the corresponding radii is the sphere having as diameter the segment bounded by the centres of similitude of the two spheres.*

This sphere shall be called the *sphere of similitude* of the two.

Theorem 21.] *If three unequal spheres be given, no two concentric, a line connecting a centre of similitude of one pair with a centre of similitude of a second pair will pass through a centre of similitude of the third pair.*

Theorem 22.] *If a sphere touch two others, the line connecting the two points of contact will pass through a centre of similitude or be parallel to the line of centres.*

Theorem 23.] *If four spheres be given, no two concentric or equal, nor with their four centres coplanar, they will determine in pairs twelve centres of similitude. These lie by sixes in planes through three centres of given spheres, and by threes on sixteen lines. Four such lines pass through each centre of similitude, four lie in each plane through the centres of three spheres, and four in each of eight other planes, whereof two pass through each of the sixteen lines. The centres of similitude lie by sixes in these twelve planes.*

To prove this theorem let the spheres be s_1, s_2, s_3, s_4. The external centre of similitude of s_i and s_j shall be C_{ij}, their internal centre C_{ij}'. Then by I. 31] the following triads are collinear:

$$C_{ij}\,C_{jk}\,C_{ki},\quad C_{ij}\,C_{jl}'\,C_{li}'.$$

Hence $C_{ij}\,C_{jk}\,C_{ki}\,C_{il}'\,C_{jl}'\,C_{kl}'$ are coplanar, and of these there are four. Similarly $C_{ij}\,C_{kl}\,C_{ik}'\,C_{il}'\,C_{jk}'\,C_{jl}'$ are coplanar, and here there are three. Lastly, $C_{ij}\,C_{ik}\,C_{il}\,C_{jk}\,C_{jl}\,C_{kl}$ are coplanar. The twelve planes may be grouped to be the face-planes of three tetrahedra. Every face-plane of one tetrahedron, and every face-plane of a second, will be coaxal with a face-plane of the third, thus giving the well-known desmic configuration of Stephanos.* Three parallel planes are here considered coaxal, and the word tetrahedron means any four planes, no three coaxal.

Theorem 24.] *The radius of the inverse of a sphere not through the centre of inversion is equal to that of the given sphere multiplied by the square of the radius of inversion, and divided by the absolute value of the power of the centre of inversion with regard to the given sphere.*

Theorem 25.] *The inverse of the centre of a sphere not through the centre of inversion is the inverse of the centre of inversion with regard to the inverse of the given sphere.*

Theorem 26.] *The inverse of the centre of a sphere through the centre of inversion is the reflection of the centre of inversion in that plane which is the inverse of the given sphere.*

Theorem 27.] *Any two non-intersecting spheres may be inverted into concentric spheres, the centre of inversion being on their line of centres.*

* 'Sur les systèmes desmiques de trois tétraèdres', *Bulletin des Sciences mathématiques*, Series 2, vol. iii, 1879, pp. 424 ff.

The three-dimensional analogue of Steiner's chain of successively tangent spheres is neither easy nor attractive except in special cases.* The criterion for five spheres tangent to a sixth is not neatly expressible except in determinant form, so that we pass it over till the next chapter. Let us turn to the relations of a tetrahedron to certain special spheres. In particular, let us search for something to correspond to the nine-point circle. It will be remembered that one method of finding that circle is to treat it as the pedal circle of two isogonally conjugate points. In a similar spirit we now take up the question of isogonal conjugates in three dimensions.

Let two half-planes be given forming a dihedral angle, but not coplanar. If P be any point not in either plane, and Q any point in that plane through the edge l of the dihedral angle, which is the reflection of the plane Pl in the bisector, then the points P and Q are said to be *isogonal conjugates* with regard to the dihedral angle; the relation between the two is clearly reciprocal. If $P_\alpha P_\beta$, $Q_\alpha Q_\beta$ be the feet of the perpendiculars from P and Q on the two planes,

$$\frac{(PP_\alpha)}{(PP_\beta)} = \frac{(QQ_\beta)}{(QQ_\alpha)}.$$

If P move parallel to the edge l till it fall into the plane $Q_\alpha Q Q_\beta$ at P', while Q moves parallel to l till it reaches Q' in the plane $P_\alpha P P_\beta$, then, by I. 66], $P_\alpha' P_\beta'$, $Q_\alpha Q_\beta$ lie on a circle whose centre is the middle point of $(P'Q)$, while $P_\alpha P_\beta$, $Q_\alpha' Q_\beta'$ lie in a circle whose centre is the middle point of (PQ'). Hence $P_\alpha P_\beta$ and $Q_\alpha Q_\beta$ lie on a sphere whose centre is the middle point of (PQ).

We next take a trihedral angle. The locus of points isogonally conjugate to a point which does not lie on any edge of this trihedral angle with regard to two of the dihedral angles, is a line through the vertex, and since the feet of the perpendiculars from two such points on the three face-planes lie on a certain sphere whose centre is half-way between them, we see that they are isogonally conjugate with regard to all three dihedral angles. Lastly, we take a tetrahedron. We see that

* Vahlen, loc. cit.

every point not on any edge-line has a definite isogonal conjugate with regard to all six dihedral angles. If, thus, we define as the *pedal sphere* of a point that which passes through the feet of the perpendiculars, thence to the face-planes of a tetrahedron, we get the interesting theorem *

Theorem 28.] *If two points be isogonally conjugate with regard to a tetrahedron, they have the same pedal sphere, whose centre is mid-way between them.*

We reached the nine-point circle as the pedal circle of the centre of the circumscribed circle and the orthocentre. Let us try a similar method here. We must first notice that, by **I.** 199] and 201], the locus of points whose powers with regard to two spheres differ only in sign is a sphere whose centre is mid-way between theirs. Let us call this their *radical sphere.*

Theorem 29.] *The locus of the centre of a sphere which is cut by one of two given spheres orthogonally, and by the other in a great circle, is their radical sphere or a portion thereof.*

Let us anticipate our future work to the extent of assuming that if four spheres be given with non-coplanar centres, there is just one point which has the same power with regard to all four. This shall be called their *radical centre.* If it be without the given spheres, it is the centre of their common orthogonal sphere; if within them, the centre of a sphere cut by all in great circles. Suppose, then, that we have given a tetrahedron, and the circumscribed sphere. We may find four spheres each having as a great circle a circle cut by a face-plane from s. The centres of these four spheres are not coplanar, there is a sphere s' which is either orthogonal to all or cut by all in great circles. If s' be orthogonal to the four spheres whose centres are in the face-planes, we see that these centres are on the radical sphere of s and s'. In any case, it is quite easy to prove trigonometrically that the centres of s and s' are isogonal conjugates; hence †

* Cf. Neuberg, 'Mémoire sur le tétraèdre', *Mémoires couronnés de l'Académie royale de Belgique*, vol. xxxvii, 1886, p. 11.

† Roberts, *On the Analogues, &c.*

Theorem 30.] *The centre of the circumscribed sphere to a tetrahedron and the radical centre of the four spheres, each of which has a great circle through three vertices of the tetrahedron, are isogonal conjugates; the feet of the perpendiculars from these points on the face-planes and the reflections of these feet in the point mid-way between the two points are co-spherical.*

Let us call this the *sixteen-point sphere.* It is the first analogue to the nine-point circle. We reach another analogue as follows.*

Let the vertices of a tetrahedron be $A_1A_2A_3A_4$. Let us first assume $A_1A_2 \perp A_3A_4$ and $A_1A_3 \perp A_2A_4$. The plane through A_1A_2 and the altitude line from A_1 will be perpendicular to A_2A_4, and meet that line at the foot of the A_2 altitude line in the $\triangle A_2A_3A_4$. In the same way the plane through A_1A_3 and the A_1 altitude line will meet $A_2A_3A_4$ in a second altitude line. Hence the altitude line through A_1 meets the opposite face-plane in the orthocentre of that face. Hence each pair of opposite edge-lines will be perpendicular in direction, each altitude will pass through the orthocentre of the corresponding face, and the four altitude lines are concurrent in a point called the *orthocentre* of the tetrahedron. Conversely, if the altitude lines be concurrent, each edge-line is perpendicular in direction to two altitude lines, and, so, to the opposite edge-line. This special case shall be called the *orthogonal tetrahedron.*

In the general case we see that if we pass a plane through any altitude line and the orthocentre of any face, we have a plane perpendicular to the plane of that face. If, further, we speak for the moment of parallel lines as meeting at infinity, and consider on the one hand the four altitude lines, and, on the other, the perpendiculars to each face-plane at the orthocentre, we see that each line of one system meets each of the other, but in neither case are all four lines parallel to one plane. In the general case the altitude lines of a tetrahedron are generators of the same system of a

* See an excellent article by Intrigila, 'Sul tetraedro', *Rendiconti della R. Accademia delle Scienze di Napoli,* vol. xxii, 1883.

hyperboloid.* Let us call this the *associated hyperboloid*, its centre C. Let us prove that the centre of gravity G of the tetrahedron is half-way between C and O, the centre of

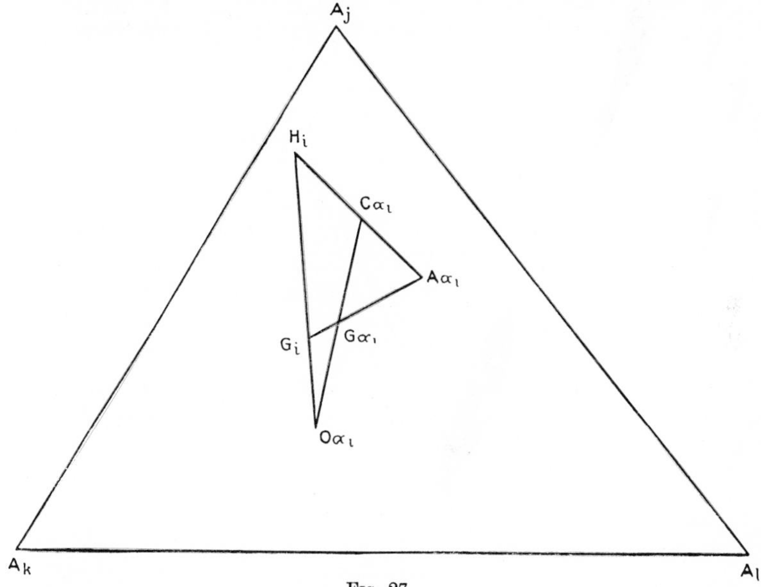

FIG. 27.

the circumscribed sphere. The orthogonal projection of P on the face-plane α_i shall be $P\alpha_i$; the orthocentre of this face shall be H_i. Remembering that the centre of the hyperboloid lies mid-way between each pair of parallel generators,

$$\frac{(\overrightarrow{A\alpha_i C\alpha_i})}{(\overrightarrow{C\alpha_i H_i})} = 1.$$

But we know, from I. 72],

$$(\overrightarrow{H_i O\alpha_i}) = -3(\overrightarrow{O\alpha_i G_i}),$$

where G_i is the centre of gravity of this face. Again, from

* It is highly unsportsmanlike to make use of a hyperboloid in elementary geometry. Frankly, the author does not know how to dispense with it in this case.

the fundamental property of the centre of gravity of a tetrahedron,

$$(\overrightarrow{A\alpha_i G\alpha_i}) = 3\,(\overrightarrow{G\alpha_i G_i}),$$

$$(\overrightarrow{A\alpha_i C\alpha_i}) \times (\overrightarrow{H_i O\alpha_i}) \times (\overrightarrow{G_i G\alpha_i}) = (\overrightarrow{H_i C\alpha_i}) \times (\overrightarrow{G_i O\alpha_i}) \times (\overrightarrow{A\alpha_i G\alpha_i}).$$

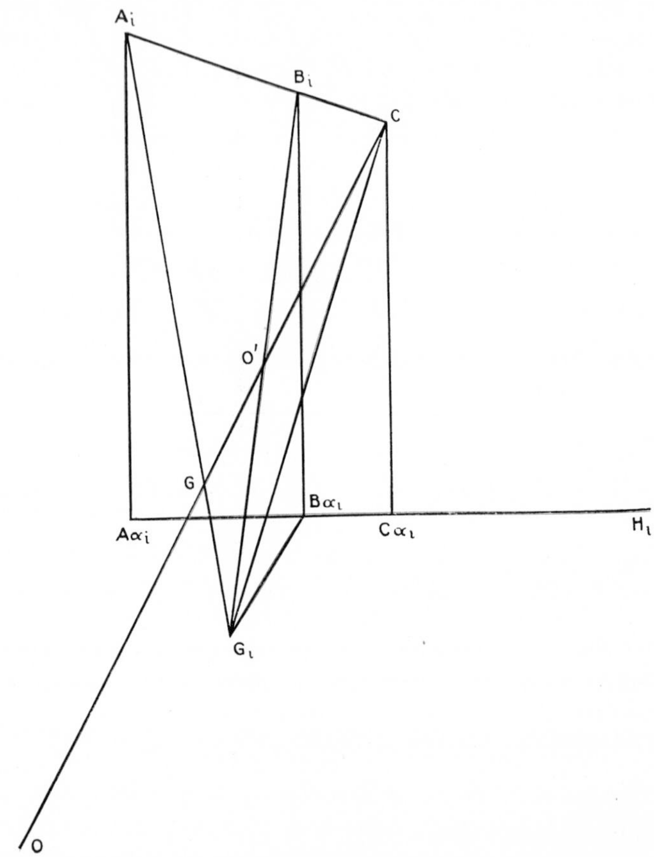

FIG. 28.

Applying Menelaus's theorem to the $\Delta H_i A\alpha_i G_i$, we see that the points $C\alpha_i$, $O\alpha_i$, $G\alpha_i$ are collinear. Reasoning in the same way for the other face-planes, COG must be collinear. Again, since $C\alpha_i$ is the middle point of $(H_i A\alpha_i)$, a line through it parallel to $A\alpha_i G_i$ meets $H_i G_i$ in the middle point of

$(H_i G_i)$, and this is the reflection of $O\alpha_i$ in G_i. Hence $G\alpha_i$ is the middle point of $(C\alpha_i O\alpha_i)$, or G is the middle point of (OC).

The tetrahedron $G_1 G_2 G_3 G_4$ is inversely similar to the tetrahedron $A_1 A_2 A_3 A_4$, the ratio of similarity being $-\frac{1}{3}$, while the centre of similitude is G. Hence O', the harmonic conjugate of O with regard to G and C, which divides (GC) in the ratio $-\frac{1}{2}$, is the centre of the sphere about $G_1 G_2 G_3 G_4$: G and C are the centres of similitude for the spheres circumscribed to our two tetrahedra. The four points, one-third of the distance from C to the vertices of the tetrahedron, lie on the new sphere. If such a point be B_i, we see that B_i and G_i are diametrically opposite on the new sphere, since $A_i B_i C O G G_i$ are in a plane through the centres of both spheres. Hence, since $\measuredangle B_i B\alpha_i C_i = \frac{\pi}{2}$, $B\alpha_i$ is on our new sphere. We also see that $B\alpha_i$ is the harmonic conjugate of H_i with regard to $C\alpha_i$ and $A\alpha_i$.

Theorem 31.] *The centres of gravity of the faces of a tetrahedron, the points on one-third of the distance from the centre of the associated hyperboloid to the vertices, and the harmonic conjugates of the orthocentre of each face with regard to the orthogonal projections on its plane of the opposite vertex, and the centre of the associated hyperboloid, are on one sphere.*

We shall call this the *twelve-point sphere* of the tetrahedron.

Let us now take up the special case of the orthogonal tetrahedron. Here C will coincide with H, the orthocentre of the tetrahedron. The points A_i, $C\alpha_i$, H_i coalesce.

Theorem 32.] *In an orthogonal tetrahedron the centres of gravity of the faces, their orthocentres, and the points one-third the distance from the orthocentre to the vertices are co-spherical.**

Let us call this the *first twelve-point sphere* of the orthogonal tetrahedron. We might naturally guess that it was identical with the sixteen-point sphere, but such is not the case. The

* Cf. Prouhet, 'Analogies du triangle et du tétraèdre', *Nouvelles Annales de Math.*, Series 2, vol. ii, 1863, p. 138.

sixteen-point sphere passes through the centres of the circles circumscribed to the face triangles, while the first twelve-point sphere passes through their orthocentres and centres of gravity, and these three points are collinear. It is true, however, that the first twelve-point sphere passes through sixteen notable points, for

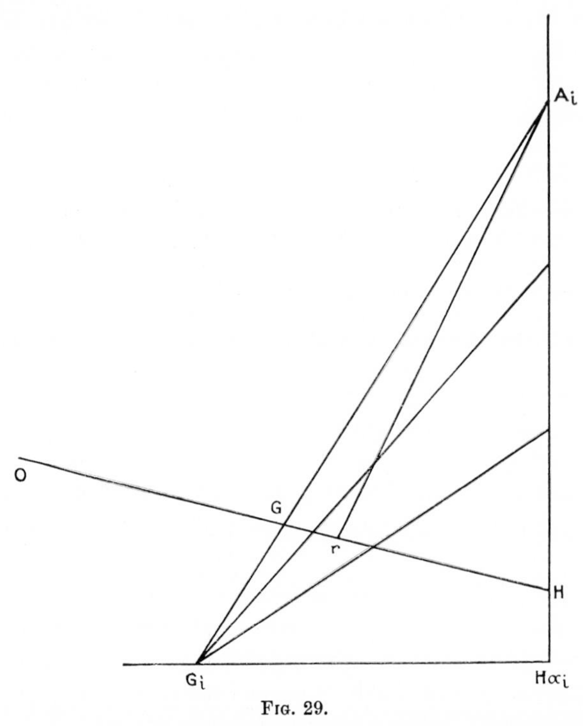

Fig. 29.

Theorem 33.] *If each altitude of an orthogonal tetrahedron be extended beyond its foot by double the distance from that foot to the orthocentre, the points so found lie on the first twelve-point sphere.*

Besides the first twelve-point sphere we have a second one reached as follows :

Theorem 34.] *The nine-point circles of the four faces of an orthogonal tetrahedron lie on the second twelve-point sphere.*

The centre of this sphere is the centre of gravity of the tetrahedron.

The four vertices and orthocentre of an orthogonal tetrahedron shall be called an *orthogonal system*: each point is the orthocentre of the tetrahedron whose vertices are the other four. We thus determine four circumscribed spheres, four twelve-point spheres of the first sort, and four of the second sort. Each face-plane is the radical plane (i.e. the locus of points having like powers) for two circumscribed spheres, and two twelve-point circles of each sort.

Theorem 35.] *Each point of an orthogonal system is the radical centre for four circumscribed spheres, four twelve-point spheres of the first sort, and four twelve-point spheres of the second sort.*

Turning especially to the twelve-point spheres of the first sort, we see that the centre of gravity of all five points lies one-fifth of the distance from G to H. The distance from G to the centre of the first twelve-point sphere associated with H is $\frac{1}{3}(GH)$, hence the distance from Γ to the centre of this sphere is $\frac{1}{6}(\Gamma H)$; Γ is the centre of similitude for the given points and the centres of the first twelve-point spheres. Lastly $(\overrightarrow{\Gamma O}) = 3(\overrightarrow{\Gamma G}) = -\frac{3}{2}(\overrightarrow{\Gamma H})$.

Theorem 36.] *The centres of the five circumscribed spheres of an orthogonal system, those of the five twelve-point spheres of the first sort, and those of the five twelve-point spheres of the second sort, form three orthogonal systems.*

There seems to be very little in the geometry of the tetrahedron which bears a close analogy to the Brocard figures; we pass therefore to the analogy of our descriptive theorems about concurrent circles and concyclic points.

We start with the figure of I. 149], three concurrent circles each through a vertex of a triangle and the marked points on the two adjacent side-lines. We invert this figure with a centre not in its plane, and we have on a certain sphere six circles concurrent by threes in eight points. Next let us take a tetrahedron, and mark one point on each edge-line.

Four spheres (or planes) may be passed, each through one vertex and the marked points of the edge-lines adjacent. If we consider how one of these spheres is met by the three planes concurrent thereon, and by the other three spheres, we have exactly the preceding figure of six circles. We thus get *

Theorem 37.] *If a point be marked on each edge-line of a tetrahedron, and a sphere be passed through each vertex and the marked points of the adjacent edge-lines, these four spheres are concurrent.*

Unfortunately, we cannot proceed immediately from this to the case of five, and so to n spheres. For if five planes be given in general position, they determine five tetrahedra and five circumscribed spheres, but these, instead of being all concurrent in one point, are concurrent by fours in the five planes.

There is an easy three-dimensional analogue to I. 166] stated as follows: †

Theorem 38.] *Given n points on a sphere, no four of which are concyclic. We may associate with them a point and a sphere as follows:*

(*a*) *The point is the centre of the associated sphere.*

(*b*) *The radius of the sphere is one-half that of the given sphere.*

(*c*) *The point lies on n spheres each associated with the systems of $n-1$ points obtained by omitting each of the given points in turn.*

(*d*) *The sphere contains the centres of these n spheres.*

We may copy closely our second proof for the above-named theorem. When $n = 5$ the centres of gravity of the five tetrahedrons are the points $-\frac{1}{4}$ of the distance from the centre of gravity of the five points, to those points, that is to say,

* The credit for this theorem is usually ascribed to Roberts, 'On certain tetrahedra specially related to four spheres', *Proceedings London Math. Soc.*, vol. xii, 1880. It is, however, implicitly given by Miquel, loc. cit.

† Intrigila, loc. cit., pp. 78, 79.

they are on a sphere whose radius is $\frac{1}{4}$ that of the given sphere. But the centre of each twelve-point sphere is $\frac{4}{3}$ the distance from the centre of the circumscribed sphere to the centre of gravity of the tetrahedrons, and the radius of the twelve-point sphere is $\frac{1}{3}$ the radius of the given sphere. Hence the centres of the five twelve-point spheres lie on a sphere of $\frac{1}{3}$ the given radius, and all pass through the centre thereof. The theorem is thus proved when $n = 5$. Assume that it is true for $n-1$ points, and that the point associated with $n-1$ points is $\frac{n-1}{3}$ of the distance from the centre of the given sphere to the centre of gravity of $n-1$ points. A centre of gravity for $n-1$ points is $-\frac{1}{n-1}$ of the distance from the centre of gravity of all n to the remaining point. These n centres of gravity will thus lie on a sphere of $\frac{1}{n-1}$, the given radius whose centre is $-\frac{1}{n-1}$ of the distance from the centre of gravity of all n points to the centre of the given sphere. Hence the associated points lie on a sphere of $\frac{1}{3}$ the given radius, whose centre is $\frac{n}{3}$ the distance from the centre of the given sphere to the centre of gravity, and this point lies on all n spheres. The centre lies $\frac{n}{3}$ of the distance from the centre of the given sphere to the centre of gravity of all n points.

§ 2. Coaxal Systems.

No part of the geometry of the sphere follows more closely the analogy of the geometry of the circle than the system of coaxal spheres, and allied systems.

Theorem 39.] *The locus of points having equal powers with regard to two non-concentric spheres is a plane perpendicular to their line of centres, and containing all points common to the two.*

This we defined as their *radical plane.* A system of spheres having a common radical plane shall be called a *coaxal system.*

Theorem 40.] *If three spheres be given, whereof no two are concentric, the radical planes which they determine two by two pass through a line or are parallel.*

Theorem 41.] *Given four spheres, whereof no two are concentric. The radical planes which they determine two by two are all parallel when the centres are collinear, they are parallel to one line when the centres are coplanar, and they are concurrent when the centres are not coplanar.*

We have already designated this point as the *radical centre* of the spheres, and noted that it was the centre of a sphere either cut orthogonally or in great circles by all four given spheres, unless indeed they all pass through that point.

Theorem 42.] *The numerical value of the difference of the powers of a point with regard to two non-concentric spheres is twice the product of its distance from their radical plane and the distance of their centres.*

Theorem 43.] *If a sphere be intersected by two others either orthogonally or in great circles, its centre lies in their radical plane. Such a sphere will be intersected either orthogonally or in a great circle by every sphere coaxal with the given two.*

Theorem 44.] *If the spheres of a coaxal system have no common points, they will have as limiting positions two point-spheres called the limiting points of the system. These are mutually inverse in every sphere of the system, and every sphere through them is orthogonal to all spheres of the system. A sphere orthogonal to any two spheres of the coaxal system will pass through the limiting points and be orthogonal to all.*

Theorem 45.] *The system of all spheres through a circle will be orthogonal to that of all spheres orthogonal to that circle.*

Theorem 46.] *The system of all spheres tangent to a given*

plane at a given point will be orthogonal to that of all spheres tangent to the normal to that plane at that point.

The system of all spheres orthogonal to two spheres or two planes, or a plane and a sphere, shall be called a *linear congruence.* The assemblage of all spheres cut by a given sphere or plane orthogonally or in great circles, or passing through a given point, shall be called a *linear complex.*

Theorem 47.] *Three non-coaxal spheres will belong to one linear congruence determined by them; four spheres of non-coplanar centres will belong to one linear complex determined by them.**

Theorem 48.] *The inverse of a coaxal system will be a coaxal system, a concentric system, a pencil of planes through a line, or a pencil of parallel planes.*

Theorem 49.] *The inverse of a linear congruence is a linear congruence, a bundle of concurrent planes, or a bundle of planes parallel to a line. The inverse of a linear complex is a linear complex, or the assemblage of all planes.*

Theorem 50.] *The assemblage of all spheres cutting orthogonally three given spheres with non-collinear centres is a coaxal system, and cuts orthogonally every member of the linear congruence determined by the three.*

Theorem 51.] *Two mutually inverse spheres are coaxal with their sphere of inversion.*

We have already named this a sphere of *antisimilitude.* Two spheres of unequal radius will always have at least one sphere of antisimilitude. It is called the *external* sphere of antisimilitude if its centre be the external centre of antisimilitude, otherwise it is the *internal* sphere.

Theorem 52.] *If four spheres be given, whereof no two are concentric and no three coaxal, nor do all belong to a linear congruence, the spheres through a given point each coaxal with two of the given spheres will belong to a linear congruence*

* Cf. Reye, *Synthetische Geometrie der Kugel*, Leipzig, 1879, p. 21.

and pass through a second common point which is inverse to the given one in the common orthogonal sphere to the four when such a sphere exists, and which coincides with the given point only when the four are concurrent.

Theorem 53.] *The locus of a point whose powers with regard to two given spheres have a constant ratio different from unity is a sphere coaxal or concentric with them.*

Theorem 54.] *Two spheres are coaxal with their sphere of similitude. If three spheres of unequal radius be given, no two concentric, their spheres of similitude are coaxal.*

Theorem 55.] *If four spheres be given with unequal radii and non-coplanar centres, the six spheres of similitude which they determine two by two belong to a linear congruence.*

Theorem 56.] *Given four spheres with non-coplanar centres. If there be a sphere orthogonal to all four, and a sphere which cuts them all in great circles, then these are coaxal with the spheres through their centres and orthogonal to the spheres of similitude which they determine two by two.*

Theorem 57.] *If a sphere so move that each of two given points has a constant power with regard to it, it traces a linear congruence.*

Theorem 58.] *If a sphere so move that each of three non-collinear points has a constant power with regard to it, it traces a coaxal system.*

Theorem 59.] *If a sphere so move that it cuts two given spheres in great circles, or cuts one in a great circle and the other orthogonally, it traces a linear congruence.*

Theorem 60.] *If a sphere so move that it cuts three spheres of non-collinear centres in great circles, or cuts two in great circles and one orthogonally, or one in a great circle and two orthogonally, it will trace a coaxal system.*

Theorem 61.] *If four mutually external spheres with non-coplanar centres be given, there is a sphere cutting each set of three orthogonally and the fourth in a great circle, and a sphere cutting any three in great circles and the fourth orthogonally.*

We have already in 29] noticed the fundamental property of the radical sphere of two given spheres. It is the sphere coaxal with them whose centre is mid-way between theirs.

Theorem 62.] *Given two non-concentric spheres. If there be a sphere coaxal with them whose centre is the reflection of the centre of the first in that of the second, then this third sphere will cut in a great circle all spheres orthogonal to the first whose centres lie on the second, and will cut orthogonally all spheres cut by the first in great circles whose centres lie on the second.*

We may sharpen our concept of the angle of two spheres, exactly as we did in the case of circles, by starting from the formula

$$\cos\theta = \frac{r^2 + r'^2 - d^2}{2\,rr'}\cdot \tag{2}$$

Theorem 63.] *If a variable sphere cut two given spheres at fixed angles, it will cut also at a fixed angle every sphere concentric or coaxal with them.*

Theorem 64.] *All spheres of a coaxal system will cut at equal or supplementary angles two spheres which cut two spheres of the system at equal or supplementary angles.*

Theorem 65.] *If a sphere intersect two others which are non-concentric and of unequal radii, the circles of intersection are in perspective from the external centre of similitude, while if it intersect them at supplementary angles, these circles are in perspective at the internal centre of similitude.*

Theorem 66.] *If a sphere intersect two others of unequal radii orthogonally, the circles of intersection are in perspective from both centres of similitude.*

Theorem 67.] *If two spheres of unequal radii intersect, all spheres cutting them at equal angles are orthogonal to the external sphere of antisimilitude, all cutting them at supplementary angles are orthogonal to the internal sphere of similitude. The first statement remains true when the spheres are mutually external, the second when one surrounds the other.*

Theorem 68.] *If each of two non-concentric and unequal spheres intersect each of two other such spheres at the same angle, the external centre of similitude of each pair lies in the radical plane of the other. If each sphere of one pair meet each of the other at supplementary angles, the internal centre of similitude of each pair lies in the radical plane of the other.*

Theorem 69.] *If three unequal spheres be given, passing through two common points, the three external spheres of antisimilitude which they determine two by two are coaxal, as are each external and the remaining two internal spheres of antisimilitude.*

Theorem 70.] *If four unequal spheres of non-coplanar centres be given, each two intersecting, the spheres cutting all at equal angles form a coaxal system, as do those cutting one in angles supplementary to the angles cut from the other three, and those cutting two in angles supplementary to the angles cut from the other two.*

Theorem 71.] *If five unequal spheres be given, no four with coplanar centres, but each two intersecting, there is at most one sphere cutting all at equal angles, five cutting one at angles supplementary to those cut from the other four, and ten cutting two at angles supplementary to those cut from the other three.*

The construction of these spheres depends on finding spheres of antisimilitude and spheres of a given coaxal system cutting a given sphere at a given angle.

Theorem 72.] *If a sphere touch two others with like contact, the line connecting the points of contact passes through the external centre of similitude, or the common centre, or is parallel to the line of centres; if it have opposite contacts with the two, it passes through the internal centre of similitude, or the common centre.*

Theorem 73.] *If two non-concentric unequal spheres touch two other such spheres, a centre of similitude of each pair will lie in the radical plane of the other.*

Theorem 74.] *If four mutually external spheres of non-coplanar centres be given, there are sixteen spheres which touch all. These fall into eight pairs. To find the points of contact of one pair with one of the given spheres we have but to connect the radical centre of the four with the pole with regard to that sphere of a plane not through three centres of given spheres, but containing six centres of similitude.**

Theorem 75.] *If two spheres be inverted from any point on their sphere of antisimilitude, but not on them, their inverses will be equal, and conversely.*

It is clear that much remains to be done in the elementary geometry of the sphere to bring it to a level with that of the circle. Leaving aside the fact that the geometry of the tetrahedron lags far behind that of the triangle, the two most important deficiencies are in the theorems about chains of concurrent spheres and cospherical points, and contact theorems. The twelve- and sixteen-point spheres are far less known than the nine-point circle; is there an analogue to Feuerbach's theorem? Above all what corresponds to the Hart systems? What is the proper analogue of Malfatti's problem, and how is it solved? These difficult but important and interesting questions offer ample scope for serious work.

The following theorems came to the Author's attention too late for insertion in place.†

Theorem 76.] *If a sphere be inscribed in a tetrahedron, the lines connecting each point of contact with the adjacent vertices make the same three angles in each case.*

Theorem 77.] *If a tetrahedron be inscribed in a sphere, the three angles made by three concurrent face-planes with the corresponding tangent planes are the same in each case.*

The proof of the first is immediate, the second comes by inversion.

* This is, of course, the analogue of Gergenne's construction.

† For a history and extension of these theorems, see Neuberg, 'Ueber die Berührungskugeln des Tetraeders', *Jahresbericht der Deutschen Mathematiker-vereinigung*, vol. 16, 1907.

CHAPTER VI

THE SPHERE IN CARTESIAN GEOMETRY

§ 1. Coordinate Systems.

ALL figures discussed in the present chapter are supposed to lie in a three-dimensional space of Euclidean measurement, rendered a perfect continuum by the adjunction of the plane at infinity; in other words, the set of points in one to one correspondence with the homogeneous complex coordinate values

$$x : y : z : t,$$

where $\frac{x}{t}, \frac{y}{t}, \frac{z}{t}$ are rectangular cartesian coordinates. Before, however, taking up the detailed study of spheres in this space, let us glance for just a moment at the application of tetrahedral coordinates to the study of the sphere. Starting with a tetrahedron of reference whose face-planes have the equations

$$\cos\alpha_i \frac{x}{t} + \cos\beta_i \frac{y}{t} + \cos\gamma_i \frac{z}{t} - \pi_i = 0, \quad i = 1, 2, 3, 4,$$

we take for our tetrahedral point coordinates the four quantities

$$p_i \equiv -\left(\cos\alpha_i \frac{x}{t} + \cos\beta_i \frac{y}{t} + \cos\gamma_i \frac{z}{t} - \pi_i\right). \tag{1}$$

The vertices of the tetrahedron being A_i, the altitudes h_i, and the edges d_{ij},

$$\sum_{i=1}^{i=4} \frac{p_i}{h_i} \equiv 1. \tag{2}$$

Let us find the equation of the circumscribed sphere.* The section of this sphere by the plane $p_l = 0$ has the trilinear equation

$$\sum_{ij} d_{ij} p_i' p_j' = 0,$$

$$\sum_{ij} d_{ij}^2 d_{jk} d_{ki} p_i' p_j' = 0,$$

$$\sum_{ij} \frac{d_{ij}^2}{(A_i Ha_i)(A_j Ha_j)} p_i' p_j' = 0.$$

The two terms in the denominator are altitudes of the triangle. But for any point in this plane

$$\frac{p_i}{p_i'} = \frac{h_i}{(A_i Ha_i)}.$$

The equation of the circle is

$$\sum_{ij} \frac{d_{ij}^2 p_i p_j}{h_i h_j}.$$

Hence we have the required equation of the circumscribed sphere,

$$\sum_{i,j=1}^{i,j=4} \frac{d_{ij}^2}{h_i h_j} p_i p_j = 0. \tag{3}$$

The equation of any sphere may thus be written

$$\sum_{i,j=1}^{i,j=4} \frac{d_{ij}^2}{h_i h_j} p_i p_j + \sum_{i=1}^{i=4} u_i p_i \sum_{i=1}^{i=4} \frac{p_i}{h_i} = 0. \tag{4}$$

The conditions that the general quadric

$$\sum_{i,j=1}^{i,j=4} a_{ij} p_i p_j = 0$$

* Salmon, loc. cit., pp. 201, 202.

should be a sphere are

$$a_{ii} = \rho\frac{u_i}{h_i}, \quad a_{ij}+a_{ji} = \rho\frac{d_{ij}^2+u_ih_i+u_jh_j}{h_ih_j},$$

$$h_i^2a_{ii}+h_j^2a_{jj}-h_ih_j(a_{ij}+a_{ji}) = kd_{ij}^2. \qquad (5)$$

If a tetrahedron be self-conjugate with regard to a sphere, the altitude lines must be concurrent, i.e. it must be an orthogonal tetrahedron, and the centre of the sphere will be the orthocentre. Conversely, if we start with an orthogonal tetrahedron, the orthocentre is orthocentre for every triangle whose vertices are two vertices of the tetrahedron, and the common foot of two face-altitudes in the opposite face-planes. Hence the product of the distances from the orthocentre to each vertex and the opposite face-plane is the same, i.e. the orthocentre is the centre of a sphere, real or imaginary, with regard to which the tetrahedron is self-conjugate.

Theorem 1.] *The sphere with regard to which an orthogonal tetrahedron is self-conjugate is a sphere of antisimilitude for the circumscribed and the first twelve-point sphere.*

We leave the subject of tetrahedral coordinates with these brief indications, and return to the homogeneous cartesian form. We shall define as a *sphere* every locus whose equation is of the type

$$x_0 i(x^2+y^2+z^2+t^2)+x_1(x^2+y^2+z^2-t^2)$$
$$+x_2(2xt)+x_3(2yt)+x_4(2zt) = 0. \qquad (6)$$

The quantities (x) may take all values, real or imaginary, provided they are not all simultaneously zero. They shall be called the *coordinates of the sphere.* Under the group of conformal collineations, we have the following types of sphere:

(*a*) proper spheres

$$(xx) \neq 0, \quad ix_0+x_1 \neq 0;$$

(*b*) non-minimal plane spheres

$$(xx) \neq 0, \quad ix_0+x_1 = 0;$$

(*c*) non-planar null spheres

$$(xx) = 0, \quad ix_0 + x_1 \neq 0:$$

these are spheres of zero radius;

(*d*) planar null spheres

$$(xx) = 0, \quad ix_0 + x_1 = 0:$$

these are planes tangent to the circle at infinity, except in the one case;

(*e*) plane at infinity

$$x_0 : x_1 : x_2 : x_3 : x_4 = i : 1 : 0 : 0 : 0.$$

The coefficients of the coordinates of the sphere in (*b*) shall be called the *special pentaspherical coordinates* of a point, or, rather, any five quantities proportional to them. Every finite point will have five such homogeneous coordinates, the sum of whose squares is zero. Conversely, if we have values (y) for which

$$iy_0 + y_1 \neq 0, \quad (yy) \equiv y_0^2 + y_1^2 + y_2^2 + y_3^2 + y_4^2 = 0,$$

we may find a corresponding finite point. The relations between homogeneous cartesian coordinates and special pentaspherical ones will be exhibited by

$$y_0 : y_1 : y_2 : y_3 : y_4$$
$$= i(x^2 + y^2 + z^2 + t^2) : (x^2 + y^2 + z^2 - t^2) : 2xt : 2yt : 2zt. \quad (7)$$

$$x : y : z : t = y_2 : y_3 : y_4 : -(iy_0 + y_1). \quad (8)$$

If our sphere (x) in (6) be non-planar, its radius will be

$$r = \frac{\sqrt{(xx)}}{ix_0 + x_1}. \quad (9)$$

We shall give to the radical such a sign, in the case of a real sphere, that this expression is positive. The special pentaspherical coordinates of the centre are

$$\rho y_0 = x_0 - \frac{i(xx)}{2(ix_0 + x_1)}.$$

$$\rho y_1 = x_1 - \frac{(xx)}{2(ix_0 + x_1)}.$$

$$\rho y_2 = x_2. \quad (10)$$

$$\rho y_3 = x_3.$$

$$\rho y_4 = x_4.$$

The special pentaspherical coordinates of a finite point are the coordinates of the null sphere whereof it is the centre or vertex. The power of the finite point (y) with regard to the non-planar sphere (x) is

$$\frac{-2(xy)}{(iy_0+y_1)(ix_0+x_1)}, \quad (xy) \equiv \sum_{i=0}^{i=4} x_i y_i. \tag{11}$$

When the sphere becomes null but not planar, this is the square of the distance between the points (x) and (y). If the sphere be proper, and we divide the power by the radius, we get

$$\frac{-2(xy)}{\sqrt{(xx)}\,(iy_0+y_1)}. \tag{12}$$

The limit of this expression as the sphere approaches the limiting form of a non-minimal plane is twice the distance from the point to that plane. Let us conserve the expression 'ratio of power to radius' even for this limiting form.

Theorem 2.] *The special pentaspherical coordinates of a point are proportional to the ratios of power to radius with regard to five mutually orthogonal not null spheres.*

If we define the cosine of the angle of two spheres as in V (2), p. 245,

$$\cos\theta = \frac{(xy)}{\sqrt{(xx)}\,\sqrt{(yy)}}, \tag{13}$$

the radicals in the denominators, in the case of real spheres, should be taken so as to give a positive sign to each radius. For mutually orthogonal spheres

$$(xy) = 0. \tag{14}$$

For tangent spheres

$$(xx)(yy) - (xy)^2 = 0. \tag{15}$$

Theorem 3.] *The assemblage of all spheres of cartesian space can be put into one to one correspondence with that of all points of a four-dimensional projective space with elliptic measurement. The angle of two not null spheres will be equal to the distance of the corresponding points. Null spheres will correspond to points of the Absolute hyperquadric. A coaxal system will correspond to points of a line, a linear congruence to points of a plane, and a linear complex to points of a hyperplane.*

If (x) and (x') be two spheres, we find one of their spheres of antisimilitude by finding (y) the sphere coaxal with them, which makes with them equal angles,

$$\rho y_i = \sqrt{(x'x')}\, x_i \pm \sqrt{(xx)}\, x_i'. \tag{16}$$

$$\sigma x_i' = (xx)\, y_i - z\,(xy)\, x_i. \tag{17}$$

The last equation will give the inverse of the sphere or point (x) in the proper sphere (y), or the reflection of (x) in the non-isotropic plane (y). If the sphere of inversion be

$$x^2+y^2+z^2 = 1,$$

the inverse of (xyz) will be

$$x' = \frac{x}{x^2+y^2+z^2}, \quad y' = \frac{y}{x^2+y^2+z^2}, \quad z' = \frac{z}{x^2+y^2+z^2}. \tag{18}$$

From these we easily find

$$\frac{dx'\,\delta x' + dy'\,\delta y' + dz'\,\delta z'}{\sqrt{dx'^2+dy'^2+dz'^2}\,\sqrt{\delta x'^2+\delta y'^2+\delta z'^2}}$$

$$= \pm \frac{dx\,\delta x + dy\,\delta y + dz\,\delta z}{\sqrt{dx^2+dy^2+dz^2}\,\sqrt{\delta x^2+\delta y^2+\delta z^2}},$$

which shows that inversion is a conformal transformation of space.

§ 2. The Identity of Darboux and Frobenius.

Suppose that we have given any two systems each of six spheres $(x)\,(y)\,(z)\,(r)\,(s)\,(t)$, $(x')\,(y')\,(z')\,(r')\,(s')\,(t')$; they will be connected by an identical relation entirely analogous to that subsisting in the case of ten coplanar circles, namely,*

* Lachlan, *On Systems of Circles*, &c., cit. Much of our work on the present identity follows this article fairly closely.

$$\begin{vmatrix} (xx') & (xy') & (xz') & (xr') & (xs') & (xt') \\ (yx') & (yy') & (yz') & (yr') & (ys') & (yt') \\ (zx') & (zy') & (zz') & (zr') & (zs') & (zt') \\ (rx') & (ry') & (rz') & (rr') & (rs') & (rt') \\ (sx') & (sy') & (sz') & (sr') & (ss') & (st') \\ (tx') & (ty') & (tz') & (tr') & (ts') & (tt') \end{vmatrix} \equiv 0. \qquad (19)$$

As a first application, let the reader prove the following:

Theorem 4.] *If five non-cospherical finite points be given whereof no four are coplanar, the sum of the reciprocals of the power of each point with regard to the sphere circumscribed to the other four is zero.*

Our formula (y) is usually more interesting when the two systems are identical. For instance, if we take five proper spheres and the plane at infinity,

$$\begin{vmatrix} 1 & \cos\measuredangle xy & \cos\measuredangle xz & \cos\measuredangle xr & \cos\measuredangle xs & \frac{1}{r_x} \\ \cos\measuredangle yx & 1 & \cos\measuredangle yz & \cos\measuredangle yr & \cos\measuredangle ys & \frac{1}{r_y} \\ \cos\measuredangle zx & \cos\measuredangle zy & 1 & \cos\measuredangle zr & \cos\measuredangle zs & \frac{1}{r_z} \\ \cos\measuredangle rx & \cos\measuredangle ry & \cos\measuredangle rz & 1 & \cos\measuredangle rs & \frac{1}{r_r} \\ \cos\measuredangle sx & \cos\measuredangle sy & \cos\measuredangle sz & \cos\measuredangle sr & 1 & \frac{1}{r_s} \\ \frac{1}{r_x} & \frac{1}{r_y} & \frac{1}{r_z} & \frac{1}{r_r} & \frac{1}{r_s} & 0 \end{vmatrix} = 0. \qquad (20)$$

The distances of any five finite points will be connected by the relation

$$\begin{vmatrix} 0 & d_{12}^2 & d_{13}^2 & d_{14}^2 & d_{15}^2 & 1 \\ d_{21}^2 & 0 & d_{23}^2 & d_{24}^2 & d_{25}^2 & 1 \\ d_{31}^2 & d_{32}^2 & 0 & d_{34}^2 & d_{35}^2 & 1 \\ d_{41}^2 & d_{42}^2 & d_{43}^2 & 0 & d_{45}^2 & 1 \\ d_{51}^2 & d_{52}^2 & d_{53}^2 & d_{54}^2 & 0 & 1 \\ 1 & 1 & 1 & 1 & 1 & 0 \end{vmatrix} = 0. \qquad (21)$$

The common orthogonal sphere to four given spheres will be

$$\rho x_i = \frac{\partial}{\partial t_i} \mid tyzrs \mid. \tag{22}$$

This will be null if

$$\begin{vmatrix} (yy) & (yz) & (yr) & (ys) \\ (zy) & (zz) & (zr) & (zs) \\ (ry) & (rz) & (rr) & (rs) \\ (sy) & (sz) & (sr) & (ss) \end{vmatrix} = 0. \tag{23}$$

(y) (z) (r) (s) will belong to one linear congruence if the following matrix have rank (3)

$$\left\| \begin{matrix} y_0 & y_1 & y_2 & y_3 & y_4 \\ z_0 & z_1 & z_2 & z_3 & z_4 \\ r_0 & r_1 & r_2 & r_3 & r_4 \\ s_0 & s_1 & s_2 & s_3 & s_4 \end{matrix} \right\|. \tag{24}$$

If (y) (z) (r) (s) (t) belong to a linear complex,

$$\mid yzrst \mid = 0.$$

Squaring, we get

$$\begin{vmatrix} (yy) & (yz) & (yr) & (ys) & (yt) \\ (zy) & (zz) & (zr) & (zs) & (zt) \\ (ry) & (rz) & (rr) & (rs) & (rt) \\ (sy) & (sz) & (sr) & (ss) & (st) \\ (ty) & (tz) & (tr) & (ts) & (tt) \end{vmatrix} = 0. \tag{25}$$

If (y), (z), (r), (s) be four non-collinear and non-concyclic finite points, (x) the sphere or plane through them, of radius r, and (t) the plane at infinity,

$$\begin{vmatrix} 0 & {d_{12}}^2 & {d_{13}}^2 & {d_{14}}^2 & \sqrt{2} & 0 \\ {d_{21}}^2 & 0 & {d_{23}}^2 & {d_{24}}^2 & \sqrt{2} & 0 \\ {d_{31}}^2 & {d_{32}}^2 & 0 & {d_{34}}^2 & \sqrt{2} & 0 \\ {d_{41}}^2 & {d_{42}}^2 & {d_{43}}^2 & 0 & \sqrt{2} & 0 \\ \sqrt{2} & \sqrt{2} & \sqrt{2} & \sqrt{2} & 0 & 1 \\ 0 & 0 & 0 & 0 & 1 & -2^2 \end{vmatrix} = 0.$$

Remembering that if V be the volume of the tetrahedron whose vertices are

$$(x_1 y_1 z_1)\ (x_2 y_2 z_2)\ (x_3 y_3 z_3)\ (x_4 y_4 z_4)$$

$$6V = \begin{vmatrix} x_1 & y_1 & z_1 & 1 \\ x_2 & y_2 & z_2 & 1 \\ x_3 & y_3 & z_3 & 1 \\ x_4 & y_4 & z_4 & 1 \end{vmatrix} = \begin{vmatrix} x_1^2+y_1^2+z_1^2 & x_1 & y_1 & z_1 & 1 \\ x_2^2+y_2^2+z_2^2 & x_2 & y_2 & z_2 & 1 \\ x_3^2+y_3^2+z_3^2 & x_3 & y_3 & z_3 & 1 \\ x_4^2+y_4^2+z_4^2 & x_4 & y_4 & z_4 & 1 \\ 1 & 0 & 0 & 0 & 0 \end{vmatrix}.$$

If we write

$$2\sigma = d_{12}d_{34} + d_{13}d_{42} + d_{14}d_{23},$$

$$r = \frac{\sqrt{\sigma(\sigma - d_{12}d_{34})(\sigma - d_{13}d_{42})(\sigma - d_{14}d_{23})}}{6V}.*$$

If our five spheres be mutually orthogonal but proper,

$$\sum_{i=1}^{i=5} \frac{1}{r_i^2} = 0. \tag{26}$$

Theorem 5.] *The sum of the squares of the reciprocals of the radii of five mutually orthogonal proper spheres is zero.*

If s_i be the ratio of power to radius with regard to the ith sphere

$$\begin{vmatrix} 1 & 0 & 0 & 0 & 0 & -s_1 \\ 0 & 1 & 0 & 0 & 0 & -s_2 \\ 0 & 0 & 1 & 0 & 0 & -s_3 \\ 0 & 0 & 0 & 1 & 0 & -s_4 \\ 0 & 0 & 0 & 0 & 1 & -s_5 \\ -s_1 & -s_2 & -s_3 & -s_4 & -s_5 & 0 \end{vmatrix} = 0.$$

$$\sum_{i=1}^{i=5} s_i^2 = 0. \tag{27}$$

Theorem 6.] *The sum of the squares of the ratios of power to radius for a finite point with regard to any five mutually orthogonal not null spheres is zero.*

* Salmon, loc. cit., p. 37.

These five ratios, or rather, any five numbers proportional to them, shall be defined as *general pentaspherical coordinates* of the point.*

Theorem 7.] *The passage from one set of pentaspherical coordinates to another is effected by means of a quinary orthogonal substitution. The equation of a sphere will be linear in every set of pentaspherical coordinates, and the expressions for the angle of two not null spheres, the inverse of one sphere in another, and the condition that a sphere should be null are invariant in form.*

If two spheres be orthogonal to three others, the line of centres of the two is orthogonal to the plane of centres of the three, since this is the radical plane of the two. Conversely, suppose that we have an orthogonal point system. Each point is the orthocentre of the tetrahedron whose vertices are the other four points, and so, as we saw recently, is the centre of a sphere with regard to which the tetrahedron is self-conjugate. Any two of these spheres will meet the plane through the centres of the other three in the circle where that plane meets the sphere whose diameter is the segment joining the two points, and the two will cut orthogonally there.

Theorem 8.] *The centres of five mutually orthogonal spheres form an orthogonal system and, conversely, every orthogonal system will yield the centres of five mutually orthogonal spheres.*

Four proper non-concurrent spheres will determine sixteen spherical tetrahedra, each having its own circumscribed sphere. If (y), (z), (s), (t) be the four spheres, (y'), (z'), (s'), (t') the vertices of such a tetrahedron, (x) a circumscribed sphere, (w) the common orthogonal sphere to the original four, while (y'') is orthogonal to (z), (s), (t), (w), we may follow exactly the steps that led to II. 12], getting

* We might, of course, take any five spheres not belonging to a linear complex and get still more general coordinates with a more complicated quadratic relation.

Theorem 9.] *The spheres circumscribed to the sixteen tetrahedra formed by four non-concurrent proper spheres cut at equal or supplementary angles the four spheres, each of which is orthogonal to three of the given spheres, and to their common orthogonal sphere.*

If five spheres touch one another externally, we have

$$\begin{vmatrix} 1 & -1 & -1 & -1 & -1 & \frac{1}{r_1} \\ -1 & 1 & -1 & -1 & -1 & \frac{1}{r_2} \\ -1 & -1 & 1 & -1 & -1 & \frac{1}{r_3} \\ -1 & -1 & -1 & 1 & -1 & \frac{1}{r_4} \\ -1 & -1 & -1 & -1 & 1 & \frac{1}{r_5} \\ \frac{1}{r_1} & \frac{1}{r_2} & \frac{1}{r_3} & \frac{1}{r_4} & \frac{1}{r_5} & 0 \end{vmatrix} = 0. \tag{28}$$

$$3\sum_{i=1}^{i=5} \frac{1}{r_1^{\,2}} = \left(\sum_{i=1}^{i=5} \frac{1}{r_i}\right)^2. \tag{29}$$

If a sphere meet five others either at $\measuredangle\,\phi$ or $\measuredangle\,\pi-\phi$,

$$\begin{vmatrix} 1 & \cos\measuredangle\,yz & \cos\measuredangle\,yr & \cos\measuredangle\,ys & \cos\measuredangle\,yt & \epsilon_1\cos\phi \\ \cos\measuredangle\,zy & 1 & \cos\measuredangle\,zr & \cos\measuredangle\,zs & \cos\measuredangle\,zt & \epsilon_2\cos\phi \\ \cos\measuredangle\,ry & \cos\measuredangle\,rz & 1 & \cos\measuredangle\,rs & \cos\measuredangle\,rt & \epsilon_3\cos\phi \\ \cos\measuredangle\,sy & \cos\measuredangle\,sz & \cos\measuredangle\,sr & 1 & \cos\measuredangle\,st & \epsilon_4\cos\phi \\ \cos\measuredangle\,ty & \cos\measuredangle\,tz & \cos\measuredangle\,tr & \cos\measuredangle\,ts & 1 & \epsilon_5\cos\phi \\ \epsilon_1\cos\phi & \epsilon_2\cos\phi & \epsilon_3\cos\phi & \epsilon_4\cos\phi & \epsilon_5\cos\phi & 1 \end{vmatrix} = 0, \tag{30}$$

$$\epsilon_1^{\,2} = \epsilon_2^{\,2} = \epsilon_3^{\,2} = \epsilon_4^{\,2} = \epsilon_5^{\,2} = 1.$$

If five spheres be tangent to a sixth we get the analogue of Casey's criterion,

$$\begin{vmatrix} 0 & t_{12}^2 & t_{13}^2 & t_{14}^2 & t_{15}^2 \\ t_{21}^2 & 0 & t_{23}^2 & t_{24}^2 & t_{25}^2 \\ t_{31}^2 & t_{32}^2 & 0 & t_{34}^2 & t_{35}^2 \\ t_{41}^2 & t_{42}^2 & t_{43}^2 & 0 & t_{45}^2 \\ t_{51}^2 & t_{52}^2 & t_{53}^2 & t_{54}^2 & 0 \end{vmatrix} = 0. \tag{31}$$

If we take four not null spheres, a sphere tangent internally to them, and a point thereon, then, if p_i be the ratio of power to radius with regard to the ith sphere,

$$\begin{vmatrix} 0 & \sin^2\frac{1}{2}\measuredangle yz & \sin^2\frac{1}{2}\measuredangle yr & \sin^2\frac{1}{2}\measuredangle ys & p_1 \\ \sin^2\frac{1}{2}\measuredangle zy & 0 & \sin^2\frac{1}{2}\measuredangle zr & \sin^2\frac{1}{2}\measuredangle zs & p_2 \\ \sin^2\frac{1}{2}\measuredangle ry & \sin^2\frac{1}{2}\measuredangle rz & 0 & \sin^2\frac{1}{2}\measuredangle rs & p_3 \\ \sin^2\frac{1}{2}\measuredangle sy & \sin^2\frac{1}{2}\measuredangle sz & \sin^2\frac{1}{2}\measuredangle sr & 0 & p_4 \\ p_1 & p_2 & p_3 & p_4 & 0 \end{vmatrix} = 0. \tag{32}$$

From this we may derive the tetrahedral equation of the inscribed sphere to a given tetrahedron. If a sphere meet four others at angles $\alpha_1, \alpha_2, \alpha_3, \alpha_4$, while its radius is r, we find

$$\begin{vmatrix} 1 & \cos\measuredangle yz & \cos\measuredangle ys & \cos\measuredangle yt & \cos\alpha_1 & \frac{1}{r_1} \\ \cos\measuredangle zy & 1 & \cos\measuredangle zs & \cos\measuredangle zt & \cos\alpha_2 & \frac{1}{r_2} \\ \cos\measuredangle sy & \cos\measuredangle sz & 1 & \cos\measuredangle st & \cos\alpha_3 & \frac{1}{r_3} \\ \cos\measuredangle ty & \cos\measuredangle tz & \cos\measuredangle ts & 1 & \cos\alpha_4 & \frac{1}{r_4} \\ \cos\alpha_1 & \cos\alpha_2 & \cos\alpha_3 & \cos\alpha_4 & 1 & \frac{1}{r} \\ \frac{1}{r_1} & \frac{1}{r_2} & \frac{1}{r_3} & \frac{1}{r_4} & \frac{1}{r} & 0 \end{vmatrix} = 0. \tag{33}$$

The equation in $\frac{1}{r}$ has real roots by II (47) if

$$\begin{vmatrix} 1 & \cos\measuredangle yz & \cos\measuredangle ys & \cos\measuredangle yt & \cos\alpha_1 \\ \cos\measuredangle zy & 1 & \cos\measuredangle zs & \cos\measuredangle zt & \cos\alpha_2 \\ \cos\measuredangle sy & \cos\measuredangle sz & 1 & \cos\measuredangle st & \cos\alpha_3 \\ \cos\measuredangle ty & \cos\measuredangle tz & \cos\measuredangle ts & 1 & \cos\alpha_4 \\ \cos\alpha_1 & \cos\alpha_2 & \cos\alpha_3 & \cos\alpha_4 & 1 \end{vmatrix}$$

$$\times \begin{vmatrix} 1 & \cos\measuredangle yz & \cos\measuredangle ys & \cos\measuredangle yt & \frac{1}{r_1} \\ \cos\measuredangle zy & 1 & \cos\measuredangle zs & \cos\measuredangle zt & \frac{1}{r_2} \\ \cos\measuredangle sy & \cos\measuredangle sz & 1 & \cos\measuredangle st & \frac{1}{r_3} \\ \cos\measuredangle ty & \cos\measuredangle tz & \cos\measuredangle ts & 1 & \frac{1}{r_4} \\ \frac{1}{r_1} & \frac{1}{r_2} & \frac{1}{r_3} & \frac{1}{r_4} & 0 \end{vmatrix} \geqq 0.$$

The second factor is

$$\frac{|\, y\, z\, s\, t\, \omega\, |}{(yy)\,(zz)\,(ss)\,(tt)}; \qquad \omega_i : \omega_1 : \omega_2 : \omega_3 : \omega_4 = i : 1 : 0 : 0 : 0,$$

and is essentially negative or zero for real spheres. Hence the condition for a real sphere cutting four real proper spheres with non-coplanar centres at given real angles is

$$\begin{vmatrix} 1 & \cos\measuredangle yz & \cos\measuredangle zs & \cos\measuredangle zt & \cos\alpha_1 \\ \cos\measuredangle zy & 1 & \cos\measuredangle zs & \cos\measuredangle zt & \cos\alpha_2 \\ \cos\measuredangle sy & \cos\measuredangle sz & 1 & \cos\measuredangle st & \cos\alpha_3 \\ \cos\measuredangle ty & \cos\measuredangle tz & \cos\measuredangle ts & 1 & \cos\alpha_4 \\ \cos\alpha_1 & \cos\alpha_2 & \cos\alpha_3 & \cos\alpha_4 & 1 \end{vmatrix} \leqq 0. \qquad (34)$$

The equation of the sphere which touches the spheres (y), (z), (r), (s), four non-concurrent proper spheres, is

$$\begin{vmatrix} 1 & \cos\measuredangle yz & \cos\measuredangle yr & \cos\measuredangle ys & \dfrac{(yx)}{\sqrt{(yy)}} \\ \cos\measuredangle zy & 1 & \cos\measuredangle zr & \cos\measuredangle zs & \dfrac{(zx)}{\sqrt{(zz)}} \\ \cos\measuredangle ry & \cos\measuredangle rz & 1 & \cos\measuredangle rs & \dfrac{(rx)}{\sqrt{(rr)}} \\ \cos\measuredangle sy & \cos\measuredangle sz & \cos\measuredangle sr & 1 & \dfrac{(sx)}{\sqrt{(ss)}} \\ \epsilon_1 & \epsilon_2 & \epsilon_3 & \epsilon_4 & 0 \end{vmatrix}$$

$$\sqrt{\begin{vmatrix} 1 & \cos\measuredangle yz & \cos\measuredangle yr & \cos\measuredangle ys & \epsilon_1 \\ \cos\measuredangle zy & 1 & \cos\measuredangle zr & \cos\measuredangle zs & \epsilon_2 \\ \cos\measuredangle ry & \cos\measuredangle rz & 1 & \cos\measuredangle rs & \epsilon_3 \\ \cos\measuredangle sy & \cos\measuredangle sz & \cos\measuredangle sr & 1 & \epsilon_4 \\ \epsilon_1 & \epsilon_2 & \epsilon_3 & \epsilon_4 & 1 \end{vmatrix}}\;\frac{|\,x\,y\,z\,r\,s\,|}{\sqrt{(yy)}\,\sqrt{(zz)}\,\sqrt{(rr)}\,\sqrt{(ss)}} = 0. \qquad (35)$$

$$\epsilon_1^{\,2} = \epsilon_2^{\,2} = \epsilon_3^{\,2} = \epsilon_4^{\,2} = 1.$$

Two spheres, tangent to four given non-concurrent proper spheres, are said to form a *couple* if they be mutually inverse in the common orthogonal sphere of the four. Evidently, in the construction given at the close of the last chapter, two such spheres will correspond to the same plane containing six centres of similitude; or, in the equation above, the spheres of a couple correspond to the same sets of values for the ϵ_i's, and differ only in the sign connecting the two terms. Let us take three spheres tangent to four not null and not concurrent spheres, no two of the three forming a couple. We easily see that the problem of finding a sphere tangent to these three and orthogonal to the common orthogonal sphere of the first four has eight solutions, corresponding to eight spheres all tangent to the inverses of the three in the common orthogonal sphere of the four.

Theorem 10.] *Any three couples of spheres tangent to four given not null and not concurrent spheres will touch four other spheres as well. There are eight such couples. There are also twelve tetrads of pairs of spheres each tangent to the given* *spheres and to the inverses of two in a sphere of antisimilitude of the other two.**

Such systems correspond in a measure to the Hart systems of the second sort of Ch. II. Does any figure in the geometry of the sphere correspond to the Hart system of the first sort? This most interesting question is still to be answered.

Let us give one theorem about cospherical points.†

Theorem 11.] *If five points, no four of which are concyclic, lying on a not null sphere, be arranged in sequence, and any five spheres be constructed, each through three successive points, the five remaining intersections, each of three successive spheres, are cospherical.*

The points shall be P_1, P_2, P_3, P_4, P_5, the original sphere s. The sphere constructed through P_i, P_j, P_k shall be s_{ijk}. The successive spheres $s_{lmi}, s_{mij}, s_{ijk}$ will meet in P_i and a second point P_i'. Consider the surface

$$\lambda_1 s_{451} s_{123} + \lambda_2 s_{512} s_{234} + \lambda_3 s_{123} s_{345} + \lambda_4 s_{234} s_{451} + \lambda_5 s_{345} s_{512} = 0.$$

This is a quartic with the circle at infinity as a double curve, and containing all ten points $P_i P_j'$. The various terms are not usually linearly dependent, as we see from a special case; hence, by varying the coefficients, we have apparently a four-parameter family of cyclics on s. Since, however, the system of cyclics through seven points have an eighth common point also by IV. 9], when eight points are fixed we still have two degrees of freedom for our surface. Hence we may choose such a value for the λ's that the surface includes s as part of itself. The remainder will be a sphere through the points P_j'. When the terms are linearly dependent we prove by continuity.

§ 3. Analytic Systems of Spheres.‡

A system of spheres whose coordinates are proportional to analytic functions of a single parameter, not all having

* Cf. Schubert, 'Eine geometrische Eigenschaft, &c.', *Zeitschrift für Mathematik und Physik*, vol. xiv, 1867, p. 506.

† E. Müller, 'Die Kugelgeometrie nach den Principien der Grassmannschen Ausdehnungslehre', *Monatshefte für Math.*, vol. iv, 1893, p. 35. Also the Author's *Circles Associated*, cit.

‡ For an admirable elementary account of systems of spheres see

constant ratios, shall be called a *series*. The simplest series is the coaxal system of pencil.

Theorem 12.] *If three non-coaxal spheres be given, the three spheres, each coaxal with two of the given spheres, and all orthogonal to a fourth sphere, are coaxal.*

We shall not waste our time in finding coordinates of the simple spheres *coaxal* with two given spheres; the formulae II (54)–(58) suffice here also. Let us rather pass to some more interesting series. An algebraic series of which two members are orthogonal to an arbitrary sphere shall be called a *conic series*. We see that all members of a conic series must be orthogonal to the spheres of a coaxal system. We may take as typical equations of a conic series

$$(ax) = (bx) = \sum_{i,j=0}^{i,j=4} a_{ij}x_i x_j = 0. \qquad (36)$$

Theorem 13.] *The spheres of a conic series are orthogonal to two distinct or coincident null spheres.*

Of course, in the usual case, the spheres pass through two distinct points. We shall mean by a *general* conic series one where this is the case, and where, also, the series is unfactorable, and four distinct solutions are obtained by combining the three equations with the identity for all null spheres.

Theorem 14.] *The assemblage of all spheres orthogonal to two not null and not tangent spheres, and to corresponding members in two projective coaxal systems with no common member, and neither containing the fixed spheres, is a general conic series.*

Theorem 15.] *The general conic series may be generated in three different ways by spheres through two fixed points, the sum of difference of whose angles with two fixed spheres through these two points is constant.*

Döhlemann, *Geometrische Transformationen*, Leipzig, 1908, vol. ii; Peschka, *Darstellende und projektive Geometrie*, Vienna, 1884, vol. iii, pp. 192–310. Also Reye and Timerding, loc. cit.

The general conic series corresponds to the general central conic in four-dimensional projective space of elliptic measurement. More interesting is the series corresponding to a circle in this space. This is a conic with double contact with the Absolute, so here we shall consider an irreducible conic series whose null spheres fall together in pairs. This shall be called a *Dupin series*. If the spheres of the series be orthogonal to a coaxal system not entirely composed of null or tangent spheres, we shall say that such a series is *general*. We may write the equations of the general Dupin series in the form

$$(c_0x_1+c_1x_1+c_2x_2)^2-(cc)\,(xx) = x_3 = x_4 = 0. \qquad (37)$$

Let us next write

$$y_0 = \rho c_0,\ y_1 = \rho c_1,\ y_2 = \rho c_2,\ y_3 = \rho c_3+\lambda,\ y_4 = \rho c_4+\mu,$$

$$\cos^2 \measuredangle xy = \frac{(xy)^2}{(xx)\,(yy)} = \frac{\rho^2\,(cx)^2}{(xx)\,(\rho)^2\,(cc)+\lambda^2+\mu^2+2\,\lambda\rho c_3+2\,\mu\rho c_4}$$

$$= \frac{\rho^2\,(cc)}{\rho^2\,(cc)+\lambda^2+\mu^2+2\,\rho\,(\lambda c_3+\mu c_4)}.$$

If, then, we make the further restriction

$$\lambda^2+\mu^2+2\,\lambda\rho c_3 2\,\mu\rho c_4 = 0,$$

we see that every sphere (y) of a certain series is tangent to every sphere of our Dupin series. This new series is a conic series, with only two null spheres, a Dupin series, since its null spheres come from $\rho = 0$.

Theorem 16.] *A Dupin series will be generated by the totality of spheres orthogonal to those of a coaxal system including distinct null spheres, and making a fixed angle different from $\frac{\pi}{2}$ with a fixed sphere not belonging to the coaxal system, and, conversely, every such series will be a Dupin series.*

Theorem 17.] *The spheres of a Dupin series are all tangent to those of a second Dupin series.**

* Strictly speaking, we have only proved this for the general case. We see by continuity, however, that it holds in the other cases.

Two such Dupin series shall be said to be *conjugate*. Suppose, conversely, that we have three spheres (y), (z), (s) which are not coaxal, nor are they all three null. If a sphere (x) be tangent to them we have

$$\sqrt{(zz)}\,(yx) - \sqrt{(yy)}\,(zx) = \sqrt{(ss)}\,(zx) - \sqrt{(zz)}\,(sx)$$
$$= (zz)\,(xx) - (zx)^2 = 0.$$

Theorem 18.] *The assemblage of all spheres tangent to three non-coaxal spheres which are not all null, and having a fixed type of contact with each, or else the exact reverse of that type of contact with each, will be a Dupin series conjugate to the Dupin series which includes the three given spheres.*

Theorem 19.] *The assemblage of all spheres tangent to three non-coaxal not null spheres is four Dupin series. The radical axis of any three spheres of one series will contain one centre of similitude of each two of the given spheres.*

The normals to any proper sphere along one of its circles generate a cone, which is a developable surface. On the other hand, by Joachimsthal's theorem, every evolute of a circle is a single point.

Theorem 20.] *The characteristic circles of the spheres of a non-coaxal series will be lines of curvature of their envelope; and, conversely, every surface, one of whose systems of lines of curvature is composed of circles, will be the envelope of a series of spheres.*

Such a surface is called an *annular* surface. The first part of the theorem suffers an exception when the characteristic circles are null. Here there will be two sets of characteristic isotropics; they will be lines of curvature on one surface or two. Conversely, any non-developable ruled surface circumscribed to the circle at infinity is the partial envelope of a family of spheres.

Suppose, next, that we have a surface where the lines of curvature of both systems are proper circles. This may be generated in two ways by a series of spheres, and all the spheres of one series will touch all of the other. The two series must be Dupin series, and the surface, when not a cone of revolution, shall be called a Dupin cyclide.*

Theorem 21.] *The only surfaces having circles for their lines of curvature of both systems are Dupin cyclides and cones of revolution. They are the envelopes of two conjugate Dupin series.*

When the null spheres of each of two conjugate Dupin series are distinct and not planar, the Dupin cyclide shall be said to be *general.* It will have four conical points at the centres of these four null spheres.

Theorem 22.] *Not more than one pair of the conical points of the general Dupin cyclide can be real, and those of one pair lie on isotropics with those of the other. The surface is of the fourth order, and has the circle at infinity as a double curve.*

To prove the latter part of the theorem we have but to notice that the Dupin series may be written parametrically,

$$y_i = \lambda^2 a_i + 2\lambda\mu b_i + \mu^2 c_i.$$

Eliminating λ and μ from

$$\left(x\frac{\partial y}{\partial \lambda}\right) = 0, \ \left(x\frac{\partial y}{\partial \mu}\right) = 0,$$

we get an equation of the second order in our special penta-spherical coordinates. The order of the surface cannot be more than four, nor can it be less, since we have two double points whose connecting line is not embedded.

Suppose, conversely, that we have a surface of the fourth order with the circle at infinity as a double curve, and two pairs of finite conical points. A plane through two such points would cut the surface in two circles, unless the line connecting the finite conical points were part of the section, in which case it would be an isotropic line. Now each of our conical points could not be on an isotropic with each other one, as we should have triangles with finite vertices and isotropic

* Dupin, *Applications de la géométrie et de la mécanique*, Paris, 1822, pp. 200 ff.

side-lines, which is an absurdity. Hence, if A and B be two conical points not connected by an isotropic, and we invert with A as centre, the inverse surface will be a cone with its vertex at the inverse of A, two generators in each plane through AB but not containing AB as a generator, i.e. a quadric cone. Since the other two conical points of our surface do not invert into conical points of the cone, they must have been on two isotropics through A. The tangent planes to the cone will invert back into a series of spheres each tangent to the surface all along a circle. A second such series may be found from the other two conical points.

Theorem 23.] *Every surface of the fourth order with the circle at infinity as double curve and four finite conical points is a Dupin cyclide.*

Since the inverse of a Dupin series is another such series,

Theorem 24.] *Every general Dupin cyclide can be inverted into a cone of revolution.*

Theorem 25.] *Every general Dupin cyclide is anallagmatic with regard to all proper spheres of two coaxal systems.*

Theorem 26.] *The locus of the centres of the spheres of a general Dupin series is a conic.*

We see, in fact, that it must be a plane curve, since the spheres of the series are orthogonal to two spheres, and also must lie on a quadric, since the sum or difference of the distances from all its points to the centres of two chosen spheres of the conjugate series is constant.

Theorem 27.] *The assemblage of all spheres orthogonal to a given sphere, and having contact of a preassigned type with each of two other given proper and not tangent spheres not coaxal therewith, or having exactly the opposite type of contact with each of these, is a Dupin series, as is the assemblage of all spheres orthogonal to two given spheres tangent to a third not coaxal with them.*

Theorem 28.] *Every general Dupin cyclide can be inverted into an anchor ring.*

We saw a moment ago that a general Dupin cyclide is anallagmatic with regard to every sphere of each of two coaxal systems. This raises the general question, what sorts of surfaces are anallagmatic with regard to an infinite number of surfaces? Every anallagmatic surface is the envelope of ∞^2 or ∞^1 spheres orthogonal to the sphere of inversion. Our given surface could not have ∞^1 systems of ∞^2 tangent spheres, for then every sphere tangent at one point would touch the surface again, and the surface could be inverted into one that touched each plane of a parallel pencil at a different point, which is quite impossible. Hence our surface is the envelope of ∞^1 spheres, i.e. annular. These generating spheres, being orthogonal to two spheres of inversion, must belong to a linear congruence. If the surface be anallagmatic in any other spheres besides those of the coaxal system determined by two, it must be doubly annular, and so a Dupin cyclide. We thus get an excellent theorem due to Hadamard.*

Theorem 29.] *The only surfaces which are anallagmatic with regard to a one-parameter family of spheres are those annular surfaces which are generated by spheres orthogonal to the spheres of a coaxal system. The only surfaces which are anallagmatic with regard to more than one one-parameter family of spheres are Dupin cyclides and their inverses. The only surfaces which are anallagmatic with regard to a two-parameter family of spheres are spheres themselves, and these are anallagmatic with regard to ∞^3 spheres.*

If a non-degenerate central conic be given, not a circle, there is a one-parameter family of quadrics confocal therewith, i.e. inscribed in the developable tangent to this and to the circle at infinity. Four quadrics of the family, considered as envelopes, degenerate into conics, whereof one is the circle at infinity. The other three lie in three mutually perpen-

* 'Recherche des surfaces anallagmatiques par rapport à une infinité de pôles d'inversion', *Bulletin des Sciences mathématiques*, Series 2, vol. xii, 1888, p. 118.

dicular planes, each piercing the plane of another in two foci of the latter. These conics are the focal conics of the confocal system of quadrics.

Theorem 30.] *If the centre of a sphere trace a central conic, while the sphere passes through a fixed point of one of the other focal conics of the confocal system determined by the given conic, then the sphere will trace a Dupin series.*

We see, in fact, that it is a conic series, whose null spheres fall together in pairs.

The characteristic circle of a sphere of a Dupin series is the locus of its points of contact with the spheres of the conjugate series. The lines from the centre of a sphere of one series to those of the spheres of the other series will generate a cone of revolution (in the limiting case two isotropic planes). Hence the deferent (i.e. locus of centres) of each series subtends a cone of revolution at each point of the deferent of the other series. The axis of revolution will be the tangent to that deferent which passes through the vertex, for it is the perpendicular on the plane of the corresponding characteristic circle. The isotropic planes through this axis touch the other deferent, hence the vertex of the cone is a double point of the developable determined by the other deferent and the circle at infinity, i.e. the conics are focal conics of a confocal system of quadrics.

Theorem 31.] *The deferents of two conjugate Dupin series whose null spheres are not planar are two central conics, focal for a confocal system of quadrics and, conversely, any two such conics will determine ∞^1 Dupin series. Each conic subtends a cone of revolution at each point of the other, the axis of revolution being tangent to the latter. The sum or difference of the distances of every point on one conic from two points of the other will depend only on the positions of the latter.**

Remembering Joachimsthal's theorem about the evolutes of curves,

* Dupin, loc. cit., pp. 207–9.

Theorem 32.] *The tangents to all lines of curvature of one system on a Dupin cyclide where they meet a line of curvature of the other system pass through a common point of the axis of the cone of revolution of normals along this same curve.*

Theorem 33.] *A sphere through a circle of curvature of a Dupin cyclide will meet the surface again in another circle of the same system. Two circles of different systems will intersect once, and only once.*

The general Dupin cyclide has two planes of symmetry, those of its two deferents. Each will cut the surface in two circles. The circles of the other system will be orthogonal to that plane, and, since they are anallagmatic in an inversion which interchanges the given circles, will meet the plane in pairs of points collinear with a determined centre of similitude of the two circles. We thus reach a neat method of constructing the cyclide due to Cayley.*

Theorem 34.] *The circles orthogonal to the plane of two proper circles and meeting them in pairs of points collinear with a fixed centre of similitude of the two will generate a Dupin cyclide.*

Enough has now been said about the conic series: we pass to the *cubic*. This may be defined as an algebraic series whereof three members are orthogonal to an arbitrary sphere. Since any four spheres have at least one common orthogonal sphere, we see that all members of a cubic series are orthogonal to at least one sphere. We shall say that the series is *general* if there be but one fixed not null sphere to which the members of the series are orthogonal. Such a series will correspond in four dimensions to a rational non-planar cubic curve in a space of three dimensions which is not tangent to the Absolute.

Theorem 35.] *A general cubic series will be generated by the spheres orthogonal to a fixed sphere and to corresponding members of three projective coaxal systems, which have no*

* 'On the Cyclide', *Quarterly Journal of Mathematics*, vol. xii, 1873, p. 150.

common sphere, and none of which includes the fixed sphere, and, conversely, every such series will be a general cubic series.

Theorem 36.] *The centres of the spheres of a general cubic series trace a rational cubic curve.*

Theorem 37.] *The general cubic series is generated by the spheres orthogonal to a not null sphere whose centres are on a rational cubic curve. If the fixed sphere be planar, the rational cubic curve lies in that plane, and vice versa.*

Theorem 38.] *When the fixed sphere for a cubic series is null but not planar, the spheres of the series may be inverted into the tangent planes to a developable of the third class.*

Theorem 39.] *The spheres of a general cubic series cut the fixed sphere in the circles of a general cubic series.*

This series was defined in Ch. II only for coplanar circles, but the definition is immediately extended to cospherical ones.

Theorem 40.] *The spheres orthogonal to sets of three successive spheres of a general cubic series, and to the fixed sphere, generate a second general cubic series. The relation between the two is reciprocal.*

Since the general cubic series is rational, we may express it in the form

$$y_i = y_i^{(3)}(r). \tag{38}$$

We get the equation of the envelope of the spheres of the series by equating to zero the discriminant of the cubic equation $(xy) = 0$.

Theorem 41.] *The spheres of a general cubic series envelop a surface of the eighth order, anallagmatic in the fixed sphere, and having the circle at infinity as a quadruple curve.*

The coordinates of the planes of the characteristic circles of the spheres of a general cubic series are easily seen to be rational quartic functions of r; these planes are all orthogonal to the fixed sphere.

Theorem 42.] *The planes of the characteristic circles of the spheres of a general cubic series generate a rational quartic cone or cylinder.*

Definition. An algebraic series of spheres whereof four are orthogonal to an arbitrary sphere shall be called a *quartic* series. If the spheres of the series be not orthogonal to any fixed sphere, the series is said to be *general.* It will correspond to a curve of the fourth order in four dimensions which lies in no space of lower dimensions. Such a series is surely rational, for each sphere of a coaxal system orthogonal to three of its members will be orthogonal to but one other member of the series.*

Theorem 43.] *The general quartic series may be generated in ∞^{12} ways by the common orthogonal spheres to the corresponding members of four projective coaxal systems, no two of which have a common member, and, conversely, every such system of projective coaxal systems will determine a general quartic series.*

Four spheres of the series, usually distinct, are planes; eight, usually distinct, are null.

Theorem 44.] *The locus of the centres of the spheres of a general quartic series is a non-planar rational quartic curve whose asymptotic directions are perpendicular to the planes of the series.*

Theorem 45.] *The common orthogonal spheres to sets of four successive spheres of a general quartic series will generate a second such series; the relation between the two is reciprocal.*

Since the coordinates of the spheres of a general quartic series are rational quartic functions of a parameter, and the discriminant of the general quartic equation is of the sixth degree,

* For an exhaustive treatment of this series by pure geometry see Timerding, loc. cit., pp. 193 ff.

Theorem 46.] *The envelope of the spheres of a general quartic series is a surface of the twelfth order with the circle at infinity as a sextuple curve. The planes of the circles of curvature of this surface generate a rational developable of the sixth class: the osculating developable to the rational quartic curve which is the locus of the centres of the spheres orthogonal to sets of four successive spheres of the given series.*

Definition. A system of spheres whose coordinates are proportional to analytic functions of two independent variables, and whose ratios also depend on two essentially independent variables, shall be called a *congruence*. When the functions involved are all algebraic, the congruence shall be said to be algebraic. Every such congruence, if irreducible, may be expressed in the form

$$x_i = f_i(rstu), \quad \phi(rstu) = 0, \qquad (39)$$

the only functions involved being homogeneous polynomials.

Definition. An algebraic congruence whereof two members are orthogonal to two arbitrary spheres shall be called a *quadric congruence*. Consider four spheres of the congruence, which, three by three, determine linear congruences. If no one of these linear congruences be included entirely in the quadric congruence, it must share therewith a conic series. A linear congruence which includes one member of each of three such conic series will meet the congruence in a conic series, and, since we may find ∞^2 spheres of our congruence in this way, we find the whole. Hence all spheres of the congruence are orthogonal to one fixed sphere orthogonal to the first four. If the congruence include in itself a linear congruence, the remainder is also a linear congruence.

Theorem 47.] *A quadratic congruence consists either in two distinct or identical linear congruences, or else all its members are orthogonal to one sphere.*

The spheres of the quadric congruence will be represented in four dimensions by the points of a quadric surface. When the sphere to which the members of the congruence are orthogonal is not null, the coaxal systems in the congruence have no common member, and the series of null spheres have no double member, we shall say that we have a *general quadric congruence*. Such a congruence will correspond to the general central quadric of three-dimensional non-Euclidean space.

Theorem 48.] *A general quadric congruence contains two families of coaxal systems. Two systems of different families have one common sphere; no two of the same family have a common sphere.*

Theorem 49.] *A general quadric congruence may be determined in* $2 \times \infty^2$ *ways by coaxal systems, each determined by corresponding members of two projective coaxal systems with no common member.**

Theorem 50.] *The locus of the centres of the spheres of a general quadric congruence is a quadric surface.*

A general central quadric in non-Euclidean space has eight sets of circular sections, a circle being a conic with double contact with the Absolute.†

Theorem 51.] *A general quadric series may be generated in eight ways by the circles of a one-parameter family of Dupin series.*

Theorem 52.] *The spheres of a general quadric congruence cut the sphere to which all are orthogonal in the circles of a general quadric congruence.*

Strictly speaking, we have only defined such congruences in the case of coplanar circles, but the definition is immediately extended to cospherical ones.

Theorem 53.] *The spheres orthogonal to sets of three successive non-coaxal members of a general quadric congruence and to the common orthogonal sphere will generate a second such congruence. The relation between the two is reciprocal.*

* Cf. Reye, 'Lehrsätze über projektive Mannigfaltigkeiten projektiver Kugelbüscheln', &c., *Annali di Matematica*, Series 3, vol. v, 1900.

† Cf. the Author's *Non-Euclidean Geometry*, cit., pp. 157, 158.

Theorem 54.] *The radical planes which the spheres of a general quadric congruence determine with a fixed sphere envelop a quadric. When the fixed sphere is that to which all spheres of the congruence are orthogonal, the planes envelop the polar reciprocal with regard to this fixed sphere of the locus of the centres of the spheres of the congruence.*

The order of the surface enveloped by the spheres of a general quadric congruence is that of the curve where the surface meets the fixed sphere. This curve is the locus of the vertices of the null spheres of a quadric congruence, and so, by IV. 2], is a cyclic.

Theorem 55.] *The spheres of a general quadric series envelop a surface of the fourth order having the circle at infinity as a double curve. It is anallagmatic with regard to the fixed sphere.*

We shall find out a great deal more about this surface in the next chapter.

Theorem 56.] *The assemblage of all spheres meeting at given angles other than $\frac{\pi}{2}$ two not null spheres will be a quadric congruence.*

Theorem 57.] *The spheres orthogonal to a not null sphere, the sum or difference of whose angles with two not null spheres is constant, will be a quadric congruence.*

Definition. The assemblage of all spheres whose coordinates are proportional to analytic functions of three independent variables, the ratios also depending on three independent variables, shall be called a *complex*. When the functions involved are algebraic, the complex is said to be algebraic. The simplest way to express an algebraic complex is by means of a single equation

$$f(x_0 x_1 x_2 x_3 x_4) = 0, \tag{40}$$

where f is a homogeneous polynomial. Next to the linear

complex already studied, the simplest algebraic complex is the quadratic one.* This has an equation of the type

$$\sum_{i,\,j=0}^{i,\,j=4} a_{ij}x_ix_j = 0, \quad a_{ij} = a_{ji}. \tag{41}$$

If we classify these complexes under the twenty-four parameter group of linear sphere transformations we have the following types:

General complex

$$| a_{ij} | \neq 0. \tag{42}$$

Simply special complex

$$| a_{ij} | = 0, \quad \frac{\partial | a_{ij} |}{\partial a_{kl}} \not\equiv 0. \tag{43}$$

Doubly special complex

$$\frac{\partial | a_{ij} |}{\partial a_{kl}} \equiv 0, \quad \frac{\partial^2 | a_{ij} |}{\partial a_{kl}\, \partial a_{mn}} \not\equiv 0. \tag{44}$$

The other cases consist in pairs of distinct or coincident linear complexes, and need not be discussed. Starting with the general quadratic complex, we may associate each sphere (y) with the linear complex

$$\sum_{i,\,j=0}^{i,\,j=4} a_{ij}y_ix_j = 0,$$

which is called the *polar* linear complex of (y). Every linear complex will be the polar of a determinate sphere called its pole sphere.

Theorem 58.] *The polar linear complex with regard to a general quadratic complex of a sphere not belonging to that complex is the totality of all spheres harmonically separated from the given sphere by pairs of spheres of the complex.*

Theorem 59.] *A linear complex will intersect a quadratic one in a quadric congruence.*

* Loria, 'Ricerche intorno alla geometria della sfera', *Memorie della R. Accademia delle Scienze di Torino*, Series 2, vol. xxxvi, 1885; Reye, 'Ueber quadratische Kugelcomplexe', *Crelle*, vol. xcix, 1885, and 'Quadratische Kugelcomplexe', &c., *Collectanea Mathematica*, Naples, 1881.

Theorem 60.] *The general quadratic complex contains* ∞^3 *coaxal systems. Each sphere of the complex belongs to* ∞^1 *such systems, and they generate the quadratic congruence common to the given complex and the polar complex of the given sphere.*

Definition. Two spheres shall be said to be *conjugate* with regard to a general quadratic complex when each belongs to the *polar linear* complex of the other.

Theorem 61.] *The assemblage of all null spheres is a general quadratic complex. Mutually orthogonal spheres are conjugate with regard to this complex, and the polar of any sphere is the complex of spheres orthogonal thereto.*

Theorem 62.] *The planes of a quadratic complex envelop a quadric.*

Theorem 63.] *The totality of spheres, each orthogonal to a sphere of a general quadratic complex and to three infinitely near spheres, the four not belonging to a linear congruence, is a second general quadratic complex. The relation between the two is reciprocal, and each may be defined as the totality of spheres orthogonal to the various spheres of the linear complexes which are polar to the spheres of the other complex.*

More generally, if we have any complex of spheres, and if we construct a sphere orthogonal to each sphere of the complex and to three infinitely near spheres thereof which do not lie with the first in a linear congruence, then, if the totality of these new spheres be actually a complex, the original one is said to be *non-developable*, and the new complex is called its *correlative.* The relation between the two is reciprocal. It is a peculiarity of the quadratic complex that we can reach the correlative by means of polar linear complexes.

If (y) be a sphere of the complex (40), the linear complex

$$\sum_{i=0}^{i=4} \frac{\partial f}{\partial y_i} x_i = 0 \tag{45}$$

shall be called the *tangent* linear complex at the sphere

(y). The correlative complex is obtained by eliminating $y_0 y_1 y_2 y_3 y_4$ from the equations

$$z_i = \frac{\partial f}{\partial y_i}, \quad f(y_0 y_1 y_2 y_3 y_4) = 0.$$

The reciprocal nature of the relation between the two appears from the fact that the equations

$$(yz) = (ydz) = 0$$

involve also $$(zdy) = 0.$$

Theorem 64.] *Two spheres of an arbitrary coaxal system belong to a given quadratic complex; two spheres of the complex have their centres at an arbitrary point.*

Let us turn for a moment to the simply special quadratic complexes (43). We may find one sphere (z) which is conjugate to every sphere in space with regard to the complex. Its coordinates will satisfy the equation

$$\sum_{j=0}^{j=4} a_{ij} x_j = 0, \quad i = 0, 1, 2, 3, 4.$$

We shall call this the *singular sphere* of the complex. Let the reader prove:

Theorem 65.] *The simply special quadratic complex contains every coaxal system determined by the singular sphere and any other sphere of the complex. All spheres of such a coaxal system have the same polar linear complex.*

This quadratic complex is the first example of a developable complex. We see

Theorem 66.] *The correlative of a simply singular quadratic complex is a quadratic congruence.*

Two quadratic complexes which have the same null spheres shall be called *homothetic*; two, whose correlatives have the same null spheres, shall be called *confocal*. If our original complex be (41), we have for a homothetic one

$$\sum_{i,j=0}^{i,j=4} a_{ij} x_i x_j + \rho(xx) = 0.$$

Theorem 67.] *A general quadratic complex will be homothetic at most and in general with five simply special complexes. The surface which is the locus of the centres of the null spheres of the complex is of the fourth order with the circle at infinity as a double curve, and may be generated in general and in five ways as the envelope of the spheres of a quadric congruence.*

We shall not now stop to define the elusive words *in general* more explicitly, as this is the surface which we have already encountered and which we have promised to discuss in detail.

The correlative to our complex (41) is

$$\sum_{i,j=0}^{i,j=4} A_{ij}x_i x_j = 0, \quad A_{ij} = \frac{\partial \mid a_{ij} \mid}{\partial a_{ij}}.$$

The general equation for a confocal complex will thus be

$$\begin{vmatrix} A_{00}+\rho & A_{01} & A_{02} & A_{03} & A_{04} & x_0 \\ A_{10} & A_{11}+\rho & A_{12} & A_{13} & A_{14} & x_1 \\ A_{20} & A_{21} & A_{22}+\rho & A_{23} & A_{24} & x_2 \\ A_{30} & A_{31} & A_{32} & A_{33}+\rho & A_{34} & x_3 \\ A_{40} & A_{41} & A_{42} & A_{43} & A_{44}+\rho & x_4 \\ x_0 & x_1 & x_2 & x_3 & x_4 & 0 \end{vmatrix} = 0 \quad (46)$$

Theorem 68.] *An arbitrary sphere will belong to four complexes confocal with a given general quadratic complex.*

There are a good many types of cubic complexes, i. e. complexes given by an equation of the third order; only one, however, is particularly interesting.* Suppose that we have three projective coaxal systems of spheres, not belonging to a linear complex, nor have any two of them a common sphere. The assemblage of linear congruences, each determined by three corresponding spheres, will determine a complex called a *rational cubic complex*. This complex will correspond in four dimensions to the hypersurface generated by planes connecting the corresponding points of three

* Discussed without proofs by Reye, *Lehrsätze*, cit.

projective ranges in general position. To justify the name of the complex, let us note that we may express it parametrically in the form

$$x_i = \rho\,[\lambda y_i + \mu z_i + \nu t_i] + \sigma\,[\lambda y_i' + \mu z_i' + \nu t_i']. \tag{47}$$

To find the order of the complex, i.e. the number of its spheres in a given coaxal system, we adjoin the three equations

$$(ux) = (vx) = (wx) = 0.$$

Substituting for (x) we get three linear homogeneous equations in the variable ρ and σ. Equating the various discriminants to zero,

$$[(uy)\lambda + (uz)\mu + (ut)\nu]\,[(vy')\lambda + (vz')\mu + (vt')\nu] \\ -[(uy')\lambda + (uz')\mu + (ut')\nu]\,[(vy)\lambda + (vz)\mu + (vt)\nu] = 0.$$

$$[(uy)\lambda + (uz)\mu + (ut)\nu]\,[(wy')\lambda + (wz')\mu + (wt')\nu] \\ -[(uy')\lambda + (uz')\mu + (ut')\nu]\,[(wy)\lambda + (wz)\mu + (wt)\nu] = 0.$$

Here we have two homogeneous quadratic equations in the variables λ, μ, ν. One solution will be

$$(uy)\lambda + (uz)\mu + (ut)\nu = (uy')\lambda + (uz')\mu + (ut')\nu = 0.$$

This must be rejected, since it will not give a solution of all three equations in ρ/σ; the three other solutions give the three spheres required.

Remembering that the spheres of a coaxal system are orthogonal to those of a linear congruence, we see

Theorem 69.] *The rational cubic complex contains all spheres orthogonal to the various coaxal systems determined by corresponding members of three projective pencils of linear complexes, which three have no common sphere, nor have any two a common coaxal system.*

We see from (47) that every sphere of the complex lies in a linear congruence obtained by giving a fixed value to ρ/σ. On the other hand, if we take two pairs of values ρ, σ and ρ', σ', and give to the other parameters first the values λ, μ, ν, then the values λ', μ', ν', and equate the corresponding expressions for (x), we have five linear homogeneous equations in the six homogeneous variables $\lambda, \mu, \nu, \lambda', \mu', \nu'$. There are thus ∞^2 double spheres, each in two linear congruences. A coaxal system determined by two double spheres must be included entirely in the complex. If this system were not composed entirely of double spheres we should have ∞^4 coaxal systems, each sphere of the complex would lie in ∞^1 of them, and so in ∞^1 of our linear congruences, which is absurd.

Theorem 70.] *A general sphere of a rational cubic complex lies in a single linear congruence of the complex; a double sphere lies in two such congruences, and the totality of double spheres is itself a linear congruence.*

Reverting to (47), if we require (x) to be a plane, we impose one linear condition; two others are imposed by fixing two points of the plane. On the other hand, each sphere of the system belongs to ∞^1 coaxal systems thereof, each plane to one pencil of planes.

Theorem 71.] *The planes of a rational cubic complex envelop a ruled surface of the third order and class. The generators of this surface are the radical axes of the linear congruences of the complex.*

The radical axis of a linear congruence is, of course, the locus of points having like powers with regard to all spheres thereof.

Theorem 72.] *The centres of the null spheres of a rational cubic complex is a surface of the sixth order with the circle at infinity as a triple curve and with a circle of double points.*

With regard to the last statement we see that the centres of the null spheres of a linear congruence must lie on a circle.

Theorem 73.] *A sphere through the double circle of the surface meets it again in a simple circle. The planes of these simple circles are those of the centres of the spheres of the linear congruences of the complex, and each circle of the surface is cospherical with the double circle.*

We see, in fact, that if we invert with a centre on the double circle we get a quartic through the circle at infinity with a double straight line, and such a surface contains no other lines or circles.

Theorem 74.] *A sphere which meets the surface in a simple circle meets it also in a cyclic. The two intersect twice on the double circle, and twice at points where the sphere touches the surface.*

It is perfectly clear that there remains a good deal to be done in the study of spheres in cartesian space. It is hard to believe that a sufficiently intelligent use of the Frobenius identity will not settle the interesting question of the existence of Hart systems, and the relation of spheres circumscribed to spherical tetrahedra and spheres tangent to other spheres. There must surely be a great deal more in the subject of tangent spheres than has yet been found. Is there a three-dimensional analogue of Malfatti's problem, and what is the solution? It seems likely that although the Dupin series is undoubtedly the most interesting of the various conic series, yet others are worthy of further investigation. The elementary metrics of four-dimensional non-Euclidean space has never been studied in great detail, and may well include many beautiful theorems of real importance in the geometry of the sphere.

CHAPTER VII

PENTASPHERICAL SPACE

§ 1. Fundamental Definitions and Theorems.

ANY set of objects which can be put into one to one correspondence with sets of essentially distinct values of five homogeneous coordinates $x_0:x_1:x_2:x_3:x_4$, not all simultaneously zero, but connected by the relation

$$(xx) \equiv x_0^2+x_1^2+x_2^2+x_3^2+x_4^2 = 0, \tag{1}$$

shall be called *points*, and their totality a *pentaspherical space.*

The assemblage of all points (x) whose coordinates satisfy a linear equation

$$(yx) \equiv y_0x_0+y_1x_1+y_2x_2+y_3x_3+y_4x_4 = 0, \tag{2}$$

where the values (y) are not all zero, shall be called a *sphere*, to which the points (x) are said to belong, or on which they lie. The coefficients (y) shall be called the *coordinates* of the sphere. If they satisfy the identity (1) the sphere is said to be *null*, the point with the coordinates (y) is called the *vertex* of the null sphere. If (y) and (z) be two not null spheres the number θ, defined by

$$\cos\theta \equiv \frac{(yz)}{\sqrt{(yy)}\sqrt{(zz)}}, \tag{3}$$

is called their *angle*. If one possible value for the angle be $\frac{\pi}{2}$, the spheres are said to be *orthogonal* or *perpendicular*, or to *cut at right angles*. The condition for this is

$$(yz) = 0, \tag{4}$$

and when this condition is satisfied we shall call the spheres orthogonal, even when one or both are null. If a possible value for the angle be 0 or π we say that the spheres are tangent. Here the condition is

$$(yy)(zz)-(yz)^2 = 0. \tag{5}$$

The assemblage of all spheres whose coordinates are linearly dependent on those of two are said to form a *coaxal system* or *pencil*. They all contain all points common to the first two, the locus of which shall be defined as a *circle*.

If (y) lie on the null sphere whose vertex is (z), and so (z) lies on the null sphere whose vertex is (y), every sphere coaxal with (y) and (z) is null. The totality of their vertices shall be called an *isotropic*. Through each point will pass ∞^1 isotropics generating the null sphere whereof this point is the vertex. The circle common to two tangent not null spheres shall be called a *null* circle; it consists in two isotropics. If two null spheres have a common isotropic this is the totality of their intersection, and shall also be classed as a null circle.

If two null spheres be coaxal with any not null sphere, every sphere through their vertices is orthogonal to this sphere. The vertices are said to be mutually *inverse* in this sphere. The inverse of the point (x) in the sphere (y) is

$$x_i \equiv (yy)\,x_i - 2\,(xy)\,y_i. \tag{6}$$

If (x) trace a sphere, (x') will also trace a sphere, and the equation will give equally well the relation between two inverse spheres (x) and (x'). The sphere (y) is called a sphere of *antisimilitude* for the two.

Two examples of pentaspherical space will at once occur to the reader. We may take a Euclidean hypersphere in a space of four dimensions. Secondly, we may start with cartesian space, that is the finite domain, and proceed as in Ch. IV. We begin with the equations VI (8)

$$x:y:z:t \equiv x_2 : x_3 : x_4 : -(x_0+x_1),$$

$$x_0 : x_1 : x_2 : x_3 : x_4 \equiv i\,(x^2+y^2+z^2+t^2) : (x^2+y^2+z^2-t^2) :$$
$$2xt : 2yt : 2zt. \tag{7}$$

Every finite point of cartesian space will correspond to a definite point of pentaspherical space, for which $ix_0 + x_1 = 0$, and conversely. If, however, we make cartesian space a perfect complex continuum by adjoining the plane at infinity, the correspondence ceases to be unique, for all infinite cartesian points not on the circle at infinity correspond to the same point of pentaspherical space. We may extend the finite cartesian domain to be a perfect pentaspherical continuum as follows:

The set of coordinates $i:1:0:0:0$ shall be said to represent the *point at infinity*. Any other set of coordinates (y) satisfying the equations

$$iy_0+y_1=(yy)=0$$

shall be taken to represent the minimal plane

$$y_2x+y_3y+y_4z+\tfrac{1}{2}(iy_0-y_1)=0.$$

The point at infinity and the totality of such minimal planes shall be called *improper points*. By adjoining them to the finite domain the cartesian space becomes once more a perfect continuum, and obeys all the laws of pentaspherical space. The definitions of sphere, circle, angle, inversion, &c., given in Ch. V for cartesian space, and here for pentaspherical space, are entirely compatible.

If we take as our pentaspherical continuum the cartesian space rendered a perfect continuum in this fashion, the following terms are synonymous:

Sphere orthogonal to point at infinity.	Plane.
Inversion in such a sphere.	Reflexion in plane.
Null sphere whose vertex is point at infinity.	Totality of minimal planes.
Null sphere containing point at infinity.	Points of minimal plane and minimal planes parallel thereto.
Isotropic not containing point at infinity.	Minimal line.
Isotropic containing point at infinity.	Pencil of parallel minimal planes.

The points of pentaspherical space on any not null sphere

will be a tetracyclide plane, and we may take over bodily for them the definitions of Ch. IV.

We shall mean by the *cartesian equivalent* of a pentaspherical figure the following. We replace the coordinates of every proper pentaspherical point by their cartesian equivalents from (7), then render the space a perfect continuum by the adjunction of the plane at infinity.

The cartesian equivalent of a surface of order n, where the point at infinity has the multiplicity k, is an algebraic surface of order $n-k$ with the circle at infinity as a curve of order $\frac{1}{2}(n-2k)$.

We mean by the *order* of an algebraic surface in pentaspherical space the number of intersections with an arbitrary circle. When the surface is given by equating to zero a homogeneous polynomial in (x), the order is twice that of the polynomial.

§ 2. Cyclides.

The definitions of series, congruences, and complexes of spheres used in the last chapter may be carried over bodily into pentaspherical space. We thus reach the fundamental locus with which we shall be occupied in the present chapter.

Definition. The locus of the vertices of the null spheres of a general quadratic complex shall be called a *cyclide*.

The equation of a cyclide may be written

$$\sum_{i,j=0}^{i,j=4} a_{ij}x_ix_j = 0, \quad a_{ij} = a_{ji} \mid a_{ij} \mid \neq 0. \tag{8}$$

The problem of classifying all cyclides under the group of quinary orthogonal transformations is the problem of classifying pairs of quinary quadratic forms, whereof one certainly has a non-vanishing discriminant. This is best done by means of the elementary divisors of Weierstrass, exactly as we classified cyclics in Ch. IV. It will be found that there are exactly twenty-six species of cyclides under this classification: an enumeration of all, with canonical forms for their equations, would lead us altogether too far afield;* we shall therefore confine ourselves to one or two types beginning with the *general* one, i.e. that characterized by the scheme of elementary divisors

$$[1\,1\,1\,1\,1].$$

The canonical form for the equation of the general cyclide will be

$$(ax^2) = 0, \quad (xx) = 0, \quad \Pi a_i(a_j - a_k) \neq 0. \tag{9}$$

Since this equation is unaltered by a change of sign of any one of the x_i's,

Theorem 1.] *The general cyclide is anallagmatic with regard to five mutually orthogonal spheres. It is a surface of the fourth order, and is the envelope of five different quadratic congruences of spheres.*

This theorem has already been proved as VI. 67].

The five spheres shall be called the *fundamental spheres* of the cyclide. The equations of the five generating congruences are easily found. An arbitrary tangent sphere at the point (x) will have the coordinates

$$\rho y_i = (\lambda + a_i)\, x_i. \tag{10}$$

In particular, if $\quad y_i = 0,$

$$x_j = \frac{y_j}{a_j - a_i}.$$

$$\sum_{j=0}^{j=4} \frac{y_j^2}{a_j - a_i} = \sum_{j=0}^{j=4} \frac{y_j^2}{(a_j - a_i)^2} = 0, \quad j \neq i. \tag{11}$$

If (r) and (s) be mutually inverse in (y),

$$\sum_{j=0}^{j=4} \frac{(r_i s_j - r_j s_i)^2}{(a_j - a_i)} = 0, \quad a_j \neq a_i.$$

Theorem 2.] *The locus of the inverse of a given point with regard to the generating spheres of one system of a general cyclide is a cyclide with the given point as a conical point.*

* Cf. Loria, *Geometria della sfera*, cit., and Segre, 'Étude des différentes surfaces de quatrième ordre à conique double', *Math. Annalen*, vol. xxiv, 1884.

Theorem 3.] *The general cartesian cyclide is a surface of the fourth order with the circle at infinity as a double curve, and every such surface is a cyclide of some sort. In the general case it may be generated in five ways by a sphere moving orthogonally to one of five mutually orthogonal spheres, while its centre lies on a central quadric.*

The words 'in general' mean that the point at infinity shall not be on a fundamental sphere of the pentaspherical cyclide, nor yet on the surface itself.

Theorem 4.] *The intersection of a not null sphere with a cyclide is a cyclic.*

The generating spheres will cut the cyclide in cyclics with two double points, i.e. in two circles. Let us show, conversely, that if any sphere have double contact with the cyclide it will be a generating sphere of one system or another. Writing that a tangent sphere at (x) is also a tangent sphere at (x'),

$$(\lambda + a_i)\, x_i \equiv (\lambda' + a_i)\, x_i'.$$

Multiplying through by x_i' and summing, also multiplying through by x_i and summing,

$$\lambda\,(xx') + \sum_{i=0}^{i=4} a_i x_i x_i' = \lambda'(xx') + \sum_{i=0}^{i=4} a_i x_i x_i' = 0.$$

This shows that $\quad \lambda = \lambda',$

$$x_i = x_i' \text{ if } \lambda + a_i \neq 0.$$

Hence λ must take one of the five values $-a_i$, which proves our result.

These facts have a good many interesting consequences which we shall develop gradually. We begin by noticing that if we define as *a focus* of a surface the vertex of any null sphere which has double contact therewith, the foci of a general cyclide come merely from the five systems of generation. Their coordinates will be given by the equations

$$z_i = z_j^2 + z_k^2 + z_l^2 + z_m^2$$
$$= \frac{z_j^2}{a_j - a_i} + \frac{z_k^2}{a_k - a_i} + \frac{z_l^2}{a_l - a_i} + \frac{z_m^2}{a_m - a_i} = 0. \quad (12)$$

Theorem 5.] *The focal curves of a general cyclide are five cyclics, one on each of the fundamental spheres. Each cyclic meets each fundamental sphere other than its own in four foci of the focal cyclic on that sphere.*

Suppose that one focal cyclic is known. Its foci and fundamental circles are known; hence the other fundamental spheres are known. On each of these spheres we know the fundamental circles of the corresponding cyclic, and four points (on the first sphere). Hence the focal cyclics are all known.

Theorem 6.] *If two general cyclides have one focal cyclic in common, they have all five focal cyclics in common.**

A cyclide contains five pairs of systems of circles. This suggests that there may be a certain number of isotropics embedded in the surface. These isotropics will not lie on the fundamental spheres, but be inverse in pairs with regard to them. Let such an isotropic be determined by the points (x) and (x'), where

$$x_i = 0, \quad (tx') = 0, \quad t_i = 0.$$

Clearly
$$(xx') = \sum_{j=0}^{j=4} a_j x_j x_j' = 0,$$

$$x_m' = \begin{vmatrix} x_j & x_k & x_l \\ a_j x_j & a_k x_k & a_l x_l \\ t_j & t_k & t_l \end{vmatrix};$$

$$\frac{x_m'}{\sqrt{a_j a_k a_l}} = \begin{vmatrix} \dfrac{x_j}{\sqrt{a_j}} & \dfrac{x_k}{\sqrt{a_k}} & \dfrac{x_l}{\sqrt{a_l}} \\ \sqrt{a_j}x_j & \sqrt{a_k}x_k & \sqrt{a_l}x_l \\ \dfrac{t_j}{\sqrt{a_j}} & \dfrac{t_k}{\sqrt{a_k}} & \dfrac{t_l}{\sqrt{a_l}} \end{vmatrix}.$$

* That erratic genius, John Casey, in an article full of interest, 'On Cyclides', &c., *Philosophical Transactions*, vol. clxi, 1871, p. 637, seems to have held the curious idea that a cyclide shared one focal curve with each of five different systems of others. He gives the equations of all five systems, failing to note that they are really identical.

$$(x'x') = (ax'^2) = 0.$$

$$x_i = (xx) = (ax^2) = a_j a_k a_l a_m \left(\frac{1}{a} x^2\right) - a_i (a^2 x^2) \equiv 0.$$

These equations give eight values for $x_j : x_k : x_l : x_m$, each corresponding to two sets of values for (x') differing in the sign of x_i'.

Theorem 7.] *A general cyclide contains sixteen isotropics, inverse in pairs with regard to the five fundamental spheres.*

The generating spheres tangent at (x) have coordinates

$$y_j = (a_j - a_i)\, x_j.$$

Four of these will have the cross ratio

$$\frac{(a_i - a_j)(a_k - a_l)}{(a_i - a_l)(a_k - a_j)}.$$

Theorem 8.] *The generating spheres of four chosen systems tangent at any point have a constant cross ratio.*

The condition that a sphere (y) should touch the cyclide is

$$\sum_{i=0}^{i=4} \frac{y_i^2}{a_i + \lambda} = 0, \quad \sum_{i=0}^{i=4} \frac{y_i^2}{(a_i + \lambda)^2} = 0. \tag{13}$$

This may be interpreted as requiring that the discriminant of the first equation, looked upon as an equation in λ, should vanish. The equation is quartic, the degree of the discriminant is six, the coefficients being linear in y_i^2.

Theorem 9.] *Twelve spheres of an arbitrary coaxal system will touch a given general cyclide.*

Theorem 10.] *The general cartesian cyclide is of class twelve, and twelve normals pass through an arbitrary point.*

We may draw still further conclusions from the first of our equations (13). Let (y) be any sphere, and (z) a point common to it and to the cyclide. A sphere tangent to the cyclide at this point will have the coordinates

$$(a_i + \lambda)\, z_i.$$

Suppose that this sphere touches the cyclic of intersection again, say at t,

$$\rho y_i + \sigma t_i + a_i t_i \equiv (a_i + \lambda) z_i.$$

$$(yz) = (zz) = (az^2) = (yt) = (tt) = (at^2) = 0.$$

$$\sigma (zt) + \sum_{i=0}^{i=4} a_i z_i t_i = 0.$$

$$\sum_{i=0}^{i=4} a_i z_i t_i + \lambda (zt) = 0.$$

$$\sigma = \lambda.$$

$$\frac{\rho y_i}{(a_i + \lambda)} + t_i = z_i.$$

$$(yz) = (yt) = 0.$$

$$\sum_{i=0}^{i=4} \frac{y_i^2}{a_i + \lambda} = 0.$$

It thus appears that the absolute invariant of our quartic in (13) gives the fundamental cross ratios for the cyclic common to the cyclide and to (y). This absolute invariant is a constant multiple of the ratio of the cube of a relative invariant of the second degree, whose vanishing gives the equiharmonic case, to the square of a relative invariant of the third degree, whose vanishing gives the harmonic case.

Theorem 11.] *In an arbitrary coaxal system are four spheres meeting a general cyclide in equiharmonic cyclics, and six meeting it in harmonic cyclics.**

We return to the tangent sphere

$$y_i = (\lambda + a_i) x_i.$$

This will be null if

$$(a^2 x^2) = 0.$$

When this equation is satisfied, every tangent sphere at that point is null, i.e. the two tangent isotropics coincide, and we have a parabolic point.

* This theorem and the three preceding are taken direct from Darboux, *Sur une classe*, cit., pp. 280 ff.

Theorem 12.] *The locus of the parabolic points of a general cyclide is the intersection with a second cyclide having the same fundamental spheres.*

It must not be supposed that the tangent isotropic to the cyclide at a point of this curve is tangent to the curve itself.

The cyclide has covariants under the quinary orthogonal group analogous to those of the cyclic. Let our orthogonal substitution be

$$x_i = \sum_{j=0}^{j=4} b_{ij} x_j', \quad \sum_{j=0}^{j=4} b_{ij}^2 = 1, \quad \sum_{j=0}^{j=4} b_{ij} b_{kj} = 0, \quad k \neq i.$$

If the corresponding cyclides be

$$(ax^2) = 0, \quad (a'x^2) = 0,$$

$$\sum_{i=0}^{i=4} a_i b_{ij} b_{kj} = 0.$$

$$a_j' = \sum_{i=0}^{i=4} a_i b_{ij}^2.$$

$$\sum_{i=0}^{i=4} a_i' = \sum_{i,j=0}^{i,j=4} a_i b_{ij}^2 = \sum_{i=0}^{i=4} a_i.$$

If thus
$$\sum a_i = 0$$
we have also
$$\sum a_i' = 0.$$

We may always suppose the first of these equations is satisfied by replacing the first of our equations (9) by a suitable linear combination of the two. If (y) be any sphere, we have the covariant *polar* sphere

$$\rho z_i = a_i y_i.$$

If (x) and (x') be any two points of the cyclide inverse with regard to (y), the other sphere orthogonal to (y) in which (x) and (x') are also inverse is orthogonal to the polar sphere. The covariance of the polar sphere is thus evident.

All spheres orthogonal to (y) will have their polar spheres orthogonal to the *antipolar* sphere of (y),

$$\rho r_i = \frac{y_i'}{a_i}.$$

This also is covariant, as we see by its definition. The locus of points whose polar spheres are null is our previous cyclide

$$(a^2 x^2) = 0.$$

The locus of points lying on their antipolar spheres is

$$\left(\frac{1}{a}x^2\right) = 0, \quad \sum_{i=0}^{i=4} a_i = 0.$$

Once more we write the tangent sphere

$$y_i = (\lambda + a_i)\,x_i.$$

If this have stationary contact, the cyclic thereon must have a cusp, the class of the corresponding cartesian cyclic will be still further reduced, and so, by the reasoning which led up to 11], the first equation (13) has three equal roots, or

$$\sum_{i=0}^{i=4} \frac{y_i^2}{a_i+\lambda} = \sum_{i=0}^{i=4} \frac{y_i^2}{(a_i+\lambda)^2} = \sum_{i=0}^{i=4} \frac{y_i^2}{(a_i+\lambda)^3} = 0. \qquad (14)$$

Now a quartic has three equal roots if the invariants of degree two and three both vanish. Hence we have an equation of the fourth degree, and one of the sixth in y_i, or,

Theorem 13.] *The congruence of stationary tangent spheres to a general cyclide is of the twenty-fourth order.*

Theorem 14.] *The locus of the centres of curvature of a general cartesian cyclide is a surface of the twenty-fourth order.**

We see that a sphere is an adjoint surface to the general

* Darboux, *Sur une classe*, cit., p. 289.

cartesian cyclide. Cospherical circles shall be said to be *residual*, hence

Theorem 15.] *If two circles of a cyclide be coresidual, every circle residual to the one is residual to the other also.*

Theorem 16.] *Two residual, or two coresidual circles of a general cyclide are orthogonal to the same fundamental sphere.*

Theorem 17.] *Two residual circles of a cyclide meet twice, two coresidual ones do not meet at all, two circles which are neither residual nor coresidual meet once.*

We have so far considered all systems of generation together; a good deal of interesting information may be obtained by fixing our attention on a single generation. We rewrite the equation

$$y_i = 0, \quad \sum_{j=0}^{j=4} \frac{y_i^2}{a_j - a_i} = 0, \quad a_j \neq a_i.$$

In an arbitrary coaxal system orthogonal to $x_i = 0$ there will be two generating spheres of this system. If (y) and (z) be orthogonal to the fundamental sphere, and if

$$\sum_{j=0}^{j=4} \frac{y_j z_j}{a_j - a_i} = 0, \quad a_j \neq a_i,$$

these two are harmonically separated by the spheres coaxal with them which are generators of the cyclide. If (z) be fixed, the spheres satisfying this equation will generate a linear congruence. The points common to the spheres of the congruence have the coordinates

$$x_i : x_j : x_k : x_l : x_m$$

$$= i \sqrt{\sum_{j=0}^{j=4} \frac{z_j^2}{(a_j - a_i)^2}} : \frac{z_j}{a_j - a_i} : \frac{z_k}{a_k - a_i} : \frac{z_l}{a_l - a_i} : \frac{z_m}{a_m - a_i}.$$

Let (x) be the point of contact of a generating sphere which is orthogonal to (y), and belongs to the present system

$$\lambda(xs) + \sum_{j=0}^{j=4} a_j s_j x_j = 0, \quad \lambda x_i + a_i x_i = 0,$$

$$a_i(xs) - \sum_{j=0}^{j=4} a_j s_j x_j = 0.$$

Theorem 18.] *The generating spheres of one system orthogonal to an arbitrary sphere touch the cyclide in the points of a cyclic anallagmatic in the corresponding fundamental sphere.*

If we keep (s) fixed and find the corresponding cyclic for another generation,

$$a_k(xs) - \sum_{j=0}^{j=4} a_j s_j x_j = 0,$$

subtracting

$$(a_i - a_k)(xs) = 0.$$

Theorem 19.] *The generating spheres of all five systems of a general cyclide orthogonal to an arbitrary sphere touch it in the points of five cyclics lying on five spheres of a coaxal system including the arbitrary sphere.*

Since the generating spheres of one system form a quadric congruence whose members may be put into one to one correspondence with the points of a cartesian quadric surface, we see that there is an immediate correspondence between such a surface and one system of generation of the cyclide. Suppose, conversely, that we have a cartesian surface covered by two networks of circles, each circle of one network being cospherical with each of the other. The axes of these circles, that is, the lines through their centres perpendicular to their planes, will generate a quadric or two pencils. If we take two circles of one network, every circle cospherical with both is orthogonal to their common orthogonal sphere, as is, also, every circle of the same network.

Theorem 20.] *The only irreducible surface which contains two networks of circles where each circle of one network is cospherical with each of the other is a cyclide.*

The correspondence between the generating spheres of one system and the points of a quadric appears very clearly in the cartesian case where the quadric is the corresponding deferent. Here we have*

Point of deferent.	Generating sphere.
Generator.	Circle of cyclide.
Residual generators.	Residual circles.
Coresidual generators.	Coresidual circles.
Conic on deferent.	Conic series.

This may also be looked upon as a means of establishing a one to one correspondence between the points of the deferent and the pairs of points of the cyclide which are mutually inverse in the corresponding fundamental sphere.

Suppose that we have the cartesian cyclide with the general pentaspherical equation (9). Eliminating x_i we have

$$\sum_{j=0}^{j=4} b_j x_j^2 = 0, \quad b_j = (a_j - a_i). \tag{15}$$

Let the condition for a planar sphere be

$$(wx) = 0.$$

If (z) be the coordinates of the centre of a sphere (s),

$$s_j = \lambda z_j + \mu w_j.$$

If (s) be a generating sphere of the present system, we get the equation of the deferent

$$s_i = 0.$$

$$\sum_{j=0}^{j=4} \frac{(w_i z_j - w_j z_i)^2}{b_j} = 0, \quad j \neq i.$$

* Cf. Moore, 'Circles orthogonal to a given sphere', *Annals of Mathematics*, Series 2, vol. viii, 1907.

An arbitrary sphere tangent thereto at (z) will be

$$y_i = \lambda' z_i - \mu' \sum_{j=0}^{j=4} \frac{w_j(w_i z_j - w_j z_i)}{b_j},$$

$$y_k = \lambda' z_k + \mu' \frac{w_i}{b_k}(w_i z_k - w_k z_i).$$

This will be a plane if it satisfy the condition of being orthogonal to (w), i. e. $\lambda' = 0$.

Calling this the plane (r), and the angle of intersection with $x_j = 0$, θ_j,

$$\cos\theta_j = \frac{r_j}{\sqrt{(rr)}}, \quad \sum_{j=0}^{j=4} b_j \cos^2\theta_j = 0, \quad j \neq i.$$

But the cosine of the angle which a plane makes with a sphere is the distance from the centre divided by the radius. This yields the curious theorem due to Casey.*

Theorem 21.] *If the equation of the general cartesian cyclide be reduced to squared terms, and if one variable be eliminated by means of the identity, the resulting form will be identical with that which gives the quadriplanar equation of the deferent corresponding to the variable eliminated, the tetrahedron of reference being that whose vertices are the centres of the four remaining fundamental spheres, and the coordinates of a plane being proportional to the distance from these centres divided by the corresponding radii.*

Let us write the tangential equations

$$\sum_{j=0}^{j=4} (a_j - a_i)\, r_j^2 = \sum_{j=0}^{j=4} (a_j - a_k)\, r_j^2 = 0,$$

subtracting

$$(a_k - a_i)\,(rr) = 0.$$

This is characteristic of isotropic or minimal planes.

* *On Cyclides*, cit., p. 598. The form there given to the theorems is not sufficiently precise. The next six theorems are from the same source.

Theorem 22.] *The five deferents of the general cartesian cyclide are confocal.*

Theorem 23.] *Given nine spheres orthogonal to a tenth, there is always at least one cyclide tangent to each at a pair of points inverse in the tenth sphere, and, conversely, if nine pairs of points be given inverse with regard to a sphere, there is at least one cyclide passing through all and inverse in the given sphere.*

Theorem 24.] *Given eight spheres orthogonal to a ninth, which is not null. There is always a one-parameter family of cyclides having double contact with these and with the spheres of a series. In special cases there may be a two-parameter family of cyclides having double contact with the eight spheres.*

Theorem 25.] *Given eight pairs of points inverse in a sphere. There is always a pencil of cyclides anallagmatic in the sphere through these points, and in special cases there may be a two-parameter family of such cyclides.*

Theorem 26.] *All cyclides having double contact with seven spheres orthogonal to a given not null sphere have double contact with an eighth sphere orthogonal thereto.*

Theorem 27.] *All cyclides passing through seven pairs of points inverse in a given sphere pass through an eighth such pair.*

Let us now turn more definitely to the cartesian cyclide. Here, in the general case, there are five deferents, confocal quadrics. To find the points of contact of any generating sphere we must drop a perpendicular from the centre of the corresponding fundamental sphere upon the tangent plane to the deferent at the centre of the generating sphere, and find where this perpendicular meets the latter sphere. This method will hold for every anallagmatic surface. When the point of contact of the generating sphere is on the circle at infinity, the tangent plane to the deferent will contain the tangent to the circle at infinity at the corresponding point. But this plane will touch the cyclide also at this infinite point, for the line connecting this point with the point of contact of the plane with the deferent should be normal to the cyclide, and the corresponding tangent plane is the plane just drawn. We are thus led to the double focal curves of our cyclide; they are the double curves of the developable of tangent planes along the (double) circle at infinity. These, unlike the focal curves, are not covariant for inversion.*

Theorem 28.] *The double focal curves of the general cartesian cyclide are the focal curves of the corresponding deferents.*

We may pass from one generation of such a cyclide to another as follows. The points where a cyclide cuts one fundamental sphere are the points of contact of the latter with the developable tangent to this sphere, and to the corresponding deferent. This developable being of the fourth class, and elliptic in type, has four conics of striction. A point on one is the centre of a sphere having double contact with the cyclide, hence

Theorem 29.] *The four quadrics confocal with the given deferent, and each passing through one conic of striction of the developable tangent to this deferent and the corresponding fundamental sphere, will be the four other deferents.*

If we consider the plane of one of the conics of striction, we see that it contains the centres of four spheres common to two generations, and so orthogonal to two fundamental spheres. It is thus a radical plane for two fundamental spheres, and so must bear a symmetrical relation to them and to the corresponding deferents.

Theorem 30.] *If two deferents be known, and the fundamental sphere corresponding to the first, that corresponding to the second is found as follows. The planes of that conic of striction of the developable determined by the first sphere and deferent which lies on the second deferent will cut the first deferent in a conic. The developable circumscribed to this conic and the second developable will touch the sphere required.*

* Darboux, *Sur une classe*, cit., for this and the two following.

We saw recently that the planes tangent to a cartesian cyclide along the circle at infinity will touch all five deferents. Through each tangent to the circle will pass two planes tangent to the deferents. These planes will fall together when, and only when, a tangent to the circle at infinity touches also a deferent; hence the five deferents and the circle at infinity touch four (usually distinct) lines. The points of contact with the circle at infinity will be points of all the focal cyclics.

Theorem 31.] *The general cartesian cyclide has four pinch-points on the circle at infinity, which are common to all the focal cyclics.*

Let us look for a normal form for the equation of a cyclide in rectangular cartesian coordinates. We begin by noticing that the locus of the centres of gravity of the intersections of a general cartesian cyclide with sets of parallel lines is a plane, the polar of the infinite point common to the lines. If a point trace a line in the plane at infinity, its polar line in each plane section through the infinite line will, by IV. 16], rotate about a point; hence its polar plane rotates about a line. Any two such lines must intersect; hence

Theorem 32.] *The polar planes of all infinitely distant points with regard to a general cartesian cyclide pass through a fixed finite point.*

This point shall be called the *centre* of the cyclide.* If we consider the plane of a focal conic of any deferent, we see that the foci of that conic are double foci of the cyclide, and of the sections thereof in that plane; hence, by IV. 16] and (20),

Theorem 33.] *The centre of the general Euclidean cyclide is the common centre of all five deferents. The planes of the focal conics of the deferents cut the plane at infinity in the side-lines of the diagonal triangle of that complete quadrangle whose vertices are the pinch-points. The tangent planes to the cyclide at the pinch-points pass through the centre.*

* Theorems 32 and 33 are from Humbert, loc. cit., p. 132.

The canonical form for the equation will thus be

$$(x^2+y^2+z^2)^2+ax^2+by^2+cz^2+ex+fy+gz+h = 0. \quad (16)$$

We now return to pentaspherical space. Before studying systems of cyclides let us look most briefly at one or two special types under the quinary orthogonal group. The general cyclide being characterized as before by [1 1 1 1 1], let us look at the type [2 1 1 1]. This notation means that in the homothetic pencil of quadratic complexes

$$\sum_{i,j=0}^{i,j=4} a_{ij}x_i x_j + \rho\,(xx) = 0,$$

two, which are simply singular, have fallen together. This gives the limiting case of the general complex when two spheres of inversion fall together. As, however, they are mutually orthogonal, in the limiting case the double sphere must be null. The vertex of this double sphere must be a conical point for the cyclide, for the surface is anallagmatic in three mutually orthogonal spheres containing this point.

Theorem 34.] *The cyclide of the type* [2 1 1 1] *in pentaspherical space has one conical point, and is anallagmatic in three mutually orthogonal spheres through that point. It is covered with eight systems of circles, residual in pairs, of which one pair of systems pass through the conical point.*

Theorem 35.] *The cartesian cyclide of type* [2 1 1 1] *may be inverted into a non-degenerate quadric surface, not a surface of revolution, unless the fundamental null sphere is planar.*

Let us next take the type [(1 1) 1 1 1]. Here there will be a doubly singular complex in the homothetic system, whose correlative is a series of spheres. We may write the general equation for our complex

$$a_0{x_0}^2+a_1{x_1}^2+a_2{x_2}^2 = 0. \quad (17)$$

Theorem 36.] *The pentaspherical cyclide of the type* [(1 1) 1 1 1] *has two conical points, and is the envelope of the spheres of a general conic series.*

A tangent sphere to our surface will have the coordinates

$$y_0 = (a_0+\lambda)\,x_0, \quad y_1 = (a_1+\lambda)\,x_1, \quad y_2 = (a_2+\lambda)\,x_2,$$
$$y_3 = \lambda x_3, \quad y_4 = \lambda x_4.$$

For the sphere orthogonal to $x_3 = x_4 = 0$ we have $\lambda = 0$. Putting

$$y_i = a_i x_i, \quad i = 0,\ 1,\ 2.$$

Our surface is the envelope of the conic series of spheres

$$\frac{1}{a_0}y_0^2 + \frac{1}{a_1}y_1^2 + \frac{1}{a_2}y_2^2 = 0, \quad y_3 = y_4 = 0.$$

Theorem 37.] *The pentaspherical cyclide of the type* $[(1\,1)\,1\,1\,1]$ *has seven systems of circles. Six are mutually residual in pairs. The circles of the seventh system all pass through two conical points and are characteristic circles of the spheres of a general conic series which envelop the cyclide.*

Theorem 38.] *The Euclidean cyclide of the type* $[(1\,1)\,1\,1\,1]$ *may be inverted into a quadric cone, not of revolution, unless the fundamental null sphere is planar.*

As a last type consider $[(1\ 1)\ (1\ 1)\ 1]$. Here there are two distinct doubly special complexes in the homothetic system; the surface may be enveloped in two ways by the spheres of a conic series.

Theorem 39.] *The cyclide of the type* $[(1\,1)\,(1\,1)\,1]$ *is a Dupin cyclide.*

The Dupin series and cyclides have only been defined in cartesian space, but the definitions carry over immediately.

We have already defined as confocal two quadratic complexes whose correlatives are homothetic; the cyclides generated by the null spheres of confocal complexes shall be defined as *confocal cyclides*. If our original cyclide have the equation (9), the general form for the confocal system will be

$$\sum_{i=0}^{i=4} \frac{x_i^2}{a_i+\lambda} = (xx) = 0. \tag{18}$$

Theorem 40.] *The cyclides which are irreducible and confocal with a general cyclide are themselves general. The five fundamental spheres, each counted twice, are the only reducible cyclides in a general confocal system.*

We mean, of course, by a *general* confocal system one composed of general cyclides. We see at once in (18) that if (x) be known, we have a cubic equation in λ.

If λ_1 and λ_2 be two roots, and we take the tangent spheres (x'), (x'') where

$$x_i' = \frac{x_i}{a_i+\lambda_1}, \quad x_i'' = \frac{x_i}{a_i+\lambda_2}.$$

$$(x'x'') = \sum_{i=0}^{i=4} \frac{x_i^2}{(a_i+\lambda_1)(a_i+\lambda_2)}$$

$$= \frac{1}{\lambda_2-\lambda_1}\left[\sum_{i=0}^{i=4} \frac{x_i^2}{a_i+\lambda_1} - \sum_{i=0}^{i=4} \frac{x_i^2}{a_i+\lambda_2}\right] = 0.$$

Theorem 41.] *Through each point of space will pass three mutually orthogonal cyclides of a general confocal system.*

The word 'space' here means 'pentaspherical space'; in cartesian space we must restrict ourselves to the finite domain. We have from the Darboux-Dupin theorem:

Theorem 42.] *The lines of curvature of a general cyclide are its intersections with confocal cyclides.*

We get immediately from our definition, or from (18),

Theorem 43.] *Confocal cyclides have the same focal curves.*

The fact that the focal curves of a cyclide are of the same type as the intersections with an arbitrary sphere leads to some curious results.* We start with the general cyclide

$$(ax^2) = (xx) = 0. \tag{9}$$

* The remaining theorems in this chapter are due to Darboux, *Sur une classe*, cit., pp. 327 ff. The proofs there given are not easy to follow.

Let (y) be an arbitrary sphere, and consider the cyclide

$$\sum_{i=0}^{i=4} \frac{y_i^2}{a_i+\lambda} \sum_{i=0}^{i=4} \frac{x_i^2}{a_i+\lambda} - \left[\sum_{i=0}^{i=4} \frac{y_i x_i}{a_i+\lambda}\right]^2 = 0. \qquad (19)$$

This will be found to be anallagmatic in (y). The tangent sphere at (x) orthogonal to (y) will be (x'), where

$$x_i' = \left[\sum_{i=0}^{i=4} \frac{y_i^2}{a_i+\lambda}\right] \frac{x_i}{a_i+\lambda} - \left[\frac{y_i}{a_i+\lambda}\right] \sum_{i=0}^{i=4} \frac{x_i y_i}{a_i+\lambda}.$$

$$\sum_{i=0}^{i=4} (a_i+\lambda) x_i'^2 = (yx') = 0.$$

The intersection of our original cyclide with (y) will be a focal curve for the new cyclide. By varying λ we get a confocal system of new cyclides, and each is tangent along a cyclic to a cyclide confocal with the original one.

Theorem 44.] *The cyclides having for one focal curve the cyclic common to a general cyclide and an arbitrary sphere are confocal, and each is tangent along a cyclic to a cyclide confocal with the original one.*

Conversely, let us take an arbitrary cyclide tangent along a cyclic to the general cyclide of our system,

$$\kappa \sum_{i=0}^{i=4} \frac{x_i^2}{a_i+\lambda} + \lambda (zx)^2 = 0.$$

If we write

$$y_i = (a_i+\lambda) z_i, \quad \kappa = -\sum_{i=0}^{i=4} \frac{y_i^2}{a_i+\lambda},$$

we fall back on (19).

Theorem 45.] *The focal curves of all cyclides touching a general cyclide along a cyclic lie on cyclides confocal with the given one.*

Theorem 46.] *If a sphere cut a general cyclide in a cyclic, that will be a focal curve for five cyclides each containing one focal curve of the original cyclide.*

We see, in fact, that, considered as envelopes, the focal curves are limiting cases of the confocal cyclides; we then apply 44].

Several ideas for continuing the geometry of pentaspherical space will occur to any one after reading the preceding chapter. We have made no mention of problems of construction; it would be easy to lead up to the solution of the problem of drawing a sphere tangent to four others exactly as we did to the corresponding problem in Ch. IV. It seems certain that some of the other cyclides deserve a more detailed study than we have given to any but the general and Dupin cyclides. The residuation theory for curves on cyclides should be easy and interesting.

CHAPTER VIII

CIRCLE TRANSFORMATIONS

§ 1. General Theory.

WE have frequently had occasion, especially in Ch. IV, to draw distinction between the cartesian and the tetracyclic planes. There is a one to one continuous correspondence between their circles, but not between their points, for they have different connectivity. In the cartesian plane we considered, besides the angles of circles, the positions of their centres and the magnitudes of their radii. In the tetracyclic plane we considered only those properties of circles which are invariant for inversion, or for quaternary orthogonal substitutions. No circle has an absolute invariant under this group, although the expression (xx) is a relative invariant. The cosine of the angle of two circles is, however, an absolute simultaneous invariant of two not null circles, and they have no other invariant independent of this.

We next observe that although we have said a good deal about this invariant we have paid next to no attention to the transformations themselves, except the inversions and conformal collineations of the cartesian plane. It is the purpose of the present chapter to discuss the various types of circle transformations and the groups thereof.*

Let us begin by defining as a *circle transformation* any analytic transformation that carries circles of a plane into circles. In circle coordinates this will be

$$x_i = x_i(x_0'x_1'x_2'x_3'x_4'),$$

* For an elaborate treatment by pure geometry see Sturm, *Theorie der geometrischen Verwandtschaften*, vol. iv, Leipzig, 1909. An admirable analytic introduction is given by Döhlemann, *Geometrische Transformationen*, cit.

where (x') represents a circle. If, further, null circles are carried into null circles,

$$(xx) = k\,(x'x')^n.$$

Here k must be a function of the coefficients of the transformation only, as otherwise not null circles of the congruence $k = 0$ would be carried into null circles.

Let us next assume that our transformation is algebraic and one to one. Such a transformation might be engendered as follows. The circles of the plane are in one to one correspondence with the points of a three-dimensional projective space; the null circles in one to one correspondence with those points of the space which lie on the Absolute quadric. If we take the most general Cremona transformation of space which leaves the quadric in place, we have the required circle transformations. Now this Absolute quadric may be stereographically projected on the projective plane, and the Cremona transformation of space in question will give a Cremona transformation of that plane. Conversely, let a Cremona transformation of the plane be given. If that be expressed in tetracyclic coordinates, it will be a transformation of projective space which leaves the Absolute quadric invariant. There remains, lastly, the question, could not two different Cremona transformations of projective space produce the same Cremona transformation of the Absolute quadric? If such were the case, the product of the one and the inverse of the other would be a Cremona transformation where all points of the quadric were invariant. Such transformations do not, however, exist. For suppose we had one,

$$\rho x_i = f_i\,(x_0' \ldots x_4').$$

Putting $x_i' = x_i$, and eliminating ρ,

$$x_i f_j - x_j f_i = 0.$$

We must then have

$$x_i f_j - x_j f_i \equiv (xx)\,\phi_{ij}.$$

Solving these equations for f_i, we find that each f_i contains

(xx) as a factor, an absurd result, as we should naturally remove such a factor at the start.

Theorem 1.] *The group of all algebraic circle transformations of the plane which carry null circles into null circles is simply isomorphic with that of all Cremona transformations of the cartesian plane.**

An interesting sub-group of these transformations is composed of those which carry tangent circles into tangent circles. We shall reserve to a subsequent chapter the discussion of these. Let us rather note that although our transformations carry points into points, and circles into circles (in the tetracyclic plane), we have not yet required that they should carry points on a circle into points on another circle. For this we require the additional restriction

$$(ax) \equiv (a'x') \text{ if } (x'x') = 0.$$

The first of these equations must be independent of the second, for $\frac{\partial x_i}{\partial x_j'} = \text{const.}$ for all values of (x') where $(x'x') = 0$, i.e.

$$\frac{\partial x_i}{\partial x_j'} = b_{ij} + k(x'x')^n.$$

This equation is not homogeneous, as it should be unless $k = 0$.

Now the most general analytic transformation of four homogeneous variables that carries a linear form into a linear form is a linear transformation, and since (xx) is covariant, we shall always have

$$\frac{(xy)}{\sqrt{(xx)}\sqrt{(yy)}} = \frac{(x'y')}{\sqrt{(x'x')}\sqrt{(y'y')}}.$$

Theorem 2.] *The most general transformation of the tetracyclic plane that carries a point into a point, and the points of a circle into points of a circle, is an orthogonal substitution.*

We shall call such transformations *circular* transformations, and study them in detail analytically presently. For the

* Nothing seems ever to have been published about these general transformations. The Author's attention was called to them by a conversation with his colleague Prof. C. L. Bouton.

moment we confine ourselves to the real cartesian plane, and approach the subject of circular transformations by pure geometry.*

We begin by returning to the domain of Ch. I, the real finite cartesian domain, and inquire what will be the nature of a transformation which is one to one, with the exception of a finite number of exceptional points, and carries points on a circle or line into concyclic or collinear points. Let such a transformation be called T, P, and P', two corresponding points. The circles through P will go into circles through P'. If we precede T by an inversion with P as centre, and follow it by an inversion with P' as centre, we have a transformation T' of the same type as T, which carries lines into lines. It is clear that parallel lines will go into parallel lines, for if two intersecting lines were carried into parallels, the ∞^2 circles through the intersection would go into ∞^2 circles meeting each parallel once; such circles do not exist in such numbers. A parallelogram will go into a parallelogram, an inscriptible parallelogram into an inscriptible parallelogram, i.e. a rectangle into a rectangle, a square into a square, since a square is the only rectangle with mutually perpendicular diagonal lines. Now a necessary and sufficient condition that a point should be between two others is that every line through this point should intersect every circle through the other two twice, and this is invariant under our transformation T'. Let T' carry the square $ABCD$ into the square $A'B'C'D'$. We may follow T' by a rigid motion of the plane and a similarity transformation which carries $A'B'C'D'$ back into $ABCD$, when corresponding orders of letters correspond to the same sense of progress about the perimeters of these squares. Where the sense of progress is opposite, we may accomplish the desired result by first reflecting in a diagonal line of one square. In any case we get a transformation T'' of the same type as T', which leaves

* The groundwork of what follows is from Möbius's Collected Works, vol. ii, p. 243, Leipzig, 1886. He defines a circular transformation as being necessarily continuous, but we have avoided that assumption by following Darboux, 'Sur la géométrie projective', *Math. Annalen*, vol. xvii, 1880, and Swift, 'On the Conditions that a Point Transformation of the Plane be a Projective Transformation', *Bulletin American Math. Soc.*, vol. x, 1904.

$ABCD$ in place. Now if a square be invariant, every square contiguous, that is, sharing one of its sides, is invariant. Moreover, if a square be invariant, the four equal contiguous squares into which it can be divided will be invariant. Hence the plane is covered with an everywhere dense network of invariant squares of sides as small as we please, and as betweenness is invariant every point is invariant. Hence T'' is the identical transformation, T' is a conformal collineation, and T is the product of such a collineation and inversions. As a matter of fact, if T be not itself a conformal collineation, it will carry straight lines into circles meeting in only one point not exceptional for the transformation, i.e. into circles through a singular point, and may be factored into the product of a conformal collineation and an inversion with the singular point as centre.

Theorem 3.] *Every circular transformation is either a conformal collineation, an inversion, or the product of the two.*

Theorem 4.] *Every circular transformation is conformal.*

We may sharpen our idea of conformal transformations by using the angular notation described on p. 20. If

$$\measuredangle ABC = \measuredangle A'B'C',$$

$$\measuredangle \overrightarrow{ABC} = \measuredangle \overrightarrow{A'B'C'} \text{ or } \measuredangle \overrightarrow{ABC} = \measuredangle \overrightarrow{C'B'A'}.$$

The first equality holds when the two directed angles have the same sense of description, the second when they have opposite senses.

Suppose now that we have a conformal collineation, and that

$$\measuredangle \overrightarrow{ABC} = \measuredangle \overrightarrow{A'B'C'}; \quad \measuredangle \overrightarrow{ABC} = \measuredangle \overrightarrow{ADC}.$$

Hence $ABCD$ are concyclic, as are $A'B'C'D'$.

$$\measuredangle \overrightarrow{A'D'C'} = \measuredangle \overrightarrow{A'B'C'} = \measuredangle \overrightarrow{ABC} = \measuredangle \overrightarrow{ADC}.$$

Here $\measuredangle ADC$, $\measuredangle A'D'C'$ are any two equal angles; we may at once extend to the case of any two commensurable angles, and so to any two angles, so that if

$$\measuredangle\,\overrightarrow{ABC} = \measuredangle\,\overrightarrow{A'B'C'},$$

then
$$\measuredangle\,\overrightarrow{HKL} = \measuredangle\,\overrightarrow{H'K'L'}.$$

Such a collineation is said to be *directly conformal*; if the sense of description be reversed in the case of one angle it will be for every angle, and the collineation is said to be *inversely conformal.* The concepts of directly and inversely conformal may be extended from collineations to conformal transformations of any sort; in the one case the sense of every angle is preserved, in the second it is reversed. Since by I. 9] an inversion is an inversely conformal transformation, if we factor a circular transformation into a collineation and an inversion, the circular transformation will be directly (inversely) conformal if the collineation be inversely (directly) conformal.

Theorem 5.] *The group of all circular transformations depends upon six parameters, and has a six-parameter sub-group of all directly conformal circular transformations, and a six-parameter sub-assemblage of all inversely conformal circular transformations.*

We may find the number of parameters by counting the amount of freedom in conformal collineation, and in an inversion, or by the number of arbitrary points presently to be determined. The sub-group is called the group of *direct* circular transformations, the sub-assemblage is composed of the *indirect* ones.

Consider a directly conformal collineation. If there be no fixed point it is a translation. If there be a fixed point, the product of this transformation and a properly chosen similarity transformation is a directly conformal collineation which keeps one, and hence all, distances invariant. Let the reader show that this must be a rotation or translation; hence

Theorem 6.] *Every direct circular transformation may*

be factored into an even number, and every indirect one into an odd number of inversions or reflections.

We shall determine the minimum values for these numbers with greater precision later. Our theorem is of importance as showing the basal rôle played by inversion in the theory of circular transformations; it is, in fact, the reason why inversion lies at the very heart of the geometry of the circle.

Theorem 7.] *Every direct or indirect circular transformation is completely determined by the fate of three points.*

We leave the proof, which is very simple, to the reader. The great use of the theorem is that it enables us to write the analytic expression for the most general circular transformation of the cartesian plane. If x and y be the cartesian rectangular coordinates of a real finite point let us put

$$z \equiv x+iy, \quad \bar{z} \equiv x-iy.$$

A real circle will have an equation of the type

$$lz\bar{z}+\mu z+\bar{\mu}\bar{z}+n=0.$$

The most general real direct circular transformation may then be written*

$$z'=\frac{\alpha z+\beta}{\gamma z+\delta}, \quad \bar{z}'=\frac{\bar{\alpha}\bar{z}+\bar{\beta}}{\bar{\gamma}\bar{z}+\bar{\delta}}, \quad (\alpha\delta-\beta\gamma)\neq 0. \tag{1}$$

The most general real indirect one will be

$$z'=\frac{\alpha \bar{z}+\beta}{\gamma \bar{z}+\delta}, \quad \bar{z}'=\frac{\bar{\alpha}z+\bar{\beta}}{\bar{\gamma}z+\bar{\delta}}, \quad (\alpha\delta-\beta\gamma)\neq 0. \tag{2}$$

Let us confine ourselves for the present to direct transformations. Suppose that the four points A, B, C, D are carried

* For a truly admirable discussion of circular transformations starting with these equations see Cole, 'Linear Functions of the Complex Variable', *Annals of Mathematics*, Series 1, vol. v. Also Döhlemann, loc. cit.

into the points A', B', C', D'. For any inversion of reflection we have by I. 4] and 7]

$$\measuredangle \overrightarrow{CBA} + \measuredangle \overrightarrow{ADC} = -\,(\measuredangle \overrightarrow{C'B'A'} + \measuredangle \overrightarrow{A'D'C'}). \qquad (3)$$

$$\frac{(AB)\,(CD)}{(AD)\,(CB)} = \frac{(A'B')\,(C'D')}{(A'D')\,(C'B')}. \qquad (4)$$

The first of these expressions is called the *double angle* of the four points, the second their *double ratio*.*

Theorem 8.] *In every direct circular transformation double angles and double ratios are invariant.*

It is worth while to verify this analytically. If four points correspond to the parameter values z_1, z_2, z_3, z_4, and the transformed values are z_1', z_2', z_3', z_4',

$$\frac{(z_1-z_2)\,(z_3-z_4)}{(z_1-z_4)\,(z_3-z_2)} = \frac{(z_1'-z_2')\,(z_3'-z_4')}{(z_1'-z_4')(z_3'-z_2')}. \qquad (5)$$

Now $(z_1-z_2) = (AB)$, and by taking the absolute values of both sides we find the equal double ratios. Again, the argument of z_1-z_2 is the angle which the line AB makes with the axis of x, so that

$$\text{argument}\ \frac{z_1-z_2}{z_3-z_2} = \measuredangle \overrightarrow{CBA}.$$

The argument of the left-hand side of the equation is thus

$$\measuredangle \overrightarrow{CBA} + \measuredangle \overrightarrow{ADC}.$$

Theorem 9.] *The modulus of the cross ratio of four values of the complex variable is the double ratio of the four corresponding points in the Gauss plane; the argument is the double angle of these four points.*

Theorem 10.] *A necessary and sufficient condition that four real points of the cartesian plane should be concyclic or collinear is that their double angle should be congruent to* 0, mod. π.

* These invariants are due to Möbius, loc. cit.

Theorem 11.] *A necessary and sufficient condition that four points should be orthocyclic is that their double ratio should have the value* 1.

Since harmonic points are both concyclic and orthocyclic, their double angle is zero and their double ratio unity.

Let the reader show that in tetracyclic coordinates

$$\frac{(xy)\,(zt)}{(xt)\,(yz)} \equiv \frac{(z_1-z_2)\,(z_3-z_4)}{(z_1-z_4)\,(z_3-z_2)} \times \frac{(\bar{z}_1-\bar{z}_2)\,(\bar{z}_3-\bar{z}_4)}{(\bar{z}_1-\bar{z}_4)\,(\bar{z}_3-\bar{z}_2)}.$$

The expression for the double ratio is thus

$$\sqrt{\frac{(xy)\,(zt)}{(xt)\,(zy)}}.$$

To find the expression for the double angle we take the special case where three of the points are the origin, the unit point of the x axis, and the infinite point (there is but one in the Gauss plane). We thus get

$$\sin \text{double} \measuredangle = \frac{i\,|\,x\,y\,z\,t\,|}{\sqrt{(xy)}\,\sqrt{(zt)}\,\sqrt{(xt)}\,\sqrt{(yz)}}.$$

Let us find the locus of a point in space forming an orthocyclic set with three given points B, C, D. We wish to find X, so that

$$\frac{(XB)\,(CD)}{(XD)\,(CB)} = 1.$$

One point of the locus will be A, the harmonic conjugate of C with regard to B and D. When A, B, C, D are collinear, we see, by elementary geometry, that the locus is the sphere on AC as diameter. Moreover, since double ratios are invariant for inversion in three dimensions, the locus is always the sphere through B and D orthogonal to the circle BCD. More generally, if

$$\frac{(XB)\,(CD)}{(XD)\,(CB)} = k,$$

we see that the locus of X is a sphere orthogonal to the circle BCD.

Two circles shall be said to be in *bi-involution* if every sphere through one be orthogonal to every sphere through the other.

This relation is clearly invariant for inversion in three dimensions, and if one circle be inverted into a straight line, that line will be the axis of the inverse of the other circle.

Let A and C lie on a circle, B and D on another in bi-involution therewith. Taking a centre of inversion on the AC circle,

$$\frac{(AB)\,(CD)}{(AD)\,(CB)} = \frac{(A'B')\,(C'D')}{(A'D')\,(C'B')} = 1,$$

since $$(A'B') = (A'D'), \quad (C'D') = (C'B').$$

Hence $$\frac{(AB)}{(AD)} = \frac{(CB)}{(CD)}.$$

Theorem 12.] *If two circles be in bi-involution the ratio of the distances of any point on one from two fixed points of the other depends merely on the position of the latter.**

We write again

$$\left| \frac{(z_1-z_2)\,(z_3-z_4)}{(z_1-z_4)\,(z_3-z_2)} \right| = \frac{(AB)\,(CD)}{(AD)\,(BC)}.$$

If our four points be on a circle, the double angle is zero or π. Assuming that A, C separate B, D, we see from the special case of points on the x axis that

$$\frac{(z_1-z_2)\,(z_3-z_4)}{(z_1-z_4)\,(z_3-z_2)} = -\frac{(AB)\,(CD)}{(CB)\,(AD)} = \lambda < 0.$$

$$\frac{(z_1-z_3)\,(z_2-z_4)}{(z_2-z_3)\,(z_1-z_4)} = \frac{(AC)\,(BD)}{(BC)\,(AD)} = 1-\lambda.$$

$$(AC)\,(BD) = (AB)\,(CD) + (AD)\,(BC).$$

This last equation proves Ptolemy's theorem by a method that surely would have surprised Ptolemy.

We easily see from 11] that the locus of points forming definitely paired orthocyclic sets with three given points and lying in their plane is a circle, hence

Theorem 13.] *A necessary and sufficient condition that a one to one transformation of the finite domain of the real*

* Möbius, loc. cit., p. 277, and Chasles, loc. cit., p. 559.

plane should be a circular transformation, is that double angles or double ratios should be invariant.

Among circular transformations inversion enjoys the advantage of being involutory. It is, however, indirect. There is a direct involutory transformation which we reach as follows. Let us start with two conjugate coaxal systems. Each point P, other than the limiting points of one coaxal system, will determine a circle of each system, and these two shall intersect again in P'. The transformation from P to P' is clearly involutory, and shall be called a *Möbius involution.* If we invert one coaxal system into a pencil of radiating lines, we see that corresponding points are harmonically separated by the limiting points of one coaxal system. The inverted transformation is clearly a circular transformation; hence we have in general a circular transformation. In the inverted case it is the product of the reflections in any two mutually perpendicular lines of the radiating set; hence

Theorem 14.] *Every Möbius involution is the product of inversions in any two mutually orthogonal circles of a determinate coaxal system through two points.*

It will be convenient to extend the term 'Möbius involution' to include the limiting case of a reflection in a point, which is the product of reflection in two mutually perpendicular lines, and from now on we understand the term to be so extended.

Theorem 15.] *A Möbius involution may be found to interchange any two pairs of points.*

If the two pairs be concyclic, the two circles orthogonal to the given circle through the two pairs of points will determine conjugate coaxal systems, or concentric circles and radiating lines through their centres, and so the involution required.

Suppose that they are not concyclic, and that P and Q are to be interchanged with P' and Q' respectively. We first invert in such a circle of antisimilitude of the circles $PP'Q$, $PP'Q'$, that Q passes to Q_1 on the circle $PP'Q'$, where Q_1 is not separated from Q' by P and P', then interchange Q_1 and Q' by inverting in a circle of the coaxal system with limiting points P and P'.

Theorem 16.] *Every involutory direct circular transformation is a Möbius involution.*

Such a transformation is surely determined when we know two pairs that are interchanged, but we may find a Möbius involution to interchange any two pairs.

Theorem 17.] *If a direct circular transformation interchange a single pair of points it is a Möbius involution.*

Suppose that such a transformation carry $ABCC'$ into $BAC'C''$. If we follow with the Möbius involution

$$BAC'C'' \sim ABC''C',$$

the product will have three fixed points and so be the identical transformation.

Theorem 18.] *Every direct circular transformation is the product of two Möbius involutions.**

Suppose that we call our transformation T and determine it by $ABC \sim A'B'C'$.

Consider the Möbius involution I which interchanges A and B', A' and B. Then under TI

$$B'A'K \sim A'B'C'.$$

Hence TI is an involution J, or $T = JI$.

Theorem 19.] *If an indirect circular transformation be involutory, it is either a reflection, an inversion, or the product of an inversion and a reflection in the centre.*

If it be a collineation, it could not be a reflection in a point, since this is direct. There can be no self-corresponding

* It is instructive to compare these last theorems and 7] with theorems 4] to 6] of ch. iv. Let the reader give the analytic reason for the similarity.

points, hence lines connecting corresponding points are all parallel, and we have a reflection in a line.

Suppose, next, that it is not a collineation. Three non-concurrent and not parallel lines will go into three concurrent circles, since the sum of the angles of the arcual triangle must be π. Any other line in the plane will go into a circle or line meeting each of the first three in only one point not singular for the transformation. Hence lines go into circles through a point O. If P and P' be two corresponding points not collinear with O, the circle OPP' and the line PP' are interchanged. The angle from (PP') to arc PP' at P would be equal to the negative of the angle at P' from the arc $P'P$ to $(P'P)$. But evidently these angles are equal both in magnitude and sign. Hence corresponding points are collinear with O. If P and P' be not separated by O every circle through P and P' is transformed into itself, and clearly we have an inversion. If P and P' be separated by O it is the product of an inversion and a reflection in O.

§ 2. Analytic Treatment.

The majority of facts so far noted about circular transformations have been reached by the methods of plane geometry. It is now time to make a more detailed study of the analytic aspect of these transformations. We shall take as our domain the real sphere, or a real tetracyclic domain such as the Gauss plane, the real finite cartesian plane made a perfect continuum by the adjunction of a single point at infinity. This may also be defined as that region of the general tetracyclic plane where x_0 is proportional to a pure imaginary number; each other x is proportional to a real number. Since the groups of circular transformations of the cartesian and tetracyclic planes are simply isomorphic, we have made no essential alteration by such a choice of domain. We express our domain parametrically in terms of the isotropic parameters. Recalling the equations of IV,

$$\dot{x}_0 = ix_0, \quad \dot{x}_1 = x_1, \quad \dot{x}_2 = x_2, \quad \dot{x}_3 = x_3. \tag{6}$$

We rewrite IV (13) in non-homogeneous form:

$$\begin{aligned} \dot{x}_0 &= z\bar{z}+1, \\ \dot{x}_1 &= z\bar{z}-1, \\ \dot{x}_2 &= z+\bar{z}, \\ \dot{x}_3 &= -i(z-\bar{z}). \end{aligned} \tag{7}$$

We get the whole tetracyclic plane by removing the restriction that z and $\bar{z}$ should have conjugate imaginary values. Real direct circular transformations will be given by (1) and indirect ones by (2). A direct transformation has the form of a complex one-dimensional projectivity, so that our theorems 16] to 19] might have been deduced from familiar theorems of projective geometry.*

The inverse of our transformation (1) will be found by interchanging α and $-\delta$; a necessary and sufficient condition for a Möbius involution is thus

$$\alpha+\delta = 0. \tag{8}$$

On the other hand, the inverse of (2) is

$$\bar{z} = \frac{-\delta z'+\beta}{\gamma z'-\alpha}. \tag{9}$$

The transformation (2) will thus be involutory if

$$\delta = -\bar{\alpha}, \quad \beta = \bar{\beta} = b, \quad \gamma = \bar{\gamma} = c. \tag{10}$$

All points of the circle

$$cz\bar{z}+\delta z+\bar{\delta}\bar{z}-b = 0 \tag{11}$$

are invariant. If this be real, we have an inversion; if self-conjugate, imaginary, the product of an inversion and a Möbius involution.

If we follow our transformation (1) by

$$z'' = \frac{\alpha' z'+\beta'}{\gamma' z'+\delta'},$$

* This point of view is emphasized by Wiener, loc. cit. Much of the following discussion is taken from an article of unusual excellence by Von Weber, 'Zur Theorie der Kreisverwandtschaften in der Ebene', *Münchener Berichte*, xxxi, 1901.

the product will be

$$z'' = \frac{(\alpha'\alpha + \beta'\gamma)\,z + (\alpha'\beta + \beta'\delta)}{(\gamma'\alpha + \delta'\gamma)\,z + (\gamma'\beta + \delta'\delta)}. \tag{12}$$

If (2) be followed by

$$z'' = \frac{\alpha'\bar{z}' + \beta'}{\gamma'\bar{z}' + \delta'},$$

the product is

$$z'' = \frac{(\alpha'\bar{\alpha} + \beta'\bar{\gamma})\,z + (\alpha'\bar{\beta} + \beta'\bar{\delta})}{(\gamma'\bar{\alpha} + \delta'\gamma)\,z + (\gamma'\bar{\beta} + \delta'\bar{\delta})}. \tag{13}$$

Let us see what fixed or self-corresponding points there may be in a circular transformation. We begin with the indirect case. Here we must have

$$\gamma z\bar{z} + \delta z - \alpha\bar{z} - \beta = 0,$$
$$\bar{\gamma} z\bar{z} + \bar{\delta}\bar{z} - \bar{\alpha} z - \bar{\beta} = 0.$$

Here, if the equations be distinct, we have two real and distinct, coincident, or conjugate imaginary fixed points. If they be identical, i.e. if

$$\gamma = \bar{\gamma}, \quad \beta = \bar{\beta}, \quad \delta = -\bar{\alpha},$$

the transformation is involutory, and we are back on a real inversion, or an inversion in a self-conjugate imaginary circle, which amounts to the product of an inversion and a Möbius involution. We pass to the more interesting direct case. Here the fixed points must be the roots of

$$\gamma z^2 + (\delta - \alpha)\,z - \beta = 0. \tag{14}$$

The discriminant of this equation is

$$(\delta + \alpha)^2 - 4\,(\alpha\delta - \beta\gamma).$$

When this vanishes, the transformation is said to be *parabolic*. If the single fixed point of the parabolic transformation correspond to the value $z = \infty$, the transformation may be written in the canonical form

$$z' = z + \beta. \tag{15}$$

Theorem 20.] *Corresponding points in a parabolic transformation lie on tangent circles through a fixed point.*

A transformation of a non-parabolic type will have distinct fixed points. It may be written in the highly suggestive form

$$\frac{z'-z_1}{z'-z_2} = re^{i\theta}\frac{z-z_1}{z-z_2}. \tag{16}$$

The expression $re^{i\theta}$ is called the *invariant* of the transformation. Let the reader show that in the cartesian plane r will give the double angle of two corresponding points and the fixed points, while θ gives the corresponding double ratio. The point of the word 'invariant' is that if we carry our transformation into an equivalent one by means of a circular transformation, the invariant does not change in value. Taking as the fixed points those which correspond to the parameter values ∞ and 0, we get the canonical form for our non-parabolic transformation

$$z' = re^{i\theta}z. \tag{17}$$

We see from this that there are three standard types of these transformations:

Hyperbolic $\theta \equiv 0 \pmod{\pi}$.

Corresponding points are concyclic with the fixed points.

Elliptic $r = 1$.

Corresponding points are orthocyclic with the fixed points.

Notice that a Möbius involution may be classified under either of these types.

Loxodromic $r \neq 1,\quad \theta \not\equiv 0 \pmod{\pi}$.

This we might naturally call the *general* case. Corresponding points will lie on the same double spiral which circulates around the two fixed points and meets at a fixed angle all circles through them.

Theorem 21.] *The only periodic circular transformations are of elliptic type.*

Suppose that we have two non-parabolic direct transformations with one common fixed point. We may take this to correspond to $z = \infty$.

$$z' - z_1 = re^{i\theta}(z - z_1).$$

$$z'' - z_2 = r'e^{i\theta'}(z' - z_2).$$

$$z'' = r'e^{i\theta'}\left[re^{i\theta}(z - z_1) + (z_1 - z_2)\right] + z_2.$$

$$(z'' - z_3) = rr'e^{i(\theta+\theta')}(z - z_3).$$

Theorem 22.] *If two non-parabolic direct transformations have one common fixed point, the invariant of their product is the product of their invariants.*

Consider a hyperbolic transformation with fixed points H and K which carries P into P'. Take any circle which has H and K as mutually inverse points, and invert. Let P' be carried into P_1. We can find a second circle of inversion interchanging H and K which carries P_1 into P'. The product of these two inversions will be a direct transformation with H and K fixed and carrying P into P', i.e. our original hyperbolic transformation. Let the reader show similarly that an elliptic transformation may be factored into the product of two inversions in circles through the fixed points, and a parabolic transformation may be factored into the product of inversions in two tangent circles. Conversely, if we have two inversions, their product will transform into themselves all circles orthogonal to the two circles of inversion.

Theorem 23.] *The hyperbolic, elliptic, and parabolic direct transformations, and these alone, are the product of two inversions.*

Theorem 24.] *A necessary and sufficient condition that the product of three inversions should be an inversion is that the three circles of inversion should be coaxal, or else the circles of two successive inversions should be orthogonal to the third circle of inversion.*

Theorem 25.] *The product of two inversions may be replaced by that of two other inversions whereof one has a circle taken at random in the coaxal system determined by the two given circles of inversion; the second inversion is uniquely determined by the first.*

There are certain problems in construction associated with direct circular transformations which should now claim our attention. The postulates assumed are those of Ch. IV.

Problem 1.] *Given two pairs of points corresponding in a Möbius involution, to find the mate of any point.*

When the two pairs are concyclic, this has already been done in Ch. IV, problem 8. If not, suppose that the involution is given by the pairs PP' and QQ', and we wish to find R' the mate of R.*

Let the harmonic conjugate of R with regard to PP' be R_1, that of R_1 with regard to QQ' shall be R_{12}; in like manner the harmonic conjugate of R with regard to QQ' shall be R_2, while that of R_2 with regard to PP' shall be R_{21}. Lastly, the harmonic conjugate of R with regard to $R_{12}R_{21}$ shall be $\bar{R}$, while R' is the required point. Let us first take the product of the two Möbius involutions with fixed points PP' and QQ'. We have a direct transformation whose fixed points are those of the given involution. If, further, we operate with a Möbius involution whose fixed points are RR', these last-found fixed points are interchanged. Hence the product of these three involutions is an involution. The product of the involutions having the successive pairs of double points PP', QQ', RR', PP', QQ' is the involution with the double points RR', but this involution will carry R_{12} into R_{21}. Hence R and R' are harmonically separated by R_{12} and R_{21}, and the problem consists in finding a succession of harmonic conjugates, and was solved in Ch. IV.

Problem 2.] *Given a direct transformation by means of three sets of corresponding points, to find the mate of any point.*

We have but to factor our transformation into two involutions by means of 18], then apply the solution of problem 1.

* Cf. Wiener, loc. cit., pp. 670, 671.

Problem 3.] *Given two pairs of points of a Möbius involution, to find the double points.*

This is a problem of the second degree. We may, by the solution of problem 1], find as many pairs of corresponding points as we please, and so construct as many pairs of corresponding circles as we like, through two chosen corresponding points. These circles will cut on any circle through one of the latter points, two ranges of points in one to one reciprocal algebraic correspondence, i.e. an involution, and the double points of this involution must be real, since they lie on real self-corresponding circles of the Möbius involution. We may thus, by Ch. IV, problem 7, find the double points of the involution on the circle, and so two self-corresponding circles of the transformation. On one of these circles find two pairs of corresponding points, and through each pair pass a circle orthogonal to the given circle. Then either these two intersect in the two self-corresponding points sought, or else those points are the limiting points of the coaxal system determined by these circles.

Problem 4.] *Given two Möbius involutions, to find their common pair.*

We find the fixed points of each, then the fixed points of that Möbius involution having them as two pairs.

Problem 5.] *Given a direct circular transformation, to find the fixed points.*

We factor the transformation into two involutions, then apply the solution of the last problem.

Let us now turn to the classification of indirect transformations. We see that the square of an indirect transformation is a direct one. The fixed points of the direct transformation were either interchanged or fixed in the indirect one. We thus get the following types of indirect transformation, the points mentioned being, when distinct, those which correspond to the parameter values ∞, 0.

Hyperbolic $$z' = re^{i\theta}\bar{z}, \quad \bar{z}' = re^{-i\theta}z'. \qquad (18)$$

Two real points are fixed, two conjugate imaginary ones interchanged. The circles through the fixed points are interchanged, two being invariant; the circles orthogonal to these are also interchanged, but no real ones stay in place.

Elliptic $$z'\bar{z} = re^{i\theta}, \quad z\bar{z}' = re^{-i\theta}. \tag{19}$$

Two real points are interchanged, two conjugate imaginary ones invariant. Circles through the interchanging points are interchanged, none invariant. Circles orthogonal to these are interchanged, one real and one self-conjugate imaginary one invariant.

Parabolic $$z' = \bar{z} + \alpha, \quad \bar{z}' = z + \bar{\alpha}. \tag{20}$$

No fixed proper circle. Members interchanged in each of two orthogonal systems of tangent circles.

Inversion $$z'\bar{z} = z\bar{z}' = k^2. \tag{21}$$

Product of inversion and Möbius involution

$$z'\bar{z} = z\bar{z}' = -k^2. \tag{22}$$

Let the reader, with the aid of 24], complete 6] as follows:

Theorem 26.] *Every indirect circular transformation may be factored into three inversions; every direct one may be factored into four inversions.*

The last statement may also be proved immediately from 18].

Let us turn aside for a moment to consider the effect of a real circular transformation upon the imaginary points of our domain. Suppose that we take an imaginary point of our tetracyclic plane, which we shall here suppose a real sphere. It will have parameter values $(z, \bar{z}')$. On each of the isotropics through this point will lie one real point, namely, the points $(z\bar{z})$, $(\bar{z}'z')$. Conversely, to each pair of real points $(z\bar{z})$, $(z'\bar{z}')$ will correspond two conjugate imaginary points $(z\bar{z}')$, $(z'\bar{z})$. The geometrical interpretation is as follows in the case of a sphere. If two real points be given, we may draw tangent planes to the sphere thereat, which planes meet in a line without the sphere. Conversely, if such a line be given, through it we may draw real tangent planes.

Suppose, next, that we have a real indirect transformation; what will be the locus of the pairs of conjugate imaginary points associated in this with corresponding pairs of real points under the transformation?

$$\bar{z}' = \frac{\alpha z + \beta}{\gamma z + \delta}, \quad z' = \frac{\bar{\alpha}\bar{z} + \bar{\beta}}{\bar{\gamma}\bar{z} + \bar{\delta}}.$$

$$\gamma z\bar{z}' + \delta\bar{z}' - \alpha z - \beta = 0, \quad \bar{\gamma}z'\bar{z} - \bar{\delta}z' - \bar{\alpha}\bar{z} - \bar{\beta} = 0.$$

Theorem 27.] *If a real indirect circular transformation be given for a real sphere, the polars with regard to that sphere of the lines connecting pairs of corresponding points will intersect two conjugate imaginary circles of the sphere. These circles will fall together when, and only when, the transformation is an inversion in a real or self-conjugate imaginary circle.**

A curious figure arises when we consider the corresponding problem for a direct transformation:

$$z' = \frac{\alpha z + \beta}{\gamma z + \delta}, \quad \bar{z}' = \frac{\bar{\alpha}\bar{z} + \bar{\beta}}{\bar{\gamma}\bar{z} + \bar{\delta}}.$$

$$\gamma z z' + \delta z' - \alpha z - \beta = 0, \quad \bar{\gamma}\bar{z}\bar{z}' + \bar{\delta}\bar{z}' - \bar{\alpha}\bar{z} - \bar{\beta} = 0.$$

We have two assemblages of points depending on two real parameters, but not on one complex parameter.

Theorem 28.] *If a real direct circular transformation be given for a real sphere, the polars with regard to that sphere of the lines connecting corresponding points will meet the sphere in pairs of points depending on two real parameters. These systems are characterized by the fact that the corresponding cross ratios of the four isotropics of the two sets through four points are conjugate imaginary.*

It is to be noted that the real domain is a special case of one of these systems.

* Von Weber, loc. cit., pp. 383 ff. See also Study, *Ausgewählte Gegenstände der Geometrie*, Part 1, Leipzig, 1911, p. 32.

We next turn our attention to the question of commutative transformations. We begin by recalling the familiar fact that in any group of transformations those which are commutative with a chosen member will form a sub-group. In fact, if

$$TA = AT, \qquad TB = BT,$$

then

$$TAT^{-1} = A, \quad TBT^{-1} = B,$$

and

$$TABT^{-1} = AB, \quad TAB = ABT.$$

It is also to be noted that if two transformations be commutative, each must leave invariant or permute all points which are invariant in the other.

A.] Two direct transformations. If neither be involutory they must have the same fixed points. Conversely, we see at once from formulae (15) and (17) that two direct transformations with the same fixed points are commutative. If one be a Möbius involution and the other not, the fixed points of the Möbius involution must be fixed for the other. Lastly, we see that harmonic pairs will determine two commutative Möbius involutions, each interchanging the other's fixed points.

Theorem 29.] *A necessary and sufficient condition that two direct circular transformations should be commutative is that they should have the same distinct or coincident fixed points, or that they should be two Möbius involutions whose fixed points separate one another harmonically.*

B.] A direct and an indirect transformation. If the indirect one be not involutory, the fixed points of the direct one must be fixed or interchanged thereby. If the indirect one be hyperbolic, the two fixed points might be interchanged if the direct one were involutory. Otherwise the fixed points and real fixed circles of the indirect transformation must be fixed for the direct one also, i. e. the direct one is hyperbolic also. If the indirect one were elliptic, the direct one might be involutory, and either keep invariant or interchange the interchanged points of the indirect one, or else the direct one might be elliptic, keeping invariant the interchanging points of the indirect one. If one were parabolic, the other would have to be parabolic, if not involutory. On the other hand, it is easy to see from equations (17) to (20) that an inversion is commutative with any direct transformation whose fixed points are either invariant or interchanged, and these same equations show us the sufficiency of our necessary conditions.

Theorem 30.]* *If a direct and an indirect circular transformation be commutative, neither being involutory, then both are hyperbolic or parabolic with the same fixed points, or the interchanging points of an elliptic indirect transformation are fixed in the elliptic direct one. If the direct transformation be involutory, its fixed points are either fixed or interchanged in the indirect one. These conditions are both necessary and sufficient.*

C.] Two indirect transformations. If neither be involutory, they will be hyperbolic, elliptic, or parabolic together, with the same fixed or interchanging points. If one be involutory, it must either keep fixed or interchange two points which are fixed or interchanged in the other, or transfer them to another pair of fixed or interchanging points. If both be inversions, they must either have the same circle of inversion (in which case they are identical) or else their circles of inversion intersect orthogonally. If one be an inversion and the other the product of an inversion and a Möbius involution, the circle of inversion must be invariant in the other transformation. Two transformations of this latter type cannot be commutative, for if two self-conjugate imaginary circles could intersect orthogonally, two planes conjugate with regard to a real sphere might both be outside of it, an impossibility.

Theorem 31.]* *If two indirect circular transformations be commutative and neither be an inversion, they must be hyperbolic, elliptic, or parabolic together with the same fixed or interchanging points. If one be an inversion, its fixed circle is fixed in the other. These conditions are necessary and sufficient.*

* The conditions given in these theorems that two circular transformations should be commutative, are necessary but not sufficient. A hyperbolic or elliptic indirect transformation will be commutative with a system

If we have two indirect transformations

$$T \quad z' = \frac{\alpha\bar{z}+\beta}{\gamma\bar{z}+\delta}, \qquad \bar{z}' = \frac{\bar{\alpha}z+\bar{\beta}}{\bar{\gamma}z+\bar{\delta}},$$

$$S \quad z' = \frac{\alpha'\bar{z}+\beta'}{\gamma'\bar{z}+\delta'}, \qquad \bar{z}' = \frac{\bar{\alpha}z+\bar{\beta}'}{\bar{\gamma}z+\bar{\delta}'},$$

they are associated respectively with the circles

$$\gamma'\bar{z}z' + \delta'z' - \alpha\bar{z} - \beta = 0,$$

$$\gamma'\bar{z}z' + \delta'z' - \alpha'\bar{z} - \beta' = 0.$$

The condition that these should be orthogonal is the polar of the condition that one should be null, i.e. the polar of the condition that the corresponding indirect transformation should be improper. We thus get

$$\alpha'\delta + \alpha\delta' - \beta'\gamma - \beta\gamma' = 0.$$

$$\bar{\alpha}'\bar{\delta} + \bar{\alpha}\bar{\delta}' - \bar{\beta}'\bar{\gamma} - \bar{\beta}\bar{\gamma}' = 0.$$

These equations tell us, however, that

$$(TS^{-1})^2 = (T^{-1}S)^2 = (ST^{-1})^2 = (S^{-1}T)^2 = 1.$$

Theorem 32.] *A necessary and sufficient condition that the product of an indirect transformation of a sphere and the inverse of a second indirect one should be an involution is that the conjugate imaginary circles determined by the polars with regard to the sphere of lines connecting corresponding points for each transformation should intersect orthogonally in pairs.*

We may treat direct transformations in the same way, and arrive at a condition which we shall call *orthogonality* for two parameter systems of complex points. We may likewise solve problems in construction associated with these systems. To a circle known by inversion will correspond a two-parameter system known by the corresponding direct circular transformation. If two circles be known, and they have intersections in the domain in question, their intersections are supposed known. So here the problem of finding the intersections of two of our two-parameter systems is the

of real ones depending on one real parameter. See Benedetti, 'Sulla teoria delle forme iperalgebriche', *Annali della R. Scuola Normale di Pisa*, vol. viii, 1899, p. 62.

problem of finding two points which correspond in two direct circular transformations, i. e. the fixed points of the product of one and the inverse of the other. This is problem 5] above.

§ 3. Continuous Groups of Transformations.

Enough attention has now been given to individual circular transformations; it is time to turn our attention to groups of such. The study of finite groups is nothing but the study of finite groups of fractional linear substitutions of the linear complex variable. It is well known that such groups are simply isomorphic with the groups of the regular solids. We may consider such groups as sufficiently familiar; in any case they are of more importance to the algebraist than to the geometer.* In the same way the study of infinite discontinuous groups would lead us into a vast field but little germane to our present purpose.† Let us rather turn to the geometrically more interesting study of continuous and mixed groups. The problem here is nothing but the problem of studying the groups of collineations of a three-dimensional space which leave a real quadric with imaginary generators in place.‡ Under a continuous group (corresponding to direct transformations) the generators of each system are permuted among themselves; in a mixed group there will be transformations where the two systems of generators are interchanged. What can we say about three-parameter groups? If such a group have an invariant two-parameter sub-group it is integrable, since every two-parameter group is integrable.§ On the other hand, if a three-parameter group of direct circular transformations had an invariant one-parameter sub-group, the two fixed points of the sub-

* The classic discussion is, of course, in Klein's *Ikosaeder*, Leipzig, 1884.

† Especially Klein-Fricke, *Theorie der automorphen Functionen*, vol. i, Leipzig, 1897.

‡ The following discussion is an amplification of Amaldi, 'I gruppi reali di trasformazioni dello Spazio', *Memorie della R. Accademia delle Scienze di Torino*, Series 2, vol. lv, 1905.

§ Cf. Lie-Scheffers, *Vorlesungen über continuirliche Gruppen*, Leipzig, 1893, p. 563.

group would be invariant throughout the three-parameter group; this is quite impossible when the points are distinct. If the one-parameter group consisted in parabolic transformations, the single fixed point would be invariant throughout the whole non-integrable three-parameter group. We shall presently see that this is impossible.

Let us begin with the study of simple three-parameter groups. A simple three-parameter collineation group in three dimensions must leave invariant either a cubic space curve, a conic and a point not in the plane thereof, one system of generators of a quadric, or a line and all points of a second line skew thereto.* In the present case, where we have real transformations leaving a real quadric with imaginary generators in place, all but the second case will be impossible. If there be a real fixed point in three dimensions, there will be a real or self-conjugate imaginary fixed circle in the tetracyclic plane. The group with a real fixed circle is simply isomorphic with the real binary projective group. It has no fixed real point, and so is simple. There will be two-parameter sub-groups with any chosen point of the fixed circle fixed, one-parameter sub-groups with two fixed points in the circle. These will be hyperbolic. There will be one-parameter elliptic sub-groups which keep invariant a pair of points mutually inverse in the circle; also a parabolic one-parameter sub-group. When the three-parameter group leaves a self-conjugate imaginary circle in place, the only real sub-groups are one-parameter elliptic ones. There are no other two-parameter sub-groups in either case, for if in such a group both fixed points might be chosen at random on the circle, the transformation would be determined by its fixed points, which is absurd.

If there were any four- or five-parameter groups of circular transformations, they would have to contain three-parameter groups. A five-parameter group would have to contain a three-parameter sub-group keeping a chosen point invariant, while

* Cf. Fano, 'Sulle varietà algebriche con un gruppo continuo non integrabile di trasformazioni in se', *Memorie della R. Accademia delle Scienze di Torino*, Series 2, vol. xlvi, 1896, p. 209.

every four-parameter group has a three-parameter sub-group.* If our three-parameter group were contained in a four-parameter group it would have to be invariant, or else have an invariant sub-group of its own.† The latter, however, is ruled out, as we are assuming for the present that the three-parameter group is simple. But if our three-parameter group were invariant, its fixed circle would be invariant in the four-parameter group, which again cannot be, as the total group with a circle fixed has but three parameters. If our simple group were in a five-parameter group, the latter, not having any fixed circle or point, would have to have a fixed circle congruence composed of the transforms of the fixed circle of the three-parameter group Keeping any one circle of the congruence fixed, we may carry any second circle into any third circle thereof, as otherwise, each circle having but one degree of freedom, we should have four-parameter groups with a fixed circle. But escaping this absurdity, we fall into the worse one of having a congruence of circles, each two of which make the same angle. Our simple three-parameter group lies thus neither in a four- nor a five-parameter one.

Let us next look at integrable groups. Every such group has a one-parameter invariant sub-group, and the fixed points of the one-parameter group must be invariant throughout. But if the integrable group be of more than two parameters, the invariant one-parameter group must be parabolic. A canonical form for such a group will be

$$z' = z + b.$$

What two-parameter groups might include our one-parameter one? The second fixed point for a transformation of such a group could not trace the whole plane. If $a \neq 1$ the transformation is hyperbolic, and the only curves carried into themselves are circles. Hence the other fixed point must lie on a circle through the point $z = \infty$. Hence the two-parameter group must either be of the type

$$z' = az + b,$$

* Lie-Scheffers, loc. cit., p. 577.

† Ibid., p. 544.

or else of the parabolic type

$$z' = z + \beta.$$

Now take an integrable three-parameter group. This has a fixed point which we may take as $z = \infty$, and we have a three-parameter sub-group of

$$z' = \alpha z + \beta.$$

It will have the two-parameter parabolic sub-group invariant. The second fixed point for a transformation of the group must be free to move over the whole plane, for if it were restricted to a certain curve, that curve would be carried into itself by the transformations of the group, while the parabolic sub-group is transitive for the whole plane except its own fixed point. Now let both points be fixed. We have a one-parameter group of the form

$$z' = re^{i\theta} z, \quad r = r(\theta),$$

$$r(\theta)\, r(\theta') = r(\theta + \theta'),$$

$$r = e^{k\theta}.$$

Here k would seem to depend on the position of the second fixed point. Such is not, however, the case. We see, in fact, that if we take two transformations with the invariants $e^{k\theta} e^{i\theta}$ and $e^{k'\theta'} e^{i\theta'}$, since, by 22], the invariant of the product is the product of the invariants

$$k\theta + k'\theta' = l(\theta + \theta'),$$

$$k = k'.$$

We thus get three-parameter groups of the form

$$z' = e^{(k+i)\theta} z + \beta,$$

where θ and β are independent variables. This equation may be written

$$z' + \frac{\beta}{e^{(k+i)\theta} - 1} = e^{(k+i)\theta}\left[z + \frac{\beta}{e^{(k+i)\theta} - 1}\right], \qquad (23)$$

and it is evident, conversely from 22], that the totality of these transformations will be a group. We shall call such a group a *Newson* group.* We characterize it geometrically by examining the significance of the constant k. Taking for simplicity a transformation of the group with fixed points ∞ and 0, we write our transformation in polar form,

$$\rho' e^{i\phi'} = e^{(k+i)\theta} \rho e^{i\phi}.$$

What sort of a double spiral will be carried into itself by this transformation? If the equation of such be

$$\rho = e^{\lambda\phi},$$

we easily find $$\lambda = k.$$

This shows that k is the tangent of the angle which the spiral makes with a circle through the fixed points: the *pitch* of the double spiral, let us say. The Newson group is thus characterized by the fact that one point is fixed, and all double spirals carried into themselves by transformations of the group have a constant pitch.

We have thus covered three-parameter groups. There are no five-parameter groups. A five-parameter group would have a three-parameter sub-group with any chosen point fixed, and such a group would be a Newson group. The pitch here must be independent of the position of the fixed point, for in any Newson group one fixed point can be chosen at random. But this leads us to another absurdity, for it is easy to show that if two loxodromic transformations have different fixed points but the same pitch, the pitch of their product is different.

If there be any four-parameter groups, and we know that there are, they must have Newson sub-groups, as we saw two pages back. If the Newson group be invariant, its fixed point will be invariant throughout the four-parameter group, and, conversely, the four-parameter group will be entirely characterized by the invariance of this point. If the Newson group were not invariant the position of one fixed point for each transformation would have to be limited to a specific

* Newson, 'Groups of Circular Transformations', *Bulletin of the American Math. Soc.*, vol. iv, 1897.

curve, for if the position of both fixed points were free the pitch would have to be constant, there being only four parameters, and we should run into our preceding contradiction. But if one fixed point lay on a certain curve, this curve must be invariant throughout the group, whereas the transformations of the Newson group are transitive. We may summarize as follows:

Theorem 33.] *There are no real five-parameter continuous groups of circular transformations.*

Theorem 34.] *The only real four-parameter groups are those with one fixed point.*

Theorem 35.] *The only real non-integrable three-parameter groups are those with a fixed real or self-conjugate imaginary circle.*

Theorem 36.] *The only real integrable three-parameter groups are the Newson groups.*

Theorem 37.] *The only real two-parameter groups are those with two fixed points, the parabolic ones with one fixed point, and those with a real fixed circle and real fixed point thereon.*

Theorem 38.] *The only real one-parameter groups are the loxodromic, hyperbolic, elliptic, and parabolic ones.*

It is doubtful whether there be room for much further investigation of the subject of real circular transformations. On the other hand, the sort of circle transformation which is obtained from a Cremona transformation of the projective plane, and was mentioned at the beginning of the chapter, is still utterly unexplored, and may well contain new theorems of interest and importance.

CHAPTER IX

SPHERE TRANSFORMATIONS

§ 1. General Theory.

THE subject of sphere transformations presents, naturally enough, many analogies to that of circle transformations. It is not so rich, however, in interesting and easily obtainable results, owing to the impossibility of representing either cartesian or pentaspherical space parametrically by means of isotropics. Thus, a large part of the theory of circular transformations which is reached through their connexion with the theory of the linear function of the complex variable is lost.

We shall mean by a *sphere transformation* any analytic transformation that carries spheres into spheres. In sphere coordinates this will be

$$x_i' = f_i(x_0 x_1 x_2 x_3 x_4),$$

where (x) represents a sphere. If, further, we require that null spheres shall be carried into null spheres,

$$(x' x') = k(xx),$$

where k depends merely on the coefficients of the transformation. We find, exactly as in the last chapter,

Theorem 1.] *The group of all one to one algebraic transformations of the spheres of pentaspherical space which carry null spheres into null spheres is simply isomorphic with that of all Cremona transformations of projective three-dimensional space.*

We shall mean by a *spherical transformation any analytic point transformation of cartesian or pentaspherical space which carries the points of a sphere or plane into points on a sphere or plane.*

We have

Theorem 2.] *The most general spherical transformation of pentaspherical space is given by a quinary orthogonal substitution.*

In the last chapter we dealt with a real tetracyclic plane. In the present one we shall deal with finite real cartesian space, as well as a second continuum which may be called a *real pentaspherical space.* We may define this as that region of the general pentaspherical space where x_0 is proportional to a pure imaginary number, and the other pentaspherical coordinates to real numbers. An example of such a domain is afforded by a real hypersphere in four-dimensional projective space with Euclidean measurement. Or we may start with the real finite domain of cartesian space and extend it to a real continuum by adjoining a single real point at infinity. We shall also fix our attention on real spherical transformations of this space.

Let us begin with a purely geometrical analysis of the cartesian case as before. If no finite point be singular for the transformation, the latter is a conformal collineation, and may be factored into translations, rotations, reflections in planes, and similarity transformations, the latter being easily factorable into two inversions, while the three preceding are factorable into reflections in planes. If one finite point be singular, the spheres through it being carried into planes, we may factor into an inversion with this point as centre, and a transformation of the preceding type.

Theorem 3.] *Every spherical transformation of real cartesian space is conformal, and may be factored into a product of inversions and reflections in planes.*

We see, incidentally, that a similar theorem holds in pentaspherical space, the word 'inversions' covering both types. We shall return to this presently; for the moment we prefer to prove the remarkable converse theorem which is due to Liouville.*

Theorem 4.] *Every conformal analytic transformation of cartesian space is a spherical transformation.*

The easiest proof is, perhaps, the following. Every conformal transformation will carry a triply orthogonal system into another such system. Hence, by the Darboux-Dupin theorem, it will carry a line of curvature into a line of curvature. It will therefore carry a surface, all of whose curves are lines of curvature, into another such surface, i.e. it will carry a sphere into a sphere or plane.

Theorem 5.] *A spherical transformation is necessarily a circular transformation, and every analytical point transformation that carries circles into circles will be a spherical transformation.*

The first part of this theorem is immediate; the second comes from the fact that the necessary and sufficient condition that two circles should be cospherical is that there should be ∞^3 circles meeting both twice.

Our formula for double ratio in Ch. I (4) yields an invariant for inversion and reflection, and so for all spherical transformations. Conversely, if we have a quadrilateral where the sum of the products of the opposite sides is equal to the product of the diagonals, we may take a centre of inversion at one vertex and transform the other three vertices into collinear points. The original four were thus concyclic. If, then,

$$(AB)(CD)+(AD)(BC) = (AC)(BD),$$

we have also

$$\frac{(BA)(CD)}{(CA)(BD)} + \frac{(AD)(BC)}{(BD)(AC)} = 1.$$

Theorem 6.] *A necessary and sufficient condition that real one to one transformation of the real finite cartesian space should be a spherical transformation is that double ratios should be invariant.*

* See his appendix to Monge's *Applications de l'analyse à la géométrie*, Paris, 1850, pp. 609 ff.

Let the reader show that in special pentaspherical coordinates the double ratio of four points may be expressed in the form

$$\frac{(AB)(CD)}{(AD)(BC)} = \sqrt{\frac{(xy)(zt)}{(xt)(zy)}}. \tag{1}$$

Since the expressions involved are covariants, we may remove the restriction that the pentaspherical coordinates should be special, and let them be any pentaspherical set. Another invariant for spherical transformations is

$$\frac{i\sqrt{\begin{vmatrix} 0 & (xy) & (xz) & (xt) \\ (yx) & 0 & (yz) & (yt) \\ (zx) & (zy) & 0 & (zt) \\ (tx) & (ty) & (tz) & 0 \end{vmatrix}}}{\sqrt{(xy)}\sqrt{(zt)}\sqrt{(xt)}\sqrt{(zy)}}. \tag{2}$$

In the case of coplanar points this reduces to the sine of the double angle.

Let us next take up the question of direct and indirect spherical transformations. We start in finite real cartesian space, and suppose that a transformation T carries a point H into a point H'. Let S be the translation that carries H' back into H, while R is the inversion with H as centre. The transformation $RSTR$, the operator being written to the left of the operand, will be a conformal collineation. Considering the effect of this transformation on the whole of projective cartesian space, we see that in the plane at infinity there will be two conjugate imaginary fixed points on the circle at infinity, and a fixed real point besides. The lines through this point are permuted by a collineation, two conjugate imaginary ones tangent to the circle at infinity are fixed, hence one finite real one is fixed. Our conformal collineation may thus be reduced to one of the following forms:

$$\begin{aligned} x &= r(\cos\theta x' - \sin\theta y'), \\ y &= r(\sin\theta x' + \cos\theta y'), \\ z &= rz' + d. \end{aligned} \tag{3}$$

$$\begin{aligned} x &= r(\cos\theta x' + \sin\theta y'), \\ y &= r(\sin\theta x' - \cos\theta y'), \\ z &= rz' + d. \end{aligned} \tag{3'}$$

Equations (3′) may be much simplified. We see that two mutually perpendicular planes are invariant. These may be taken as fundamental in the coordinate system, so that we may write transformations of this type in the simple form

$$\begin{aligned} x &= rx', \\ y &= -ry', \\ z &= rz+d. \end{aligned} \tag{4}$$

Such a transformation is the product of one of type (3) and a reflection, so that we may confine ourselves mainly to type (3). Here, if $r > 0$, we have the product of a translation parallel to the fixed axis, a rotation about that axis and a similarity transformation, that is, the product of four reflections and two inversions. When $r < 0$, if we change θ into $\pi+\theta$, we fall back on the other form. When the number of reflections and inversions is even, we may pass by a continuous change of parameters from the given transformation to the identical one. When the number is odd, we may pass continuously to a single inversion, but not to the identity. We see, in fact, that if we take a reflection in a plane, the sense of each trihedral angle is reversed, and we cannot pass continuously from a transformation which alters the senses of trihedral angles to one that does not. As for the number of parameters involved, any not-null sphere may be carried into any other such, which uses up four degrees of freedom; when one sphere is fixed we have as many free parameters left as there are in a circular transformation.

Theorem 7.] *The group of all spherical transformations of pentaspherical space depends upon ten parameters. It has a ten-parameter sub-group of direct transformations, and a ten-parameter sub-assemblage of indirect ones. A direct transformation may be factored into an even number of inversions and reflections, and may be continuously changed into the identical transformation; an indirect transformation may be factored into an odd number of inversions and reflections, and may be continuously changed into a single inversion, but not into the identical transformation.*

Let us look at the fixed points of a direct transformation. We shall confine ourselves to real transformations, and begin in cartesian space. First take a conformal collineation. If we consider the whole of cartesian space there will be one fixed real point in the plane at infinity, and one fixed real line through it; or all infinite points are fixed. Taking the cases in order, if all the points of the fixed line are themselves fixed, we have in (3) $d = 0, r = 1$. We have a rotation; corresponding points lie on circles with the line of fixed points as axis, i. e. on circles in bi-involution therewith.

Suppose, next, that but one finite point of the fixed lines is invariant. Here, if we take this point for the origin, $d = 0$. If $\theta \neq 0$, $r \neq 1$, corresponding points lie on non-circular isogonal trajectories of the generators of cones of revolution whose common vertex is the origin and whose common origin is the z axis. If $\theta = 0$, corresponding points are collinear with the origin. If $r = 1$ we fall back on the preceding case.

Let us, thirdly, assume that no finite point of the fixed lines is fixed. Here $r = 1$, $\theta \neq 0$. Corresponding points are on circular helices, i.e. isogonal not circular trajectories of the generators of cylinders of revolution with a common axis.

There then remains the case where all infinite points are invariant. Here $\theta = 0$. If $r \neq 1$ we have essentially the next to the last case; if $r = 1$ we have a translation, and corresponding points lie on lines of given direction.

Suppose, now, that we have any direct spherical transformation T. If it have a finite fixed point, and I be the inversion with this point as centre, we see that ITI is a conformal collineation which, under our spherical group, is equivalent to the given transformation. The only spherical transformations not equivalent to conformal collineations under our spherical group are those with no finite fixed point, if any such exist.

A real spherical transformation will appear in four-dimensional projective space as a real collineation, and will there leave at least one real point invariant in four dimensions. Corresponding to this point will be a real or self-conjugate imaginary sphere. If this sphere be real we may assume it not null, as otherwise we might assume its vertex in the domain which amounts to the finite domain of cartesian space, and fall back on our previous types. The real spherical transformation will produce on the fixed real not null sphere a real circular transformation, which must be indirect, as otherwise there would be a real fixed point, the case we wish to avoid. It must either be elliptic, or an inversion in a self-conjugate imaginary circle. Taking the cases in order, when the circular transformation is elliptic one real circle of the sphere is invariant, as is the circle orthogonal to the sphere through the interchanging points. This is equivalent to a transformation of cartesian space, which leaves invariant the z axis and the unit circle of the z plane, i.e.

$$x=\frac{(\cos\theta x'-\sin\theta y')}{x'^2+y'^2+z'^2},\quad y=\frac{(\sin\theta x'+\cos\theta y')}{x'^2+y'^2+z'^2},$$
$$z=\frac{-z'}{x'^2+y'^2+z'^2}. \qquad (5)$$

We see, in fact, that the transformation so written is the product of a rotation, a reflection in a plane, and an inversion, and so direct. It has no finite fixed point. Corresponding points are on the non-circular isogonal trajectories of the generators of cones of revolution whose common vertex is the origin, and whose axis is the axis of z.

We must not forget the possibility of an inversion on our real sphere. This may be thrown back to the previous case with $\theta=\pi$. It is thus an inversion in a self-conjugate imaginary sphere. If we mean by a *Möbius involution* in three dimensions the transformation where corresponding points are harmonically separated by two given points, we see that this transformation is the product of an inversion and a Möbius involution.

There remains but one possible case, that where the one real fixed point in four dimensions corresponds to a self-conjugate imaginary sphere in three dimensions. Let this sphere be $x_0=0$. We get the real transformations of the sort desired by keeping x_0 invariant, and subjecting the other four coordinates to a quaternary orthogonal substitution; the direct transformations will come from those transformations where the determinant is positive, for here, and here alone, we may pass continuously to the identical transformation. Each set of isotropics of the fixed sphere will be permuted among themselves. For a real transformation there are two possibilities; either two conjugate imaginary isotropics of each set are invariant, or all are invariant in one set, while but two of the other remain in place. Taking these in turn, we have the canonical equations for a transformation of the first sort:

$$\begin{aligned} \rho x_0 &= x_0', \\ \rho x_1 &= x_1 \cos\phi - x_2 \sin\phi, \\ \rho x_2 &= x_1 \sin\phi + x_2 \cos\phi, \\ \rho x_3 &= x_3 \cos\theta - x_4 \sin\theta, \\ \rho x_4 &= x_3 \sin\theta + x_4 \cos\theta. \end{aligned} \tag{6}$$

Conversely, it is clear that this transformation fulfils all the requirements. No real point or sphere is invariant; corresponding points lie on the non-circular isogonal trajectories of the generators of the Dupin cyclides

$$x_1^2 + x_2^2 + \kappa x_0^2 = 0,$$

or of the Dupin cyclides

$$x_3^2 + x_4^2 + \lambda x_0^2 = 0.$$

Lastly, it is possible that every isotropic of one set of the fixed sphere is in place. Such transformation belongs to one of the two three-parameter invariant sub-groups of the quaternary orthogonal group. It may be written admirably in quaternion form. Following Hamilton, we write

$$i^2 = j^2 = k^2 = ijk = -1. \tag{7}$$

Our transformation will have one of the two forms

$$\begin{aligned} &\rho x_0 = x_0', \quad \rho(x_1 + ix_2 + jx_3 + kx_4) \\ &\qquad = \frac{(a + bi + cj + dk)}{\sqrt{a^2 + b^2 + c^2 + d^2}}(x_1' + ix_2' + jx_3 + kx_4'), \\ &\rho x_0 = x_0', \quad \rho(x_1 + ix_2 + jx_3 + kx_4) \\ &\qquad = (x_1' + ix_2' + jx_3' + kx_4')\,\frac{(a + bi + cj + dk)}{\sqrt{a^2 + b^2 + c^2 + d^2}}. \end{aligned} \tag{8}$$

We see, in fact, that these equations give invariant three-parameter sub-groups of the total group which leaves $x_0 = 0$ invariant. The groups leaving the one or the other system of isotropics invariant are also invariant three-parameter sub-groups. Hence each three-parameter sub-group in one pair must have invariant sub-groups of its own composed of what it has in common with the one or the other invariant group of the other pair. But the groups leaving all isotropics of one system invariant are merely the binary projective group applied to the isotropics of the other group, and have no invariant sub-groups. Hence the two methods of dividing into three-parameter sub-groups must be the same, and we have indeed the transformations desired.

The transformation (8) will be involutory if

$$a = 0, \quad b^2 + c^2 + d^2 = 1.$$

It may also be infinitesimal. Corresponding points lie on circles orthogonal to the fixed sphere intersecting the two fixed isotropics of that set where only two are fixed. Two of these circles cannot intersect, for if they intersected once they would do so again in the inverse of the first point with regard to $x_0 = 0$, and so be cospherical. But the two isotropics which intersect both would have to lie on this sphere; the latter would meet $x_0 = 0$ in two skew isotropics, which is quite impossible.

Take an infinitesimal transformation of the present type. Corresponding points lie on circles meeting two chosen isotropics of the same set on $x_0 = 0$, and there will be ∞^2 such circles which are carried into themselves by all transformations of the one-parameter group generated by the infinitesimal transformation. If we anticipate our future work to the extent of assuming that any two circles are cut twice orthogonally by at least one third circle, we see that any two circles of the present system, though not cospherical, are cut orthogonally twice by ∞^1 circles. Two circles so related shall be said to be *paratactic*.*

* These will correspond to paratactic or Clifford parallel lines of non-Euclidean space already mentioned on p. 164. The present type of spherical transformations will correspond to translations of elliptic space. See the Author's *Non-Euclidean Geometry*, cit., p. 99.

Theorem 8.] *There are nine types of real direct spherical transformations of real pentaspherical space:*

(*a*) *Inversions in self-conjugate imaginary spheres. A transformation of this sort is the product of an inversion and a Möbius involution whose fixed points are mutually inverse in the sphere of inversion. Corresponding points are concyclic with the fixed points of the involution.*

(*b*) *Rotatory transformations. Corresponding points lie on circles in bi-involution with a real circle of fixed points.*

(*c*) *Loxodromic transformations. Corresponding points are on non-circular isogonal trajectories of the circles of curvature of a Dupin cyclides with given double points which are fixed.*

(*d*) *Hyperbolic transformations. Corresponding points concyclic with two fixed points.*

(*e*) *Loxodromo-parabolic transformations. These are identical with the loxodromic except that two conical points of the Dupin cyclides fall together.*

(*f*) *Parabolic transformations. Corresponding points lie on tangent circles through a fixed point.*

(*g*) *Semi-elliptic transformations. Corresponding points lie on non-circular isogonal trajectories of Dupin cyclides with given conical points, whereof two are real and are interchanged in the transformation.*

(*h*) *Loxodromo-elliptic transformations. These are like the loxodromo-hyperbolic transformations, but there are no real fixed points or real fixed sphere. The conical point of the Dupin cyclides are two pairs of conjugate imaginaries.*

(*i*) *Paratactic transformations. No real fixed spheres. Corresponding points on paratactic circles.*

Theorem 9.] *There are infinitesimal spherical transformations of every type but inversions and semi-elliptic ones.*

Theorem 10.] *The only involutory transformations are of the hyperbolic rotatory, and paratactic types. The first are inversions in self-conjugate imaginary spheres, and may be factored into the products of inversions and Möbius involutions; the second are inversions in fixed circles.*

Let us look for types of indirect transformations. Starting with (4) we see that if $d \neq 0$ no finite point is fixed and the axis of z is transformed into itself. If $d = 0$, the origin and point at infinity are fixed and lines through the origin are interchanged, whereas in a reflection in the origin these lines are all invariant.

An indirect transformation with no real fixed point but with a real fixed sphere is obtained from (6) by changing the sign of z. There are no real indirect transformations with no real fixed sphere. Such a transformation would be given by a quaternary orthogonal substitution with negative discriminant, and permute the isotropics of the two sets on the fixed sphere. There would thus be two fixed conjugate imaginary points, and two interchanging conjugate imaginary points (as in the case of an indirect circular transformation). The transformation could be written

$$\rho x_0 = x_0'.$$
$$\rho x_1 = x_1' \cos \phi + x_2' \sin \phi.$$
$$\rho x_2 = x_1' \sin \phi - x_2' \cos \phi.$$
$$\rho x_3 = x_3' \cos \theta - x_4 \sin \theta.$$
$$\rho x_4 = x_3' \sin \theta + x_4 \cos \theta.$$

But we see at once here that two real spheres $x_1 + \lambda x_2 = 0$ are invariant.

Theorem 11.] *There are but six types of indirect real spherical transformations:*

(*a*) *Inversions.*

(*b*) *Two real fixed points, circles through them fixed.*

(*c*) *Two real fixed points, circles through them interchanged.*

(*d*) *Adjacent fixed points, circles all tangent, at one point interchanged.*

(*e*) *Adjacent fixed points, circles tangent at one point invariant.*

(*f*) *No real fixed points, points of real circle interchanged, real fixed sphere.*

Theorem 12.] *A necessary and sufficient condition that two inversions should be commutative is that the spheres of inversion should be mutually orthogonal.*

Theorem 13.] *The product of inversions in two mutually orthogonal spheres is the inversion in their common circle, the product of three such inversions is the Möbius involution whose fixed points are common to the three mutually orthogonal spheres; the product of four such inversions is the inversion in the self-conjugate imaginary sphere orthogonal to the four given mutually orthogonal spheres.*

Theorem 14.] *The only real indirect involutory spherical transformations are inversions and Möbius involutions.*

Theorem 15.] *If two points be interchanged in a spherical transformation, either the transformation is real and semi-elliptic, the two points lying on a circle of interchanging points, or else the transformation is involutory.*

Theorem 16.] *A single direct spherical transformation may be found to carry any three points, whereof no two are on an isotropic, and any not null sphere through them into any other three points and sphere similarly arranged.*

Theorem 17.] *The product of two inversions may be replaced by that of two others. One of the new spheres of inversion may be taken at random in the coaxal system determined by the original two; the other is thereby uniquely determined.*

Theorem 18.] *Any direct spherical transformation of pentaspherical space may be factored into the product of four inversions.*

The proof of this important theorem is as follows. The transformation being called T, take two corresponding spheres s and s'. Let us find an inversion I_4 to interchange the spheres s' and s. Then I_4T leaves s invariant. Let it carry three points A, B, C into three points A', B', C'. We next find an indirect circular transformation on s which carries A', B', C' into A, B, C, and, by VIII. 26], factor into three inversions.

Through the circles of inversion pass spheres orthogonal to s, and let I_1, I_2, I_3 be the inversions in these spheres. The transformation $I_1 I_2 I_3 I_4 T$ will be a direct transformation with a fixed sphere and three fixed points thereon, i.e. the identical transformation. Hence

$$T = I_4 I_3 I_2 I_1.$$

Theorem 19.] *Any direct spherical transformation may be factored into the product of two circular inversions.*

The proof comes from 13], 17], and 18], and is left to the reader.

Theorem 20.] *Any indirect spherical transformation of pentaspherical space may be factored into the product of five inversions.*

It would be tedious to discuss the various cases where pairs of spherical transformations might be commutative. We can foresee the answer for the general case from what we have already done in the case of circular transformations.

§ 2. Continuous Groups.

The classification of all real continuous groups of spherical transformations is a long and laborious task which would lead us altogether too far afield. We shall therefore content ourselves with noting the results which others have found *

Theorem 21.] *The group of all direct spherical transformations depends upon ten essential parameters, and has the following real sub-groups:*

One of seven parameters.

Three of six parameters.

One of five parameters.

Six of four parameters.

* Cf. Lie-Engel, *Theorie der Transformationsgruppen*, vol. iii, pp. 219 ff., Leipzig, 1893. In more detail Standen, *Invariante Flächen und Kurven bei konformen Gruppen des Raumes*, Dissertation, Leipzig, 1899. In the text we follow this enumeration, although it is not clear whether it will check up exactly with that of Amaldi, loc. cit.; in other words, it is not apparent whether Amaldi undertook to find all groups.

Eight of three parameters.

Six of two parameters.

Seven of one parameter.

Let the reader note that the existence of seven one-parameter groups agrees with theorem 9].

Before leaving spherical transformations let us return to the pentaspherical notation, or rather, to sphere coordinates, and try to find a parametric representation for the general direct case. We shall follow the classic method of Cayley.* We begin with the twin equations

$$x_i = \sum_{j=0}^{j=4} b_{ij} z_j, \quad x_i' = \sum_{j=0}^{j=4} b_{ji} z_j, \quad b_{ii} = b_{jj} = b,$$
$$b_{ij} = -b_{ji}, \quad j \neq i. \tag{9}$$

We then find at once

$$x_i + x_i' = 2bz_i,$$
$$b(zz) = (zx) = (zx'),$$
$$(xx) = (x'x').$$

We have thus, indeed, an orthogonal substitution. Solving for z_j,

$$|\, b_{ij} \,| z_j = \sum_{k=0}^{k=4} B_{jk} x_k'.$$

This gives our substitution in final form

$$|\, b_{ij} \,| x_i = \sum_{j=0}^{j=4} a_{ij} x_j'.$$
$$a_{ii} = 2bB_{ii} - |\, b_{ij} \,|, \quad a_{ij} = 2bB_{ij}, \quad j \neq i. \tag{10}$$

Unfortunately, it is not possible to express all quinary orthogonal substitutions in this form.† We have, however, ten independent parameters, so that we have a ten-parameter assemblage, and this contains no indirect transformations, for

* 'Sur quelques propriétés des déterminants gauches', *Crelle's Journal*, vol. xxxii, 1846. Cf. Pascal, *Die Determinanten*, Leipzig, 1900, pp. 159 ff.

† Cf. Netto, 'Über orthogonale Substitutionen', *Acta Mathematica*, vol. ix, 1887, p. 295.

we may pass continuously to the identical transformation. To get the real transformations of the assemblage, since a real point must have its first coordinate proportional to a pure imaginary number, and the others proportional to real numbers, we must have every a_{ij} in the first row and column of the matrix a pure imaginary with the exception of a_{ii}, which, like the other a_{jk}'s, must be real. In other words, every b_{ij} which has the subscript $_0$ appearing once only is pure imaginary; the other b_{ij}'s are real.

The theory of spherical transformations as here outlined is far behind that of circular transformations in completeness. There seems room for various interesting investigations connected therewith. A direct circular transformation has one invariant; how about a spherical transformation? For instance, there is always a fixed not null sphere. On this we have a circular transformation whose invariant must be invariant for the spherical transformation. How many of these invariants are there? What is their geometrical meaning? The detailed study of commutative transformations might be worth while. The various continuous groups must have interesting geometrical characteristics not yet discovered. The subject of spherical transformations must also have an important relation to certain systems of oriented circles in space. There is ample room for much valuable geometrical work on any of these questions.

CHAPTER X

THE ORIENTED CIRCLE

§ 1. Elementary Geometrical Theory.

We have occasionally found, in the work done so far, that the concept of the angle of two circles is lacking in precision. For instance, we saw in I. 212] that if a variable circle cut two others at given angles, it will cut at either of two supplementary angles every circle coaxal or concentric with the two. To remove this ambiguity we defined the angle of two circles in the form

$$\cos\theta = \frac{r^2 + r'^2 - d^2}{2\,rr'}, \tag{1}$$

and were thus enabled to avoid the confusion as to which of two supplementary angles given circles made with each other.

It is possible to reach even greater precision in handling the angles of circles by the interesting device of assuming that the radius of a real circle may be either *positive* or *negative*, with a similar extension for complex circles.* This is accomplished analytically by the introduction of a redundant coordinate. Let us, however, postpone such a method for a short while, and begin geometrically in the finite cartesian plane of elementary geometry, the word *circle* having the restricted significance allowed in Ch. I. When a positive or negative sign has been assigned to the radius of such a circle it shall be said to be *oriented.* Let us assume, after the positive aspect of the plane has been chosen, that every circle of positive radius is described by a point moving about the circumference in the positive or counter-clockwise

* The Author has the impression that this idea is due to Cayley.

direction, when viewed from the positive side of the plane. An oriented circle of negative radius shall similarly be looked upon as generated by a clockwise moving point. Or again, we shall assume that the normal to a circle of positive radius is oriented towards the centre, while a normal to a circle of negative radius is oriented outwards.

As a circle is oriented, so a straight line may be oriented also. We may either assume it generated by a point moving in the one or the other sense, or by assuming that it divides the plane into a positive and a negative region, and that the normals to it are oriented from the negative to the positive region of the plane.

The angle of two oriented circles shall be defined as that of their oriented normals, and a similar definition shall hold for the angle of two oriented lines. This form of definition in terms of normals has the advantage of being easily extended to three dimensions. The cosine of the angle of two oriented lines or circles is thus single valued, and agrees, in the latter case, with (1).

Two oriented lines shall be said to be *properly parallel* when they have the same system of oriented normals; when the normals to one have the opposite orientation to those of the other, they are said to be *improperly parallel.* An oriented line and circle shall be said to be *properly tangent* when they touch, and have the same oriented normal at the point of contact. When there is still contact, but the normals have opposite orientation, they are said to be *improperly tangent.* Two oriented circles are said to be properly tangent when their angle is $\equiv 0 \pmod{2\pi}$. They will be both properly tangent to the same oriented line at the same point. When their angle is $\equiv \pi \pmod{2\pi}$ they are said to be *improperly tangent*: the proper tangent to one is an improper tangent to the other. Let the reader show that when two oriented circles are properly tangent they touch internally if their radii have like signs, externally if the signs are unlike.

The fundamental concepts developed in Ch. I were *point, circle, power*, and *inversion.* It is the object of the present chapter to show that these concepts have duals in the geometry of the oriented line and circle, and that a corresponding duality extends to a large number of theorems.*

Suppose that we have an oriented circle c of radius r and an oriented line l. The centre of the circle shall be C. Let

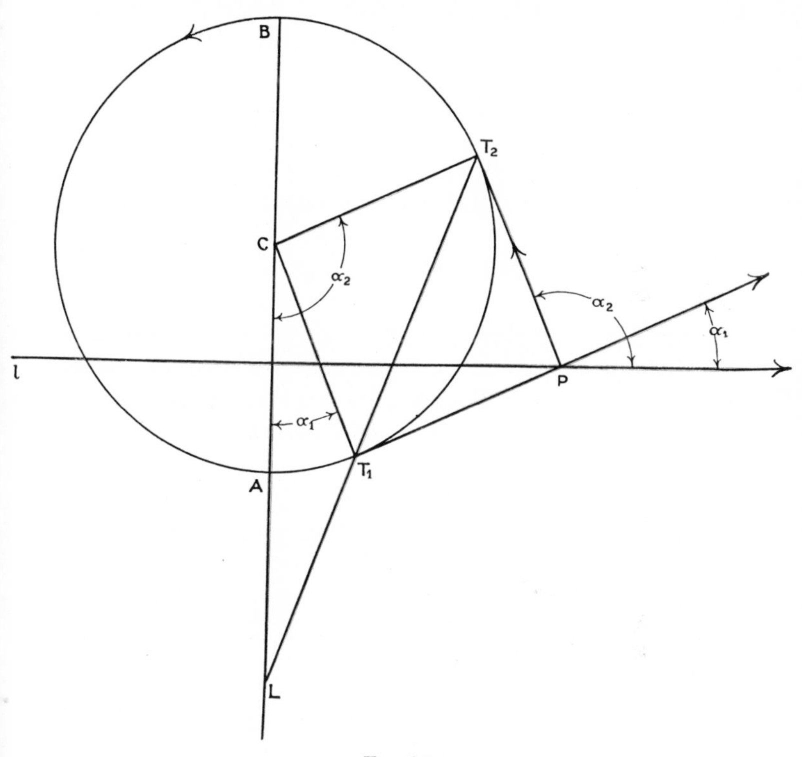

FIG. 30.

P be a point of l outside of c, and T_1, T_2 the points of contact of the oriented tangents to P from C, α_1 and α_2 being the angles which these oriented lines make with l. The line T_1T_2 passes through a fixed point L (the pole of l) as P

* The idea of this duality was certainly present to the mind of that excellent geometer Laguerre. An admirable exposition is found in Epstein, 'Die dualistische Ergänzung des Potenzbegriffes', *Zeitschrift für mathematischen Unterricht*, vol. xxxvii, 1906.

traces l. Let the line CL which is $\perp l$ meet C in A and B so that

$$\measuredangle ACT_1 = \alpha_1, \quad \measuredangle ACT_2 = \alpha_2,$$

$$(LT_1)^2 = (CL)^2 + r^2 - 2(CL)\,r\cos\alpha_1,$$

$$\tan^2\frac{\alpha_1}{2} = \frac{[(CL)-r]^2-(LT_1)^2}{-[(CL)+r]^2+(LT_1{}^2)} = \frac{-(AL)^2+(LT_1{}^2)}{(BL)^2-(LT_1)^2}.$$

In the same way we have

$$\tan^2\frac{\alpha_2}{2} = \frac{-(AL)^2+(LT_2)^2}{(BL)^2-(LT_2)^2}.$$

But
$$(LT_1)(LT_2) = (AL)(BL). \tag{2}$$

Hence
$$\tan\frac{\alpha_1}{2}\tan\frac{\alpha_2}{2} = \frac{(AL)}{(LB)}.$$

Theorem 1.] *If, from all points outside an oriented circle and lying on an oriented line, oriented lines be drawn properly tangent to the oriented circle, the product of the tangents of the halves of the angles which they form with the given oriented line is constant.*

This constant shall be called the *power* of the oriented line with regard to the oriented circle. It will be positive when the oriented line intersects the oriented circle in real points, negative when there is no common point. When there is proper contact the power is zero; improper contact produces an infinite power.

Theorem 2.] *Given two pairs of oriented lines concurrent on a given oriented line, but not concurrent or parallel with one another. A necessary and sufficient condition that they should be properly tangent to one same oriented circle is that the product of the tangents of the halves of the angles which one pair make with the fifth line should be equal to the corresponding product for the other pair.*

Two non-oriented circles which are non-concentric and unequal in radius have two centres of similitude. Two oriented circles shall be defined as having at most and in general one centre of similitude, namely, the external centre when they are non-concentric and have unequal radii of like sign, the internal centre when their radii are of unlike sign.

Theorem 3.] *An oriented line having like powers with regard to two oriented circles of unequal radius passes through their centre of similitude, and every line through this centre has like powers with regard to the two. When the radii are equal the line is parallel to the line of centres, and every such line has the same power with regard to each circle.*

We have so far established the following duality:

Oriented line.	Point.
Oriented circle.	Circle.
Power.	Power.
Centre of similitude.	Radical axis.
Circles with common centre of similitude.	Coaxal circles.

Suppose that we have an oriented line l; we may transform other oriented lines not parallel thereto as follows. Corresponding oriented lines shall be concurrent on l, and the product of the tangents of the halves of their angles with l shall be a given constant. Such a transformation shall be called a *Laguerre inversion*; let us show that it carries an oriented circle into another oriented circle.*

To begin with, the transform of every oriented line not parallel to l is uniquely determined. By (1) there are ∞^2 oriented circles which are transformed into themselves; anallagmatic let us say. Every oriented line parallel to the line is properly tangent to ∞^1 of these circles. The remainder of the envelope of these oriented circles is a second oriented line parallel to l which we define as the transform of the first oriented line. The transformation is thus one to one for all oriented lines; the oriented line l is reversed.

We next observe that, if we take a point on the radical axis of two circles and draw one tangent to each, the line connecting the points of contact meets the two circles at

* Laguerre, 'Sur la transformation par semi-droites réciproques, *Nouvelles Annales de Math.*, Series 3, vol. i, 1882, pp. 542 ff.

equal angles, and so passes through a centre of similitude, or may be parallel to the line of centres when the radii are equal. The tangent to one circle where it meets again the line connecting the points of contact is parallel to the tangent to the other circle. The two tangents to the first circle meet on the polar of the centre of similitude; hence the product of the tangents of the halves of their angles therewith is constant Hence the product of the tangents of the halves of the angles which the given tangents, properly oriented, make with the oriented radical axis is also constant. If one circle and radical axis be given, the other circle may be found so that this product shall take any desired value. Hence in a Laguerre inversion an oriented circle goes into an oriented circle, l being the radical axis. Be it noticed that as a common proper tangent to two oriented circles goes into a common proper tangent to their transforms, the two meeting on the common radical axis of each circle of the first pair with its mate in the second, then the common proper tangential segment of two oriented circles is equal to that for their transforms under a Laguerre inversion. The Laguerre inversion is the simplest type of what we shall later study under the general name of *equilong* transformation.

Laguerre inversion.	Inversion.
Oriented line to oriented line.	Point to point.
Oriented circle to oriented circle.	Circle to circle.
An oriented circle properly tangent to corresponding oriented lines anallagmatic.	Circle through two mutually inverse points anallagmatic.
Proper tangency of oriented circles invariant.	Tangency of circles invariant.
Common proper tangential segment of two oriented circles invariant.	Angle of intersection of two circles invariant.
Corresponding circles have the fundamental line as radical axis.	Corresponding circles have the fundamental point as centre of similitude.

There are a number of simple theorems concerning oriented lines which are duals to the point theorems, I. 149–63]. The algebraic proofs are, however, so much simpler than the geometric ones that we postpone these for the moment. We make an exception in favour of the following.

Let us start with four oriented lines l_1, l_2, l_3, l_4, which we suppose so related that no two are parallel. We shall mean by the bisector of the angle of two intersecting oriented lines the locus of the centres of circles properly tangent to one and improperly tangent to the other. The bisector of the angle of l_i with the *opposite to* l_j shall be l_{ij}, which is the locus of the centres of circles properly tangent to l_i and l_j. Let C_i be the centre of the circle properly tangent to $l_j l_k l_l$. Let C_i, C_k be on the same side of C_j, C_l, and let

$$\measuredangle\, \overrightarrow{C_j\,.\,C_i\,.\,C_l} > 0, \quad \measuredangle\, \overrightarrow{C_j C_k C_l} > 0,$$

$$\pi - \tfrac{1}{2}[\measuredangle\, \overrightarrow{l_j l_k} + \measuredangle\, \overrightarrow{l_k l_l}] = \measuredangle\, \overrightarrow{C_j C_i C_l} \quad \text{or} \quad \pi - \measuredangle\, \overrightarrow{C_j C_i C_l},$$

$$\pi - \tfrac{1}{2}[\measuredangle\, \overrightarrow{l_j l_i} + \measuredangle\, \overrightarrow{l_i l_l}] = \measuredangle\, \overrightarrow{C_j C_k C_l} \quad \text{or} \quad \pi - \measuredangle\, \overrightarrow{C_j C_k C_l}.$$

Subtracting

$$\begin{aligned} 0 &= \tfrac{1}{2}[\measuredangle\, \overrightarrow{l_i l_j} + \measuredangle\, \overrightarrow{l_j l_k} + \measuredangle\, \overrightarrow{l_k l_l} + \measuredangle\, l_l l_i] \\ &= \measuredangle\, \overrightarrow{C_j C_i C_l} - \measuredangle\, \overrightarrow{C_j C_k C_l}, \quad \text{or} \\ &= \measuredangle\, \overrightarrow{C_j C_i C_l} + \measuredangle\, \overrightarrow{C_j C_k C_l} \pm \pi. \end{aligned}$$

We see in a special case that the first hypothesis is right, and so, by continuity, it is always right.

Theorem 4.] *The centres of the oriented circles each properly tangent to three out of four given oriented lines, whereof no two are parallel, are concyclic.*

§ 2. Analytic Treatment.

It is now time to take up the analytic treatment of oriented lines and circles, as thus, naturally, we shall obtain a far greater wealth of results than from purely geometric methods. The domain shall be the complex cartesian plane, including the line at infinity. We shall slightly alter the traditional form for the equation of a circle, writing

$$x_0(x^2+y^2)+x_2(2xt)+x_3(2yt)+x_1(2t^2)=0. \tag{3}$$

$$-2x_0x_1+x_2^2+x_3^2+x_4^2\equiv 0. \tag{4}$$

The radius will have the value

$$r=\frac{ix_4}{x_0}. \tag{5}$$

We see thus that we have for every oriented circle five homogeneous coordinates (x) connected by the identity (4). Conversely, we shall define as an oriented circle every locus which satisfies the equations (3) and (4). We have the following types:

1) Proper oriented circles $x_0x_4 \neq 0.$

2) Non-linear null circles $x_4=0, \quad x_0\neq 0.$

3) Oriented lines $x_0=0.$

4) Minimal lines $x_0=x_4=0.$

The line at infinity is included in this latter class. This may also be looked upon as the class of all oriented lines which are identical with their opposites. If we have two proper oriented circles (x) and (y), the length of their proper common tangential segment will be

$$\sqrt{d^2-(r-r_1)^2}=\sqrt{\frac{2(-x_0y_1-x_1y_0+x_2y_2+x_3y_3+x_4y_4)}{-x_0y_0}}. \tag{6}$$

For their common angle we have the expression

$$\sin^2\frac{\theta}{2}=\frac{-x_0y_1-x_1y_0+x_2y_2+x_3y_3+x_4y_4}{2x_4y_4}. \tag{7}$$

By continuity this formula will hold even when $x_0y_0=0$. The condition of proper tangency will be

$$-x_0y_1-x_1y_0+x_2y_2+x_3y_3+x_4y_4=0. \tag{8}$$

The point of contact may be a circular point at infinity; the radii are then equal and the centres on a minimal line.

It must be specially noted that this is the identity polarized.

A point in the tetracyclic plane has four homogeneous coordinates connected by a quadratic identity; the same is true for an oriented line.* Fundamentally important, however, is this difference, that whereas the discriminant of the quadratic identity for tetracyclic coordinates does not vanish, the oriented line identity has a vanishing discriminant. The two may not therefore be put into a one to one analytic correspondence. Still, the circle has a linear equation in tetracyclic coordinates; so, too, a linear relation among the coordinates of an oriented line will usually give us the proper tangents to a circle. Let us begin by writing

$$-a_0x_1+a_2x_2+a_3x_3+a_4x_4=0. \tag{9}$$

When $a_0\neq 0$ we see that (x) is properly tangent to the oriented circle

$$\left(a_0,\quad \frac{a_2^{\,2}+a_3^{\,2}+a_4^{\,2}}{2a_0},\quad a_2,\quad a_3,\quad a_4\right).$$

When $a_0=0,\ a_2^{\,2}+a_3^{\,2}\neq 0$ the slope will be

$$-\frac{x_2}{x_3}=\frac{a_2a_3\pm ia_4\sqrt{a_2^{\,2}+a_3^{\,2}+a_4^{\,2}}}{a_2^{\,2}+a_4^{\,2}}.$$

The oriented line makes a fixed angle with a fixed oriented line. An equation of the first degree will give either the oriented lines properly tangent to a fixed circle or two distinct or coincident pencils of parallels. The power of the oriented line (x) with regard to the oriented circle (y) will be

$$\frac{-x_1y_0+x_2y_2+x_3y_3+x_4y_4}{x_1y_0-x_2y_2-x_3y_3+x_4y_4}. \tag{10}$$

If (a) be a fixed non-minimal oriented line, and (x) an arbitrary oriented line, let us write the following transformation:

$$\begin{aligned} x_i'&=(\alpha_4^{\,2}-a_4^{\,2})\,x_i-2\,(a_2x_2+a_3x_3+\alpha_4x_4)\,a_i, \quad i\neq 4,\\ x_4'&=(\alpha_4^{\,2}-a_4^{\,2})\,x_4-2\,(a_2x_2+a_3x_3+\alpha_4x_4)\,\alpha_4. \end{aligned} \tag{11}$$

* The comparison of the two is well brought out by Müller, 'Die Geometrie orientierter Kugeln', *Monatshefte für Math.*, vol. ix, 1898, pp. 288 ff.

Here, regardless of the value of the parameter α_4, we have

$$(x_2'^2+x_3'^2+x_4'^2) = (a_2^2+a_3^2+\alpha_4^2)^2\,(x_2^2+x_3^2+x_4^2).$$

The oriented lines (x) and (x') are concurrent on the line (a), the product of the tangents of half their angles therewith is $\frac{\alpha_4-a_4}{\alpha_4+a_4}$. We have thus, in general, i.e. when $\alpha_4^2\neq a_4^2$, a Laguerre inversion, and every Laguerre inversion may be thrown into this form. In the limiting case where $a_2=a_3=0$, we replace each oriented line by its opposite moved through a fixed distance.

It seems almost axiomatic that there must be a number of interesting theorems about oriented lines arising from the adaptation of the Frobenius identity. The present Author has been somewhat disappointed in the results obtained in this fashion. There is one interesting theorem which is perhaps more easily proved in this way than in any other. Let us take four proper oriented circles $c^{(1)}$, $c^{(2)}$, $c^{(3)}$, $c^{(4)}$, whereof each is properly tangent to the preceding and the succeeding in the natural cyclic order. The common proper oriented tangents shall be $l^{(12)}$, $l^{(23)}$, $l^{(34)}$, $l^{(41)}$. We write the identity

$$\begin{vmatrix} c_0^{(1)} & c_1^{(1)} & c_2^{(1)} & c_3^{(1)} & c_4^{(1)} & 0 \\ c_0^{(2)} & c_1^{(2)} & c_2^{(2)} & c_3^{(2)} & c_4^{(2)} & 0 \\ 0 & l_1^{(12)} & l_2^{(12)} & l_3^{(12)} & l_4^{(12)} & 0 \\ 0 & l_1^{(23)} & l_2^{(23)} & l_3^{(23)} & l_4^{(23)} & 0 \\ 0 & l_1^{(34)} & l_2^{(34)} & l_3^{(34)} & l_4^{(34)} & 0 \\ 0 & l_1^{(41)} & l_2^{(41)} & l_3^{(41)} & l_4^{(41)} & 0 \end{vmatrix} \times \begin{vmatrix} -c_1^{(1)} & -c_0^{(1)} & c_2^{(1)} & c_3^{(1)} & c_4^{(1)} & 0 \\ -c_1^{(2)} & -c_0^{(2)} & c_2^{(2)} & c_3^{(2)} & c_4^{(2)} & 0 \\ -c_1^{(3)} & -c_0^{(3)} & c_2^{(3)} & c_3^{(3)} & c_4^{(3)} & 0 \\ -c_1^{(4)} & -c_0^{(4)} & c_2^{(4)} & c_3^{(4)} & c_4^{(4)} & 0 \\ -y_1 & -y_0 & y_2 & y_3 & y_4 & 0 \\ 1 & 0 & 0 & 0 & 0 & 0 \end{vmatrix} = 0.$$

Here (y) is supposed to be properly tangent to $l^{(12)}, l^{(23)}, l^{(34)}$. Divide each of the first two rows of the first determinant by its first member; divide each of the first five rows of the second determinant by its second member. Multiply the two determinants together by rows. For simplicity write $T(ab)$ for half the square of the common tangential segment of the circles (a) and (b).

$$\begin{vmatrix} 0 & 0 & T(c^{(1)}c^{(3)}) & 0 & T(c^{(1)}y) & 1 \\ 0 & 0 & 0 & T(c^{(2)}c^{(4)}) & T(c^{(2)}y) & 1 \\ 0 & 0 & \alpha & \beta & 0 & 0 \\ \gamma & 0 & 0 & \delta & 0 & 0 \\ \epsilon & \zeta & 0 & 0 & 0 & 0 \\ 0 & \eta & \theta & 0 & A & 0 \end{vmatrix} = 0.$$

Now $$T(c^{(1)}y) = T(c^{(2)}y).$$

Taking a constant multiple of the members of the last column from the corresponding ones of the next to the last, we find the product of two factors is equal to zero. The second factor is found not to vanish in a special case, the first is

$$A = \frac{-l_1^{(41)}y_0 + l_2^{(41)}y_2 + l_3^{(41)}y_3 + l_4^{(41)}y_4}{y_0}.$$

Theorem 5.] *If four proper oriented circles be so arranged in cyclic order that each is properly tangent to its two next neighbours, then the four common proper tangents to pairs of successive circles are properly tangent to a fifth oriented circle or are properly parallel in pairs.**

Since the two tangential segments to two circles from a point on their radical axis are equal, we may construct a circle tangent to each at its point of contact with one of the given circles.

Theorem 6.] *From each of two points on the radical axis of two non-concentric oriented circles two oriented proper tangents are drawn to each circle. Take two of the tangents to each circle, not intersecting on the radical axis. These four lines, if not properly parallel in pairs, may be so oriented as to be properly tangent to an oriented circle.*

A number of interesting theorems follow from the fact that the condition for proper contact between an oriented line and circle is linear. For example

* Müller, 'Einige Gruppen von Sätze, &c.', *Jahresbericht der deutschen Mathematiker-Vereinigung*, vol. xx, 1911, p. 181.

Theorem 7.] *If four proper oriented circles meet a fixed oriented line at a fixed angle, and if a common proper tangent be drawn to each two, then, if no two of these tangents be parallel, four new oriented circles may be found, each properly tangent to three of these tangents which do not touch one particular circle. These four new oriented circles will also meet a fixed oriented line at a fixed angle,* or be the limiting form of four such circles.*

The proof is as follows. If an oriented circle meet a fixed line at a fixed angle we see by (7) that all of the coordinates but the second are connected by a linear relation, and conversely, if such a relation exist, the circle will usually meet a fixed line at a fixed angle. The exceptions to this last statement are easily noted. Secondly, if three oriented lines be given, all of the coefficients but the second of the oriented circle properly tangent to them are given by the three rowed determinants of the matrix formed by the coordinates of the given oriented lines, the first coefficient having the reverse of the natural sign. These facts premised, let our oriented circles be $c^{(1)}$, $c^{(2)}$, $c^{(3)}$, $c^{(4)}$, the common tangents $l^{(ij)}$ as before. Finding all the coordinates but the second of the oriented circles each tangent properly to three of these lines which lack a common index, the statement that these new oriented circles meet a fixed oriented line at a fixed angle will lead to the equations

$$|\, a \; l^{(34)} \; l^{(42)} \; l^{(23)} \,| = 0,$$

$$|\, a \; l^{(34)} \; l^{(41)} \; l^{(12)} \,| = 0,$$

$$|\, a \; l^{(34)} \; l^{(41)} \; l^{(13)} \,| = 0,$$

$$|\, a \; l^{(23)} \; l^{(31)} \; l^{(12)} \,| = 0.$$

Eliminating (a)

$$\begin{vmatrix} 0 & |\, l^{(34)} \; l^{(42)} \; l^{(23)} \; l^{(31)} \,| & |\, l^{(34)} \; l^{(42)} \; l^{(23)} \; l^{(12)} \,| \\ |\, l^{(34)} \; l^{(41)} \; l^{(12)} \; l^{(23)} \,| & |\, l^{(34)} \; l^{(41)} \; l^{(12)} \; l^{(31)} \,| & 0 \\ |\, l^{(34)} \; l^{(41)} \; l^{(13)} \; l^{(23)} \,| & 0 & |\, l^{(34)} \; l^{(41)} \; l^{(13)} \; l^{(12)} \,| \end{vmatrix} = 0.$$

* Ibid., p. 183.

This last equation will be unaltered if we replace each $l^{(ij)}$ by the complementary $l^{(kl)}$, which has the effect of replacing each of our new circles by the corresponding old circle. By hypothesis, the four original circles met a fixed oriented line at a fixed angle. Hence the four new ones will do the same, or at least be the limits of four circles doing so.

Suppose that we have three oriented circles $c^{(1)}$, $c^{(2)}$, $c^{(3)}$ properly tangent to an oriented line l. The remaining common proper tangents shall be $l^{(ij)}$, $l^{(i)}$ shall be an arbitrary proper tangent to $c^{(i)}$, and the oriented circle properly tangent to $l^{(i)}$, $l^{(j)}$, $l^{(ij)}$ shall be $\bar{c}^{(k)}$. Consider the one-parameter family of quadratic oriented line envelopes given by the equation

$$\lambda c^{(2)}\,\bar{c}^{(2)} + \mu c^{(3)}\,\bar{c}^{(3)} = 0.$$

All of these will share the proper tangents l, $l^{(1)}$, $l^{(12)}$, $l^{(13)}$ with $c^{(1)}$. We may therefore so choose λ/μ that this envelope shall touch another proper tangent to $c^{(1)}$, i.e. include $c^{(1)}$ as part of itself. The remainder of the envelope will be another oriented circle which is properly tangent to $l^{(2)}$, $l^{(3)}$, $l^{(23)}$. It will then be identical with $\bar{c}^{(1)}$, which latter must touch the remaining proper common tangent to $\bar{c}^{(2)}$ and $\bar{c}^{(3)}$.

Theorem 8.] *If three oriented circles be properly tangent to an oriented line, and an arbitrary proper tangent be drawn to each, then the three oriented circles, each properly tangent to two of these arbitrary tangents and to the remaining common proper tangent of the corresponding original circles, are themselves properly tangent to one oriented line or are the limit of such circles.*

This theorem is dual to our fundamental I. 149]. It may be somewhat generalized by a contact transformation of circles, as the previous one was generalized by inversion. The result is, however, rather involved; it is better to draw corollaries from the proposition as it stands. As a first, let the reader show

Theorem 9.] *If four oriented tangents to an oriented circle be taken in cyclic order, and four oriented circles be drawn each properly tangent to two successive oriented tangents, then the remaining common proper tangents to successive oriented* *circles of the sequence are themselves tangent to an oriented circle or are properly parallel in pairs.**

The number of simple results which can be deduced from this is almost absurd. We first get a precise wording for a well-known theorem due to Plücker.†

Theorem 10.] *From two points, one on each common proper tangent to two oriented circles, the remaining proper tangents to these circles are drawn. The two tangents to one circle and the opposites to those to the other touch a circle or are properly parallel in pairs.*

Theorem 11.] *Three oriented circles $c^{(1)}$, $c^{(2)}$, $c^{(3)}$ are so arranged that one common proper tangent to $c^{(1)}$ and $c^{(2)}$ and one to $c^{(1)}$ and $c^{(3)}$ are concurrent on an improper common tangent to $c^{(2)}$ and $c^{(3)}$; then the remaining common proper tangents of $c^{(1)}$, $c^{(2)}$ and $c^{(1)}$, $c^{(3)}$ are concurrent on the remaining improper tangent of $c^{(2)}$, $c^{(3)}$ or are parallel thereto.*

Theorem 12.] *Four oriented concurrent lines are arranged in cyclic order and an oriented circle drawn properly tangent to each two successive lines. Then the remaining common proper tangents to successive circles are properly tangent to one oriented circle or are properly parallel in pairs.*

Theorem 13.] *Given two oriented non-parallel lines and four oriented circles, the first properly tangent to both, the second properly tangent to the first and improperly tangent to the second, the third improperly tangent to both, and the fourth improperly tangent to the first and properly tangent to the second. The remaining common proper tangents of successive circles are properly tangent to an oriented circle or are properly parallel in pairs.*

We get a special case of this when the first and second circles differ only in the sign of the radius from the third and fourth.

Theorem 14.] *Given two non-concentric circles and the*

* Ibid., p. 173. The five following are from the same source.

† 'Analytisch-geometrische Aphorismen', *Crelle's Journal*, vol. xi, 1834, p. 117.

two common direct or transverse common tangents. Then, if any circle be taken touching these two tangents, but not with its centre collinear with those of the given circles, its remaining common tangents with these circles are tangent to another circle.

Theorem 15.] *Let four oriented lines be drawn properly tangent to an oriented circle, and six oriented circles be drawn each properly tangent to two of these oriented lines. These six may be arranged in three sequences of four, each sequence determining a new oriented circle by Theorem* 9]. *The three new oriented circles are properly tangent to the same two oriented lines.*

The proof is entirely analogous to II. 22] though demanding a little more care. We leave the details to the reader and continue the process of drawing corollaries from 8] even as we drew some from I. 149].

Theorem 16.] *Given four oriented circles properly tangent to an oriented line. If four other oriented circles can be found, each properly tangent to the three remaining common proper tangents to three pairs of the given circles, these four new oriented circles are also properly tangent to a common line.*

Theorem 17.] *Given five oriented circles properly tangent to a common line. With each set of four we may, by* 16], *associate another oriented line, and these five oriented lines are properly tangent to an oriented circle.*

The proof is entirely analogous to that of I. 160]. We leave the details to the reader, as well as the task of proving the following, which is dual to I. 162].

Theorem 18.] *Given n oriented circles properly tangent to an oriented line. If n be odd we may associate therewith an oriented circle, if n be even an oriented line, in such a way that the circle (line) is properly tangent to the n oriented lines (circles) associated with the n sets of oriented circles obtained by omitting each of the original ones in turn.*

In the same way we have the dual to I. 163].

Theorem 19.] *Given n oriented circles properly tangent to one oriented line, and an oriented tangent common to each and to a given oriented circle. If n be odd we may associate therewith an oriented line, and if it be even an oriented circle, in such a way that the line (circle) is properly tangent to the oriented circles (lines) associated with the n sets of $(n-1)$ oriented circles obtained by omitting each of the original circles in turn.*

It is to be noted that in 17] to 19] lines properly tangent to an oriented circle will, if two be properly parallel, be replaced by lines of given direction.

§ 3. Laguerre Transformations.

In our development of the analogy between the geometry of the oriented line and elementary circle geometry we have pointed out the three following analogies :

Oriented line.	Point.
Oriented circle.	Circle.
Tangential segment of two oriented circles.	Angle of two circles.

We are immediately led to the idea that there must be a whole theory of transformations which carry oriented lines into oriented lines, and oriented circles into oriented circles. An example of such a transformation was the Laguerre inversion, under which the common tangential segment of two oriented circles is invariant. The question arises immediately, Is this segment invariant under every transformation that carries an oriented line into an oriented line, and the oriented lines properly tangent to an oriented circle into proper tangents to another circle ? The answer to this question is ‘ No ’. At the same time the total group of such transformations has an important sub-group where tangential segments do retain their size. This group presents interesting analogies to the conformal group of circular transformations, and we find invariants analogous to the double angle and double ratio. Before attacking this group directly, let us give a space representation of our oriented lines and circles which is of capital importance in our subsequent work with them.*

We begin with the equations

$$\begin{aligned} T^2 &\equiv \rho x_0, \\ \tfrac{1}{2}(X^2+Y^2+Z^2) &\equiv \rho x_1, \\ -XT &\equiv \rho x_2, \\ -YT &\equiv \rho x_2, \\ -ZT &\equiv \rho x_4. \end{aligned} \tag{12}$$

Here $(XYZT)$ are supposed to indicate homogeneous rectangular cartesian coordinates. The null sphere with centre $(XYZT)$ will have the equation

$$(Tx-Xt)^2+(Ty-Yt)^2+(Tz-Zt)^2 = 0.$$

It will meet the plane $z = 0$ in the circle

$$x_0(x^2+y^2)+x_2(2xt)+x_3(2yt)+x_1(2t^2) = 0. \tag{3}$$

$$-2x_0x_1+x_2{}^2+x_3{}^2+x_4{}^2 = 0. \tag{4}$$

This is the oriented circle with centre (XYT) and radius

$$\frac{ix_4}{x_0} = -\frac{iZ}{T}.$$

If now we make our three-dimensional space a perfect pentaspherical continuum by adjoining improper points, as in Ch. VII, the improper point

$$x_0 = x_2{}^2+x_3{}^2+x_4{}^2 = 0,$$

being interpreted as the minimal plane

$$x_2x+x_3y+x_4z+x_1t = 0,$$

we see that the latter cuts our plane $z = 0$ in the oriented line

$$x_2x+x_3y+x_1t = 0.$$

This transformation shall be called a *minimal projection.*

* Cf. Klein, *Höhere Geometrie*, Göttingen, 1893, vol. i, pp. 473 ff.; Lie-Scheffers, *Berührungstransformationen*, Leipzig, 1896, pp 428 ff.; Scheffers, 'Bestimmung aller Berührungstransformationen des Kreises in der Ebene', *Leipziger Berichte*, vol. li, 1899, p. 145.

Theorem 20.] *If finite cartesian space be made a perfect pentaspherical continuum by adjoining a single point at infinity, and the totality of minimal planes as improper points, there is a perfect one to one correspondence between the points of such a space and the oriented circles of the plane. In this correspondence each proper point corresponds to that proper oriented circle whose centre is the orthogonal projection of the given point on the plane, and whose radius is the algebraic distance from the point to the plane multiplied by $-i$, a definite square root of -1; each improper point is represented by the line where the corresponding minimal plane meets the given plane, with an orientation rationally dependent upon the coordinates of the improper point.*

If $(XYZT)$ and $(X'Y'Z'T')$ be two proper points, their distance will be

$$\sqrt{\frac{(XT'-TX')^2+(YT'-TY')^2+(ZT'-TZ')^2}{T^2T'^2}}$$

$$\equiv\sqrt{\frac{2\left(-x_0y_1-x_1y_0+x_2y_2+x_3y_3+x_4y_4\right)}{-x_0y_0}}. \quad (13)$$

Theorem 21.] *In a minimal projection the distance between two proper points is equal to the common tangential segment of the corresponding proper oriented circles.*

This last theorem is also simply proved by elementary geometry. Two proper points whose distance is null determine an isotropic, the locus of points at a null distance from both. If a proper point lie in a minimal plane, there is one isotropic through the point lying in the plane. Parallel minimal planes determine one tangent to the circle at infinity. Our correspondence is as follows:

Plane π.	Pentaspherical space Σ.
Oriented circle.	Point.
Proper oriented circle.	Proper point.
Non-linear null circle.	Point of plane π.
Oriented line.	Minimal plane.
Minimal line.	Minimal plane parallel to normals to π.

Common tangential segment of proper oriented circles.	Distance of proper points.
Pencil of properly tangent oriented circles.*	Isotropic in finite domain.
Pencil of properly parallel oriented lines.	Pencil of parallel minimal planes.
Transformation carrying oriented line into oriented line, and proper tangents to oriented circle into proper tangents to oriented circle.	Conformal collineation.

Theorem 22.] *The group of transformations which carry oriented lines into oriented lines, and the proper tangents to an oriented circle into those of another circle, depends upon seven parameters. Every transformation of the group will multiply the common tangential segment of two proper oriented circles by some constant. There is a six-parameter sub-group that keeps such tangential segments invariant.*

Seven-parameter group of oriented lines and circles as envelopes of such.	Seven-parameter group of conformal collineations.

Theorem 23.] *The six-parameter group which carries oriented lines into oriented lines, oriented circles which are the envelopes of such lines into other such circles, and keeps invariant the common tangential segment of two proper oriented circles, is mixed. It has a six-parameter continuous sub-group, and a six-parameter continuous sub-assemblage.*

Six-parameter group keeping tangential segments invariant.	Six-parameter congruent group.
Six-parameter continuous sub-group.	Six-parameter group of motions.
Six-parameter sub-assemblage.	Six-parameter assemblage of symmetry transformations.

* Proper circles of equal radii whose centres lie on a minimal line must be considered as properly tangent.

The mixed six-parameter group for the oriented line shall be called the group of *Laguerre transformations*, or the *Laguerre group*. We may reach this group from another point of view which possesses the highest interest. Let us define as *equilong* any analytic transformation of the oriented lines of a plane which keeps invariant the distance between the points of contact of each line with any two oriented envelopes properly tangent to it.* This is the natural dual to a conformal transformation. Even as in determining the most general conformal transformation of the Gauss plane we express each point by means of a single complex coordinate, so here we shall introduce complex coordinates of another sort for an oriented line.† We begin by writing

$$\zeta \equiv \xi + \epsilon\eta, \quad \bar{\zeta} \equiv \xi - \epsilon\eta, \quad \epsilon^2 = 0,$$
$$\zeta \equiv \frac{x_4 + \epsilon x_1}{x_2 + i x_3}. \tag{14}$$

Every oriented line not passing through a specified circular point at infinity will thus have a complex coordinate ζ. When $x_2 + ix_3 = 0$ $\zeta = \infty$, and a whole parallel pencil of isotropics correspond to the single value ∞. Conversely, suppose that we have given $\zeta = \xi + \epsilon\eta$, we may write

$$\rho x_1 = 2\eta, \quad \rho x_2 = (1 - \xi^2), \quad \rho x_3 = -i(1 + \xi^2), \quad \rho x_4 = 2\xi, \tag{15}$$

and thus find a determinate oriented line corresponding to our complex value ζ. Suppose now that we have an equilong transformation. It will give rise to equations of the type

$$\xi' = \xi'(\xi\eta), \quad \eta' = \eta'(\xi\eta).$$

If an oriented line touch an envelope at infinity, the same will hold for the transformed line and envelope, hence in-

* Cf. Scheffers, 'Isogonalkurven, Equitangentialkurven, und complexe Zahlen', *Math. Annalen*, vol. lx, 1905.

† First done by Scheffers, ibid., p. 528. We follow the treatment of Blaschke, 'Untersuchungen über die Geometrie der Speere in der Euklidischen Ebene', *Monatshefte für Math.*, vol. xxi, 1910. We shall lean heavily on this excellent article in the present chapter. See also Grünwald, 'Duale Zahlen in der Geometrie', ibid., vol. xvii, 1906.

finitely near parallel lines go into other such lines, i.e. ξ' is a function only of ξ. The invariance of the distance of the intersections of a line with two others infinitely near requires that the expression

$$\left[\frac{d\eta\,\delta\xi - d\xi\,\delta\eta}{d\xi\,\delta\xi}\right]^2$$

shall be invariant. The corresponding expression for $\xi'\eta'$ will give the value

$$\frac{d\eta'\delta\xi' - d\xi'\delta\eta'}{d\xi'\delta\xi'}$$

$$\equiv \pm \frac{\left(\frac{\partial\eta'}{\partial\xi}d\xi + \frac{\partial\eta'}{\partial\eta}d\eta\right)\left(\frac{\partial\xi'}{\partial\xi}\delta\xi + \frac{\partial\xi'}{\partial\eta}\delta\eta\right) - \left(\frac{\partial\eta'}{\partial\xi}\delta\xi + \frac{\partial\eta'}{\partial\eta}\delta\eta\right)\left(\frac{\partial\xi'}{\partial\xi}d\xi + \frac{\partial\xi'}{\partial\eta}d\eta\right)}{\left(\frac{\partial\xi'}{\partial\xi}d\xi + \frac{\partial\xi'}{\partial\eta}d\eta\right)\left(\frac{\partial\xi'}{\partial\xi}\delta\xi + \frac{\partial\xi'}{\partial\eta}\delta\eta\right)}$$

$$\equiv \pm \frac{\left(\frac{\partial\eta'}{\partial\eta}\right)}{\left(\frac{\partial\xi'}{\partial\xi}\right)}\frac{(d\eta\,\delta\xi - d\xi\,\delta\eta)}{d\xi\,\delta\xi}.$$

On the other hand, if we have either function of the complex variable,

$$(\xi' + \epsilon\eta') = f(\xi + \epsilon\eta), \quad (\xi' + \epsilon\eta') = f(\xi - \epsilon\eta),$$

the differential equations corresponding to the Cauchy-Riemann equations for the usual complex variable are

$$\frac{\partial\eta'}{\partial\eta} = \pm\frac{\partial\xi'}{\partial\xi}, \quad \frac{\partial\xi'}{\partial\eta} = 0, \tag{16}$$

and these give an equilong transformation above. Conversely, when these equations are satisfied ζ' is an analytic function of ζ or of $\bar{\zeta}$.

Theorem 24.] *The most general equilong transformation of the plane is obtained by taking ζ' as an analytic function of ζ or of $\bar{\zeta}$, and, conversely, every such function will give an* *equilong transformation. The group of all equilong transformations is mixed, having a sub-group which is continuous and depends upon an arbitrary function, and also a continuous sub-assemblage depending on an arbitrary function.*

Every equilong transformation of the sub-group shall be called *direct*, those of the sub-assemblage *indirect*. Every indirect transformation is the product of a direct transformation, and of the indirect one

$$\zeta' = -\bar{\zeta},$$

which reverses the orientation of every line. We may pass continuously from any direct transformation to the identical one; not so for an indirect transformation.

The Laguerre transformation is a special case of the equilong. The analytical expressions for the direct and indirect transformations are

$$\xi' + \epsilon\eta' = \frac{(\alpha + \epsilon\alpha')(\xi + \epsilon\eta) + (\beta + \epsilon\beta')}{(\gamma + \epsilon\gamma')(\xi + \epsilon\eta) + (\delta + \epsilon\delta')}. \tag{17}$$

$$\xi' + \epsilon\eta' = \frac{(\alpha + \epsilon\alpha')(\xi - \epsilon\eta) + (\beta + \epsilon\beta')}{(\gamma + \epsilon\gamma')(\xi - \epsilon\eta) + (\delta + \epsilon\delta')}. \tag{18}$$

The Laguerre inversion is an indirect transformation, for it is involutory and keeps invariant all circles which meet an oriented line at a fixed angle, and so corresponds to an involutory congruent transformation of space with all points of a plane invariant, i.e. a reflection in a plane.

Plane π.	Space Σ.
Laguerre inversion.	Reflection in a plane.

Four oriented lines will have an absolute complex invariant under the group of direct Laguerre transformations, namely

$$\frac{(\zeta_1 - \zeta_2)(\zeta_3 - \zeta_4)}{(\zeta_1 - \zeta_4)(\zeta_3 - \zeta_2)}.$$

An indirect transformation will carry this into its conjugate. The part independent of η is seen at once to be

$$\frac{(\xi_1-\xi_2)(\xi_3-\xi_4)}{(\xi_1-\xi_4)(\xi_3-\xi_2)}=\frac{\sqrt{\sum\limits_{i=2}^{i=4} x_i^{(1)}x_i^{(2)}}\sqrt{\sum\limits_{i=2}^{i=4} x_i^{(3)}x_i^{(4)}}}{\sqrt{\sum\limits_{i=2}^{i=4} x_i^{(1)}x_i^{(4)}}\sqrt{\sum\limits_{i=2}^{i=4} x_i^{(3)}x_i^{(2)}}}$$

$$=\frac{\sin\frac{1}{2}\measuredangle\theta_{12}\sin\frac{1}{2}\measuredangle\theta_{34}}{\sin\frac{1}{2}\measuredangle\theta_{14}\sin\frac{1}{2}\measuredangle\theta_{32}}. \quad (19)$$

This is the cross ratio of the points of contact with a fixed oriented proper circle of proper tangents properly parallel to the given oriented lines. The coefficient of ϵ is more complicated: it may be written

$$\frac{\sum\limits_{i=1}^{i=4}\pm\eta_i(\xi_j-\xi_k)(\xi_k-\xi_l)(\xi_l-\xi_j)}{(\xi_1-\xi_4)^2(\xi_3-\xi_2)^2}=\frac{-i\,|\,x_1^{(1)}x_2^{(2)}x_3^{(3)}x_4^{(4)}|}{2\sum\limits_{i=2}^{i=4}x_i^{(1)}x_i^{(4)}\sum\limits_{i=2}^{i=4}x_i^{(3)}x_i^{(2)}}.$$

The ratio between the two invariants is a third invariant

$$\frac{-i\,|\,x_1^{(1)}x_2^{(2)}x_3^{(3)}x_4^{(4)}\,|}{\sqrt{\sum\limits_{i=2}^{i=4}x_i^{(1)}x_i^{(2)}}\sqrt{\sum\limits_{i=2}^{i=4}x_i^{(3)}x_i^{(4)}}\sqrt{\sum\limits_{i=2}^{i=4}x_i^{(1)}x_i^{(4)}}\sqrt{\sum\limits_{i=2}^{i=4}x_i^{(3)}x_i^{(2)}}} \quad (20)$$

We get the meaning as follows. Let the four oriented lines be arranged in cyclic order, and four oriented circles be taken, each properly tangent to two successive lines of the system. The difference between the sum of the first and third sides of the oriented (perhaps re-entrant) quadrilateral determined by the given lines in order, and the sum of the second and fourth sides, is equal to the corresponding expression among the common tangential segments of the four circles, each circle corresponding to the vertex of the quadrilateral which is the intersection of the two oriented lines which it touches. This expression among the common tangential segments is an invariant for a Laguerre transformation; hence the difference between the sums of pairs of opposite sides of the quadrilateral is also invariant. Reverting to our expression (2), let us take for our oriented lines (x_1, x_2, x_3, x_4) $(0,1,0,i)$ $(0,a,b,i\sqrt{a^2+b^2})$ $(0,0,1,i)$, as any four may be reduced to these by a suitable Laguerre transformation. The quadrilateral has for its sides three sides of a right triangle and an infinitesimal side passing through the vertex of the right angle. The invariant is thus the difference between the sum of the two legs and the hypotenuse. But our expression (26) reduces to

$$\frac{x_1(a+b-\sqrt{a^2+b^2})}{2\sqrt{x_2+ix_4}\sqrt{x_3+ix_4}\sqrt{a-\sqrt{a^2+b^2}}\sqrt{b-\sqrt{a^2+b^2}}}$$
$$=\frac{x_1}{x_2+x_3+ix_4}$$
$$=\frac{x_1(x_2+x_3-ix_4)}{2x_2x_3},$$

which is one-half the difference in question.

Theorem 25.] *The complex invariant of four oriented lines in chosen order is made up of two parts. The first part is the corresponding cross ratio of the points of contact with any proper oriented circle of four oriented proper tangents, properly parallel to the given lines; the ratio of the two parts is one-half the difference between the sums of the pairs of opposite sides of the quadrilateral determined by the given lines in the given order.**

The invariants of a Laguerre transformation and the simplification of figure by these transformations lead us to certain properties which might not suggest themselves naturally. Let us say that three pairs of oriented lines belong to *an involution* when this is true of the points of contact with a fixed proper oriented circle of proper tangents properly

* Blaschke, loc. cit., p. 17, confuses the second part of the invariant with the ratio of the two parts. He courteously acknowledged the mistake when it was shown to him.

parallel to them. This condition is invariant for a Laguerre transformation.* We thus get

Theorem 26.] *If three proper oriented circles have each a pair of proper common tangents, these will be three pairs of an involution.*

We see, in fact, that there is a motion of our three-dimensional space Σ which will bring any three proper points to lie in a plane $\parallel \pi$; hence there is a Laguerre transformation of π which will carry three proper circles into three others all of equal radius. The common tangents to these latter are improperly parallel in pairs, and so clearly are pairs of an involution.

Suppose that we have $l_1 l_1'$, $l_2 l_2'$, $l_3 l_3'$, three pairs of a non-parabolic involution. Since we may find a motion of Σ to bring any two non-isotropic lines to be parallel to the plane π, so we may find a Laguerre transformation carrying $l_1 l_1'$ and $l_2 l_2'$ into two pairs of parallel lines, and owing to the existence of the involution $l_3 l_3'$ become a third pair of parallel lines also. Let the triangle whose side-lines are $\bar{l}_1 \bar{l}_2 \bar{l}_3$ be marked according to the standard notation of Ch. I, A_i being opposite to $\bar{l}_i$. Let $\bar{l}_i'$ meet $\bar{l}_j \bar{l}_k$ in points whose distances from A_i are $r_i a_j$, $r_i a_k$. Under these circumstances it is an easy matter to calculate the lengths of the common tangential segments of the oriented circles properly tangent to the triads of oriented lines; then, exercising great care as to signs, we apply Casey's criterion, I. 47]. We thus reach an admirable theorem due to Bricard.†

Theorem 27.] *If $l_1 l_1'$, $l_2 l_2'$, $l_3 l_3'$ be three pairs of oriented lines in involution, the four oriented circles properly tangent respectively to the triads $l_1 l_2 l_3$, $l_1 l_2' l_3'$, $l_1' l_2 l_3'$, $l_1' l_2' l_3$ are properly tangent to a fifth oriented circle.*

Numerous corollaries follow immediately. The side-lines of a triangle and those of the middle point triangle, when

* This definition is due to Laguerre, *Collected Works*, vol. ii, p. 597.

† 'Sur le problème d'Apollonius', *Nouvelles Annales de Math.*, Series 4, vol. vii, 1907, p. 503.

properly oriented, are pairs of an involution. The four circles in this case are the inscribed circle (or an escribed circle) and the middle points of the sides. We thus get Feuerbach's theorem, I. 49]. Again, consider the circle inscribed in a triangle and the tangent thereto anti-parallel to one side-line. We see that the line connecting the points of contact of the anti-parallel tangents is parallel to the bisector of the corresponding angle of the triangle. But if through each point of contact with a side of the triangle we draw a parallel to the bisector of the opposite angle, we have three concurrent lines, as we see by applying Ceva's theorem to the triangle whose vertices are the points of contact. We thus find

Theorem 28.] *The side-lines of a triangle and three anti-parallel transversals when properly oriented are pairs of an involution.*

Since the tangents to the circumscribed circle at the vertices are respectively anti-parallel to the corresponding side-lines, we have

Theorem 29.] *The inscribed circle to a triangle and three circles escribed to the triangles, each having as side-lines a side-line of the given triangle and the tangents to the circumscribed circle at the corresponding vertices, are tangent to a fifth circle.**

We return to the Laguerre group.† There are three types of involutory transformation, corresponding to the involutory congruent collineations of Σ.

Plane π.	Space Σ.
Laguerre inversion.	Reflection in plane.
Corresponding oriented lines properly tangent to the same oriented circle which touches properly two fixed	Reflection in line.

* 'Sur le problème d'Apollonius', *Nouvelles Annales de Math.*, Series 4, vol. vii, 1907, p. 505.

† Cp. Bricard, 'Sur la géométrie de direction', *Nouvelles Annales de Mathématiques*, Series 4, vol. vi, 1906.

oriented lines, while the points of contact make two harmonic pairs.	
Each oriented line is reflected in the properly parallel tangent to a fixed oriented circle.	Reflection in a point.

Every direct Laguerre transformation corresponds to a motion in Σ; every motion is either a rotation, a translation or a screw, or a limiting case of these. The first and third of these leave just two minimal planes invariant; the translation leaves invariant all minimal planes parallel to a given line.

Theorem 30.] *In a direct Laguerre transformation there will be invariant either two distinct oriented lines, or else all oriented lines making a fixed angle with a given line will be invariant.*

Let us look at this analytically. We see from (17) that the invariant lines correspond to roots of the equations

$$\gamma\xi^2+(\delta-\alpha)\epsilon-\beta = \gamma'\xi^2+2\gamma\xi\eta+(\delta'-\alpha')\xi+(\delta-\alpha)\beta-\beta' = 0.$$

If
$$(\delta-\alpha) = \beta = \gamma = 0,$$

these equations are satisfied regardless of η. Our indirect transformation (18) will be involutory if

$$\beta' = \gamma' = 0,\ \delta+\alpha = \delta'-\alpha' = 0,$$

or
$$\beta = \gamma = 0,\ \delta-\alpha = \delta'+\alpha' = 0.$$

It will be the first of these which gives the Laguerre inversion, as the special Laguerre inversion

$$\zeta' = -\bar{\zeta}$$

is included therein.

It is clear that a direct and close analogy exists between a considerable number of theorems concerning circular transformations which were developed in Ch. VIII and theorems concerning Laguerre transformations. One reason for this may be seen in the fact that the circular group is that of congruent transformations of three-dimensional non-Euclidean space, while the Laguerre group is that of congruent Euclidean transformations. In order to exhibit as clearly as possible this analogy, we shall give in parallel columns the corresponding theorems, appending to the theorem on circular transformations the number which it had in Ch. VIII. We start with the well-known fact that every motion of Σ can be factored into four plane reflections.

Laguerre group.	Circular group.
Theorem 31.] *Every direct Laguerre transformation can be factored into four Laguerre inversions.*	Theorem 26.] *Every direct circular transformation can be factored into four inversions.*
Theorem 32.] *Every direct Laguerre transformation is completely determined by the fate of three oriented lines.*	Theorem 7.] *Every direct circular transformation is completely determined by the fate of three points.*

In (17), if there be a single pair of values ζ and ζ' which are interchanged,

$$(\alpha + \epsilon\alpha') = -(\delta + \epsilon\delta').$$

Every pair of corresponding values are changed in an involutory manner.

Theorem 33.] *If a direct Laguerre transformation interchange a single pair of oriented lines it is involutory.*	Theorem 17.] *If a direct circular transformation interchange a single pair of points it is involutory.*

There is a fundamental theorem due to Wiener whereby every motion can be factored into the product of the reflection in two lines.*

Theorem 34.] *Every direct Laguerre transformation is the product of two involutory transformations.*	Theorem 18.] *Every direct circular transformation is the product of two Möbius involutions.*

* 'Zur Theorie der Umwendungen', *Leipziger Berichte*, vol. xlii, 1890, p. 20.

Every real motion of Σ is a translation, a rotation, or a screwing about a real axis. Unfortunately, the real motions of Σ will not usually give real Laguerre transformations in π and vice versa. Among the imaginary congruent transformations of Σ there are those where the only fixed point is on the circle at infinity. These will correspond to parabolic Laguerre transformations where the fixed oriented lines form a parallel pencil. Moreover, such transformations can be real, as we see if we replace ix_4 by x_4, and the denominator of ζ by x_2-x_3. A real oriented line will then have real coordinates ξ and η, and it is easy to find a real parabolic transformation.

A proper choice of our (old) complex coordinates will enable us to write any parabolic transformation in the form

$$\zeta' = \frac{\zeta}{(\gamma+\epsilon\gamma')\zeta+1}.$$

Let the reader show that under this transformation every circle

$$-\gamma x_1+k(x_2-ix_3)+\gamma' x_4 = 0$$

is invariant.

Theorem 35.] *Corresponding oriented lines in a parabolic Laguerre transformation touch properly the same oriented circle of a properly tangent pencil.*

Theorem 20.] *Corresponding points in a parabolic circular transformation lie on tangent circles through a fixed point.*

A Laguerre transformation which corresponds to a screw in Σ shall be called *loxodromic*. The indefinite repetition of an infinitesimal transformation of this sort will carry into itself a one-parameter family of oriented envelopes. There is also a one-parameter invariant family of oriented circles (not a one-parameter family of invariant circles) properly tangent to the two oriented lines which are fixed in the transformation, and these are permuted by the given transformation. If an envelope be carried into itself it will be repeatedly so transformed when the transformation is indefinitely repeated.

Every screw motion is a member of a continuous one-parameter group of screw motions all carrying into themselves the same systems of curves. We may then inquire what sort of an envelope will be carried into itself by a one-parameter family of loxodromic Laguerre transformations. When we remember that a Laguerre transformation is equilong, we see that all such envelopes must be equitangential of this system of circles, i.e. each envelope determines a common tangential segment of constant length with each circle of the system. Lastly, this tangential length will be the same for all the equitangentials, for each transformation of the loxodromic group will be seen to be commutative with every Laguerre transformation having the same two fixed oriented lines. Thus let T be our loxodromic transformation, S a Laguerre transformation with the same fixed oriented lines, k an equitangential envelope of T,

Let
$$Sk = k',$$
$$TSk = Tk' = k'',$$
$$S^{-1}TSk = Tk = k = S^{-1}k'',$$
$$k = S^{-1}k',$$
$$k' = k'',$$
$$Tk' = k',$$

and k' is an equitangential of our group. All these equitangentials will thus correspond to the same fixed length. We characterize our group analytically as follows. We write the transformation
$$\frac{\zeta'-\zeta_1}{\zeta'-\zeta_2} = (p+\epsilon q)\frac{\zeta-\zeta_1}{\zeta-\zeta_2}.$$

For a one-parameter group
$$q = f(p);$$
following with a second such transformation we have
$$pq' + p'q = f(pp'),$$
$$pf(p') + p'f(p) = f(pp'),$$
$$f(p) = k\log p.$$

The fixed tangential segment will thus be a function of k.

The Laguerre transformation corresponding to a rotation of space shall be called *hyperbolic* if the fixed oriented lines be real, *elliptic* if they be conjugate imaginary. The Laguerre transformation corresponding to a translation of space shall be called a *parallel* transformation. All oriented lines of two distinct or coincident parallel pencils are invariant.

Real Laguerre transformations.	Real circular transformations.
Loxodromic. Two types. Corresponding oriented lines touch the same equitangential of a system of oriented circles properly tangent to two real or conjugate imaginary oriented lines.	*Loxodromic.* Corresponding points lie on the same isogonal trajectory of a system of circles through two real points.
Hyperbolic. Corresponding oriented lines touch properly the same oriented circle properly tangent to two real lines.	*Hyperbolic.* Corresponding points are concyclic with two real points.
Elliptic. Corresponding oriented lines touch properly the same oriented circle properly tangent to conjugate imaginary lines.	*Elliptic.* Corresponding points are concyclic with two conjugate imaginary points, and orthocyclic with two real points.
Parabolic. Corresponding oriented lines touch properly the same oriented circle touching properly a fixed oriented circle at a fixed point.	*Parabolic.* Corresponding points are on the same circle touching a fixed circle at a fixed point.
Parallel. Three cases. Corresponding oriented lines are properly parallel. Two real and distinct coincident or conjugate imaginary pencils of parallel invariant lines.	

Theorem 36.] *Every periodic direct Laguerre transformation is elliptic.*

Theorem 21.] *Every periodic direct circular transformation is elliptic.*

If we write a non-parabolic Laguerre transformation in the form

$$\frac{\zeta'-\zeta_1}{\zeta'-\zeta_2} = (p+\epsilon q)\frac{\zeta-\zeta_1}{\zeta-\zeta_2},$$

the expression $(p+\epsilon q)$ is called the *invariant* of the transformation. It is the complex invariant of the fixed lines and any two corresponding lines.

Theorem 37.] *If two direct non-parabolic Laguerre transformations have a common fixed oriented line, the invariant of their product is the product of their invariants.*

Theorem 22.] *If two direct non-parabolic circular transformations have a common fixed point, the invariant of their product is the product of their invariants.*

The product of the reflection in two planes may never be a screw motion, but may be any of the other kinds of motion.

Theorem 38.] *The hyperbolic, elliptic, parabolic, and parallel Laguerre transformations, and these alone, are the products of two inversions. The necessary and sufficient condition that two inversions should be commutative is that the system of oriented circles invariant in the one should be an invariant system in the other.*

Theorem 23.] *The hyperbolic, elliptic, and parabolic circular transformations, and these alone, are the product of two inversions. When the circles of inversion are mutually orthogonal the two inversions are commutative.*

Let us return for a moment to the indirect Laguerre transformation. The square of an indirect transformation is a direct one, with the same fixed elements. We see also that this square cannot be a loxodromic transformation.

Theorem 39.] *Every indirect Laguerre transformation may be factored into three inversions, every direct one into four inversions.*

Theorem 26.] *Every indirect circular transformation may be factored into three inversions, every direct one into four inversions.*

§ 4. Continuous Groups.

The problem of finding real continuous groups of Laguerre transformations may be handled like the similar problem for circular transformations. We have the advantage of starting from familiar facts concerning Euclidean motions, but, as already remarked, the question of reality requires delicate handling, for real motions do not usually give real direct Laguerre transformations. Our Laguerre group is simply isomorphic with the six-parameter group, which leaves invariant one real non-degenerate conic. We begin with three-parameter groups, and find that our previous reasoning holds in the matter of integrable and non-integrable groups.

Theorem 40.] *The only real non-integrable three-parameter groups of Laguerre transformations are those with one real fixed oriented circle.*

Theorem 35.] *The only real non-integrable three-parameter groups of circular transformations are those with one real or self-conjugate imaginary fixed circle.*

It should be noticed at this point that whereas a real tetracyclic equation will give a real or self-conjugate imaginary circle, the latter type of circle having a real centre but a pure imaginary radius will correspond to a real point in Σ. The group in Σ will be that of a real fixed point, and will not carry into one another those points whose distances from π are pure imaginary, i.e. will not give a real Laguerre group.

An integrable three-parameter group of Laguerre transformations might have as its invariant sub-group a two-parameter group of parallel transformations. In Σ there are two such integrable groups, corresponding to screw motions of fixed pitch about axes of given directions and the group of translations.* The first of these will correspond to a three-parameter Laguerre group with fixed direction for the invariant lines, and a fixed segment for the invariant equi-tangentials. The second will give the parallel Laguerre group. We have also the Newson group with fixed real line and fixed tangential segment.

With regard to four- and five-parameter groups we may pursue our previous reasoning whereby these must contain three-parameter sub-groups. As before, we see that there are no five-parameter groups. Four-parameter groups will fall into two classes—those with one fixed real oriented line, and those with fixed directions for the fixed lines. The two-parameter groups are composed of those with two real and distinct or conjugate imaginary invariant oriented lines, and those composed of parallel transformations. We see, in fact, that every two-parameter group is integrable. If the one-parameter invariant sub-group have two distinct fixed oriented lines, these will be invariant throughout the whole two-parameter group. On the other hand, if the one-parameter group be a parallel group, there are ∞^1 invariant parallel oriented lines which must be permuted by every transformation of the two-parameter group ; hence every such transformation must be a parallel or parabolic one. Let us exhibit in parallel columns the corresponding groups in the two systems.

Real Laguerre transformations.	Real circular transformations.
Five-parameters.	
None.	None.

* Cf. Study, 'Von den Bewegungen und Umlegungen', *Math. Annalen*, vol. xxxix, 1891, pp. 485 ff. Our enumeration of groups of Laguerre transformations differs from his for motions precisely in this, that the real domains are not in correspondence.

Four-parameters.

Real fixed oriented line.	Real fixed point.
Fixed lines have two real and distinct, coincident, or conjugate imaginary directions.	

Three-parameters.

Fixed real oriented circle.	Fixed real or self-conjugate imaginary circle.
Newson.	Newson.
Constant real or conjugate imaginary directions for fixed lines, and fixed tangential lengths for invariant equitangential curves.	

Two-parameters.

Two real fixed oriented lines.	Two real fixed points.
Conjugate imaginary fixed lines.	
Lines of one direction fixed.	Coincident real fixed points.
Fixed real tangent to fixed real circle.	Fixed real point on fixed real circle.

One-parameter.

Loxodromic, with real or conjugate imaginary fixed lines.	Loxodromic.
Hyperbolic.	Hyperbolic.
Elliptic.	Elliptic.
Parabolic.	Parabolic.
Parallel.	

§ 5. **Hypercyclics.**

We have now studied sufficiently the simplest oriented envelope, the oriented circle. It is time to pass on to oriented envelopes of a more complicated sort.

Let us define as a *hypercyclic* * every oriented envelope having an equation of the type

$$\sum_{i,j=1}^{i,j=4} a_{ij}x_ix_j = 0, \quad x_2^2+x_3^2+x_4^2 = 0. \tag{21}$$

We may, without restriction, assume that the discriminant of the first of these quadratic forms is different from zero. Let us seek to reduce our equations to a canonical form under the Laguerre group. A transformation of the coordinates of a circle will be of the form

$$\begin{aligned}
y_0' &= c_{00}y_0,\\
y_2' &= c_{20}y_0+c_{22}y_2+c_{23}y_3+c_{24}y_4,\\
y_3' &= c_{30}y_0+c_{32}y_2+c_{33}y_3+c_{34}y_4,\\
y_4' &= c_{40}y_0+c_{42}y_2+c_{43}y_3+c_{44}y_4.
\end{aligned}$$

The coefficients c_{ij}, $i \neq 0$, $j \neq 0$ have the form of a ternary orthogonal substitution. Reverting to our previous hypercyclic, we write the equation

$$\begin{vmatrix} a_{22}-\rho & a_{23} & a_{24} \\ a_{32} & a_{33}-\rho & a_{34} \\ a_{42} & a_{43} & a_{44}-\rho \end{vmatrix} = 0.$$

Let us suppose that this has three distinct roots. We may in this case perform such an orthogonal substitution on the coordinates x_2, x_3, x_4 that the transformed equation lacks the terms x_2x_3, x_2x_4, x_3x_4. We still have the parameters c_{00}, c_{20}, c_{30}, c_{40} free, and we make use of them to destroy the terms x_1x_2, x_1x_3, x_1x_4. Our hypercyclic has thus the canonical equations

$$(ax^2) = x_2^2+x_3^2+x_4^2 = 0. \tag{22}$$

We mean by a *general* hypercyclic one where this reduction is possible, and where the envelope is not transformed into

* The more usual name is *hypercycle*. See Laguerre, *Collected Works*, vol. ii, and Blaschke, loc. cit. The word is here modified to accentuate the comparison with the cyclic.

itself by reversing the orientation of every line. We have the following correspondence.

Plane π.	Space Σ.
Hypercyclic.	Focal developable of quadric.
General hypercyclic.	Focal developable of central quadric.

Theorem 41.] *The general hypercyclic is transformed into itself by a group of eight involutory transformations including the identical one. Three others are Laguerre inversions, and three are direct transformations.*

These transformations will, of course, correspond to the reflections of various sorts which carry a central quadric into itself.

If an oriented envelope be anallagmatic in a given Laguerre inversion, the anallagmatic circle of that inversion which touches the envelope at any point will also have proper contact at another point. The envelope will thus be generated by a one-parameter family of anallagmatic oriented circles. These circles will correspond to the points of a focal curve of the corresponding focal developable. In the case of the hypercyclic, this focal curve is a conic, whose orthogonal projection on the plane π is another conic, the deferent of the corresponding generation of the hypercyclic.

Theorem 42.] *The general hypercyclic may be generated in three ways by an oriented circle which meets a fixed oriented line at a fixed angle, while its centre traces a central conic.*

Remembering the fundamental property of the focal conics of a central quadric, VI. 31],

Theorem 43.] *If two fixed oriented circles be taken in one generation of a general hypercyclic, the sum or difference of their common tangential segments with all proper circles of a second generation is constant.*

Since every oriented circle having double proper contact

with a general hypercyclic will correspond to a focal point of the corresponding quadric,

Theorem 44.] *The only oriented circles having double proper contact with a general hypercyclic are generating circles of one system or another.*

Reverting to the oriented circles of one generation of a hypercyclic, we see that the oriented tangents at the points of contact meet on the fundamental line of the corresponding Laguerre inversion, so that this fundamental line passes through the centre of similitude of two successive generating circles. If the angles which these tangents make with the fundamental line be α_1 and α_2,

$$\tan\frac{\alpha_1}{2}\tan\frac{\alpha_2}{2} = k.$$

Now let θ be the angle which a normal to the deferent makes with a line perpendicular to the fundamental line, while θ' is the angle which it makes with a normal to the hypercyclic, i.e. the line connecting corresponding points of deferent and hypercyclic,

$$\frac{\sin\theta}{\sin\theta'} = \frac{\cos\dfrac{\alpha_1+\alpha_2}{2}}{\cos\dfrac{\alpha_1-\alpha_2}{2}} = \frac{1-k}{1+k}.$$

Theorem 45.] *The general hypercyclic is an anticaustic by refraction of the deferent of each generation, the incident rays being supposed to come in a direction orthogonal to the fundamental line of the corresponding inversion.*

In the last equation we may not have $k(k+1) = 0$. The first would not give a Laguerre inversion at all, the second would give an inversion which merely replaced each oriented line by its opposite, excluded by the definition of the general hypercyclic.

There are two tangents to the deferent perpendicular to each real or imaginary direction that is unaltered by the inversion. These will meet the fundamental line in points where the hypercyclic has four-point contact with the generating circle. They will correspond in Σ to the points where the focal conic intersects the quadric. The hypercyclic will meet the fundamental line at these four points, and at the two intersections with the deferent, which are double points with distinct tangents, since the radius of a generating circle and the distance of its centre from the fundamental line are infinitesimals of the same order. The hypercyclic will have no other intersections with the fundamental line.

Theorem 46.] *The general hypercyclic meets the fundamental line of each Laguerre inversion in two double points, in four points where the corresponding generating circle has four-point contact, and in no other points.*

The condition of passing through a fixed point is linear in the coordinates of an oriented line, hence

Theorem 47.] *The general hypercyclic is a curve of the eighth order and fourth class.*

The extremities of the asymptotes of a deferent will not usually correspond to self-corresponding directions for the Laguerre inversion, so that each asymptote of the deferent gives two distinct asymptotes of the hypercyclic. Conversely, each asymptote of the hypercyclic must correspond to one of the deferent in each generation, unless the point of contact happen to be a circular point at infinity. Remembering that the hypercyclic must meet the line at infinity eight times, we have

Theorem 48.] *The general hypercyclic has each circular point at infinity as a double point.*

Theorem 49.] *The general hypercyclic has four double foci which are also the foci of the four deferents.*

The proof for this last is identical with that given for the analogous theorem for the cyclic, IV. 15].

Suppose that we know one generation of a hypercyclic, we pass to another as follows. First of all we know all the asymptotes and double foci. The other fundamental lines are the remaining diagonals of the complete quadrilateral of the asymptotes. The asymptotes of the deferents bisect the angles of the asymptotes of the hypercyclic. The deferents are all known because we know their foci and asymptotes, the fundamental lines are known, and the asymptotes determine the constants of the Laguerre inversions. It is to be noted that we have here taken no account of the difference between real and imaginary.

The tangents to the general hypercyclic are in one to one correspondence with the planes of the focal developable of a central quadric, which is a developable elliptic. Thus, since the points and tangents of any curve are in one to one correspondence, the hypercyclic must be an elliptic curve. This fact, combined with its order and class, will enable us to find all of its Plückerian characteristics.

Theorem 50.] *The general hypercyclic is of order eight, class four, and deficiency one. It has eight nodes, twelve cusps, two double tangents, and no inflexions.*

Suppose that we have a Laguerre inversion characterized by the equation

$$\tan\frac{\alpha_1}{2}\tan\frac{\alpha_2}{2} = k.$$

Let us transform it by inversion with the same fundamental line, and the equation

$$\tan\frac{\alpha}{2}\tan\frac{\alpha'}{2} = \frac{1}{i\sqrt{k}}.$$

The result will be the inversion

$$\tan\frac{\alpha_1'}{2}\tan\frac{\alpha_2'}{2} = -1,$$

which merely has the effect of reversing every oriented line. A hypercyclic which is anallagmatic under this new inversion must be a conic counted twice.

Theorem 51.] *There are six Laguerre inversions which carry a general hypercyclic into a non-oriented conic.*

Four oriented lines shall be said to form a *harmonic set* when their complex invariant has the value -1. Under these circumstances they touch one oriented circle and, if the latter be proper, their points of contact form a harmonic set. A given oriented line will have a single harmonic conjugate with regard to two given oriented lines. If these two be opposite lines, the other pair of the harmonic set are the reflections of one another therein. If we reflect all the tangents to a conic in a straight line we get tangents to another conic, hence

Theorem 52.] *The oriented envelope of the harmonic conjugates of a given non-minimal oriented line with regard to the pairs of oriented tangents to a general hypercyclic which correspond in one Laguerre inversion, is a second hypercyclic anallagmatic in the same inversion.*

The theorems so far developed for the hypercyclic correspond very closely to theorems developed in Ch. IV for the cyclic. The reader would do well to turn back and compare one by one. The correspondence may be brought into an even stronger light as follows. The general hypercyclic may be written

$$a_1x_1^2+a_2x_2^2+a_3x_3^2+a_4x_4^2=x_2^2+x_3^2+x_4^2=0.$$

A general cyclic may be written

$$(b_2-b_1)x_2^2+(b_3-b_1)x_3^2+(b_4-b_1)x_4^2=x_1^2+x_2^2+x_3^2+x_4^2=0.$$

Theorem 53.] *There is a perfect one to one correspondence between the points of a general cyclic and the oriented tangents to a general hypercyclic. Concyclic points of the first will correspond to oriented tangents to the other, which either touch one oriented circle or meet a fixed oriented line at a fixed angle.*

This theorem may be used to verify the calculation of the Plückerian characteristics of the hypercyclic given in 50]. Suppose that we have four oriented circles properly tangent to two oriented lines. Every oriented circle tangent to these lines will have coordinates of the form

$$(\lambda x_0' + \mu x_0'',\quad x_1,\quad \lambda x_2' + \mu x_2'',\quad \lambda x_3' + \mu x_3'',\quad \lambda x_4' + \mu x_4''),$$

where (x') and (x'') are two oriented circles of the system. The cross ratio of the centres of four such will be

$$\frac{|\lambda\mu'| \cdot |\lambda''\mu'''|}{|\lambda\mu'''| \cdot |\lambda''\mu'|},$$

an absolute invariant for every Laguerre transformation. We thus get from IV. 21]

Theorem 54.] *The centres of four oriented circles properly tangent to a general hypercyclic and to two fixed oriented tangents thereto, have a cross ratio which is independent of the choice of the oriented tangents, and is an invariant of the hypercyclic for every Laguerre transformation.*

Theorem 55.] *If three generating circles be properly tangent to a general hypercyclic at the same point, the distances from the centre of the first to those of the second and third have a fixed ratio which is an invariant of the hypercyclic.*

The general cyclic is anallagmatic in four inversions and has four systems of generating circles. These will correspond to the three systems of generating circles of the hypercyclic and to a one-parameter family of pencils of parallel lines, two members of each pencil being properly tangent to the hypercyclic and being the reflections of one another in the properly parallel proper tangent to a fixed proper circle. They correspond in the remaining involutory transformation which carries the hypercyclic into itself. The circles of four-point contact with the cyclic will correspond to the generating circles which have four-point contact and to the asymptotes.

It is at once apparent that our various residuation theorems, IV. 27–34], have exact duals in the case of the hypercyclic. We give some of the most interesting.

Theorem 56.] *If a, b, c, d be the four proper tangents which a hypercyclic shares with a circle, a second oriented circle touches the proper tangents to the hypercyclic a, b, c_1, d_1, a third touches the proper tangents a_1, b_1, c, d, then the four oriented lines a_1, b_1, c_1, d_1 touch a circle or make a fixed angle with a fixed oriented line.*

Theorem 57.] *The envelope of pairs of oriented lines each touching a common oriented circle with each of three given pairs of oriented lines is a hypercyclic.*

There are many special cases of this which we shall not investigate further.

Theorem 58.] *The osculating oriented circles to a hypercyclic corresponding to the oriented tangents which the hypercyclic shares with a given oriented circle, will each determine with the hypercyclic one other common proper tangent, and these will also be properly tangent to an oriented circle or meet a fixed line at a fixed angle.*

Theorem 59.] *If three of the proper tangents common to a general hypercyclic and an oriented circle correspond to generating circles having four-point contact, the same is true of the fourth common tangent, or else the latter is an asymptote.**

Theorem 60.] *If three oriented circles of one generation of a general hypercyclic share with the curve the pairs of proper tangents aa', bb', and cc' respectively, and if a, b, c, d touch one oriented circle, while a', b', c', d' touch another, then dd' will be a pair of corresponding proper tangents in this generation.*

§ 6. The Oriented Circle treated directly.

We have so far, in the present chapter, treated the oriented circle almost exclusively as an envelope of oriented lines.

* Cf. Blaschke, loc. cit., p. 59. Much of what we have given in the present chapter, both in connexion with the Laguerre group and the hypercyclic, is from this excellent memoir.

Let us now change our point of view, and examine the oriented circle directly. We repeat the previous analogies.

Plane π.	Space Σ.
Proper oriented circle.	Proper point.
Common tangential segment.	Distance.
Ten-parameter group of contact transformations of oriented circles.	Ten-parameter group of spherical transformations.

This last truly admirable correspondence is due to Lie.* It is proved by noticing the isomorphism of the conformal group of Ch. IX with the collineation group in S_4, that leaves invariant the quadratic form

$$-2x_0x_1+x_2^2+x_3^2+x_3^2+x_4^2=0.$$

A one-parameter system of oriented circles, that is, a system whose coordinates are proportional to analytic functions of one independent variable whose ratios are not all constants, shall be called a *series*. A system whose coordinates are proportional to analytic functions of two independent variables, the ratios being not all constants, nor functions of one same variable, shall be a *congruence*. Among congruences the simplest are the linear ones, determined by equations of the type

$$-a_1x_0-a_0x_1+a_2x_2+a_3x_3+a_4x_4=0. \qquad (23)$$

They have the following interpretations with the aid of (7) and (6):

a) $$-2a_0a_1+a_2^2+a_3^2\neq 0,$$

$$\frac{-a_1x_0-a_0x_1+a_2x_2+a_3x_3+\sqrt{2a_0a_1-a_2^2-a_3^2}\,x_4}{2\sqrt{2a_0a_1-a_2^2-a_3^2}\,x_4}$$

$$=\frac{1}{2}\left[1-\frac{a_4}{\sqrt{2a_0a_1-a_2^2-a_3^2}}\right].$$

* 'Ueber diejenige Theorie des Raumes', etc., *Göttingische Nachrichten*, 1871.

The circles of the congruence meet an oriented circle at a fixed angle which will be null if

$$-2a_0a_1+a_2^2+a_3^2+a_4^2 = 0.$$

When this last equation holds the congruence is said to be *special*, even when the inequality above does not hold. It consists in the oriented circles properly tangent to one oriented circle.

b) $$a_0 \neq 0.$$

$$\frac{2\left[\frac{(a_2^2+a_3^2+a_4^2)}{-2a_0}x_0-a_0x_1+a_2x_2+a_3x_3+a_4x_4\right]}{a_0x_0} = 2\left[\frac{a_2^2+a_3^2+a_4^2}{-a_0^2}-\frac{a_1}{a_0}\right].$$

The circles have a fixed common tangential segment with a fixed oriented circle.

c) $$-2a_0a_1+a_2^2+a_3^2 = 0, \quad a_0 = 0.$$

Circles invariantly related to a minimal line.

One or two special cases deserve notice.

d) $$a_0 = 0, \quad a_2^2+a_3^2 \neq 0.$$

Circles meet an oriented line at a fixed angle.

e) $$a_1 = a_2 = a_3 = 0.$$

Circles have a given radius which is null if $a_4 = 0$.

$$a_0 = a_2 = a_3 = a_4 = 0.$$

Congruence of all line circles.

If the coordinates of an oriented circle be connected by a linear relation, it will thus in the general case, where all of our inequalities hold, meet one circle at a fixed angle, and have a fixed tangential segment in common with another. If we mean by 'in general' that these inequalities shall be in force, we see

Theorem 61.] *The assemblage of all oriented circles whose common tangential segments with two proper circles bear* *a fixed ratio, will meet at a fixed angle an oriented circle coaxal with the given two, or be the limit of such an assemblage.**

Theorem 62.] *The ratio of the common tangential segments of all circles of a coaxal system with any two of the conjugate coaxal system is constant.*

Theorem 63.] *The assemblage of all oriented circles common to two linear congruences is that of all oriented circles properly tangent to two oriented circles, or to a given oriented circle at a given point.*

The proof is immediate and left to the reader. We pass to the general consideration of two linear congruences. These will have an invariant under the contact group of oriented circles, namely

$$\frac{-a_0b_1-a_1b_0+a_2b_2+a_3b_3+a_4b_4}{\sqrt{-2a_0a_1+a_2^2+a_3^2+a_4^2}\ \sqrt{-2b_0b_1+b_2^2+b_3^2+b_4^2}} \equiv I. \quad (24)$$

Let us restrict ourselves to the general case where

$$(-2a_0a_1+a_2^2+a_3^2+a_4^2)(-2b_0b_1+b_2^2+b_3^2+b_4^2) \neq 0.$$

Under these circumstances, if ϕ_1 and ϕ_2 be the angles associated with the definition of the congruences, and θ the angle of their fundamental circles,

$$\frac{\cos\phi_1\cos\phi_2-\cos\theta}{\sin\phi_1\sin\phi_2} = I.$$

Two linear congruences shall be said to be *in involution* when I vanishes, i.e.

$$-a_0b_1-a_1b_0+a_2b_2+a_3b_3+a_4b_4 = 0.$$

If one of the complexes be special, its fundamental circle is a member of the other congruence; if both be special, their fundamental circles are tangent.

* This theorem and the next are from rather a poor article by Sobotka, 'Eine Aufgabe aus der Geometrie der Bewegung', *Monatshefte für Math.*, vol. vii, 1896, p. 347.

Plane π.	Space Σ.
Linear congruence.	Sphere.
Invariant of two linear congruences.	Cosine of angle of spheres.
Congruences in involution.	Orthogonal spheres.

This analogy of linear congruence with sphere leads us to consider that oriented circle transformation which corresponds to an inversion in Σ and which contains the Laguerre transformation as a special case.* We start with a non-special linear congruence $(a_0 a_1 a_2 a_3 \alpha_4)$, and define as an *inversion* therein the following transformation, which is seen at once to correspond to inversion in space.

$$x_i' = (\alpha_4^2 - a_4^2)x_i - 2(-a_1 x_0 - a_0 x_1 + a_2 x_2 + a_3 x_3 + \alpha_4 x_4)a_i, \; i \neq 4,$$

$$x_4' = (\alpha_4^2 - a_4^2)x_i - 2(-a_1 x_0 - a_0 x_1 + a_2 x_2 + a_3 x_3 + \alpha_4 x_5)\alpha_4, \quad (25)$$

$$-2a_0 a_1 + a_2^2 + a_3^2 + a_4^2 = 0,$$

$$(-2x_0' x_1' + x_2'^2 + x_3'^2 + x_4'^2) = (\alpha_4^2 - a_4^2)^2 (-2x_0 x_1 + x_2^2 + x_3^2 + x_4^2).$$

When $a_0 = 0$ we fall back upon the Laguerre inversion (11).

The oriented circles (x) and (x') are coaxal with the circle (a). The product of the tangents of the halves of their angles therewith is

$$\tan\frac{\theta}{2}\tan\frac{\theta'}{2} = \frac{\alpha_4 - a_4}{\alpha_4 + a_4} = \tan^2\frac{\phi}{2},$$

where ϕ is the fundamental angle associated with the linear congruence.

Theorem 64.] *If a general linear congruence consist in oriented circles meeting a fixed oriented circle at a fixed angle not zero, then two oriented circles are mutually inverse in the congruence when, and only when, they are coaxal with the fixed circle, and the product of the tangents of half their*

* This transformation seems to be due to Smith, 'On a Transformation of Laguerre', *Annals of Math.*, Series 2, vol. i, 1900, and 'On Surfaces enveloped by spheres belonging to a linear Complex', *Transactions American Math. Soc.*, vol. i, 1900.

angles therewith is equal to the square of the tangent of half the angle associated with the congruence.

We shall not enter into the special or limiting cases of this transformation. Let the reader prove the five following theorems:

Theorem 65.] *When the angle associated with a linear congruence of oriented circles is* $\frac{\pi}{2}$, *inversion in the congruence is circular inversion.*

Theorem 66.] *Inversion in the linear congruence of all null circles is the reversal of the sign of the radius of each oriented circle.*

Theorem 67.] *Inversion in the congruence of all oriented circles of given radius will change each oriented circle into a concentric one whose radius differs from the negative of that of the given circle by a constant.*

Theorem 68.] *A necessary and sufficient condition that the product of the inversions in two linear congruences should be commutative is that the congruences should be in involution.*

Theorem 69.] *Every contact transformation of oriented circles can be factored into the product of five or less inversions in linear congruences.*

Before making any further study of series or congruences of oriented circles it will be wise to make a slight alteration of notation. This amounts essentially to determining each circle by the special pentaspherical coordinates of the representing point in Σ, that is, in (12) we replace the homogeneous cartesian coordinates X, Y, Z, T by special pentaspherical ones. We write, therefore,

$$\begin{aligned} -\sqrt{2}\,x_0 &\equiv \rho\,(X_0+iX_1),\\ \sqrt{2}\,x_1 &\equiv \rho\,(X_0-iX_1),\\ x_2 &\equiv \rho X_2,\\ x_3 &\equiv \rho X_3,\\ x_4 &\equiv \rho X_4. \end{aligned} \tag{26}$$

An oriented circle has thus five homogeneous coordinates (X) connected by the relation

$$(XX) = 0. \tag{27}$$

The condition of proper contact of two oriented circles (X) and (Y) is

$$(XY) = 0. \tag{28}$$

In Ch. VI we passed from special to general pentaspherical coordinates which were related to five mutually orthogonal spheres. Let five congruences in involution be

$$(A^{(0)}), (A^{(1)}), \ldots, (A^{(4)}),$$

the angle associated with the congruence $(A^{(i)})$ shall be ϕ_i, the angle at which an arbitrary circle cuts the fundamental circle of $(A^{(i)})$ shall be θ_i. Let us then write

$$X_i' \equiv \frac{\cos\phi_i - \cos\theta_i}{\sin\phi_i} = \frac{(A^{(i)}X)}{X_4\sqrt{(A^{(i)}A^{(i)})}}. \tag{29}$$

$$| A_0^{(0)} A_1^{(1)} A_2^{(2)} A_3^{(3)} A_4^{(4)} \; 0 |^2 = 0.$$

$$\begin{vmatrix} (A^{(0)}A^{(0)}) & 0 & 0 & 0 & 0 & (A^{(0)}X) \\ 0 & (A^{(1)}A^{(1)}) & 0 & 0 & 0 & (A^{(1)}X) \\ 0 & 0 & (A^{(2)}A^{(2)}) & 0 & 0 & (A^{(2)}X) \\ 0 & 0 & 0 & (A^{(3)}A^{(3)}) & 0 & (A^{(3)}X) \\ 0 & 0 & 0 & 0 & (A^{(4)}A^{(4)}) & (A^{(4)}X) \\ (A^{(0)}X) & (A^{(1)}X) & (A^{(2)}X) & (A^{(3)}X) & (A^{(4)}X) & 0 \end{vmatrix} = 0.$$

$$\begin{vmatrix} 1 & 0 & 0 & 0 & 0 & X_0' \\ 0 & 1 & 0 & 0 & 0 & X_1' \\ 0 & 0 & 1 & 0 & 0 & X_2' \\ 0 & 0 & 0 & 1 & 0 & X_3' \\ 0 & 0 & 0 & 0 & 1 & X_4' \\ X_0' & X_1' & X_2' & X_3' & X_4' & 0 \end{vmatrix} = 0.$$

$$(X'X') = 0. \tag{30}$$

Theorem 70.] *If five linear congruences in involution of oriented circles consist in the circles meeting five fundamental oriented circles at five fundamental angles, then we may take for the coordinates of every proper oriented circle five quantities each proportional to the quotient of the difference of the cosine of a fundamental angle and the cosine of the angle made with the corresponding fundamental circle, divided by the sign of that fundamental angle. The sum of the squares of these coordinates will in every case be zero.*

Four oriented circles have an absolute invariant for all contact transformations.

$$\frac{(XY)\,(ZT)}{(XT)\,(ZY)} \equiv \frac{\sin^2 \frac{1}{2}\measuredangle\, XY \sin^2 \frac{1}{2}\measuredangle\, ZT}{\sin^2 \frac{1}{2}\measuredangle\, XT \sin^2 \frac{1}{2}\measuredangle\, ZY}. \tag{31}$$

We pass to the general study of systems of oriented circles. A general congruence may be expressed in the form*

$$f(X_0 \dots X_4) = 0. \tag{32}$$

Or else in the form

$$X_i \equiv X_i(uv). \tag{33}$$

Let us look for series of oriented circles in the congruence which osculate their envelopes. If (X) and $(X+dX)$ make an angle which is infinitesimal to higher order we have

$$\left(\frac{\partial X}{\partial u}\,\frac{\partial X}{\partial u}\right)du^2 + 2\left(\frac{\partial X}{\partial u}\,\frac{\partial X}{\partial v}\right)du\,dv + \left(\frac{\partial X}{\partial v}\,\frac{\partial X}{\partial v}\right)dv^2 = 0. \tag{34}$$

The solutions of this differential equation give the two one-parameter families of osculating series; an arbitrary circle will usually belong to two different series; they coalesce if

$$\left(\frac{\partial X}{\partial u}\,\frac{\partial X}{\partial u}\right)\left(\frac{\partial X}{\partial v}\,\frac{\partial X}{\partial v}\right) - \left(\frac{\partial X}{\partial u}\,\frac{\partial X}{\partial v}\right)^2 = 0. \tag{35}$$

* The following study of congruences and series of circles is based on two interesting but highly unreliable articles by Snyder, 'Geometry of some Differential Expressions in Hexaspherical Coordinates', *Bulletin American Math. Soc.*, vol. iv, 1897, and 'On the Geometry of the Circle', ibid., vol. vi, 1899. On p. 464 the author corrects several misstatements in the preceding paper; certain erroneous results appearing in the earlier paper have not, apparently, ever been corrected.

Again, if the adjacent circles (X) and $(X+dX)$ be tangent to one another,

$$(XX) = (XdX) = (dXdX) = 0,$$

$(X+dX)$ belongs to the properly tangent pencil determined by the mutually tangent circles (X) and (dX). These two both belong to the linear congruences

$$(XX') = \left(\frac{\partial f}{\partial X} X'\right) = 0.$$

To find the special linear congruences linearly dependent on these we must solve the quadratic equation

$$\lambda^2 (XX) + 2\lambda\mu \left(X \frac{\partial f}{\partial X}\right) + \mu^2 \left(\frac{\partial f}{\partial X} \frac{\partial f}{\partial X}\right) = 0. \qquad (36)$$

Assuming first

$$\left(\frac{\partial f}{\partial X} \frac{\partial f}{\partial X}\right) \neq 0,$$

the only solutions of the quadratic are $\mu = 0$. The series determined by the two linear equations consist in oriented circles properly tangent to (X) at either of two given points, its points of contact with its envelopes in the osculating series. We next assume

$$\left(\frac{\partial f}{\partial X} \frac{\partial f}{\partial X}\right) \equiv 0. \qquad (37)$$

Here the state of affairs is entirely different. Let us first assume that the congruence is algebraic, i.e. f is a homogeneous polynomial.* Representing our oriented circles by means of points in a projective four-dimensional space we see that, as in this case, $\left(\frac{\partial f}{\partial X} \frac{\partial f}{\partial X}\right)$ vanishes identically with (XX) and $\left(X \frac{\partial f}{\partial X}\right)$; we may apply Nöther's fundamental theorem and write

$$\left(\frac{\partial f}{\partial X} \frac{\partial f}{\partial X}\right) \equiv (XX)\,\phi + \left(X \frac{\partial f}{\partial X}\right)\psi,$$

* Cf. Klein, 'Ueber einige in der Liniengeometrie auftretenden Differentialgleichungen', *Math. Annalen*, vol. v, 1872, p. 288.

$$f\left(X+\lambda\frac{\partial f}{\partial X}\right) \equiv \frac{\lambda^2}{2\,!}\sum_{i,j=0}^{i,j=4}\frac{\partial^2 f}{\partial X_i\,\partial X_j}\cdot\frac{\partial f}{\partial X_i}\frac{\partial f}{\partial X_j}+\ldots\ldots$$

$$2\sum_{i=0}^{i=4}\frac{\partial f}{\partial X_i}\frac{\partial^2 f}{\partial X_i\,\partial X_j} \equiv (XX)\frac{\partial\phi}{\partial X_j}+\left(X\frac{\partial f}{\partial X}\right)\frac{\partial\psi}{\partial X_j} + 2X_j\phi + n\frac{\partial f}{\partial X_j}\psi,$$

$$2\sum_{i,j=0}^{i,j=4}\frac{\partial^2 f}{\partial X_i\,\partial X_j}\frac{\partial f}{\partial X_i}\frac{\partial f}{\partial X_j} \equiv (XX)\left(\frac{\partial f}{\partial X}\frac{\partial\phi}{\partial X}\right) + \left(X\frac{\partial f}{\partial X}\right)\left(\frac{\partial f}{\partial X}\frac{\partial\psi}{\partial X}\right)+2\,nf\phi+n\psi\left(\frac{\partial f}{\partial X}\frac{\partial f}{\partial X}\right).$$

It appears thus that the first term in the above expansion of $f\left(X+\lambda\frac{\partial f}{\partial X}\right)$ vanishes. We may continue thus and show that every term vanishes, i.e. every circle properly tangent to the properly tangent circles (X) and $\left(\frac{\partial f}{\partial X}\right)$ belongs to the congruence. The latter must consist in the circles tangent to one curve. Conversely, suppose that we have a congruence of circles properly tangent to an oriented envelope. The osculating circle being (Z) we may write

$$Y_i = Z_i(u)+vZ_i'(u),$$

$$(ZZ) = (ZZ') = (Z'Z') = (ZZ'') = (Z'Z'') = 0,$$

$$\left(T\frac{\partial f}{\partial Y}\right) = \left(Z\frac{\partial f}{\partial Y}\right) = \left(\frac{\partial Y}{\partial u}\frac{\partial f}{\partial Y}\right) = \left(\frac{\partial Y}{\partial v}\frac{\partial f}{\partial Y}\right) = 0,$$

$$\rho\frac{\partial f}{\partial Y_i} = |\,T_j\,Z_k\,Z_l'\,Z_m''\,|,$$

$$\left(\frac{\partial f}{\partial Y}\frac{\partial f}{\partial Y}\right) \equiv 0.$$

When the equation of the congruence is not algebraic we may express it approximately as closely as we please by a development in Taylor's series, the theorem holding for each approximation. Hence, it is universally true that (37) expresses the necessary and sufficient condition for a congruence of oriented circles properly tangent to an envelope.

Suppose that we have two algebraic congruences

$$f=0, \quad \phi=0.$$

The oriented circle (X) and its immediate neighbours belong to the linear congruences

$$\left(\frac{\partial f}{\partial X}X'\right)=\left(\frac{\partial \phi}{\partial X}X'\right)=0.$$

The special linear congruences linearly dependent upon them will be given by the roots of the quadratic equation

$$\lambda^2\left(\frac{\partial f}{\partial X}\frac{\partial f}{\partial X}\right)+2\lambda\mu\left(\frac{\partial f}{\partial X}\frac{\partial \phi}{\partial X}\right)+\mu^2\left(\frac{\partial \phi}{\partial X}\frac{\partial \phi}{\partial X}\right)\equiv 0. \qquad (38)$$

When the roots of this equation are identically equal, i.e. when

$$\left(\frac{\partial f}{\partial X}\frac{\partial f}{\partial X}\right)\left(\frac{\partial \phi}{\partial X}\frac{\partial \phi}{\partial X}\right)-\left(\frac{\partial f}{\partial X}\frac{\partial \phi}{\partial X}\right)^2\equiv 0,$$

we have a series of osculating circles, by II. 25].

Reverting to our congruence (32), that series for which

$$\left(\frac{\partial f}{\partial X}\frac{\partial f}{\partial X}\right)=0 \qquad (39)$$

is said to be composed of *singular circles.* The infinitely near circles tangent to (X) touch it where it touches the oriented circle $\left(\frac{\partial f}{\partial X}\right)$. That part of the envelope of the singular series which comes from this point of contact is called the *singular curve.*

This hazy talk about neighbouring or infinitely near circles attains a clear meaning when we consider our minimal projection.

Plane π.	Space Σ.
Congruence of oriented circles.	Surface.
Congruence of oriented lines properly tangent to one envelope.	Isotropic ruled surface.

Pencils determined by adjacent circles properly tangent to given circle.	Isotropic directions in surface.
Singular circle.	Parabolic point.
Series of osculating circles.	Minimal curve.

Let us apply these general principles to the study of the quadratic congruence, that is, the congruence given by equations

$$\sum_{i,j=0}^{i,j=4} a_{ij} X_i X_j = (XX) = 0. \tag{40}$$

We shall mean by a *general* quadratic congruence one for which the equation

$$\begin{vmatrix} a_{00}-\rho & a_{01} & a_{02} & a_{03} & a_{04} \\ a_{10} & a_{11}-\rho & a_{12} & a_{13} & a_{14} \\ a_{20} & a_{21} & a_{22}-\rho & a_{23} & a_{24} \\ a_{30} & a_{31} & a_{32} & a_{33}-\rho & a_{34} \\ a_{40} & a_{41} & a_{42} & a_{43} & a_{44}-\rho \end{vmatrix} = 0$$

has distinct roots.

Plane π.	Space Σ.
Quadratic congruence.	Cyclide.
General quadratic congruence.	General cyclide.

We may, as we well know, find a contact circle transformation to carry the equations of the general quadratic congruence to the canonical form

$$(aX^2) = (XX) = 0. \tag{41}$$

Theorem 71.] *The general quadratic congruence is carried into itself by an inversion in any one of five linear congruences in involution.*

Let the condition for a null circle in our present coordinates be

$$(UX) = 0,$$

while that for a line circle is

$$(VX) = 0.$$

Theorem 72.] *The null circles of a general quadratic congruence generate a cyclic, its line circles a hypercyclic.*

Let us look for the singular circles of the congruence

$$(aX^2) = (a^2X^2) = (XX) = 0.$$

If (Z) be the point of contact with the singular curve, i.e. the null circle whose vertex is there, while (Z') is the oriented tangent,

$$Z_i = (a_i+\lambda)X_i, \quad Z_i' = (a_i+\lambda')X_i'.$$

$$\sum_{i=0}^{i=4}\frac{Z_i^2}{a_i+\lambda} = \sum_{i=0}^{i=4}\frac{Z_i^2}{(a_i+\lambda)^2} = 0; \quad \sum_{i=0}^{i=4}\frac{Z_i'^2}{a_i+\lambda'} = \sum_{i=0}^{i=4}\frac{Z_i'^2}{(a_i+\lambda)^2} = 0.$$

The first pair of equations are a cubic in λ and its derivative. The elimination of λ amounts to setting the discriminant to zero, and is a quartic in Z_i^2; another such quartic will come from the other pair of equations. If we combine with

$$(UZ) = (AZ) = (ZZ) = 0, \quad (VZ') = (A'Z') = (Z'Z') = 0,$$

we get sixteen solutions.

Theorem 73.] *The singular curve of a general quadratic congruence is of the sixteenth order and class with a multiplicity eight at each circular point at infinity.**

This curve may be generated in various ways as an envelope of circles. We see, in fact, that a properly tangent oriented circle belonging to one fundamental linear congruence is doubly tangent.

Plane π.	Space Σ.
Five generations of singular curve of general quadratic congruence.	Five focal cyclics of cyclide.

The centre of one of the generating circles will be the orthogonal projection on Σ of a focus of the cyclide; we thus see

* Blaschke, loc. cit., p. 55.

Theorem 74.] *The singular curve of a general quadratic congruence of oriented circles may be generated in five ways by an oriented circle which meets a fixed oriented circle at a fixed angle, while its centre traces a binodal quartic. It will have a node at each finite intersection of one of these quartics with the corresponding fixed circle.*

The deficiency of this curve, being that of a cyclic, is unity. We have thus a sufficient number of facts to enable us to calculate the Plückerian characteristics of the curve; we find

Theorem 75.] *The singular curve of a general quadratic congruence has the equivalent of* 88 *nodes and double tangents,* 16 *inflexions and cusps.*

The singular circles which satisfy all the relations

$$(XX) = (aX^2) = (a^2X^2) = (a^3X^2) = 0 \tag{42}$$

are said to be singular of the *second sort.* Every oriented circle of the system

$$\lambda X_i + \mu a_i X_i$$

will belong to the congruence. Our correspondence is

Plane π.	Space Σ.
Sixteen singular circles of the second sort of general quadratic circle congruence.	Sixteen isotropics of general cyclide.

The one-parameter family of quadratic congruences

$$\sum_{i=0}^{i=4} \frac{X_i^2}{a_i + \lambda} = 0 \tag{43}$$

are said to be *cosingular.* They have, in fact, the same singular curve. We have

Plane π.	Space Σ.
Cosingular quadratic circle congruences.	Confocal cyclides.

Long as has been our present chapter, it is clear that we have by no means exhausted the subject of the oriented circle. We began with the study of these circles by means of elementary geometry, and there is no doubt that we merely scratched the surface. The analogy between the geometry of inversion and the geometry of direction could be pushed much further. Our next task was to study the Laguerre group, and, though we carried this far, yet much remains to bring it to an equality with the group of circular transformations. Thirdly, be it noticed that we have made no mention of any of the special forms of quadratic circle congruence. Several of these must be of importance, especially those which correspond to two-horned and Dupin cyclides. Lastly, it would seem as if the coordinates of the oriented circle offered an ideal method of studying Hart systems. The Author must confess that his success in this last line has been disappointing, yet he has not lost his conviction that much might be done.

CHAPTER XI

THE ORIENTED SPHERE

§ 1. Elementary Geometrical Theorems.

WE saw in the last chapter what a profound change was introduced into the geometry of the circle in the plane, or, for that matter, the circle on the sphere, by giving a sign to the radius. Exactly similar changes occur in the geometry of the sphere when a like orientation is introduced.

We start, as before, with the real finite cartesian domain, the domain of Euclidean geometry. The radius of a sphere shall be considered as positive when each normal is oriented inwards; outwardly oriented normals shall correspond to a negative radius. The angle of intersection of two spheres shall be defined as that of their oriented normals at a point of intersection; its cosine will be

$$\cos\theta = \frac{r^2 + r'^2 - d^2}{2rr'}. \qquad (1)$$

When this expression is equal to unity the spheres are said to be *properly tangent*; when it is equal to -1, *improperly tangent*. Every plane (we are in the real domain) may be oriented by requiring all of its normals to point from one of the two regions into which the plane divides space to the other. Two planes shall be *properly parallel* when they have the same system of oriented normals; when the normals to one are opposite to those to the other they are said to be *improperly parallel*. The angle between two intersecting planes shall be defined as that of their oriented normals at a point of intersection, and this definition shall be extended to include the angle of any two oriented surfaces. If at any non-singular point they have the same oriented normal, they are said to be *properly tangent*; when the normal to one is opposite to that to the other, *improperly tangent*.

A number of simple theorems about oriented spheres may be obtained immediately from the elementary theorems about oriented circles. Let the reader prove:

Theorem 1.] *If, through all lines of an oriented plane which lie outside of an oriented sphere, pairs of properly tangent planes be drawn to the sphere, the product of the tangents of half the angles which they make with the given oriented plane is constant.*

This constant shall be called the *power* of the oriented plane with regard to the oriented sphere. It may take any value between $-\infty$ and ∞ according to the position of the plane. Let us note that two oriented spheres, like two oriented circles, have but one centre of similitude; we have then

Theorem 2.] *All oriented planes having equal powers with regard to two non-concentric oriented spheres of unequal radii pass through the centre of similitude; when the spheres have equal radii, these planes are parallel to the line of centres.*

We get at once from the Laguerre inversion in the plane

Theorem 3.] *If oriented planes be transformed in such a way that corresponding ones are coaxal with a fundamental oriented plane, while the product of the tangents of the halves of the angles which they form therewith is constant, then an oriented sphere is transformed into an oriented sphere, and corresponding oriented spheres have the given fundamental plane as their radical plane. The common tangential segment of two oriented spheres will be invariant under this transformation.*

We mean, of course, by the 'common tangential segment' of two oriented spheres the distance of the points of contact with any properly tangent oriented plane. Our transformation shall be called a *Laguerre inversion*. It is a special case of the general equilong transformation, to be studied later.

§ 2. Analytic Treatment.

It is hardly worth while to delay any longer on the elementary geometry of the oriented plane and sphere, as much more interesting material lies beyond. We pass, therefore, to cartesian space rendered a perfect complex continuum by the adjunction of the plane at infinity, and write the equations

$$x_0(x^2+y^2+z^2)+x_2(2xt)+x_3(2yt)+x_4(2zt)+x_1(2t^2)=0; \quad (2)$$

$$-2x_0x_1+x_2{}^2+x_3{}^2+x_4{}^2+x_5{}^2\equiv 0. \quad (3)$$

Every oriented surface corresponding to these two equations shall be called an *oriented sphere*. When the first coefficient does not vanish, we shall define as its radius the expression

$$r=\frac{ix_5}{x_0}. \quad (4)$$

We have the following types of oriented sphere:

a) Proper oriented spheres $x_0x_5\neq 0$.

b) Non-planar null spheres $x_0\neq 0,\quad x_5=0$.

c) Non-minimal oriented planes $x_0=0,\quad x_5\neq 0$.

d) Minimal oriented planes $x_0=x_5=0$.

The plane at infinity is included in the latter category. The common tangential segment of two non-planar spheres (x) and (y) will be

$$t=\sqrt{\frac{2(-x_0y_1-x_1y_0+x_2y_2+x_3y_3+x_4y_4+x_5y_5)}{-x_0y_0}}. \quad (5)$$

For the angle of intersection of two not null spheres we have

$$\sin^2\frac{\theta}{2}=\frac{-x_0y_1-x_1y_0+x_2y_2+x_3y_3+x_4y_4+x_5y_5}{2x_5y_5}. \quad (6)$$

We shall leave till a later stage of the present chapter the further direct discussion of the oriented sphere, and take up for the present a detailed study of the oriented plane. This has five homogeneous coordinates (x) where

$$x_2 x + x_3 y + x_4 z + x_1 t = 0,$$
$$x_2^2 + x_3^2 + x_4^2 + x_5^2 = 0. \tag{7}$$

Suppose that these coordinates are limited by a single linear equation

$$-a_0 x_1 + a_2 x_2 + a_3 x_3 + a_4 x_4 + a_5 x_5 = 0. \tag{8}$$

If $a_0 \neq 0$ the plane is properly tangent to the oriented sphere

$$\left(a_0, \quad \frac{a_2^2 + a_3^2 + a_4^2 + a_5^2}{-2a_0}, \quad a_2, \ a_3, \ a_4, \ a_5\right).$$

If $a_0 = 0$ we see that the extremities of the normals to the planes will trace in the plane at infinity a conic having double or four-point contact with the circle at infinity, i.e. the oriented planes make a fixed angle with a fixed oriented line or is invariantly related to an isotropic. Every oriented sphere has a linear equation in oriented plane coordinates; every such equation will represent either an oriented sphere, or the planes making a fixed angle with an oriented line, or invariantly related to a minimal line. This latter corresponds to the case

$$a_0 = a_2^2 + a_3^2 + a_4^2 = 0.$$

The power of the oriented plane (x) with regard to the proper oriented sphere (y) is

$$\frac{-y_0 x_1 + y_2 x_2 + y_3 x_3 + y_4 x_4 + y_5 x_5}{y_0 x_0 - y_1 x_1 - y_2 x_2 - y_3 x_3 - y_4 x_4 + y_5 x_5}.$$

The formula for a Laguerre inversion in the oriented plane will be

$$x_1' = (\alpha_5^2 - a_5^2) x_i - 2(a_2 x_2 + a_3 x_3 + a_4 x_4 + \alpha_5 x_5) a_i, \quad i \neq 5,$$
$$x_5' = (\alpha_5^2 - a_5^2) x_5 - 2(a_2 x_2 + a_3 x_3 + a_4 x_4 + \alpha_5 x_5) \alpha_5.$$

The geometry of the oriented sphere furnishes simple duals to various residuation theorems, which we found in our sphere geometry. We start with V. 37]. Suppose that we have four oriented spheres s_1, s_2, s_3, s_4 properly tangent to a plane π. Let π_i be the other oriented plane properly tangent to the three spheres s_j, s_k, s_l, while π_{ij} is an arbitrary oriented plane properly tangent to s_k and s_l. The oriented sphere properly tangent to π_i, π_{ij}, π_{ik}, π_{il} shall be s_i'. We write the equation

$$\lambda s_1 s_1' + \mu s_2 s_2' + \nu s_3 s_3' = 0.$$

This oriented envelope will determine with an arbitrary proper oriented sphere a developable of the fourth order; the points of contact will trace a cyclic. The terms of this equation are linearly independent, the planes of the system properly tangent to s_4 would seem to determine thereon a two-parameter family of cyclics through the seven points of contact with π, π_1, π_2, π_3, π_{12}, π_{13}, π_{23}. Through these points there will pass but a one-parameter family of cyclics. Hence we may choose λ, μ, ν so that s_4 itself shall be part of the envelope. The rest will be a second oriented sphere which will share with s_4' the proper tangent planes π_4, π_{14}, π_{24}, π_{34}. The points of contact are not concyclic in the general case, hence this sphere must be identical with s_4'; the latter must touch the remaining common proper tangent plane to s_1', s_2', s_3'.

Theorem 4.] *If four oriented spheres be given, all properly tangent to the same oriented plane, but no three properly tangent to three oriented planes, and if the remaining common proper tangent planes be taken to each set of three, as well as one arbitrary plane properly tangent to each two, then the four oriented spheres each properly tangent to a plane touching three of the original ones and to the three arbitrary planes each tangent to two of these three spheres are themselves properly tangent to one oriented plane.*

This theorem may be somewhat generalized by a contact transformation of spheres; the statement, which is certainly bad enough at present, becomes altogether too involved to be worth while. Let the reader prove by methods analogous to those which were used in the case of VI. 11].

Theorem 5.] *Given five oriented spheres properly tangent to an oriented plane, no three properly tangent to more than two oriented planes, and all arranged in cyclic order. Let the common proper tangent plane be constructed to each set of three successive spheres. And let five oriented spheres be constructed each properly tangent to three successive oriented planes just found. Then the remaining common proper tangent planes to the five sets of three successive spheres in the new sequence will themselves touch a sphere, or make a fixed angle with a fixed line, or be invariantly related to a minimal line.*

It is intuitively clear that the geometry of the oriented sphere may be treated by a minimal projection exactly as was that of the oriented circle. We begin with the fundamental equations

$$\begin{aligned} T^2 &\equiv \rho x_0, \\ \tfrac{1}{2}(X^2+Y^2+Z^2+W^2) &\equiv \rho x_1, \\ -XT &\equiv \rho x_2, \\ -YT &\equiv \rho x_3, \\ -ZT &\equiv \rho x_4, \\ -WT &\equiv \rho x_5. \end{aligned} \tag{9}$$

The finite point in four-dimensional space Σ_4 with the homogeneous rectangular cartesian coordinates $(X:Y:Z:W:T)$ will be the centre of a null hypersphere which will cut the hyperplane $W=0$ in the sphere

$$x_0(x^2+y^2+z^2)+x_2(2x)+x_3(2y)+x_4(2z)+x_5(2t)=0. \tag{2}$$

The radius of this sphere will be

$$r=\frac{ix_5}{x_0}=-\frac{iW}{T}. \tag{10}$$

We shall make our Σ_4 a perfect continuum by adjoining a single point at infinity with the coordinates $x_1=1$, $x_i=0$, and ∞^3 other improper points whose coordinates satisfy the equations

$$x_0=x_2^2+x_3^2+x_4^2+x_5^2=0. \tag{11}$$

We shall define such points as the minimal hyperplanes

$$x_2X+x_3Y+x_4Z+x_5W+x_1T=0,$$

which cut the hyperplane Σ_3 in the oriented planes

$$x_2x+x_3y+x_4z+x_1t=0.$$

Theorem 6.] *If a four-dimensional cartesian space of Euclidean measurement be made a perfect hexaspherical continuum by the adjunction of a single point at infinity, and the totality of minimal hyperplanes as improper points, there is a perfect one to one correspondence between the points of such a space and the oriented spheres of a three-dimensional cartesian space. The correspondence may be effected by letting each proper point of Σ_4 correspond to the sphere in Σ_3 whose centre is the foot of the perpendicular on Σ from the given point, while the radius is $-i$ times the algebraic distance from that centre to the given point. Each improper point is represented by the plane where Σ_3 meets the corresponding minimal hyperplane of Σ_4, with an orientation rationally determined by the coordinates of the improper point. The distance of two proper points will be the tangential distance of the corresponding oriented spheres.*

Σ_3.	Σ_4.
Oriented sphere.	Point.
Proper oriented sphere.	Proper point.
Non-planar null sphere.	Point of Σ_3.
Non-minimal plane.	Minimal hyperplane whose point of contact with the sphere at infinity is not in Σ_3.
Minimal plane.	Minimal hyperplane whose point of contact is in Σ_3.
Common tangential segment of proper spheres.	Distance of two proper points.
Pencil of properly tangent spheres.	Isotropic line.
Pencil of properly parallel planes.	Pencil of parallel minimal hyperplanes.

Transformation carrying oriented planes properly tangent to an oriented sphere into oriented planes also tangent to a sphere.	Conformal collineation.

Theorem 7.] *There is an eleven-parameter group of transformations which carry oriented planes into oriented planes, and those tangent to an oriented sphere into others tangent to another sphere. Every transformation of the group will multiply the common tangential segments of two proper spheres by a constant, characteristic of the transformation.*

Σ_3.	Σ_4.
Eleven-parameter group of oriented planes and spheres.	Eleven-parameter group of conformal collineations.

If we use temporarily x_1, x_2, x_3, x_4 for rectangular cartesian coordinates in Σ_4, and put p_i for $\frac{\partial f}{\partial x_i}$, the Lie symbol for this eleven-parameter group is *

$$[p_i, \ x_i p_j - x_j p_i, \ \sum_{i=1}^{i=4} x_i p_i].$$

The transformations of this group which multiply tangential segments by the factor ± 1 shall be called *Laguerre transformations.*

Theorem 8.] *The Laguerre group in space is a ten-parameter mixed group, with a ten-parameter continuous subgroup, and a ten-parameter continuous sub-assemblage.*

The Laguerre transformations of the sub-group shall be

* Cf. Page, 'On the Primitive Groups of Transformations in a Space of Four Dimensions', *American Journal of Math.*, vol. x, 1888, p. 345.

called *direct*, those of the sub-assemblage *indirect*. The Lie symbol for the continuous group is

$$[p_i,\ x_i p_j - x_j p_i].$$

Σ_3.	Σ_4.
Laguerre group.	Congruent group.
Continuous direct sub-group.	Group of motions.
Sub-assemblage of indirect transformations.	Sub-assemblage of symmetry transformations.
Laguerre inversion.	Reflection in hyperplane.

It would be natural to assume that this ten-parameter group was simply isomorphic with that of all conformal transformations of three-dimensional space. Such is not, however, the case. The conformal group appears in four dimensions as a collineation group keeping invariant a hyperquadric of non-vanishing discriminant; the Laguerre group is a sub-group of the eleven-parameter group which leaves in place, not a hyperquadric, but a quadric surface.

Before resolving our Laguerre transformations into factors, it is well to approach the subject from another point of view, exactly as we did in the plane. Let us define as *equilong* any analytic transformation of oriented planes which keeps invariant the distance of the points of contact of an arbitrary oriented plane with any two envelopes. We seek the general analytic expression for such a transformation. We first write our oriented plane

$$ax+by+cz = p. \tag{12}$$

Here a, b, c are supposed to be the direction cosines of the directed normal to the plane. We may write this same equation also in Bonnet coordinates *

$$\frac{u+v}{uv+1}x+i\frac{u-v}{uv+1}y+\frac{uv-1}{uv+1}z = \frac{w}{uv+1}. \tag{13}$$

* Cf. Darboux, *Théorie générale des surfaces*, vol. i, Paris, 1887, pp. 243 ff.

Let the reader establish the relations between these and our previous oriented plane coordinates, namely

$$\begin{aligned} \rho x_1 &= -w, & u &= -\frac{x_2 - i x_3}{x_4 + i x_5}, \\ \rho x_2 &= u + v, & & \\ \rho x_3 &= i(u - v), & v &= \frac{x_2 + i x_3}{x_4 + i x_5}, \\ \rho x_4 &= uv - 1, & & \\ \rho x_5 &= i(wv + 1), & w &= \frac{x_1}{x_4 + i x_5}; \end{aligned} \tag{14}$$

u and v are the isotropic parameters of the spherical representation of the plane. In an equilong transformation parallel planes must go into parallel planes, exactly as in the case of the plane equilong transformation, hence

$$U = U(u, v), \quad V = V(u, v).$$

Moreover, two planes which intersect in a minimal line, whereon all distances are null, must correspond to two other such planes, hence

either $$U = U(u), \quad V = V(v),$$
or else $$U = U(v), \quad V = V(u).$$

Evidently, if we prove that we have transformations of both types, the first of these will be direct and the second indirect. In order that the transformation be real we must have

$$v = \bar{u}, \quad V = \bar{U}.$$

Let us now look at the triangle in the plane (u, v, w) formed by the adjacent planes

$$(u + du,\ v + dv,\ w + dw), \quad (u + d'u,\ v + d'v,\ w + d'w),$$
$$(u + d''u,\ v + d''v,\ w + d''w).$$

Since our equilong transformation is to carry this into an equal triangle (the sides of one are, by definition, equal to those of the other), its area must be invariant, at least if we disregard the sign. Conversely, suppose that we have a transformation of oriented planes which is conformal for their spherical representations and leaves the areas of such triangles invariant. Every such triangle is carried into a similar and equivalent triangle, i. e. into an equal one, and the transformation is equilong. Reverting to the first form for the equation of the plane (12), we find the area of a triangle from the volume of a tetrahedron, and the distance from a vertex to the opposite face-plane. Thus the square root of the area of the triangle in question will differ by a constant factor from

$$\frac{\begin{vmatrix} a & b & c & p \\ da & db & dc & dp \\ d'a & d'b & d'c & d'p \\ d''a & d''b & d''c & d''p \end{vmatrix}}{\sqrt{\begin{vmatrix} a & b & c \\ da & db & dc \\ d'a & d'b & d'c \end{vmatrix}}\sqrt{\begin{vmatrix} a & b & c \\ d''a & d''b & d''c \\ da & db & dc \end{vmatrix}}\sqrt{\begin{vmatrix} a & b & c \\ d'a & d'b & d'c \\ d''a & d''b & d''c \end{vmatrix}}}$$

In expanding in terms of the first row the coefficient of p vanishes, since

$$(ada+bdb+cdc)=(ad'a+bd'b+cd'c)=(ad''a+bd''b+cd''c)=0.$$

$$\begin{vmatrix} a & b & c \\ da & db & dc \\ d'a & d'b & d'c \end{vmatrix}=\sqrt{\begin{vmatrix} (da^2+db^2+dc^2) & (dad'a+dbd'b+dcd'c) \\ (dad'a+dbd'b+dcd'c) & (d'a^2+d'b^2+d'c^2) \end{vmatrix}}$$

$$=\frac{\pm 2i\,(du\,d'v-dv\,d'u)}{(uv+1)},$$

$$dp=\frac{dw}{uv+1}-\frac{w\,(udv+vdu)}{(uv+1)^2}.$$

Our invariant expression thus becomes, except for a constant factor,

$$\frac{\begin{vmatrix} du & dv & dw \\ d'u & d'v & d'w \\ d''u & d''v & d''w \end{vmatrix}}{\sqrt{\pm(du\,d'v-dv\,d'u)}\;\sqrt{\pm(d'u\,d''v-d'v\,d''u)}\;\sqrt{\pm(d''u\,dv-d''u\,du)}}.$$

To carry through the direct transformation we put this equal to the corresponding form, remembering that

$$dU = U'du,\ dv = V'dv,\ dW = \frac{\partial W}{\partial u}du + \frac{\partial W}{\partial v}dv + \frac{\partial W}{\partial w}dw.$$

Our expression above thus becomes equal to

$$\frac{\dfrac{\partial W}{\partial w}U'V'\begin{vmatrix} du & dv & dw \\ d'u & d'v & d'w \\ d''u & d''v & d''w \end{vmatrix}}{\sqrt{\pm(U'V')^3}\sqrt{\pm(dud'v-dvd'u)}\sqrt{\pm(d'ud''v-d''ud'v)}\sqrt{\pm(d''udv-d''vdu)}}.$$

$$\frac{\partial W}{\partial w} = \sqrt{\pm U'V'}.$$

We thus get the fundamental equations for direct and indirect equilong transformations,*

$$\begin{gathered} U = U(u),\ V = V(v),\ W = w\sqrt{\pm U'V'} + F(u, v), \\ U = U(v),\ V = V(u),\ W = w\sqrt{\pm U'V'} + F(u, v). \end{gathered} \tag{15}$$

Let us now see what form these various functions take if the oriented planes properly tangent to an oriented sphere go into other such tangents. The equation of an oriented sphere

* The fact that these transformations depend upon arbitrary functions was first suggested by Study, 'Ueber mehrere Probleme der Geometrie, welche der konformen Abbildung analog sind', *Sitzungsberichte der niederrheinischen Gesellschaft für Natur- und Heilkunde*, December 5, 1904. The formulae with the upper sign were first published by the Author, 'The Equilong Transformations of Space', *Transactions American Math. Soc.*, vol. ix, 1908. The complete form given above was published independently a year or two later by Blaschke, 'Ueber einige unendliche Gruppen von Transformationen orientierter Ebenen im Euklidischen Raume', *Grunerts Archiv*, Series 3, vol. xvi, 1910. Between these two articles appeared the dissertation of Löhrl, *Ueber konforme und äquilonge Transformationen im Raume*, Würzburg, 1910. Here we find no such formulae. On the contrary, the author's object is to establish the correspondence between the conformal and equilong groups. In consequence the whole dissertation rests upon the erroneous idea (p. 27) that the equilong group depends upon ten parameters. The proof which he offers for this incorrect theorem is geometrical, and contains an obvious error.

is linear in the quantities u, v, $uv-1$. Hence U and V must be linear functions, W must be a fraction whose denominator is the product of the denominators of U and V, while its numerator is linear in u, v, w, $(uv-1)$. Our direct Laguerre transformation will thus have the form

$$U=\frac{\alpha u+\beta}{\gamma u+\delta},\quad V=\frac{\alpha' v+\beta'}{\gamma' v+\delta'},$$

$$W=\frac{\sqrt{\pm(\alpha\delta-\beta\gamma)(\alpha'\delta'-\beta'\gamma')}\,w+puv+qu+rv+s}{(\gamma u+\delta)(\gamma' v+\delta')}. \tag{16}$$

The general indirect one will be

$$U=\frac{\alpha v+\beta}{\gamma v+\delta},\quad V=\frac{\alpha' u+\beta'}{\gamma' u+\delta'},$$

$$W=\frac{\sqrt{\pm(\alpha\delta-\beta\gamma)\cdot(\alpha'\delta'-\beta'\gamma')}\,w+puv+qu+rv+s}{(\gamma' u+\delta')(\gamma v+\delta)}. \tag{17}$$

The ten independent parameters are well set in evidence in these equations. Four oriented planes will have an absolute invariant under the Laguerre group

$$\frac{(u_1-u_2)(u_3-u_4)}{(u_1-u_4)(u_3-u_2)}.$$

Recalling theorem 9] and equation (5) of Ch. VIII, we see that the modulus and argument of this complex expression will give the double ratio and double angle of the stereographic projection of the points of contact of the four oriented planes with their common proper tangent sphere. We find a more direct interpretation for the modulus as follows. Consider the real oriented planes $(u_1\bar{u}_1w_1)(u_2\bar{u}_2w_2)$.

$$\sum_{i=2}^{i=5} x_i^{(1)}x_i^{(2)} \equiv -2(u_1-u_2)(\bar{u}_1-\bar{u}_2).$$

$$\sqrt{\frac{(u_1-u_2)(u_3-u_4)}{(u_1-u_4)(u_3-u_2)}}\sqrt{\frac{(\bar{u}_1-\bar{u}_2)(\bar{u}_3-\bar{u}_4)}{(\bar{u}_1-\bar{u}_4)(\bar{u}_3-\bar{u}_2)}}$$

$$\equiv \pm\sqrt{\frac{\sum_{i=2}^{i=5} x_i^{(1)}x_i^{(2)}\sum_{i=2}^{i=5} x_i^{(3)}x_i^{(4)}}{\sum_{i=2}^{i=5} x_i^{(1)}x_i^{(4)}\sum_{i=2}^{i=5} x_i^{(3)}x_i^{(2)}}}$$

$$\equiv \frac{\sin\frac{1}{2}\measuredangle 12\sin\frac{1}{2}\measuredangle 34}{\sin\frac{1}{2}\measuredangle 14\sin\frac{1}{2}\measuredangle 32}. \tag{18}$$

This invariant will take the value unity when the points of contact form an orthocyclic set. Another absolute invariant is

$$\frac{\begin{vmatrix} 0 & \sum_{i=2}^{i=5} x_i^{(1)}x_i^{(2)} & \sum_{i=2}^{i=5} x_i^{(1)}x_i^{(3)} & \sum_{i=2}^{i=5} x_i^{(1)}x_i^{(4)} \\ \sum_{i=2}^{i=5} x_i^{(2)}x_i^{(1)} & 0 & \sum_{i=2}^{i=5} x_i^{(2)}x_i^{(3)} & \sum_{i=2}^{i=5} x_i^{(2)}x_i^{(4)} \\ \sum_{i=2}^{i=5} x_i^{(3)}x_i^{(1)} & \sum_{i=2}^{i=5} x_i^{(3)}x_i^{(2)} & 0 & \sum_{i=2}^{i=5} x_i^{(3)}x_i^{(4)} \\ \sum_{i=2}^{i=5} x_i^{(4)}x_i^{(1)} & \sum_{i=2}^{i=5} x_i^{(4)}x_i^{(2)} & \sum_{i=2}^{i=5} x_i^{(4)}x_i^{(3)} & 0 \end{vmatrix}}{\sum_{i=2}^{i=5} x_i^{(1)}x_i^{(2)}\sum_{i=2}^{i=5} x_i^{(3)}x_i^{(4)}\sum_{i=2}^{i=5} x_i^{(4)}x_i^{(4)}\sum_{i=2}^{i=5} x_i^{(3)}x_i^{(2)}}$$

$$\equiv \frac{\Pi\left(\sin\frac{1}{2}\measuredangle 12\sin\frac{1}{2}\measuredangle 34 \pm \sin\frac{1}{2}\measuredangle 13\sin\frac{1}{2}\measuredangle 42 \pm \sin\frac{1}{2}\measuredangle 14\sin\frac{1}{2}\measuredangle 23\right)}{\sin^2\frac{1}{2}\measuredangle 12\sin^2\frac{1}{2}\measuredangle 34\sin^2\frac{1}{2}\measuredangle 14\sin^2\frac{1}{2}\measuredangle 32}. \tag{19}$$

This vanishes when the points of contact are concyclic.

A real indirect transformation will be involutory if it have the form

$$U = \frac{\alpha v + b}{cv - \bar{\alpha}}, \quad V = \frac{\bar{\alpha} u + b}{cu - \alpha},$$

$$W = \frac{-(\alpha\bar{\alpha} + bc)\, w + luv + \mu u + \bar{\mu} v + n}{(cv - \bar{\alpha})(cu - \alpha)}.$$

$$b = \bar{b},\ c = \bar{c},\ l = \bar{l},\ n = \bar{n},\ -bl + \mu\alpha + \bar{\mu}\bar{\alpha} + uc = 0.$$

Let us consider the real Laguerre group a little more closely. Instead of minimal projection, let us consider the analog of Fiedler's cyclographic method developed in Ch. IV. If the centre of a real sphere be xyz, and its radius r, we represent it by the point $xyzr$ of four-dimensional space. We define as the distance of two points in this space the expression

$$\sqrt{(x-x')^2 + (y-y')^2 + (z-z')^2 + (r-r')^2}.$$

The Laguerre group is simply isomorphic with the group in our four-dimensional space S_4, leaving invariant a real quadric with imaginary generators in the hyperplane at infinity, i.e. the Lorenz group of the modern theory of relativity.* The hyperplane at infinity shall be called S_3, the quadric S_2^2. The group in S_3 is that of Ch. VIII. As we may pass continuously from a direct transformation to the identical transformation, we have in S_3 a collineation permuting among themselves the generators of each set of S_2^2. We have the following types of fixed elements in S_3:

1) Two real and two conjugate imaginary fixed points of S_2^2, no other fixed point in S_3.

2) One real line of fixed points, and two real or conjugate imaginary fixed points where its polar intersects S_2^2.

* Cf. Wilson and Lewis, 'The Space-Time Manifold of Relativity', *Proceedings American Academy of Arts and Sciences*, vol. xlviii, 1912. The nature of the group of relativity was, of course, known from the first; its geometric exposition is nowhere more elaborately discussed than in this article.

3) Limiting cases of 1) and 2), where pairs of real fixed points tend to fall together.

4) All points of S_3 fixed.

In a transformation of type 1) there will be two real fixed lines in S_3, mutually polar in S_2^2. The totality of planes through each will be a two-dimensional projective manifold which is subject to a real collineation. Two fixed planes are the tangent ones to S_2^2 through the given line; there will be a third not in S_3. We thus get two real fixed planes not lying in a real space, and so intersecting in a finite fixed point. Each of the fixed planes through this point is the locus of non-Euclidean perpendiculars to the other thereat (lines conjugate with regard to S_2^2). The transformation of S_4 is the product of successive rotations about these two planes, and each rotation can be factored into the product of reflections in two hyperplanes.

2) In a transformation of this type, if there were a finite fixed point there would be a pencil of fixed lines through it to the fixed points in S_3, the transformation would have to be a non-Euclidean rotation about the plane of the pencil, and so the product of two reflections. If there were no finite fixed point, but a finite fixed plane as before, we should have a screwing about this plane which, again, could be factored into four reflections in hyperplanes.

3) Since these transformations are limiting cases of the previous, they also can be factored into four reflections, for every limiting form of a hyperplane reflection which is not a degenerate transformation is still a reflection.

4) There can be no finite fixed point, for then every point would be fixed; we must have a translation.

Theorem 9.] *Every direct Laguerre transformation can be factored into four Laguerre inversions, every indirect one into five inversions.*

The Laguerre inversion is not the only type of involutory Laguerre transformation. We have transformations which correspond to reflections in planes, lines, and hyperplanes.

The nature of these is seen by choosing the simplest case, and noticing the behaviour of the invariants.* We thus get

Σ_3.	Σ_4.
Laguerre inversion.	Reflection in hyperplane.
Corresponding oriented planes touch properly ∞^1 oriented spheres properly tangent to two fixed oriented planes, the points of contact being in every case a harmonic set.	Reflection in plane.
Corresponding oriented planes properly tangent to same oriented sphere which touches properly ∞^1 fixed oriented planes; points of contact with corresponding planes inverse in circle of contact with fixed planes.	Reflection in line.
Corresponding oriented planes reflections of one another in oriented plane properly parallel to them, and properly tangent to fixed proper oriented sphere.	Reflection in a point.

It should be noticed that the second and fourth of these transformations are direct, while the first and third are indirect.

§ 3. The Hypercyclide.

Let us next take up the oriented surface which corresponds in three dimensions to the plane hypercyclic. We shall define

* First studied by Laguerre, *Collected Works*, vol. ii, pp. 432 ff. See also Smith, *On a Transformation*, cit., p. 165; Blaschke, *Geometrie der Speere*, cit., p. 55; Müller, 'Geometrie orientierter Kugeln', cit., pp. 295 ff.; Bricard, 'Sur la géométrie de direction dans l'espace', *Nouvelles Annales de Math.*, Series 4, vol. vi, 1906.

as a *hypercyclide* every oriented envelope given by oriented plane equations of the form

$$\sum_{i,j=1}^{i,j=5} a_{ij}x_ix_j = 0, \quad x_2^2+x_3^2+x_4^2+x_5^2 = 0. \tag{20}$$

Σ_3.	Σ_4.
Hypercyclide.	Focal hyperdevelopable of hyperquadric.

Theorem 10.] *The hypercyclide is an oriented envelope of the fourth class.*

Suppose that our hypercyclide is of such structure that the following equation has four distinct roots,

$$\begin{vmatrix} a_{22}-\rho & a_{23} & a_{24} & a_{25} \\ a_{32} & a_{33}-\rho & a_{34} & a_{35} \\ a_{42} & a_{43} & a_{44}-\rho & a_{45} \\ a_{52} & a_{53} & a_{54} & a_{55}-\rho \end{vmatrix} = 0,$$

and that it is not transformed into itself by a reversal of the orientation of every plane, it shall be called a *general* hypercyclide.

$$(ax^2) = x_2^2+x_3^2+x_4^2+x_5^2 = 0. \tag{21}$$

Σ_3.	Σ_4.
General hypercyclide.	Focal hyperdevelopable of central hyperquadric.

Theorem 11.] *The general hypercyclide is anallagmatic in four Laguerre inversions and carried into itself by a group of ten involutory Laguerre transformations, including the identical transformation.*

In each hyperplane of symmetry of a central hyperquadric of Σ_4 will lie a focal central quadric, a double surface of the focal hyperdevelopable. The orthogonal projection of this is also a central quadric; hence

Theorem 12.] *The general hypercyclide may be generated in four ways by an oriented sphere whose centre traces a central quadric, while it meets a fixed oriented plane at a fixed angle.*

Theorem 13.] *The general hypercyclide is an anticaustic by refraction of each of its deferents, the rays of light being supposed to come in a direction perpendicular to the fundamental plane of the corresponding Laguerre inversion.*

The proof of this is identical with that for the corresponding theorem about the hypercyclic, and so is omitted. We pass to the determination of the order of the general hypercyclide. We see, first of all, that the intersection of the fundamental plane of each Laguerre inversion, with the corresponding deferent, is a double curve of the hypercyclide (cf. Ch. X, theorem 46). It will also meet this plane simply along a curve which is the envelope of lines through which pass tangent planes to the deferent which are perpendicular to anallagmatic planes through the same lines. This curve is the intersection of the fundamental plane with the developable which touches the deferent and a conic in the plane at infinity which has double contact with the circle at infinity. The order of the curve is eight, for it is of the fourth class with deficiency one, and has two double tangents. (It is dual to the cone from an arbitrary point to the curve of intersection of two quadrics.)

Theorem 14.] *The general hypercyclide is of the twelfth order. It meets the fundamental plane of each Laguerre inversion when the plane is finite, in a double conic, the intersection with the corresponding deferent, and in a simple curve of the eighth order and fourth class which is a line of curvature of the surface.*

The truth of the last statement is evident from the fact that the tangent generating spheres have stationary contact. Each infinite point of the deferent in general position will give two distinct asymptotic planes of the hypercyclide. Conversely, each asymptotically tangent plane to the hypercyclide whose point of contact is not on the circle at infinity will give an asymptotically tangent plane to the deferent. How about the circle at infinity, is that also on the hypercyclide? If we take a line in a fundamental plane tangent to the focal developable of the corresponding deferent, one tangent plane to the deferent is minimal, hence one tangent plane to the hypercyclide touches it on the circle at infinity, but in no other case will a finite line in the plane give a point of the hypercyclide on the circle at infinity. Through each point of the infinite line of a fundamental plane will pass four tangent planes to the focal developable of the corresponding deferent, but only two tangents to the circle at infinity; the latter must be a double curve of the hypercyclide.

Theorem 15.] *The general hypercyclide has the circle at infinity as a double conic, it has no proper foci, its double foci are those of the deferents.*

It is easy to see that in four dimensions there are two hyperplane reflections which carry any non-minimal hyperplane into any other not parallel to it. When they are parallel there is but one.

Theorem 16.] *If no fundamental plane for a Laguerre inversion in which a general hypercyclide is anallagmatic lie at infinity, there are eight Laguerre inversions which will carry it into a central quadric counted doubly.*

Since every transformation of the Laguerre group is a contact transformation of oriented spheres, it will carry strips of curvature of a surface into other such strips. We thus get from the familiar properties of the lines of curvature of a central quadric

Theorem 17.] *The lines of curvature of a general hypercyclide are algebraic, and are determined by the common generating spheres of one system which it shares with a one-parameter family of hypercyclides, all anallagmatic in the same set of Laguerre inversions.*

Theorem 18.] *The oriented tangent planes to a general hypercyclide may be put into one to one correspondence with the points of a general cyclide in such a way that the planes properly tangent to an oriented sphere, or making a fixed angle with a fixed oriented line, or invariantly related to a fixed minimal line, will correspond to points on a sphere.*

The common oriented developable of a general hypercyclide and oriented sphere will touch the latter at the points of a cyclic. The cyclic degenerates into two circles in the case of a generating sphere of one system or another, and in that case alone, for there are no other spheres with double proper contact. The common developable becomes in this case two cones of revolution. Two such cones circumscribed to the hypercyclide and to the same oriented sphere are said to be residual; two cones residual to the same cone are coresidual.

Theorem 19.] *If two cones of revolution properly circumscribed to the same general hypercyclide be coresidual, every such cone residual to the one is residual to the other.*

The vertex of a properly circumscribed cone of revolution, being the centre of a generating null sphere, is on a focal conic of the deferents, hence

Theorem 20.] *The cones of revolution properly circumscribed to a general hypercyclide fall into four systems, each containing two series. The cones of each system are anallagmatic in one of the fundamental Laguerre inversions, their vertices trace the corresponding double conic, each point of the conic being the vertex of a cone of each series belonging to the system. Two cones of the same series are coresidual, two of the same system but different series residual.*

§ 4. The Oriented Sphere Treated Directly.

We turn now from the consideration of oriented planes and their envelopes to the direct discussion of oriented spheres; in other words, we pass from the Laguerre group to the fifteen-parameter contact group of oriented spheres. A system of such spheres whose coordinates are analytic functions of three parameters, their ratios not being all functions of a lesser number of parameters, shall be called a *complex*. The simplest complex is the linear one composed of spheres whose coordinates satisfy a linear equation of the type

$$-a_1x_0-a_0x_1+a_2x_2+a_3x_3+a_4x_4+a_5x_5=0. \tag{22}$$

The discussion of the special cases is exactly analogous to that carried out in the last chapter:

$$-2a_0a_1+a_2^2+a_3^2+a_4^2 \neq 0.$$

$$\frac{-a_1x_0-a_0x_1+a_2x_2+a_3x_3+a_4x_4+\sqrt{2a_0a_1-a_2^2-a_3^2-a_4^2}x_4}{2\sqrt{2a_0a_1-a_2^2-a_3^2-a_4^2}x_5}$$

$$=\frac{1}{2}\left[1-\frac{a_5}{\sqrt{2a_0a_1-a_2^2-a_3^2-a_4^2}}\right]. \quad (23)$$

The oriented spheres of the complex meet a fixed oriented sphere at a fixed angle. If we have

$$-2a_0a_1+a_2^2+a_3^2+a_4^2+a_5^2=0, \quad (24)$$

our oriented spheres are properly tangent to a fixed oriented sphere, and the complex shall be said to be *special*.

$$a_0 \neq 0.$$

$$\frac{2\left[\dfrac{a_2^2+a_3^2+a_4^2+a_5^2}{-2a_0}x_0-a_0x_1+a_2x_2+a_3x_3+a_4x_4+a_5x_5\right]}{-a_0x_0}$$

$$=2\left[\frac{a_1}{-a_0}+\frac{a_2^2+a_3^2+a_4^2+a_5^2}{2a_0^2}\right].$$

The oriented spheres have a fixed common tangential segment with a given sphere.

$$-2a_0a_1+a_2^2+a_3^2+a_4^2=a_0(a_2^2+a_3^2+a_4^2+a_5^2)=0.$$

Oriented spheres invariantly related to a fixed minimal plane.

$$a_0=0, \quad a_2^2+a_3^2+a_4^2 \neq 0.$$

Oriented spheres meeting a fixed plane at a fixed angle.

$$a_1=a_2=a_3=a_4=0.$$

Complex or oriented spheres of given radius, which is null if $a_5=0$.

$$a_0=a_2=a_3=a_4=a_5=0.$$

Complex of all oriented planes.

Theorem 21.] *The assemblage of all oriented spheres whose common tangential segments with two fixed oriented spheres bear to one another a fixed ratio is, in general, that of all oriented spheres meeting at a fixed angle a fixed oriented sphere coaxal with the given spheres.*

Two linear complexes (a) and (b), neither of which is special, have an absolute invariant under the fifteen-parameter group, namely

$$\frac{-a_0b_1-a_1b_0+a_2b_2+a_3b_3+a_4b_4+a_5b_5}{\sqrt{-2a_0a_1+a_2{}^2+a_3{}^2+a_4{}^2+a_5{}^2}\,\sqrt{-2bd_1+b_2{}^2+b_3{}^2+b_4{}^2+b_5{}^2}}\equiv I. \quad (25)$$

In the general case where the complexes consist in oriented spheres meeting fixed spheres at the angles ϕ_1 and ϕ_2 respectively, while θ is the angle of these fixed spheres, we have

$$\frac{\cos\phi_1\cos\phi_2-\cos\theta}{\sin\phi_1\sin\phi_2}=I. \quad (26)$$

When the numerator of this expression vanishes, the complexes are said to be in *involution*.

Theorem 22.] *Two linear complexes of oriented spheres, which consist in the assemblages of all spheres meeting two fixed oriented spheres at fixed angles, will be in involution when, and only when, the product of the cosines of these fixed angles is equal to the cosine of the angle of the fixed spheres.*

Theorem 23.] *A special linear complex is in involution with every linear complex which includes its fundamental sphere.*

Theorem 24.] *If a linear complex consisting in oriented spheres meeting a fixed sphere at a fixed angle be in involution with that of all null spheres, the angle is* $\frac{\pi}{2}\cdot$

Theorem 25.] *If the fixed sphere of a linear complex be planar, the complex is in involution with that of all oriented planes.*

The transformation of inversion in a linear complex is

analogous to the corresponding transformation in the plane. The analytical expression will be

$$x_i' = (\alpha_5^2 - a_5^2)\, x_i$$
$$- 2\,(-a_1 x_0 - a_0 x_1 + a_2 x_2 + a_3 x_3 + a_4 x_4 + \alpha_5 x_5)\, a_i, \quad i \neq 5.$$

$$x_5' = (\alpha_5^2 - a_5^2)\, x_i$$
$$- 2\,(-a_1 x_0 - a_1 x_0 + a_2 x_2 + a_3 x_3 + a_4 x_4 + a_5 x_5)\, \alpha_5. \quad (27)$$

$$-2\, a_0 a_1 + a_2^2 + a_3^2 + a_4^2 + a_5^2 = 0.$$

$$(-2\, x_0' x_1' + x_2'^2 + x_3'^2 + x_4'^2 + x_5'^2)$$
$$= (\alpha_5^2 - a_5^2)^2\, (-2\, x_0 x_1 + x_2^2 + x_3^2 + x_4^2 + x_5^2).$$

Theorem 26.] *Two oriented spheres are mutually inverse in a linear complex consisting in the totality of oriented spheres which meet a fixed sphere at a fixed angle, if they be coaxal with the fixed sphere, and the product of the tangents of the halves of their angles therewith is equal to the square of the tangent of half the fixed angle.**

Theorem 27.] *When the fixed angle of a linear complex is* $\frac{\pi}{2}$, *inversion in that complex is inversion in its fundamental sphere.*

Theorem 28.] *Inversion in a linear complex of oriented spheres of given radius is a dilatation.*

Theorem 29.] *Inversion in the linear complex of all null spheres reverses the orientation of every sphere.*

We may easily find such a linear complex, called temporarily the *inverting* complex, that inversion therein will carry the complex of all null spheres into any other non-spacial linear complex. We see also that our geometrical definition for inversion in a linear complex breaks down in the case of null

* Cf. Smith, *Transformation of Laguerre*, cit., *On the surfaces enveloped*, cit., and 'Geometry within a Linear Spherical Complex', *Transactions American Math. Soc.*, vol. ii, 1901.

spheres. In this inversion corresponding spheres are coaxal with the fixed sphere, and the oriented spheres of the inverting complex properly tangent to the one are properly tangent to the other. The points of the fundamental sphere of the complex transform into themselves. The points of a minimal line will be transformed into a pencil of properly tangent oriented spheres whose point of contact is on the fundamental sphere.

Theorem 30.] *Each oriented sphere of a linear complex is properly tangent to a pencil of spheres of the complex at each point of a circle.*

A surface considered as a point locus will transform into a congruence of oriented spheres in a linear complex. The spheres tangent to a point surface at a given non-singular point will fall into two pencils, according to their orientation, and will be transformed into two pencils of properly tangent oriented spheres to the oriented envelope of the spheres of the congruence. When the point of contact in the first place is on the fundamental sphere of the inverting complex, the points of contact of the two transformed pencils fall together.

Theorem 31.] *If a point surface be inverted in a linear complex with a fundamental sphere, the curve of intersection with this fundamental sphere will be a double curve of the envelope of the oriented spheres which correspond to the points of the surface.*

Lines of curvature of the point surface will correspond to strips of curvature of the corresponding envelope, the curve of contact with the focal developable will correspond to a strip of contact along a line of curvature of the envelope which lies on the fundamental sphere of the inverting complex, and confocal surfaces will correspond to surfaces with the same strip of curvature along such a curve.*

* These ideas are developed in detail by Smith, *Geometry within a Complex*, cit., pp. 238 ff. He finds interesting analogues to various classical theorems, such as those of Joachimsthal and Darboux-Dupin.

We revert to the minimal projection

Σ_3.	Σ_4.
Linear complex.	Hypersphere.
Complexes in involution.	Orthogonal hyperspheres.
Special linear complex.	Null hypersphere.
Fifteen parameter group of contact transformations of oriented spheres.	Fifteen parameter group of hyperspherical transformations.

Every hyperspherical transformation of Σ_4 may be carried, by an inversion in a hypersphere, into a conformal collineation, and this will be accomplished by any inversion with a fixed finite point of the hyperspherical transformation as centre. We speak loosely here about hyperspherical inversions and collineation of Σ_4, leaving to the reader the simple task of defining these analytically according to the analogy of what was done in two and three dimensions. The ratio of similitude in this transformation will not be independent of the radius of the hypersphere of inversion; hence we may choose this radius so that the ratio of similitude shall have the value ± 1. Our conformal collineation is thus a congruent one; we ha ve from 9]:

Theorem 32.] *Every contact transformation of oriented spheres may be factored into the product of inversions in six or seven linear complexes.**

When we come to the study of complexes of oriented spheres of a more complete structure there is advantage in changing slightly the form of our coordinates, exactly as we did in the case of the oriented circle. Let us write

$$\begin{aligned} -\sqrt{2}\,x_0 &\equiv \rho\,(X_0+iX_1), \\ \sqrt{2}\,x_1 &\equiv \rho\,(X_1-iX_1), \\ x_2 &\equiv \rho X_2. \\ x_3 &\equiv \rho X_3, \\ x_4 &\equiv \rho X_4, \\ x_5 &\equiv \rho X_5. \end{aligned} \tag{28}$$

* Smith, *Surfaces Enveloped*, cit., p. 380.

We have a one to one correspondence between our oriented spheres, and systems of coordinates (X) connected by the identical relation

$$(XX) \equiv 0. \tag{29}$$

The condition for proper contact of two oriented spheres (X) and (Y) will be

$$(XY) = 0. \tag{30}$$

More generally, if we have six linear non-special complexes in involution, where ϕ_i is the fundamental angle of the ith complex, and θ_i is the angle which a given sphere makes with the fundamental sphere of that complex, then we may take as the homogeneous coordinates of that sphere the six quantities

$$\rho X_i = \frac{\cos\phi_i - \cos\theta_i}{\sin\theta_i}.$$

Equations (29) and (30) will subsist and retain their meanings for this more general system of coordinates.

§ 5. The Line-sphere Transformation.

The system of oriented sphere coordinates just explained lead in the most natural way to one of the most beautiful transformations in the whole field of geometry, the line-sphere transformation of Sophus Lie.* We begin by taking two points of complete cartesian (projective) three-dimensional space with the homogeneous coordinates (ξ) and (η). Their line has the Plücker coordinates

$$p_{ij} = \xi_i \eta_j - \xi_j \eta_i$$

The condition of intersection of two lines is

$$\sum_{i,j} p_{ij} q_{kl} = 0, \quad (i-k)(i-l)(\gamma-k)(\gamma-l) \neq 0.$$

* There are innumerable accounts of this transformation. It was first published by Lie in his article, 'Ueber Complexe, insbesondere Linien und Kugelcomplexe', *Math. Annalen*, vol. v, 1872.

From these we pass to what are sometimes called the Klein coordinates, as follows:

$$p_{01} \equiv X_0 + iX_3, \qquad p_{23} \equiv X_0 - iX_3,$$
$$p_{02} \equiv X_1 + iX_4, \qquad p_{32} \equiv X_1 - iX_4,$$
$$p_{03} \equiv X_2 + iX_5, \qquad p_{12} \equiv X_2 - iX_5.$$
$$(XX) \equiv 0. \tag{29}$$

The condition for the intersection of the lines (X) and (Y) will be

$$(XY) = 0 \tag{30}$$

Let us further call the linear complex

$$X_5 = 0$$

the *notable complex*, while the line with the coordinates $(1, i, 0, 0, 0, 0)$ shall be the *notable* line. We have, then, the following correspondence.*

Sphere space.	Line space.
Oriented sphere.	Line.
Properly tangent oriented spheres.	Intersecting lines.
Null spheres.	Lines of notable complex.
Spheres differing only in orientation.	Polar lines in notable complex.
Plane at infinity.	Notable line.
Oriented planes.	Lines intersecting notable line.
Minimal planes.	Lines of notable complex intersecting notable line.
Pencil of properly tangent oriented spheres.	Pencil of lines.
Oriented surface element.	Surface element.

There is one special case of the last correspondence which should be mentioned. Two spheres of equal radius whose

* The form here given is that followed by the Author, 'Metrical Aspect of the Line-sphere Transformation', *Transactions American Math. Soc.*, vol. xii, 1911. Cf. also Snyder, *Ueber die linearen Complexe der Lieschen Kugelgeometrie*, Dissertation, Göttingen, 1895.

centres are on the same minimal line fulfil the analytic requirements of contact and will correspond to intersecting lines; a pencil of lines may thus correspond to a system of spheres of given radius whose centres lie on a minimal line. The surface element here is at infinity at the end of the minimal line; the tangent plane is the corresponding minimal plane. But the point of contact and tangent plane are independent of the magnitude of the constant radius assigned to all the spheres, so that the correspondence of surface elements in the natural sense is not one to one for such cases. When we speak of a surface element as in general position we shall mean that this case does not arise.*

Points of minimal line.	Pencil of lines of notable complex meeting notable line.
Pencil of properly parallel planes.	Pencil of lines meeting notable line, but not belonging to notable complex.
Pencil of parallel minimal planes.	Pencil of lines of notable complex meeting notable line.
Spheres containing minimal line.	Point and polar plane in null system of notable complex.
Group $g_{15}h_{15}$ of all contact transformations of oriented spheres.	Group $g_{15}h_{15}$ of all collineations and correlations.
Group g_{15} of all contact transformations factorable into an even number of inversions.	Group g_{15} of all collineations.
Inversion in linear complex.	Polarization in null system.
Linear complexes in involution.	Linear complexes in involution.
Oriented spheres properly tangent to oriented spheres.	Linear congruence with distinct directrix lines.
Dupin series.	Regulus.

* The Author's attention was first called to this exceptional case by a conversation with Professor Study.

Not null circle.	Regulus in notable complex.
Involutory transformation where corresponding members are properly tangent to same two members of two conjugate Dupin series.	Polarization in quadric of non-vanishing discriminant.
Six linear complexes in involution.	Six linear complexes in involution.
Group of thirty-two involutory transformations interchanging these.	Group of sixteen collineations and correlations interchanging these.

It is clear that the properties of six linear line complexes in involution will lead to a number of simple theorems about six linear oriented sphere complexes in involution. It would be tedious to carry through the results, in the case of the line complexes, as they are familiar enough ;* we have but to translate into sphere geometry, as follows:

Theorem 33.] *Six linear complexes in involution determine, by fours, fifteen pairs of oriented spheres, and, by threes, twenty Dupin series, forming ten pairs of conjugate series. Each pair of spheres determined by four complexes belongs to four Dupin series; each such series contains three of the fifteen pairs of spheres. If a pair of spheres do not belong to a Dupin series or its conjugate, they correspond in the involutory transformation determined by the series. Each of the thirty spheres is properly tangent to six others constituting three pairs.*

We shall define as a *series* of spheres a system whose coordinates are proportional to analytic functions of a single variable, the ratios being not all constants. A *congruence*, likewise, shall be a system whose coordinates are proportional to analytic functions of two independent variables, the ratios not being all functions of one variable. The envelope of a series of spheres is an annular surface, in the general case. If, however, adjacent spheres tend towards contact, i.e. the

* Cf. Koenigs, *La Géométrie réglée*, Paris, 1895, pp. 99–125.

difference between the radii of any two spheres of the series is equal to the negative of the corresponding arc of the deferent, the spheres trace two isotropic ruled surfaces. Conversely, if an isotropic ruled non-developable surface be given, adjacent generators determine a sphere. A one-parameter family of spheres is thus determined, part of whose envelope is the given surface. The remainder will be another surface of the same sort, which we shall speak of as *coupled* with the first.*

Coupled isotropic ruled surfaces.	Developable and its polar in notable complex.

Let us now show that not only do surface elements correspond to surface elements, but surfaces to surfaces; in other words, we have a contact transformation. Consider a continuously oriented non-developable surface in sphere space, i. e. a non-developable surface where the orientation of the normal is analytically determined—the envelope of ∞^2 oriented planes. The surface elements can be assembled in two ways into a one-parameter family of curvature strips, each element belonging to two strips, and the point will have two usually distinct lines of advance in the plane, so that the plane is tangent to the locus of the point. Corresponding to these we shall have ∞^2 surface elements which can be assembled in two ways into ∞^1 developable strips, each element belonging to two strips, and once more the point has two lines of advance in the plane. Hence again the envelope of the planes is the locus of the point.

Minimal developable.	Developable in notable complex.
Oriented surface not minimal developable.	Non-developable surface.
Congruence.	Congruence.
Envelope of oriented spheres of congruence.	Focal surface of congruence.

* The Author has been told that this idea of coupling ruled minimal surfaces dates back to Monge; he has not, however, been able to verify the statement.

This correspondence of surface element to surface element is subject to the 'in general' restriction mentioned on p. 436.

Theorem 34.] *In a general congruence of oriented spheres, the strip of contact with the envelope of those whose surface elements of contact fall together is a strip of curvature.*

In a general line congruence the strip of contact with the focal surface of those lines whose focal points and planes fall together is an asymptotic strip.

Theorem 35.] *The spheres of curvature of one system of a surface have no other envelope.*

The asymptotic lines of one system of a surface are tangent to no other surface.

The two theorems on the right are familiar enough in differential line geometry, arising from the fact that the developable surfaces of a line congruence determine two conjugate systems of curves on the two nappes of the focal surface; the theorems on the left come from those on the right by our transformation.

The focal surface of a line congruence of the second order and class is the Kummer quartic surface with sixteen conical points and sixteen planes of conical contact.* Since every such congruence is contained in a linear complex, let us assume that we have a congruence in our notable linear complex. It will be the total intersection of this complex and a quadratic one, and correspond under our transformation to the points of a cyclide. Now, by VII. 42], the lines of curvature of a cyclide are its complete intersection with the confocal cyclides, and are space curves of the eighth order.

Theorem 36.] *The Kummer quartic surface has algebraic asymptotic lines of the sixteenth order, being curves of contact with doubly enveloping ruled surfaces of the eighth order.*

* Cf. Jessop, *Treatise on the Line Complex*, Cambridge, 1903, pp. 101 and 296.

§ 6. Complexes of Oriented Spheres.

The general complex of oriented spheres shall be indicated by an equation

$$f(X_1 \dots X_0) = 0, \tag{31}$$

or else in the parametric form

$$X_i \equiv X_i(u, v, w). \tag{32}$$

If a sphere adjacent to (X) in the complex be properly tangent thereto,

$$(XX) = (XdX) = (dXdX) = 0.$$

The oriented spheres (X) and (dX) belong to the linear complexes

$$(XX') = \left(\frac{\partial f}{\partial X} X'\right) = 0.$$

To find the special linear complexes linearly dependent on these, we must solve the quadratic equation

$$\lambda^2 (XX) + 2\lambda\mu \left(X \frac{\partial f}{\partial X}\right) + \mu^2 \left(\frac{\partial f}{\partial X} \frac{\partial f}{\partial X}\right) = 0. \tag{33}$$

Assuming first $\left(\frac{\partial f}{\partial X} \frac{\partial f}{\partial X}\right) \neq 0,$

the only solutions of the quadratic are $\lambda = 0$; hence

Theorem 37.] *The oriented spheres of a complex infinitely near and properly tangent to an arbitrary sphere thereof, touch it in the points of a circle.**

We next assume that

$$\left(\frac{\partial f}{\partial X} \frac{\partial f}{\partial X}\right) \equiv 0.$$

We may repeat exactly our reasoning in the last chapter and show that the spheres of the complex are properly tangent to an oriented surface or curve. The converse is better proved as follows. Consider the corresponding question for a line complex. Equation (33) in line space means that the lines of

* Lie, 'Ueber Complexe', cit., p. 207.

a complex infinitely near one line thereof and intersecting it will usually belong to a linear congruence with one directrix line, i.e. those which meet it at any point lie in a plane connected with that point by a process called a normal correlation. When, however, the identity is satisfied, the lines of the complex infinitely near (X) pass through a fixed point, or lie in a fixed plane.* It is this, and not the other, which must happen for a complex of tangents; hence the identity must be satisfied, and it will likewise be satisfied for a complex of oriented spheres properly tangent to a surface.

Suppose, thirdly, that the identity is not satisfied, but (X) is such a sphere that

$$\left(\frac{\partial f}{\partial X}\frac{\partial f}{\partial X}\right) = 0.$$

It is then said to be a *singular* sphere of the complex. The oriented spheres (X) and $\left(\frac{\partial f}{\partial X}\right)$ are properly tangent. The envelope of their surface element is called the *singular surface.* The point of the name is seen as follows. If (Y) be an oriented sphere through an isotropic common to (X) and $\left(\frac{\partial f}{\partial X}\right)$, so that

$$(XY) = \left(\frac{\partial f}{\partial X}Y\right) = 0,$$

then in finding the oriented spheres of the tangent pencil $\lambda(Y)+\mu(X)$ which belong to the complex, we see that two members fall together in (X), the series of spheres through this isotropic, and in the complex will have (X) as a double member. Conversely, suppose that (X) belongs to the complex, and there is such an isotropic thereon that (X) counts as a double member of the series through this isotropic and in the complex. If (Y) and (Z) be any two oriented spheres through this isotropic, they touch (X) and one another.

$$(XY) = (XZ) = (YZ) = \left(\frac{\partial f}{\partial X}Y\right) = \left(\frac{\partial f}{\partial X}Z\right) = \left(\frac{\partial f}{\partial X}X\right) = 0.$$

* Koenigs, loc. cit., p. 36.

The linear complex $\left(\frac{\partial f}{\partial X} X'\right) = 0$ having three mutually tangent members, not linearly dependent, must be special, and our equation is satisfied.*

Let us next consider a congruence

$$f = \phi = 0.$$

The spheres of the congruence infinitely near an arbitrary sphere and tangent to it are given by the equations

$$(XX') = \left(\frac{\partial f}{\partial X} X'\right) = \left(\frac{\partial \phi}{\partial X} X'\right) = 0.$$

There are but two special linear complexes linearly dependent on those given by these equations; they fall together if

$$\left(\frac{\partial f}{\partial X} \frac{\partial f}{\partial X}\right)\left(\frac{\partial \phi}{\partial X} \frac{\partial \phi}{\partial X}\right) - \left(\frac{\partial f}{\partial X} \frac{\partial \phi}{\partial X}\right)^2 = 0. \tag{34}$$

This equation will be identically satisfied when, and only when, the spheres of the congruence are properly tangent to but one surface or curve. Next take a series

$$f = \phi = \psi = 0.$$

The spheres of the series infinitely near (X) belong to the linear complexes

$$(XX') = \left(\frac{\partial f}{\partial X} X'\right) = \left(\frac{\partial \phi}{\partial X} X'\right) = \left(\frac{\partial \psi}{\partial X} X'\right) = 0.$$

These equations have usually two distinct solutions. They fall together if

$$\begin{vmatrix} \left(\frac{\partial f}{\partial X} \frac{\partial f}{\partial X}\right) & \left(\frac{\partial f}{\partial X} \frac{\partial \phi}{\partial X}\right) & \left(\frac{\partial f}{\partial X} \frac{\partial \psi}{\partial X}\right) \\ \left(\frac{\partial \phi}{\partial X} \frac{\partial f}{\partial X}\right) & \left(\frac{\partial \phi}{\partial X} \frac{\partial \phi}{\partial X}\right) & \left(\frac{\partial \phi}{\partial X} \frac{\partial \psi}{\partial X}\right) \\ \left(\frac{\partial \psi}{\partial X} \frac{\partial f}{\partial X}\right) & \left(\frac{\partial \psi}{\partial X} \frac{\partial \phi}{\partial X}\right) & \left(\frac{\partial \psi}{\partial X} \frac{\partial \psi}{\partial X}\right) \end{vmatrix} = 0, \tag{35}$$

* The matter of singular elements is more luminous when regarded from the point of view of line space. Transforming our reasoning above about isotropics, we see that a singular line of a complex is a double line of the cone of the complex whose vertex is any point of the line.

and this equation is identically verified when, and only when, the envelope is not an annular surface, but two coupled isotropic ruled surfaces.

Let us apply these general methods to the particular case of the quadratic complex. This will be defined by the equations

$$\sum_{i,j=0}^{i,j=5} a_{ij}X_iX_j = (XX) = 0. \tag{36}$$

We shall limit ourselves to the study of the *general* quadratic complex, that giving distinct roots to the equation *

$$\begin{vmatrix} a_{00}-\rho & a_{01} & a_{02} & a_{03} & a_{04} & a_{05} \\ a_{10} & a_{11}-\rho & a_{12} & a_{13} & a_{14} & a_{15} \\ a_{20} & a_{21} & a_{22}-\rho & a_{23} & a_{24} & a_{25} \\ a_{30} & a_{31} & a_{32} & a_{33}-\rho & a_{34} & a_{35} \\ a_{40} & a_{41} & a_{42} & a_{43} & a_{44}-\rho & a_{45} \\ a_{50} & a_{51} & a_{52} & a_{53} & a_{54} & a_{55}-\rho \end{vmatrix} = 0.$$

We may find a contact transformation of oriented spheres to reduce the equation of such a general complex to the canonical forms

$$(aX^2) = (XX) = 0. \tag{37}$$

Theorem 38.] *The general quadratic complex of oriented spheres is anallagmatic in six linear complexes in involution.*

Theorem 39.] *The null spheres of a quadratic complex generate a cyclide, its planar spheres a hypercyclide.*

The singular spheres of the general quadratic complex have the equations

$$(a^2X^2) = (aX^2) = (XX) = 0. \tag{38}$$

The condition that an oriented sphere shall be either null or planar is linear in our present coordinate system. If Z be

* For an elaborate discussion of the various types of quadratic complex, see Moore, 'Classification of the Surfaces of Singularities of the Quadratic Spherical Complex', *American Journal of Math.*, vol. xxii, 1905.

either the point or plane of contact of a singular sphere with the singular surface, we have

$$Z_i = (a_i + r) X_i.$$

$$(ZZ) = \sum_{i=0}^{i=5} \frac{Z_i^2}{(a_i + r)} = \sum_{i=0}^{i=5} \frac{Z_i^2}{(a_i + r)^2} = 0.$$

Of these last two equations, the first is quartic in r, the second its derivative. The result of eliminating r will be to equate to zero the discriminant of the quartic, an expression of the sixth degree in Z^2. Considered with the first equation in Z and three linear equations, we get twenty-four solutions.

Theorem 40.] *The singular surface of the general quadratic complex of oriented spheres is of the twenty-fourth order and class with the circle at infinity as a curve of the twelfth order. There are six spheres which meet it in a double cyclic and a line of curvature of the sixteenth order. It is the envelope of six quadratic congruences of oriented spheres, each contained in one linear complex with regard to which the given quadratic complex is anallagmatic.**

We may also reach this surface from the focal surface of the general quadratic line complex, the Kummer surface with sixteen conical points and sixteen planes of conical contact.†

Sphere space.	Line space.
General quadratic complex.	General quadratic complex.
Singular surface.	Kummer surface.

We next turn to the lines of curvature of the singular surface. They are connected with the singular spheres of the *second order* determined by the equations

$$(XX) = (aX^2) = (a^2X^2) = (a^3X^2) = 0. \qquad (39)$$

* Smith, *Surfaces Enveloped*, &c., p. 387, and Blaschke, *Geometrie der Speere*, cit., p. 59, incorrectly by Snyder, *Some Differential Expressions*, cit., p. 150.

† Jessop, *Line Complex*, cit., pp. 97 ff.

The easiest proof comes from considering the corresponding question in line space. Here every line of the pencil $\lambda(X_i)+\mu a_i X_i$ belongs to the complex, and this pencil consists of tangents to the Kummer surface which is the singular surface of quadratic line complex. The line connecting the centres of two infinitely near pencils of this sort is tangent to the singular surface, and belongs to both pencils. Hence, as the centre of such a pencil proceeds along the surface, its plane rolls about the tangent to the curve, and we are following an asymptotic direction which corresponds to a direction of curvature of the singular surface of the sphere complex. Thus one line of curvature is given by the singular spheres of the second order. More generally, consider the quadratic complex

$$(X'X') = \sum_{i=0}^{i=5} \frac{X_i'^2}{a_i+\lambda} = 0. \tag{40}$$

Let us write further

$$(a_i+\lambda)\,X_i' = X_i.$$

Then, if (X) be a singular sphere of the complex (37), (X') is a singular sphere of the complex (40), and, since (X') belongs to the tangent pencil determined by (X) and $a_i X_i$, the two quadratic complexes have the same singular surfaces. We get other lines of curvature of our surface from the series

$$(X'X') = \sum_{i=0}^{i=5} \frac{X_i'^2}{a_i+\lambda} = \sum_{i=0}^{i=5} \frac{X_i'^2}{(a_i+\lambda)^2} = \sum_{i=0}^{i=5} \frac{X_i'^2}{(a_j+\lambda)^3} = 0.$$

Since the lines of curvature are algebraic, and the two sets are not rationally separable, we get all of our lines of curvature in this way. To find the order of one such line or the class of the enveloping developable, let (Z') be the point of contact, or the properly tangent plane

$$Z_i = (a_i+\rho)\,X_i,$$

$$(ZZ) = (aZ^2) = \sum_{i=0}^{i=5} \frac{Z_i^2}{(a_i+\rho)} = \sum_{i=0}^{i=5} \frac{Z_i^2}{(a_i+\rho)^2} = 0.$$

We may eliminate ρ exactly as in previous cases, and find

Theorem 41.] *The lines of curvature of the singular surface of the general quadratic complex of oriented spheres are of the thirty-second order; the tangent developable along such a line is of the thirty-second class.**

We saw in 40] that the singular surface can be generated in six ways by an oriented sphere belonging to a linear complex. Such a sphere will usually meet a fixed sphere at a fixed angle. These generating spheres are the minimal projections of one focal surface of the hypersurface in four-dimensional space which corresponds to the complex; which surface is a pentaspherical cyclide. The locus of the centres of the generating spheres will be the orthogonal projection of this cyclide. A hyperplane will meet this cyclide in a spherical cyclic whose orthogonal projection is a binodal quartic. The projection of the cyclide will be a surface of the fourth order with two double points in an arbitrary plane:

Theorem 42.] *The singular surface of a general quadratic complex of oriented spheres may be generated in six ways by an oriented sphere which meets a fixed sphere at a fixed angle, while its centre traces a surface of the fourth order with a double conic.*

A slight modification must be made to this theorem when the linear complex does not consist in spheres meeting a fixed sphere at a fixed angle.

There are few parts of our whole subject where more remains to be done than in connexion with the oriented sphere. It is impossible not to believe that a sufficiently ingenious use of our plane and sphere coordinates will settle the interesting question of whether there be any systems of spheres which correspond to the Hart systems of circles of the first sort. Again, the group of euclidean motions in four-dimensional space and the allied group which leaves a real

* Deduced differently by Smith, *Surfaces Enveloped*, cit., p. 387.

quadric invariant are deserving of the same sort of careful study that has been bestowed on three-dimensional motions, for their own sakes, for the light thereby thrown on the oriented sphere, and for their relation to the theory of relativity. The general equilong three-dimensional transformation has never received any more attention than we have here given to it; surely there must be much of interest to be found in this connexion. Lastly, there must still remain a number of interesting undiscovered properties of the simplest linear and quadratic systems of oriented spheres.

CHAPTER XII

CIRCLES ORTHOGONAL TO ONE SPHERE

§ 1. Relations of Two Circles.

We have at various times caught glimpses of curious relations which can exist between circles which are not on one sphere. In particular, in Ch. VIII. 12] we met two circles in an interesting relation which we called bi-involution, while in Ch. IX we met two circles so situated that they were each cut twice perpendicularly by an infinite number of circles. The time has now come to make a detailed study of circles which are not cospherical, and the rest of the present work will be chiefly devoted to this purpose. The space in question is pentaspherical space. We begin with a couple of elementary theorems.

Theorem 1.] *Any two circles will have one common orthogonal sphere, and only one unless they be cospherical, in which case they are orthogonal to a coaxal system.*

Theorem 2.] *If a circle be cospherical with two others which are not cospherical with one another, it is orthogonal to their common orthogonal sphere.*

We leave the proofs of these simple theorems to the reader.

A most fundamental element in the study of the circle in pentaspherical space is its *focus*. Suppose that we have a circle determined by two spheres (x') and (y'). Let us see whether there be any null spheres through this circle. Such a sphere will be linearly dependent on (x') and (y'), and have coordinates (x), where

$$x_i \equiv \lambda x_i' + \mu y_i'.$$

Substituting in the fundamental identity for pentaspherical coordinates,

$$\lambda^2 (x'x') + 2\lambda\mu (x'y') + \mu^2 (y'y') = 0.$$

If, now, these spheres (x') and (y') do not touch one another, the discriminant of this quadratic equation is not zero, there are two null spheres in the coaxal system and their vertices shall be called the *foci* of the circle.

Theorem 3.] *A necessary and sufficient condition that two not null circles should be cospherical is that their foci should be concyclic.*

Theorem 4.] *A necessary and sufficient condition that two not null circles should touch is that their foci should lie on two intersecting isotropics, whose intersection is not a common focus of the two circles.*

If a sphere be orthogonal to a circle, and so to the spheres through it, it passes through the foci of that circle when they are distinct, and vice versa. Two circles shall be said to be in *involution* when each is orthogonal to a sphere through the other. If a circle c' be orthogonal to a sphere S through a circle c, its foci, if distinct, lie on s. Every sphere through c' will be orthogonal to s, since the two null spheres through c' are orthogonal to it, and the foci of c are clearly mutually inverse in every sphere through c. Hence a sphere through c' and one focus of c goes through the other focus.

Theorem 5.] *If one not null circle be cospherical with the foci of a second, then the second is cospherical with the foci of the first, and the two are in involution.*

We have defined two circles as being in bi-involution when every sphere through one is orthogonal to the other. Let the reader prove

Theorem 6.] *If the foci of one not null circle lie on a second such circle, then the foci of the second lie on the first, and the two are in bi-involution.*

Theorem 7.] *A necessary and sufficient condition that two not null circles should be in bi-involution is that each should contain the foci of the other.*

The common orthogonal sphere of two non-cospherical not null circles is that through their foci.

Theorem 8.] *A necessary and sufficient condition that two non-cospherical not null circles should be in involution is that each should be cospherical with the circle orthogonal to their common orthogonal sphere which is in bi-involution with the other.*

Let the reader show that the word *each* may be replaced by the word *one.*

Let us next seek a common perpendicular to two non-cospherical circles. The first shall have the foci $(x)\,(y)$, the second the foci $(x')\,(y')$. If such a common perpendicular, and we mean thereby a circle cospherical and orthogonal to both, be determined by the spheres $\lambda\,(x)+\mu\,(y)$, $\lambda'\,(x')+\mu'\,(y')$, there must be some sphere through each circle that is orthogonal to each of these spheres. This will require that $\lambda\,(x)-\mu\,(y)$ be orthogonal to $\lambda'(x')+\mu'(y')$, and $\lambda'(x')-\mu'(y')$ orthogonal to $\lambda\,(x)+\mu(y)$. We thus get the equations

$$\lambda\lambda'(xx')-\mu\mu'(yy') = 0,$$

$$\mu\lambda'(yx')-\lambda\mu'(xy') = 0.$$

Eliminating λ'/μ',

$$\lambda^2(xx')\,(xy')-\mu^2(yx')\,(yy') = 0.$$

The roots of this equation differ only in sign, and give two mutually orthogonal spheres, and the same would be true of the corresponding equation in λ'/μ'.

Theorem 9.] *Two non-cospherical not null circles are usually cospherical and orthogonal to two circles in bi-involution and no others.*

We must now find the exact meaning to attach to the word 'usually'. The discriminant of this quadratic equation is

$$-4\,(xx')\,(xy')\,(yx')\,(yy').$$

Theorem 10.] *A necessary and sufficient condition that two not null and non-cospherical circles should be cospherical and orthogonal to two and only two circles is that no focus of one should lie on an isotropic with a focus of the other.*

Suppose, next, that a single pair of foci lie on an isotropic, say

$$(xx') = 0, \quad (xy')\,(yx')\,(yy') \neq 0.$$

If they had a common focus they would be cospherical, which we exclude. We therefore cannot have $(x) \equiv (x')$; we must have $\mu = \mu' = 0$, and the isotropic connecting (x) and (x') is the only circle to fit the conditions.

Next suppose

$$(xx') = (yx') = 0.$$

Here λ/μ is entirely indeterminate, but the system of circles found are all null. Thirdly, let

$$(xx') = (yy') = 0.$$

Here, since the circles are not cospherical, their foci lie in pairs of two skew isotropics. There are ∞^1 sets of values for λ/μ; the circles, though not cospherical, are cut twice orthogonally by ∞^1 circles, i. e. they are paratactic.

Theorem 11.] *A necessary and sufficient condition that two not null and non-cospherical circles should be cospherical and orthogonal to a single circle is that just one focus of one should lie on an isotropic with one focus of the other.*

Theorem 12.] *A necessary and sufficient condition that two not null and non-cospherical circles should be paratactic is that their foci should lie in pairs on two skew isotropics.*

Lastly, suppose

$$(xx') = (xy') = (yx') = 0.$$

If $(yy') \neq 0$ the common orthogonal circles are all null; if $(yy') = 0$ the given circles are in bi-involution.

Theorem 13.] *A necessary and sufficient condition that two not null circles should be cospherical and orthogonal to a two-parameter family of circles is that they should be in bi-involution.*

We shall define as the *non-Euclidean angles* of two circles those of two pairs of spheres, the spheres of each pair being determined by the given circles and a circle cospherical and orthogonal to both. If we represent each sphere of our pentaspherical space by a point in four-dimensional projective space of elliptic measurement, the angles of two circles will correspond to the distance of the corresponding lines, or the angles of these lines.* The pairs of spheres are

$$\sqrt{(yx')}\,\sqrt{(yy')}\,(x) \pm \sqrt{(xx')}\,\sqrt{(xy')}\,(y)\,;$$

$$\sqrt{(y'x)}\,\sqrt{(y'y)}\,(x') \pm \sqrt{(x'x)}\,\sqrt{(x'y)}\,(y').$$

If θ_1 and θ_2 be the angles of these circles, we find

$$\cos^2\theta_1 + \cos^2\theta_2 = \frac{2\left[(xx')(yy') + (yx')(xy')\right]}{(xy)(x'y')},$$

$$\cos^2\theta_1 \cos^2\theta_2 = \frac{\left[(xx')(yy') - (yx')(xy')\right]^2}{(xy)^2 (x'y')^2}. \tag{1}$$

The condition that these angles should be equal or supplementary is

$$(xx')(xy')(yx')(yy') = 0.$$

This will involve either $(xx') = (yy') = 0$,

or else $$(xy') = (yx') = 0,$$

as otherwise the spheres making these angles are null and the angles meaningless.

Theorem 14.] *A necessary and sufficient condition that two non-cospherical not null circles should be paratactic is that their non-Euclidean angles should be equal or supplementary.*

Suppose that we have two paratactic circles whose common orthogonal sphere is not null. Take any circle cospherical

* Cf. the Author's *Non-Euclidean Geometry*, cit., pp. 111 and 113.

with both and the ∞^1 circles into which it is transformed by the one-parameter group of spherical transformations which leave the given circles invariant (cf. IX, p. 344). The foci of ∞^1 circles must all lie on the same pair of generators of one set of the fixed sphere; any two of the ∞^1 circles are thus paratactic. There will thus be a second one-parameter group of spherical transformations leaving all of the ∞^1 circles in place, and carrying the two original ones into ∞^1 others generating the same surface as the first group of ∞^1 circles; hence

Theorem 15.] *Two paratactic circles are generators of ∞^1 cyclides, each having two conjugate generations composed of paratactic circles.*

Strictly speaking, we have only proved this in the case where the common orthogonal sphere of the first two circles is not null. When it is null we reach the same theorem by continuity, or, in cartesian space, by inverting into a non-Euclidean hyperboloid with paratactic generators.

§ 2. Circles Orthogonal to one Sphere.

The theorems so far developed in the present chapter were of a general character for circles in pentaspherical space; from this point on we shall limit ourselves to the discussion of circles orthogonal to one fixed not null sphere, which for definiteness we shall take as $x_4 = 0$. Every sphere orthogonal to the fundamental sphere will lack the last coordinate. If two such spheres (x) and (y) be given, we may determine their common circle by the following six homogeneous coordinates called the *Plücker* coordinates of the circle.

$$\begin{aligned}\rho p_{01} &\equiv x_0 y_1 - x_1 y_0, \quad \rho p_{02} \equiv x_0 y_2 - x_2 y_0, \quad \rho p_{03} \equiv x_0 y_3 - x_3 y_0, \\ \rho p_{23} &\equiv x_2 y_3 - x_3 y_2, \quad \rho p_{31} \equiv x_3 y_1 - x_1 y_3, \quad \rho p_{12} \equiv x_1 y_2 - x_2 y_1.\end{aligned} \tag{2}$$

These coordinates being homogeneous are essentially unaltered when the original spheres are replaced by any two others coaxal with them. They depend, therefore, on the circle, and not on the individual spheres. They are, moreover, connected by the fundamental quadratic identity

$$|\, xyxy \,| \equiv 2\,(p_{01}\,p_{23}+p_{02}\,p_{31}+p_{03}\,p_{12}) = 0.$$

Introducing (merely for the purposes of the present chapter) the symbolism

$$(p/q) \equiv (p_{01}q_{23}+p_{02}q_{31}+p_{03}q_{12}+p_{23}q_{01}+p_{31}q_{02}+p_{12}q_{03}), \quad (3)$$

our identity becomes

$$(p/p) = 0. \quad (4)$$

Suppose, conversely, that we have a system of homogeneous values, not all zero, which satisfy (4); to be specific, suppose $p_{23} \neq 0$, we then write

$$\begin{aligned} p_{23}x_1+p_{31}x_2+p_{12}x_3 &= 0,\\ -p_{23}x_0 \qquad\qquad +p_{03}x_2-p_{02}x_3 &= 0,\\ -p_{31}x_0-p_{03}x_1 \qquad\qquad +p_{01}x_3 &= 0,\\ -p_{12}x_0+p_{02}x_1-p_{01}x_2 \qquad\qquad &= 0. \end{aligned}$$

We see by a simple elimination that every solution of the first two equations is also a solution of the last two; the four will represent a coaxal system of spheres. Take the spheres of the system orthogonal to two arbitrary spheres (u) and (v)

$$\rho y_i = \frac{\partial}{\partial t_i}\begin{vmatrix} t_0 & t_1 & t_2 & t_3 \\ 0 & p_{23} & p_{31} & p_{12} \\ -p_{23} & 0 & p_{03} & -p_{02} \\ u_0 & u_1 & u_2 & u_3 \end{vmatrix}, \quad \sigma z_i = \frac{\partial}{\partial t_i}\begin{vmatrix} t_0 & t_1 & t_2 & t_3 \\ 0 & p_{23} & p_{31} & p_{12} \\ -p_{23} & 0 & p_{03} & -p_{02} \\ v_0 & v_1 & v_2 & v_3 \end{vmatrix}.$$

The Plücker coordinates of the circle of intersection will be

$$\rho\sigma\,(y_i z_j - y_j z_i) = -p_{23}\begin{vmatrix} 0 & p_{33} & p_{31} & p_{12} \\ -p_{23} & 0 & p_{03} & -p_{02} \\ u_0 & u_1 & u_2 & u_3 \\ v_0 & v_1 & v_2 & v_3 \end{vmatrix} p_{ij}.$$

Here neither coefficient of p_{ij} will vanish, so that there is indeed a circle with these coordinates. Lastly, if we know the coordinates of a circle, we may easily find, rationally, in terms of them the coordinates of the sphere through the circle orthogonal to an arbitrary sphere, so that two different circles could not have the same Plücker coordinates.

The condition that two circles of our system should be cospherical is

$$(p/q) = 0. \tag{5}$$

If (p) and (q) be two circles of our system, then

$$\lambda (p) + \mu (q)$$

will represent a circle when, and only when, the two are cospherical, and, by varying λ/μ in this case we get all circles coaxal with (p) and (q). Similarly, if we have three circles cospherical two by two, then either all pass through two points, or are on one sphere orthogonal to the fundamental sphere. The circles

$$\lambda (p) + \mu (q) + \nu (r)$$

will in the first case be those through the two points, in the second those of our system lying on the particular sphere.

If (x) (y) and (x') (y') be respectively the foci of two circles, the condition that the two should be in involution is

$$\begin{vmatrix} (xx')\,(xy') \\ (yx)\,(yy') \end{vmatrix} = \sum_{i,j=0}^{i,j=3} p_{ij}p_{ij}' = 0.* \tag{6}$$

For bi-involution we shall have

$$(xx) = (yy) = (xx') = (xy') = (yx') = (yy') = (x'x') = (y'y') = 0.$$

$$\rho x_i = \begin{vmatrix} x_j & x_k & x_l \\ x_j' & x_k' & x_l' \\ y_j' & y_k' & y_l' \end{vmatrix}, \qquad \sigma y_i = \begin{vmatrix} y_j & y_k & y_l \\ x_j' & x_k' & x_l' \\ y_j' & y_k' & y_l' \end{vmatrix}.$$

$$\rho\sigma p_{ij} = |\, x\, y\, x'\, y' \,|\, p_{kl}'. \tag{7}$$

* In every expression of the form

$$\sum_{i,j=0}^{i,j=3} q_{ij}\, p_{ij},$$

the summation is meant to include the six terms given by (2).

We therefore get the circle of our system in bi-involution with a given circle by replacing each Plücker coordinate by its complementary. With regard to the non-Euclidean angles of two circles we easily find

$$\sum_{i,j=0}^{i,j=3} p_{ij}p_{ij}' = [(xx')(yy') - (yx')(xy')];$$

$$\sum_{i,j=0}^{i,j=3} p_{ij}^2 \sum_{i,j=0}^{i,j=3} p_{ij}'^2 = (xy)^2(x'y')^2. \tag{8}$$

$$(p/p')^2 = \begin{vmatrix} 0 & (xy) & (xx') & (xy') \\ (yx) & 0 & (yx') & (yy') \\ (x'x) & (x'y) & 0 & (x'y') \\ (y'x) & (y'y) & (y'x') & 0 \end{vmatrix}. \tag{9}$$

The cosines of the angles of the circles will be found from the equation

$$\sum_{i,j=0}^{i,j=3} p_{ij}^2 \sum_{i,j=0}^{i,j=3} p_{ij}^2 \cos^4\theta$$
$$+\left[(p/p')^2 - \left(\sum_{i,j=0}^{i,j=3} p_{ij}p_{ij}'\right)^2 - \sum_{i,j=0}^{i,j=3} p_{ij}^2 \sum_{i,j=0}^{i,j=3} p_{ij}'^2\right]\cos^2\theta$$
$$+\left(\sum_{i,j=0}^{i,j=3} p_{ij}p_{ij}'\right)^2 = 0. \tag{10}$$

The condition that the circles should be paratactic is

$$\left\{\left[(p/p') + \sum_{i,j=0}^{i,j=3} p_{ij}p_{ij}'\right]^2 - \sum_{i,j=0}^{i,j=3} p_{ij}^2 \sum_{i,j=0}^{i,j=3} p_{ij}'^2\right\}$$
$$\left\{\left[(p/p') - \sum_{i,j=0}^{i,j=3} p_{ij}p_{ij}'\right]^2 - \sum_{i,j=0}^{i,j=3} p_{ij}^2 \sum_{i,j=0}^{i,j=3} p_{ij}'^2\right\} = 0. \tag{11}$$

It will be found, in fact, that the first factor vanishes if

$(yx') = (xy') = 0$, while the second vanishes if $(xx') = (yy') = 0$. Conversely, when this equation is satisfied, we find

$$(xx')\,(xy')\,(yx')\,(yy') = 0.$$

The most important invariant for two circles under the spherical sub-group which leaves the fundamental sphere invariant is the product of the cosines of their non-Euclidean angles. This is

$$I(pp') \equiv \cos\theta_1 \cos\theta_2 = \frac{\sum\limits_{i,j=0}^{i,j=3} p_{ij}p_{ij}'}{\sqrt{\sum\limits_{i,j=0}^{i,j=3} p_{ij}^2}\sqrt{\sum\limits_{i,j=0}^{i,j=3} p_{ij}'^2}}. \quad (12)$$

When we speak in general of the *invariant* of two not null circles we shall mean this product.

A great flood of light is thrown upon the system of circles orthogonal to one sphere when we compare the formulae here developed with those of line geometry in the usual Plücker coordinates. We have the following correspondence:

Spheres orthogonal to fixed sphere.	Points of projective space.
Circles orthogonal to fixed sphere.	Lines.
Cospherical circles.	Intersecting lines.
Plücker coordinates.	Plücker coordinates.
Coaxal system.	Pencil of lines.
Point-pair inverse in fixed sphere.	Plane.
Angle of cospherical circles.	Angle of intersecting lines in elliptic measurement.
Non-Euclidean angles of circles.	Angle of skew lines in elliptic measurement.

This correspondence, which we have reached by purely algebraic means,* may also be derived directly by geometrical

* Cf. Forbes, *Geometry of Circles Orthogonal to a given Sphere*, Dissertation, New York, 1904.

considerations.* Let us imagine that our circles lie in cartesian space of Euclidean measurement, and that the fundamental sphere is not planar. Each circle orthogonal to this sphere may be set into correspondence to its axis, i.e. the line through the centre orthogonal to the plane of the sphere. Every proper circle and every null circle whose centre is finite and on the sphere will correspond to a determinate line. A diametral line will correspond to the line in the plane at infinity, and in the planes perpendicular to the given diametral line. Conversely, if any line be given, there is one plane through the centre of the sphere orthogonal to it. If the line be not an isotropic through the centre, the plane meets it in a definite point, the centre of a circle in the plane orthogonal to the sphere. An isotropic through the centre will correspond to a parabolic circle, i.e. a parabola touching the circle at infinity, with the line as axis. Here the correspondence is not one to one. Intersecting lines will correspond to cospherical circles.

We pass to the consideration of the simplest systems of circles orthogonal to one sphere. We begin with the *linear complex*, defined as the totality of circles whose coordinates satisfy an equation of the type

$$\sum_{i,j=0}^{i,j=3} a_{ij}\, p_{ij} = 0. \qquad (13)$$

If
$$(a/a) = 0,$$

the complex consists in the totality of circles cospherical with the fixed circle $q_{ij} = \rho a_{kl}$. When this equation does not hold, we have a more complicated system, which we shall call the non-special case, the sphere being not null.

Theorem 16.] *A non-special linear complex of circles orthogonal to a fixed sphere will share a coaxal system with every sphere orthogonal to this sphere, and with every point-pair anallagmatic therein. It will set up such a one to one correspondence between the spheres orthogonal to the fundamental*

* Cf. Moore, 'Circles Orthogonal to a given Sphere', *Annals of Math.*, Series 2, vol. viii, 1907.

*sphere and the point-pairs, that each point-pair lies on the corresponding sphere and each sphere passes through the corresponding point-pair; the spheres through a circle orthogonal to the fixed sphere will correspond to point-pairs on such a circle, and vice versa. The two circles so defined bear a reciprocal relation, and will fall together in the case of the circles of the complex, and in no other case.**

This long theorem may be immediately deduced from the fundamental properties of the linear line complex, or else proved immediately by a simple analysis. A linear complex has an absolute invariant, under the sub-group of spherical transformations which leaves the fundamental sphere invariant, namely

$$H \equiv \frac{\sum\limits_{i,j=0}^{i,j=3} a_{ij}^2}{(a/a)}. \tag{14}$$

Let us see what will be the meaning of this. We note, to begin with, that circles orthogonal to the fundamental sphere, and in bi-involution with those of the given complex, will generate a second complex

$$(a/p) = 0. \tag{15}$$

The circles common to the two complexes will satisfy the equation

$$\lambda \sum_{i,j=0}^{i,j=3} a_{ij} p_{ij} + \mu (a/p) = 0,$$

regardless of the value of λ/μ. We shall thus get two circles, usually distinct, by requiring the complex whose equation was last written to be special, i.e.

$$(\lambda^2 + \mu^2)(a/a) + 2\lambda\mu \sum_{i,j=0}^{i,j=3} a_{ij}^2 = 0.$$

The resulting circles shall be

$$\rho\alpha_{ij} = (-H + \sqrt{H^2-1})\, a_{ij} + a_{kl}.$$

$$\sigma\alpha_{ij}' = a_{ij} + (-H + \sqrt{H^2-1})\, a_{kl}.$$

* Cf. Forbes, loc. cit., pp. 19 ff.

It is immediately evident that these two are in bi-involution. If p be any circle orthogonal to our fundamental sphere, the ratio of its invariants with regard to these two will be

$$\frac{I(\alpha p)}{I(\alpha' p)} = \frac{(-H+\sqrt{H^2-1})\sum\limits_{i,j=0}^{i,j=3} a_{ij}p_{ij} + (a/p)}{\sum\limits_{i,j=0}^{i,j=3} a_{ij}p_{ij} + (-H+\sqrt{H^2-1})(a/p)}.$$

For a circle of our complex this ratio reduces to

$$-H-\sqrt{H^2-1}.$$

Conversely, when the ratio so reduces, the circle (p) must belong to our complex.

If
$$H^2 \neq 1,$$
the two circles (α), (α') are distinct.

We shall find it convenient from now on to speak of two circles in bi-involution as forming a *cross*, and the cross formed by the circles (α) and (α') shall be the *axial cross* of the complex. Restricting the word *general* to those linear complexes where (α) and (α') are distinct, we have

Theorem 17.] *The circles orthogonal to a fixed sphere, whose invariants with the two circles of a cross orthogonal to this sphere bear a fixed ratio which is finite and different from* 0 *or* ± 1, *will generate a linear complex with the given cross as axial cross, and, conversely, every general linear complex may be generated in this way.*

When the fundamental sphere is null we must prove by continuity, or by inverting into a linear line complex in non-Euclidean sphere.

The existence of the axial cross leads us to the canonical form for the equation of the linear complex. In fact, if we take the fundamental circles of our pentaspherical coordinate system as passing two by two through the circles of the axial cross, the equation of the linear complex may be written

$$a_{01}p_{01} + a_{23}p_{23} = 0, \quad H = \frac{a_{01}{}^2 + a_{23}{}^2}{2\,a_{01}\,a_{23}}. \tag{16}$$

A thorough discussion of all special cases of the linear complex, under the quaternary orthogonal group of substitutions, would lead us too far afield. We merely note that if

$$H^2 = 1,$$

the two circles of the cross fall together in an isotropic. On the other hand, if the complex be in bi-involution with itself, i.e. if

$$a_{ij} = \rho a_{kl}, \quad \rho^2 = 1,$$

any two circles which correspond by means of the complex according to the description in (16) may be taken to give the axial cross.

The assemblage of all circles orthogonal to a fixed sphere, and common to two different linear complexes, shall be called a *linear congruence*. Let the reader show that usually this consists in the assemblage of circles cospherical with two distinct or adjacent circles orthogonal to the fundamental sphere. Let him also show that the totality of circles common to three linear complexes will usually be one generation of a cyclide.

Before leaving altogether the linear complex, and systems of linear complexes, let us look again for a moment at the transformation mentioned in 16]. If the complex be that given by (13), the transformation will be

$$q_{ij}' = (a/a)\, q_{ij} - 2\,(a/q)\, a_{kl}. \tag{17}$$

We see, in fact, that this will permute circles orthogonal to the fundamental sphere, leaving invariant only such as belong to the given complex. It will also carry cospherical circles (orthogonal to the sphere) into cospherical ones. The transformation is involutory, but is not a spherical transformation, and will carry a sphere orthogonal to the fundamental sphere into a point-pair anallagmatic with regard thereto. A circle of the complex cospherical with (q) is cospherical with (q') also. We shall call this transformation a *polarization* in the linear complex. The necessary and sufficient condition that the linear complex

$$\sum_{i,j=0}^{i,j=3} b_{ij} p_{ij} = 0$$

should be carried into itself by polarizing in our given complex is

$$(a/b) = 0.$$

Let the reader show that this is also a necessary and sufficient condition that the product of successive polarizations in the two linear complexes should be commutative.*

It is easy to find a system of six linear complexes, each of which bears this relation to the five others. They will determine ten cyclides each having the given sphere (when not null) as a fundamental one, and a group of sixteen spherical transformations which carry the whole figure over into itself.

If we define as a *complex* of circles orthogonal to our fixed sphere a system where the coordinates of each member are proportional to analytic functions of three independent variables, the ratios not being all functions of two variables, then a general complex may be written

$$f(p) = 0.$$

Remembering that our circles are in one to one correspondence with the lines in projective space, and they in turn with oriented spheres, we find at once, if

$$\left(\frac{\partial f}{\partial p} \Big/ \frac{\partial f}{\partial p}\right) \equiv 0,$$

the complex consists in circles bitangent to a surface anallagmatic in the fixed sphere, or meeting a curve in pairs of anallagmatic points. When this expression does not vanish identically, we get the *singular* circles of the complex by equating it to zero. Each singular circle determines a sphere orthogonal to the fundamental sphere, and an anallagmatic point-pair thereon; the locus of the point-pairs is the envelope of the spheres. The circles of the complex through the point-pair will generate a surface having the singular circle as

* The reader familiar with line geometry might naturally expect us to speak of two such linear complexes as being in *involution*. Such a locution might, however, lead to confusion, for if the complexes were special, $(a/a) = (b/b) = 0$, the circles (a) and (b) would be in involution if

$$\sum_{i,j=0}^{i,j=3} a_{ij} b_{ij} = 0, \text{ not if } (a/b) = 0.$$

a double circle, and the singular circle will be double among the circles of the system on the sphere.

A system of circles whose coordinates are proportional to analytic functions of two independent variables, their ratios not being all functions of one variable, shall be called a *congruence.* We determine a congruence frequently by two equations

$$f(p) = \phi(p) = 0.$$

An arbitrary circle of the congruence will, in general, be cospherical with two adjacent circles thereof, that is to say, the circles may be assembled in two ways into generators of a one-parameter family of annular surfaces. These will reduce to a single system if

$$\begin{vmatrix} \left(\frac{\partial f}{\partial p} \Big/ \frac{\partial f}{\partial p}\right) & \left(\frac{\partial f}{\partial p} \Big/ \frac{\partial \phi}{\partial p}\right) \\ \left(\frac{\partial \phi}{\partial p} \Big/ \frac{\partial f}{\partial p}\right) & \left(\frac{\partial \phi}{\partial p} \Big/ \frac{\partial \phi}{\partial p}\right) \end{vmatrix} \equiv 0.$$

A system of circles whose coordinates are proportional to analytic functions of one independent variable, their ratios not being all constants, shall be called a *series.* If a series be given by the equations

$$f(p) = \phi(p) = \psi(p) = 0,$$

the necessary and sufficient condition that the surface generated should be annular is

$$\begin{vmatrix} \left(\frac{\partial f}{\partial p} \Big/ \frac{\partial f}{\partial p}\right) & \left(\frac{\partial f}{\partial p} \Big/ \frac{\partial \phi}{\partial p}\right) & \left(\frac{\partial f}{\partial p} \Big/ \frac{\partial \psi}{\partial p}\right) \\ \left(\frac{\partial \phi}{\partial p} \Big/ \frac{\partial f}{\partial p}\right) & \left(\frac{\partial \phi}{\partial p} \Big/ \frac{\partial \phi}{\partial p}\right) & \left(\frac{\partial \phi}{\partial p} \Big/ \frac{\partial \psi}{\partial p}\right) \\ \left(\frac{\partial \psi}{\partial p} \Big/ \frac{\partial f}{\partial p}\right) & \left(\frac{\partial \psi}{\partial p} \Big/ \frac{\partial \phi}{\partial p}\right) & \left(\frac{\partial \psi}{\partial p} \Big/ \frac{\partial \psi}{\partial p}\right) \end{vmatrix} \equiv 0.$$

If the series be given in the parametric form

$$p = p(u),$$

the condition for an annular surface will be

$$\left(\frac{\partial p}{\partial u} \Big/ \frac{\partial p}{\partial u}\right) \equiv 0.$$

§ 3. Systems of Circle Crosses.

We have now given in outline the most important facts concerning systems of circles orthogonal to one sphere. Before leaving these circles altogether, let us revert to that striking figure, the circle cross, for a good deal of interest comes to light when we take the cross rather than the individual circle as a space element.* We start with the linear complex

$$\sum_{i,\,j=0}^{i,\,j=3} a_{ij}p_{ij} = 0. \tag{13}$$

Let us then write

$$a_{01}+a_{23} \equiv \rho X_1, \quad a_{01}-a_{23} \equiv \sigma \dot{X}_1.$$
$$a_{02}+a_{31} \equiv \rho X_2, \quad a_{02}-a_{31} \equiv \sigma \dot{X}_2.$$
$$a_{03}+a_{12} \equiv \rho X_3, \quad a_{03}-a_{12} \equiv \sigma \dot{X}_3.$$

If the axial cross be determinate, neither set of coordinates $(X)\,(\dot{X})$ will vanish identically. The coordinates of the circles of the axial cross will have the form

$$\alpha_{ij} = la_{ij} + ma_{kl}.$$

If, now, we replace (X) and $(\dot{X})$ by $\lambda(X)$ and $\mu(\dot{X})$ respectively, we get

$$a_{0i}'+a_{jk}' = \lambda\,(a_{0i}+a_{jk}), \quad a_{0i}'-a_{jk}' = \mu\,(a_{0i}-a_{jk}),$$

and the new linear complex (a') will have the same axial cross as the original one. Conversely, suppose that two linear complexes have the same axial cross,

$$\sum_{i,\,j=0}^{i,\,j=3} a_{ij}'p_{ij} = p\sum_{i,\,j=0}^{i,\,j=3} a_{ij}p_{ij} + q(a/p),$$

$$(a'/p) = r\sum_{i,\,j=0}^{i,\,j=3} a_{ij}p_{ij} + s\,(a/p),$$

$$r = q, \quad s = p,$$

$$X_i' = (p+q)\,X_i, \quad \dot{X}_i' = (p-q)\,X_i.$$

* The remainder of the present chapter will be found in the Author's *Study of the Circle Cross*, cit.

We see thus that (X) and $(\dot{X})$ may be taken as two separately homogeneous triads of coordinates to determine the cross.*

Two crosses, orthogonal to the same not null sphere, will have certain invariants under the quaternary orthogonal group. Let the circles of the cross have the coordinates (α) and (α'), where

$$\alpha_{ij} = \rho\alpha'_{kl}.$$

In the same way let a second cross be determined by the circles (β) and (β'). If θ_1, θ_2 be the angles of the circles (α) and (β), we have by (12)

$$\cos\theta_1\cos\theta_2 = \frac{\sum_{i,j=0}^{i,j=3} \alpha_{ij}\beta_{ij}}{\sqrt{\sum_{i,j=0}^{i,j=3}\alpha_{ij}^2}\sqrt{\sum_{i,j=0}^{i,j=3}\beta_{ij}^2}}.$$

In like manner we find without difficulty

$$\sin\theta_1\sin\theta_2 = \frac{(\alpha/\beta)}{\sqrt{\sum_{i,j=0}^{i,j=3}\alpha_{ij}^2}\sqrt{\sum_{i,j=0}^{i,j=3}\beta_{ij}^2}},$$

$$\cos(\theta_1 \mp \theta_2) = \frac{(XY)}{\sqrt{(XX)}\sqrt{(YY)}},$$

$$\cos(\theta_1 \pm \theta_2) = \frac{(\dot{X}\dot{Y})}{\sqrt{(\dot{X}\dot{X})}\sqrt{(\dot{Y}\dot{Y})}}.$$

The condition that the circles (α) and (β) should be cospherical, or that one should be cospherical with the mate of the other in a circle cross, is that

$$\frac{(XY)}{\sqrt{(XX)}\sqrt{(YY)}} = \pm\frac{(\dot{X}\dot{Y})}{\sqrt{(\dot{X}\dot{X})}\sqrt{(\dot{Y}\dot{Y})}}. \tag{18}$$

* The idea of taking a line cross as a space element is due to Study, and plays a fundamental rôle in his *Geometrie der Dynamen*, Leipzig, 1903. See also, for the case of non-Euclidean space, the Author's *Non-Euclidean Geometry*, cit., pp. 124 ff.

They will be cospherical and in involution, i.e. intersect twice orthogonally, if

$$(XY) = (\dot{X}\dot{Y}) = 0. \tag{19}$$

Theorem 18.] *If the circles of two crosses be orthogonal to the same sphere, then, if one circle of the first cross be cospherical with one of the second, the same will be true of the remaining circles of the two crosses.*

Suppose that (α) is cospherical with the two circles (β) and (β') of a certain cross. It is then, by 2], orthogonal to their common orthogonal sphere. Every sphere through (β) will be orthogonal to the sphere (α) (β'), hence (β) cuts (α) twice at right angles. Our equation (19) is thus satisfied. It will also be satisfied if we replace (α) by (α').

Theorem 19.] *If a not null circle be cospherical with two circles of a cross, it will cut each of them twice orthogonally, as will the circle orthogonal to the common orthogonal sphere of these three circles and in bi-involution with the first one.*

Theorem 20.] *If two circles, neither of which is null, intersect twice orthogonally, then each intersects twice orthogonally the circle in bi-involution with the other which is orthogonal to the common orthogonal sphere of the two.*

We shall say that two crosses *intersect orthogonally* when each circle of one meets twice orthogonally each circle of the other. The condition for this is given by (19). The condition for parataxy will be, from (11),

$$[(XX)(YY)-(XY)]^2\,[(\dot{X}\dot{X})(\dot{Y}\dot{Y})-(\dot{X}\dot{Y})^2] = 0.$$

This may be written

$$\left[\sum_{i=1}^{i=3}\begin{vmatrix} X_j & X_k \\ Y_j & Y_k \end{vmatrix}^2\right]\left[\sum_{i=1}^{i=3}\begin{vmatrix} \dot{X}_j & \dot{X}_k \\ \dot{Y}_j & \dot{Y}_k \end{vmatrix}^2\right] = 0.$$

The only real solutions will be

$$Y_i = \rho X_i \text{ or } \dot{Y}_i = \sigma \dot{X}_i. \tag{20}$$

Theorem 21.] *If two not null circles be paratactic, each is paratactic with the circle orthogonal to the common orthogonal sphere which is in bi-involution with the other.*

It will be proper here to speak of the crosses themselves as *paratactic.*

The simplest one-parameter families of crosses are those given by equations of the form

$$X_i \equiv \lambda Y_i + \mu Z_i, \quad \dot{X}_i \equiv \lambda \dot{Y}_i + \mu \dot{Z}_i. \tag{21}$$

These will be the axial crosses of the pencil of linear complexes

$$a_{ij} \equiv \lambda b_{ij} + \mu c_{ij}.$$

There are four important varieties in this system

(*a*) $$Y_i \equiv \rho Z_i, \quad \dot{Y}_i \equiv \sigma \dot{Z}_i.$$

The system consists in but a single cross

(*b*) $$Y_i \equiv \rho Z_i, \quad \dot{Y}_i \not\equiv \sigma \dot{Z}_i.$$

Here all crosses of the system are paratactic. Let $(\dot{X}')$ satisfy the equations

$$(\dot{X}'\dot{Y}) = (\dot{X}'\dot{Z}) = 0,$$

then $$(\dot{X}'\dot{X}) = 0.$$

On the other hand (X') shall merely be required to satisfy the equation

$$(X'Y) = 0.$$

We may then write

$$X_i' \equiv \lambda' Y_i' + \mu' Z_i', \quad \dot{X}_i' \equiv \dot{Y}_i' = \dot{Z}_i'.$$

We thus have a second system of crosses exactly like the first, each cross of one system cutting each of the other orthogonally. We have, in fact, two residual generations of the same cyclide, each not null circle of one generation cutting each of the other orthogonally, while each two proper circles of the same generation are paratactic.

(c) $$Y_i \not\equiv \rho Z_i, \quad \dot{Y}_i \equiv \sigma Z_i.$$

This is not essentially different from the last.

(d) $$Y_i \not\equiv \rho Z_i, \quad \dot{Y}_i \not\equiv \sigma \dot{Z}_i.$$

Here we have something quite different. Let us write

$$rU_i \equiv Y_j Z_k - Y_k Z_j, \quad s\dot{U}_i \equiv \dot{Y}_j \dot{Z}_k - \dot{Y}_k \dot{Z}_j.$$

The crosses of our system cut this fixed cross orthogonally. The surface generated by these crosses shall be called a *pseudo-cylindroid.** It has one highly remarkable property. Suppose that we have a congruence of circles orthogonal to our fundamental sphere. We may express them parametrically

$$X_i \equiv X_i(u, v), \quad \dot{X}_i \equiv \dot{X}_i(u, v).$$

Let us assume that

$$\left| X \frac{\partial X}{\partial u} \frac{\partial X}{\partial v} \right| \times \left| \dot{X} \frac{\partial \dot{X}}{\partial u} \frac{\partial \dot{X}}{\partial v} \right| \not\equiv 0.$$

The cross cutting orthogonally the adjacent crosses $(X)(\dot{X})$ and $(X+dX)(\dot{X}+d\dot{X})$ will be determined by the equations

$$\rho Y_i \equiv \begin{vmatrix} X_j & X_k \\ \frac{\partial X_j}{\partial u} & \frac{\partial X_k}{\partial u} \end{vmatrix} du + \begin{vmatrix} X_j & X_k \\ \frac{\partial X_j}{\partial v} & \frac{\partial X_k}{\partial v} \end{vmatrix} dv;$$

$$\sigma \dot{Y}_i \equiv \begin{vmatrix} \dot{X}_j & \dot{X}_k \\ \frac{\partial \dot{X}_j}{\partial u} & \frac{\partial \dot{X}_k}{\partial u} \end{vmatrix} du + \begin{vmatrix} \dot{X}_j & \dot{X}_k \\ \frac{\partial \dot{X}_j}{\partial v} & \frac{\partial \dot{X}_k}{\partial v} \end{vmatrix} dv.$$

We shall mean by the *general position* of a cross in our congruence one where the left-hand side of the identity (or rather non-identity) does not vanish.

Theorem 22.] *If a congruence of circles be given orthogonal to a fixed sphere of such a nature that its circles cannot be assembled into a one-parameter family of series of paratactic*

* It is carried by our line-circle transformation into that surface in non-Euclidean space which corresponds to the Euclidean cylindroid. See the Author's *Non-Euclidean Geometry*, cit., pp. 128 and 219.

circles, then the circles cospherical and orthogonal to a circle of the congruence in general position and its adjacent circles generate a pseudo-cylindroid.

We may find a canonical form for the equations of the pseudo-cylindroid as follows. The cross which is orthogonal to the various crosses of the series shall be (1, 0, 0) (1, 0, 0). There will be two crosses of the series which intersect orthogonally. Let us, in fact, undertake to solve the simultaneous equations

$$\lambda\lambda'(YY)+(\lambda\mu'+\mu\lambda')(YZ)+\mu\mu'(ZZ)=0,$$
$$\lambda\lambda'(\dot{Y}\dot{Y})+(\lambda\mu'+\mu\lambda')(\dot{Y}\dot{Z})+\mu\mu'(\dot{Z}\dot{Z})=0.$$

Eliminating λ'/μ' we get a quadratic in λ/μ, which is seen to be the Jacobian of (XX) and $(\dot{X}\dot{X})$ looked upon as quadratic forms in λ/μ. If our crosses $(Y)(\dot{Y})$ and $(Z)(\dot{Z})$ be real, the quadratic equations $(XX)=0$, $(YY)=0$ have conjugate imaginary roots. The two quadratics can have no common root unless they are identical. Hence the invariant cannot vanish, and the quadratic in λ/μ has distinct roots. Let us take the crosses corresponding to them as (0, 1, 0) (0, 1, 0), (0, 0, 1) (0, 0, 1). Our canonical form then becomes

$$X_1=\dot{X}_1=0,$$
$$X_2=\beta\dot{X}_2,$$
$$X_3=\gamma\dot{X}_3,\quad \beta,\ \gamma \text{ constant.}$$

To find the equation of the surface generated, let (p) be the Plücker coordinates of a circle of our system.

$$p_{01}=p_{23}=0,$$
$$\beta(p_{02}-p_{31})(p_{03}+p_{12})=\gamma(p_{02}+p_{31})(p_{03}-p_{12}).$$

A point (x) will lie on this circle if

$$x_2p_{02}+x_3p_{03}=0,$$
$$x_0p_{02}+x_1p_{12}=0,$$
$$\rho x_0=-p_{12},\quad \rho x_1=p_{02},\quad \sigma x_2=p_{03},\quad \sigma x_3=-p_{02},$$
$$(p/p)=0.$$

$$p_{12}=x_0x_3,\quad p_{02}=-x_1x_3,\quad p_{03}=x_1x_2,\quad p_{31}=x_0x_2,$$
$$\beta(x_0x_2+x_3x_1)(x_0x_3+x_1x_2)-\gamma(x_0x_2-x_3x_1)(x_0x_3-x_1x_2)=0. \quad (22)$$

Theorem 23.] *The general pseudo-cylindroid is a surface of the eighth order, which, in cartesian space, has the circle at infinity as a quadruple curve, and two circles in bi-involution as double curves.*

Among congruences of crosses the simplest are those which come from the axial crosses of a three term linear system of linear complexes. Here we have

$$X_i \equiv aY_i + bZ_i + cT_i; \quad \dot{X}_i \equiv a\dot{Y}_i + b\dot{Z}_i + c\dot{T}_i. \tag{23}$$

(*a*) $$|\,YZT\,| \equiv |\,\dot{Y}\dot{Z}\dot{T}\,| \equiv 0.$$

This gives the crosses, cutting a given cross orthogonally,

(*b*) $$|\,YZT\,| \equiv 0, \quad |\,\dot{Y}\dot{Z}\dot{T}\,| \not\equiv 0,$$

$$T_i \equiv \lambda Y_i + \mu Z_i,$$

$$X_i \equiv (a+\lambda c)\,Y_i + (b+\mu c)Z_i.$$

Let $$(a+\lambda c) = p, \quad (b+\mu c) = q.$$

$$c = \frac{bp-aq}{\lambda q - \mu p}.$$

$$X_i = pY_i + qZ_i, \quad \dot{X}_i = a\left[\dot{Y}_i - \frac{q\dot{Z}_i}{\lambda q - \mu p}\right] + b\left[\dot{Z}_i + \frac{p\dot{T}_i}{\lambda q - \mu p}\right].$$

The congruence contains one paratactic generation of each of ∞^1 cyclides corresponding to different values for $\frac{p}{q}$. The residual generations will generate a second congruence of like sort.

(*c*) $$|\,YZT\,| \times |\,\dot{Y}\dot{Z}\dot{T}\,| \not\equiv 0.$$

Here we may solve one system of equations for a, b, c, and substitute in the other, getting

$$\rho \dot{X}_i = \sum_{j=1}^{j=3} a_{ij} X_j. \tag{24}$$

The crosses intersecting orthogonally pairs of crosses of this congruence will generate a second congruence of like sort called the *conjugate* to the first.

$$\sigma U_i = \sum_{j=1}^{j=3} a_{ji} U_j. \tag{25}$$

The relation between the two is reciprocal. If we seek a cross that belongs at once to both systems, we shall fall upon three equations of the type

$$\tau X_i = \sum_{j=1}^{j=3} b_{ij} X_j.$$

The condition for compatibility gives a cubic in τ whose discriminant does not vanish identically. If, now, we give the name *general chain congruence* only to a congruence of our present type where this cubic has distinct roots, we see that it shares with its conjugate three crosses, and only three. The common perpendicular to each two of these crosses which belongs to both congruences must be the third cross; hence each two intersect orthogonally, and we may take them as fundamental in our coordinate system. Our chain congruence and its conjugate may thus be reduced to the canonical form

$$\rho \dot{X}_i = a_i X_i, \quad \sigma U_i = a_i \dot{U}_i, \tag{26}$$

$$\prod_{i=1}^{i=3} a_i \neq 0, \quad \prod_{i=1}^{i=3} (a_j - a_k) \neq 0.$$

Theorem 24.] *The crosses cutting orthogonally pairs of crosses of a general chain congruence generate a second such congruence. The relation between the two is reciprocal, and they have in common three orthogonally intersecting crosses.*

The condition that two circles of our two systems should be cospherical, or that one should be cospherical with the circle in bi-involution with the other, is

$$(UX)\left[\sqrt{(aX^2)}\sqrt{\left(\frac{1}{a}U^2\right)} \pm \sqrt{(XX)}\sqrt{(UU)}\right] = 0.$$

If the first factor vanish, $(UX) = (U\dot{X}) = 0$, the two cut orthogonally. If the second factor vanish, every circle of a cross of the first congruence for which

$$q(aX^2) = p(XX)$$

is cospherical with a circle of every cross of the second for which

$$p\left(\frac{1}{a}U^2\right) = q(UU).$$

Theorem 25.] *The circles of a general chain congruence cospherical with one circle of the conjugate congruence and orthogonal thereto generate a pseudo-cylindroid; those cospherical with a circle of the conjugate congruence but not orthogonal thereto generate a cyclide; the residual generation will be composed of circles of the conjugate congruence.*

If we have a cyclide one of whose fundamental spheres is our given sphere $x_4 = 0$ it is clear that the circles of either corresponding generation form a rational series which can be expressed in the parametric form

$$X_i = t^2 Y_i + tZ_i + T_i; \quad \dot{X}_i = t^2 \dot{Y}_i + t\dot{Z}_i + \dot{T}_i.$$

Eliminating t^2, t, and 1 we fall back upon three equations of the type (24). The series of circles is thus surely contained in one of our chain congruences.

Theorem 26.] *The circles which are cospherical and orthogonal to pairs of circles of one generation of a cyclide will, in general, generate a chain congruence whose conjugate includes this generation of the cyclide.*

The words 'in general' here indicate that there are a number of possible exceptions. We shall not, however, stop to investigate them. If we take the equation of our congruence in the form (24), we may assume that the sphere $x_0 = 0$ bears no special relation thereto. Now all circles of this sphere satisfy the relation

$$\dot{X}_i = \rho X_i.$$

Hence three circles of the sphere (orthogonal to the fundamental sphere) lie in our congruence.

Theorem 27.] *Three circles of the general chain congruence lie on an arbitrary sphere orthogonal to that sphere which is orthogonal to all circles of the congruence, and three circles of the conjugate congruence also lie thereon. Each spherical triangle formed by the three circles of one congruence is polar to each formed by the circles of the other congruence.*

Our congruence is traced by pairs of circles in bi-involution. If a circle generate a sphere orthogonal to the fixed sphere, the circle in bi-involution therewith passes through two fixed points, the foci of the circle common to the two spheres, and vice versa.

Theorem 28.] *Through each pair of points anallagmatic in the sphere, which is orthogonal to the circles of a general chain congruence, will pass three circles of the congruence, and three of the conjugate congruence. Each circle of one congruence through these points cuts orthogonally two circles of the other congruence.*

It is doubtful whether there remains a great deal in the subject of circles orthogonal to a fixed sphere which is worth protracted study. Of course it would be possible to develop the subject until we had an explicit counterpart for every known theorem in line geometry, metrical or projective, but most of these results would be of mediocre interest, and easily found by any one who needed them. The pseudo-cylindroid might well be given some further attention, and also certain complexes of circles, notably those corresponding to the quadratic line complex. The methods to be employed are perfectly obvious, and the results can be at once predicted.

CHAPTER XIII

CIRCLES IN SPACE, ALGEBRAIC SYSTEMS

§ 1. Coordinates and Identities.

THE greatest aid to the study of circles in space is the correspondence established in Ch. VI between the spheres of pentaspherical space and the points of a four-dimensional projective space of elliptic measurement. Let us write down again such facts in regard to this correspondence as will be particularly useful to us in the present chapter.

Pentaspherical space S_3.	Projective space S_4.
Sphere.	Point.
Point.	Point of S_3^2.
Circle.	Line.
Null circle.	Line on or tangent to S_3^2.
Point-pair as locus of spheres.	Plane.
Spheres orthogonal to given sphere.	Hyperplane.
Mutually orthogonal spheres.	Points conjugate with regard to S_3^2.
Coaxal circles.	Pencil of lines.
Conjugate generations of cyclide.	Conjugate generations of quadric.
Intersecting circles.	Lines whose hyperplane touches S_3^2.
Cospherical circles.	Intersecting lines.
Circles in involution.	Lines, each of which intersects the polar hyperplane of the other with regard to S_3^2.

Circles in bi-involution.	Lines, each of which lies in polar hyperplane of the other with regard to S_3^2.
Circles meeting twice orthogonally.	Intersecting lines conjugate with regard to S_3^2.
Twenty-four - parameter group carrying spheres to spheres and circles to circles.	Group of all collineations.
Ten-parameter group of spherical transformations.	Ten-parameter collineation group leaving S_3^2 invariant.

The basis for the algebraic study of circles will be the Plücker coordinates *

$$\rho p_{ij} \equiv x_i y_j - x_j y_i, \quad \begin{matrix} i = 0 \dots 4 \\ j = 0 \dots 4 \end{matrix}. \tag{1}$$

Of these there are clearly twenty-five. They are connected by the following identities:

$$p_{ij} \equiv -p_{ji}, \quad p_{kk} \equiv 0 \tag{2}$$

$$\begin{aligned} \tfrac{1}{2}\Omega_0(pp) &\equiv p_{12}p_{34} + p_{13}p_{42} + p_{14}p_{23} \equiv 0. \\ \tfrac{1}{2}\Omega_1(pp) &\equiv p_{02}p_{43} + p_{04}p_{32} + p_{03}p_{24} \equiv 0. \\ \tfrac{1}{2}\Omega_2(pp) &\equiv p_{01}p_{34} + p_{04}p_{41} + p_{04}p_{13} \equiv 0. \\ \tfrac{1}{2}\Omega_3(pp) &\equiv p_{01}p_{42} + p_{04}p_{21} + p_{02}p_{14} \equiv 0. \\ \tfrac{1}{2}\Omega_4(pp) &\equiv p_{01}p_{23} + p_{02}p_{31} + p_{03}p_{12} \equiv 0. \end{aligned} \tag{3}$$

The last five arise from the obvious equation

$$\Omega_i(pp) \equiv \frac{\partial}{\partial t_i} \mid t\,x\,y\,x\,y \mid.$$

The polar of the form $\Omega_i(pp)$ shall be written $\Omega_i(pq)$.

Since the totality of circles in space depends upon six independent parameters, it is clear that the identities above cannot all be independent. Suppose that we have twenty-five homogeneous quantities (p) which satisfy merely the equations

$$p_{ij} \equiv -p_{ji}. \tag{2}$$

* The first writer to use these seems to have been Stephanos, 'Sur une configuration remarquable de cercles dans l'espace', *Comptes rendus*, vol. xciii, 1881.

We easily find

$$\sum_{n=0}^{n=4} p_{in}\,\Omega_n(pp) \equiv 0, \quad i = 0 \ldots 4.$$

If thus

$$p_{ij} \neq 0, \quad \Omega_k(pp) = \Omega_l(pp) = \Omega_m(pp) = 0.$$

We have also $\qquad \Omega_i(pp) = \Omega_j(pp) = 0.$

Every circle will have thus ten essentially distinct Plücker coordinates which satisfy our identities (3). When we say *essentially distinct* we mean that no two, in the general case, differ merely in sign, and none vanishes automatically. Let us show, conversely, that just one circle will correspond to each set of homogeneous values not all zero which satisfy these equations. Assuming first $p_{ij} \neq 0$.

Let us take the two spheres

$$\sum_{n=0}^{n=4} p_{in}\,x_n = \sum_{n=0}^{n=4} p_{jn}x_n = 0. \tag{4}$$

Multiply the first equation by p_{jk}, the second by p_{ki}, and add, we get

$$-p_{ij}\sum_{n=0}^{n=4} p_{kn}x_n = 0.$$

The points (x) which satisfy these equations will lie on one circle. To find the Plücker coordinates of this circle we have but to take the spheres (4)

$$\rho\, q_{rs} = \begin{vmatrix} p_{ir} & p_{is} \\ p_{jr} & p_{js} \end{vmatrix} = -p_{ij}\,p_{rs}.$$

There will surely be one circle corresponding to each set of Plücker coordinates. Let the reader show that there cannot be more than one.

The coordinates of a circle are occasionally determined in another manner. Suppose we start with the knowledge that our circle is orthogonal to the three spheres (r), (s), (t). The spheres through this circle orthogonal to (u) and (v) are

$$x_i = \frac{\partial}{\partial l_i} | \, l\, u\, r\, s\, t \, |, \quad y_i = \frac{\partial}{\partial l_i} | \, l\, v\, r\, s\, t \, |.$$

$$| \, u\, v\, r\, s\, t \, | \equiv \Delta \neq 0.$$

$$\rho p_{ij} \equiv \Delta \begin{vmatrix} r_k & r_l & r_m \\ s_k & s_l & s_m \\ t_k & t_l & t_m \end{vmatrix}. \tag{5}$$

Two circles will be in involution, when, and only when, a sphere through one is among the spheres orthogonal to the other. This gives

$$| \, x\, y\, r'\, s'\, t' \, | \equiv \Sigma \, p_{ij} \, p_{ij}' = 0.* \tag{6}$$

The only circles in involution with themselves are null circles. The condition for a null circle will thus be

$$\Sigma \, p_{ij}^2 = 0. \tag{7}$$

Two circles will be cospherical when there is linear dependence between the spheres through the one and those through the other. The condition is

$$\Omega_i (pp') \equiv 0, \quad i = 0, \dots 4. \tag{8}$$

These five conditions are not independent. They amount to requiring that the rank of the matrix

$$\left\| \begin{matrix} x_0 & x_1 & x_2 & x_3 & x_4 \\ y_0 & y_1 & y_2 & y_3 & y_4 \\ x_0' & x_1' & x_2' & x_3' & x_4' \\ y_0' & y_1' & y_2' & y_3' & y_4' \end{matrix} \right\|$$

shall not exceed three, and this imposes two conditions only. When the circles (p) and (p') are not cospherical, the coordinates of their common orthogonal sphere will be

$$\rho z_i = \Omega_i (pp'). \tag{9}$$

* The expression $\Sigma \, p_{ij} \, q_{ij}$ must be understood in the present chapter to cover ten values of p, essentially distinct in the sense explained above.

This will be null, and the circles will intersect if

$$\sum_{i=0}^{i=4} \Omega_i^2 (pp') = 0. \tag{10}$$

The left-hand side of this equation is a relative invariant of the two circles for all spherical transformations. More interesting is their absolute invariant which we found in Ch. XII. (12).

$$\cos\theta_1 \cos\theta_2 = \frac{\Sigma p_{ij} p_{ij}'}{\sqrt{\Sigma p_{ij}^2}\sqrt{\Sigma p_{ij}'^2}}. \tag{11}$$

There is a second meaning which may be attached to this invariant, and which is of not a little interest. We first ask, when will the circle (p) be tangent to the sphere (x')? An arbitrary sphere orthogonal to (x), (y), (x') may be written

$$\rho z_i = \lambda \frac{\partial}{\partial l_i} \mid lux'xy \mid + \mu \frac{\partial}{\partial l_i} \mid lvx'xy \mid.$$

Here (u) and (v) are two arbitrary spheres not linearly dependent on (x), (y), (x').

We also introduce the notation

$$P_{ij} \equiv \sum_{n=0}^{n=4} p_{in} p_{jn} \equiv \sum_{n=0}^{n=4} p_{ni} p_{nj}.$$

The condition for tangency is the condition that there should be but one null sphere orthogonal to (x'), (x), (y). We thus write the condition that the equation in λ/μ, $(zz) = 0$, should have equal roots, and apply the identity

$$\Delta \frac{\partial^2 \Delta}{\partial a_{ii}\, \partial a_{jj}} \equiv \frac{\partial \Delta}{\partial a_{ii}} \frac{\partial \Delta}{\partial a_{jj}} - \frac{\partial \Delta}{\partial a_{ij}} \frac{\partial \Delta}{\partial a_{ji}},$$

$$\begin{vmatrix} (xx) & (xy) & (xx') \\ (yx) & (yy) & (yx') \\ (x'x) & (x'y) & (x'x') \end{vmatrix} = 0.$$

$$(x'x')\, \Sigma p_{ij}^2 = \sum_{i,j=0}^{i,j=4} p_{ij} x_i' x_j'.$$

Let us next define as the *orthogonal projection* of a circle on a sphere the intersection of that sphere with a second orthogonal to it, passing through the circle. This becomes indeterminate only when the circle is itself orthogonal to the sphere. Let us, then, take two not null circles (p), (p'). Let (x') be a sphere through (p') tangent to (p); we seek the cosine of the angle of (p') with the orthogonal projection of (p) on (x'). If (y') be the sphere through (p') orthogonal to (x'), we need to find the angle of (y') with the sphere (y) through (p) orthogonal to (x'). We easily find

$$y_i = \sum_{j=0}^{j=4} p_{ij} x_j'.$$

$$\cos\theta = \frac{(yy')}{\sqrt{(yy)}\sqrt{(y'y')}} = \frac{\Sigma p_{ij} p_{ij}}{\sqrt{(y'y')}\sqrt{(x'x')}\,\Sigma p_{ij}^2}$$

$$= \frac{\Sigma p_{ij} p_{ij}'}{\sqrt{\Sigma p_{ij}^2}\sqrt{\Sigma p_{ij}'^2}} = I(pp') \qquad (12)$$

Theorem 1.] *The product of the cosines of the non-Euclidean angles of two not null circles is the cosine of the angle which the one makes with the orthogonal projection of the other on a sphere through the first tangent to the second.*

We shall define this angle as the *angle* of the two circles. It will be found that such a definition will agree with the usual one when the circles are cospherical, or reduce to straight lines of cartesian space.*

The condition that two circles should be paratactic is reached by rewriting XII. (11) in invariant form

$$\left[\Sigma p_{ij}^2\,\Sigma p_{ij}'^2 + (\Sigma p_{ij} p_{ij}')^2 - \sum_{i=0}^{i=4} \Omega_i^2(pp')\right]^2$$

$$= 4[\Sigma(p_{ij} p_{ij}')]^2\,\Sigma p_{ij}^2\,\Sigma p_{ij}'^2.$$

* This invariant is due to Koenigs. See his remarkable article 'Contributions à la théorie du cercle dans l'espace', *Annales de la Faculté des Sciences de Toulouse*, vol. ii, 1888. The interpretation as the product of the cosines of two angles first appeared in the Author's *Study of the Circle Cross*, cit., p. 155. The interpretation as the cosine of a single angle is due to Dr. David Barrow.

Our last two equations can be put into particularly neat form by introducing an additional circle coordinate: let us write

$$\Sigma p_{ij}^2 + p^2 \equiv \Sigma p_{ij}'^2 + p'^2 \equiv 0. \tag{13}$$

$$I(pp') = \frac{\Sigma p_{ij} p_{ij}'}{pp'}. \tag{14}$$

For parataxy

$$\sum_{i=0}^{i=4} \Omega_i^2 (pp') = [\Sigma p_{ij} p_{ij}' \pm pp']^2.$$

When the given circles are real this last equation must amount to two distinct equations.

Assuming that the paratactic circles are determined by their foci,

$$(xx) = (yy) = (x'x') = (y'y') = (xx') = (yy') = 0.$$

$$P_{ij} = -(xy)(x_i y_j + x_j y_i).$$

The conditions for parataxy become *

$$p \sum_{n=0}^{n=4} p_{in} P_{in}' + p' \sum_{n=0}^{n=4} p_{in}' P_{in} \equiv pp' \sum_{n=0}^{n=4} p_{in} p_{in}' + \sum_{n=0}^{n=4} P_{in} P_{in}' \equiv 0.$$

$$p \sum_{n=0}^{n=4} p_{in} P_{in}' - p' \sum_{n=0}^{n=4} p_{in}' P_{in}$$

$$\equiv pp' \sum_{n=0}^{n=4} p_{in} p_{in}' - \sum_{n=0}^{n=4} P_{in} P_{in}' \equiv 0, \quad i = 0, \dots 4. \tag{15}$$

The sphere through a given circle orthogonal to a sphere (r) has the coordinates

$$\rho z_i = \sum_{n=0}^{n=4} p_{in} r_n. \tag{16}$$

* Cf. von Weber, 'Zur Geometrie des Kreises im Raume', *Grunerts Archiv*, Series 3, vol. vii, 1904, p. 292.

Two circles will be in bi-involution when every sphere through one will be orthogonal to every sphere through the other. The condition is

$$\sum_{n=0}^{n=4} p_{in} p_{jn}' \equiv 0, \quad i, j = 0 \ldots 4. \tag{17}$$

We have now a sufficient analytic basis for the study of systems of circles. Before taking up continuous systems we shall discuss a very curious figure formed by five circles, and an analogous one involving fifteen. We shall approach these figures by means of our four-dimensional representation.*

Suppose that we have three lines a, b, c in four-dimensional projective space, no two coplanar, nor are the three in one hyperplane. There are ∞^3 planes which intersect all of them. Let an arbitrary hyperplane π be taken. It will contain one plane of the system, that which joins its intersections with the three lines. We assume that the hyperplane does not contain the single line d' that intersects a, b, c. The cross ratio of the four points where d' meets a, b, c, π will be that of the four hyperplanes through any plane of the system, and through a, b, c and the point $d'\pi$, or the cross ratio of the four planes through the line of intersection of this plane with π, and the four fixed points $a\pi, b\pi, c\pi, d'\pi$. This brings out the important fact that the planes of our system meet an arbitrary hyperplane in the lines of a tetrahedral complex, which is, of course, of the second order.

Suppose, next, that we have four lines a, b, c, d, no two coplanar, no three in one hyperplane, nor does one lie in a plane which meets the other three. They will be intersected in threes by the four lines a', b', c', d', a similar system Let us fix our attention on the hyperplane (dd'). Every line in this hyperplane which intersects d' will lie in a plane meeting a, b, c.

The linear complex (d') will thus split off from the quadratic complex. The residue will be a linear complex which does

* Cf. Segre, 'Sull' incidenza di retti e di piani nello spazio a quattro dimensioni', *Rendiconti del Cercolo matematico di Palermo*, vol. ii, 1888.

not contain (d'), so that the lines thereof which meet (d) will also meet a line e. In other words, every plane which intersects a, b, c, d will also intersect e. The linear complex could not be a special one with e as directrix, for then every plane meeting a, b, c would meet e, an absurdity, since through each point of the plane common to the hyperplanes (ab) and (ce) there will pass but one line to meet ab and one to meet ce. Hence d, e are mutually polar in the linear complex, and are symmetrically related to a, b, c. No other line f can meet all the planes which meet a, b, c, d. For if it were in the hyperplane (dd') it would meet all lines of the linear congruence with directrices d, d', which is impossible; but if not in this hyperplane, the linear congruence with directrices d, f would not lie in the tetrahedral complex determined by planes cutting a, b, c. Our five lines will thus bear to one another a symmetrical relation: any plane meeting four will intersect the fifth.

Theorem 2.] *If four circles be given, no two cospherical, no three orthogonal to the same sphere, and no one containing a point-pair cospherical with each of the other three, then every circle in involution with these four circles is in involution with a fifth. The relation connecting the five is reciprocal, each being uniquely determined by the other four.*

This figure is the famous *pentacycle* of Stephanos.* To find the construction let us notice that e lies in the hyperplane (dd'); by symmetry it is in the hyperplanes (aa'), (bb'), (cc'). Four circles fulfilling the restrictions mentioned in 2] shall be said to be *in general position.*

Theorem 3.] *If four circles be given in general position, there are four others each of which is cospherical with three of the given ones. The common orthogonal spheres to each two non-cospherical circles, one from each system, will pass through the foci of the circle completing a pentacycle with the four original circles.*

* *Sur une configuration remarquable*, cit., p. 579.

We next note that e plays the same rôle with regard to a', b', c', d' as it does with regard to a, b, c, d. Draw the line connecting $(ab')(a'b)$. This lies in the hyperplanes (aa'), (bb'), and so is coplanar with e. In the same way the line $(cd')(c'd)$ is coplanar with e. But the line $(cd')(c'd)$ lies in the hyperplanes (cd), $(c'd')$, which are identical with the hyperplanes $(a'b')$, (ab), so that $(cd')(c'd)$ intersects $(ab')(a'b)$.

Theorem 4.] *Given four circles* c_1, c_2, c_3, c_4 *in general position, and four others* c_1', c_2', c_3', c_4' *so placed that* c_i *and* c_j' *are cospherical if* $i \neq j$. *The circle of intersection of the sphere* $(c_i c_j')$ *with the sphere* $(c_i' c_j)$ *and the circle of intersection of the sphere* $(c_k c_l')$ *with the sphere* $(c_k' c_l)$ *are cospherical. The three spheres so obtained pass through a circle* c_5' *which completes a pentacycle with* c_1, c_2, c_3, c_4 *and with* c_1', c_2', c_3', c_4'.

This method of construction leads to an extension of the pentacyclic figure which is of much interest. Let the circles c_1, c_2, c_3, c_4, c_5' be renamed 01, 02, 03, 04, 05 respectively, while $c_1', c_2', c_3', c_4', c_5'$ become 15, 25, 35, 45, 05. The circle determined by the spheres (01, 25), (02, 15) shall be 34; it is cospherical with the circle of the spheres (03, 45), (04, 35) which shall be 12. We have the following fifteen circles:

01	02	03	04	05
21	12	13	14	15
31	32	23	24	25
41	42	43	34	35
51	52	53	54	45

The necessary and sufficient condition that two of these should be cospherical is that their symbols should have no common digit. They are cospherical by threes on fifteen spheres. The circles with the digit 0 and those with the digit 5 form two pentacycles. Consider the circles 01, 21, 31, 41. Each three are cospherical with one of the circles 05, 25, 35, 45. 15 completes a pentacycle with the latter four, hence it does with the first four also. Next consider two triads (ij, jk, ki), (lm, mn, nl). Each circle of one triad is cospherical with each of the other, but no two of the same triad are cospherical. We thus reach a beautiful theorem due to Stephanos.*

Theorem 5.] *Given four circles in general position. We may associate therewith eleven other circles as follows. The fifteen circles lie by threes on fifteen spheres, each circle belonging to three spheres. They may be grouped by fives in six pentacycles, each circle belonging to two pentacycles. They may, lastly, be grouped by threes in ten pairs of conjugate triads, each such pair of triads belonging to conjugate generations of the same cyclide.*

Let us call such a system *associated circles.* We get a construction for them by reverting to four dimensions. Suppose that we have six points P_1, P_2, P_3, P_4, P_5, P_6, whereof no five are in one hyperplane. We have fifteen triads of lines such as (P_1P_2), (P_3P_4), (P_5P_6). These will not lie in a hyperplane, so that there will be one line $l_{12,\,34,\,56}$ intersecting all, and of these lines there will be fifteen. They are concurrent three by three as follows: $l_{ij,\,kl,\,mn}$, $l_{ij,\,km,\,nl}$, $l_{ij,\,kn,\,ml}$ meet in the intersection of (P_iP_j) with the hyperplane $(P_kP_lP_mP_n)$. In no other case will two of the lines meet. Our lines are thus concurrent by threes in fifteen points, hence †

Theorem 6.] *If six spheres be given, no five having a common orthogonal sphere, they may be divided, in fifteen ways, into three groups of two each, and a circle found cospherical with the circle of each group of two spheres. The resulting figure will be fifteen associated circles.*

It will be noticed that the pentacycle and the figure of fifteen associated circles are carried into like figures by any linear sphere transformation.

* *Sur une configuration*, cit., and *Sur une configuration de quinze cercles dans l'espace*, ibid., p. 633.

† This figure is discussed by Richmond, 'On the figure of six points in a space of four dimensions', *Quarterly Journal of Math.*, vol. xxxi, 1899. The most complete discussion is in a long article by Weitzenböck, 'Projektive Geometrie des R_4', *Wiener Berichte*, vol. cxxi, 1912.

Let us now see what can be done towards an algebraic study of the pentacycle. We start with four circles (p), (q), (r), (s). If (p') be in involution with them we have

$$\Sigma p_{ij} p_{ij}' = \Sigma q_{ij} p_{ij}' = \Sigma r_{ij} p_{ij}' = \Sigma s_{ij} p_{ij}' = 0.$$

$$\Sigma (\lambda p_{ij} + \mu q_{ij} + \nu r_{ij} + \rho s_{ij}) p_{ij}' = 0.$$

For how many values of $\lambda : \mu : \nu : \rho$ will the coordinates (p') satisfy our identities (3)? If we substitute in three identities we have three quadratic equations in $\lambda : \mu : \nu : \rho$ giving eight solutions. Let us show that only five of these will satisfy the other identities. We write

$$\lambda p_{ij} + \mu q_{ij} + \nu r_{ij} + \rho s_{ij} = l_{ij}.$$

Let us try to satisfy the equations

$$l_{34} = \Omega_0 (ll) = \Omega_1 (ll) = 0.$$

One solution will be

$$l_{23} = l_{24} = l_{34} = 0.$$

Otherwise, we have, by (2), three solutions for the equations

$$l_{34} = \Omega_0 (ll) = \Omega_1 (ll) = \Omega_2 (ll),$$

and for these

$$\Omega_3 (ll)\, \Omega_4 (ll) \neq 0.$$

There are thus five values of $\lambda : \mu : \nu : \rho$ for which $\Omega_i (ll) \equiv 0$, and our theorem 2] is proved algebraically.* To find the coordinates of the circle (s') which is cospherical with (p), (q), (r), let these latter be determined by the pairs of spheres $(x)\,(y)$, $(x')\,(y')$, $(x'')\,(y'')$. Let (u), (v), (w) be three spheres orthogonal to (s'). Then, by (5),

$$s_{ij}' = \begin{vmatrix} u_k & u_l & u_m \\ v_k & v_l & v_m \\ w_k & w_l & w_m \end{vmatrix}.$$

* The actual equation of the fifth degree, on whose solution the problem depends, was exhibited by Weitzenböck at the fifth International Congress of Mathematicians in Cambridge, 1912, and will be found ibid., p. 2574.

We may take (u), (v), (w) as orthogonal to $(q)\,(r)$, $(r)\,(p)$, and $(p)\,(q)$ respectively

$$s_{ij}' = \begin{vmatrix} \Omega_k\,(qr) & \Omega_l\,(qr) & \Omega_m\,(qr) \\ \Omega_k\,(rp) & \Omega_l\,(rp) & \Omega_m\,(rp) \\ \Omega_k\,(pq) & \Omega_l\,(pq) & \Omega_m\,(pq) \end{vmatrix}. \qquad (18)$$

The fifth circle forming a pentacycle with (p), (q), (r), (s) is orthogonal to the common orthogonal spheres of $(p)\,(p')$, $(q)\,(q')$, $(r)\,(r')$, hence

$$t_{ij} = \begin{vmatrix} \Omega_k\,(pp') & \Omega_l\,(pp') & \Omega_m\,(pp') \\ \Omega_k\,(qq') & \Omega_l\,(qq') & \Omega_m\,(qq') \\ \Omega_k\,(rr') & \Omega_l\,(rr') & \Omega_m\,(rr') \end{vmatrix}. \qquad (19)$$

If (x), (y), (z), (r), (s), (t) be six spheres, no five having a common orthogonal sphere, the fifteen associated circles thereby determined will be the intersections of pairs of spheres such as

$$|\,x\,z\,r\,s\,t\,|\,(y) - |\,y\,z\,r\,s\,t\,|\,(x); \quad |\,x\,y\,r\,s\,t\,|\,(z) - |\,x\,y\,z\,s\,t\,|\,(r).$$

These expressions may be much simplified. We may find such multipliers for the homogeneous coordinates (x), (y), (z), (r), (s), (t) that

$$x_i + y_i + z_i + r_i + s_i + t_i \equiv 0, \quad i = 0 \ldots 5.$$

Otherwise written

$$(Xx) + (Xy) + (Xz) + (Xr) + (Xs) + (Xt) \equiv 0.$$

Our circles will be determined by pairs of equations such as

$$(Xx) + (Xy) = (Xz) + (Xr) = (Xs) + (Xt) = 0.$$

These simple formulae lead us to another property of the system of fifteen associated circles. Let us write the equation

$$(Xx)^3 + (Xy)^3 + (Xz)^3 + (Xr)^3 + (Xs)^3 + (Xt)^3 = 0. \qquad (20)$$

This will be a surface containing all fifteen circles, hence

Theorem 7.] *Every system of fifteen associated circles will* *lie on a surface of the sixth order which, in cartesian space, has the circle at infinity as a triple curve.**

§ 2. Linear Systems.

A system of circles whose coordinates are proportional to functions of five independent variables, the ratios not being all functions of four variables, shall be called a *hypercomplex*. The simplest hypercomplex is the *linear* one, determined by an equation of the form

$$\Sigma a_{ij} p_{ij} = 0. \tag{21}$$

Theorem 8.] *Nine arbitrary circles will belong to one, and, in general, only one, linear hypercomplex.*

Theorem 9.] *The assemblage of all circles in involution with a given circle is a linear hypercomplex.*

Manifestly this will not give the general linear hypercomplex, in fact the necessary and sufficient condition that (21) should represent a linear hypercomplex of this sort is

$$\Omega_i (aa) \equiv 0, \quad i = 0 \dots 4.$$

A linear hypercomplex has two important invariants under the group of spherical transformations, namely

$$I = \Sigma a_{ij}^2, \quad J = \sum_{i=0}^{i=4} \Omega_i^2 (aa). \tag{22}$$

The vanishing of these invariants and of various expressions dependent on them will lead to special types of linear hypercomplex, some of which we shall investigate. Let us write (21) at length

$$\Sigma a_{ij} (x_i y_j - x_j y_i) = 0.$$

The sphere (y) being fixed, (x) is orthogonal to (z) where

$$z_i = \sum_{n=0}^{n=4} a_{in} y_n. \tag{23}$$

* Stephanos, *Quinze cercles*, cit., p. 634.

It is to be noticed that, as a result of our identities (2),

$$\sum_{n=0}^{n=4} \Omega_n (aa)\, z_n = 0.$$

Theorem 10.] *If a linear hypercomplex be not composed of circles in involution with a fixed circle, then the circles of the system on an arbitrary sphere are those which are orthogonal to the intersection with a second sphere orthogonal to the first, and to a sphere uniquely determined by the hypercomplex.*

Consider the circles of the system through two fixed points (η), (ζ). If (ξ) be any point on such a circle,

$$\Sigma\, a_{ij} \begin{vmatrix} \eta_k & \eta_l & \eta_m \\ \zeta_k & \zeta_l & \zeta_m \\ \xi_k & \xi_l & \xi_m \end{vmatrix} = 0,$$

so that (ξ) traces a sphere.

Theorem 11.] *The circles of a linear hypercomplex which pass through two arbitrary points will usually trace a coaxal system.*

We shall see presently exactly what meaning to attach to the elusive word 'usually'. The sphere (x), where

$$\rho x_i = \Omega_i\, (aa), \tag{24}$$

shall be called the *central sphere* for the linear hypercomplex. In (23) we get the same value for (z) if we replace (y) by $\lambda\,(y) + \mu\,(x)$.

Theorem 12.] *The circles of a linear hypercomplex which lie on the spheres of a coaxal system including the central sphere are orthogonal to one sphere.*

Theorem 13.] *All circles of a linear hypercomplex which are cospherical with a circle of the central sphere are orthogonal to one sphere.*

Theorem 14.] *All circles of the central sphere of a linear hypercomplex belong to the hypercomplex.*

We may find a spherical transformation to carry our central sphere, when not null, into $x_4 = 0$. The equation of the linear hypercomplex will then lack all terms with the subscript $_4$; equation (23) will give the same values for (z) if we replace (y) by either null sphere through the circle common to (y) and the central sphere. It will thus give a relation between each point-pair anallagmatic in the central sphere, and a corresponding sphere. This, however, is nothing in the world but the polarization in the linear complex of circles of our hypercomplex which are orthogonal to the central sphere, a process described on pp. 461, 462. We see by continuity, or non-Euclidean line geometry, that this relation holds even when the central sphere is null.

Theorem 15.] *The circles of a linear complex lying on an arbitrary sphere are orthogonal to the polar in the linear complex of the circles of the system which are orthogonal to the central sphere, of the foci of the circle common to the given sphere and to the central sphere.*

When the equation of the linear hypercomplex is written in the form (21) we may assume $x_4 = 0$ to be an arbitrary sphere. Circles orthogonal to this will lack the subscript $_4$, hence

Theorem 16.] *The circles of a linear hypercomplex orthogonal to an arbitrary sphere will generate a linear complex.*

Are there any exceptions to theorems 10] and 16]? For an exception to 10] we must have in (23) $z_n = \rho y_n$. If $\rho \neq 0$, $(yy) = 0$, and the sphere is null. But on a null sphere there are but ∞^2 circles orthogonal thereto, so that we have no exception. In the second case (y) is the central sphere, and this will constitute the only exception to 10] when there is a central sphere. When the hypercomplex consists in circles in involution with a given circle every sphere orthogonal thereto will be exceptional. As for 16] if all circles orthogonal to any sphere belonged to a linear hypercomplex, we might choose this (when not null) (0, 0, 0, 0, 1); in the equation of the hypercomplex there would be no terms except those involving the subscript $_4$, and the hypercomplex would be without a central sphere. By continuity this will hold even when the sphere in question is null. When the hypercomplex consists in circles in involution with a fixed circle, every sphere through the fixed circle will be an exception to 16].

We next look for a canonical form for the equation of our hypercomplex. The hypercomplex shall be said to be *general* if

$$J(I^2-J) \neq 0.$$

Whenever the first factor does not vanish there is a not null central sphere which we may take as $x_4 = 0$. We have then

$$\Omega_0(aa) = \Omega_1(aa) = \Omega_2(aa) = \Omega_3(aa) = 0, \quad \Omega_4(aa) \neq 0.$$

$$a_{01}p_{01} + a_{02}p_{02} + a_{03}p_{03} + a_{23}p_{23} + a_{31}p_{31} + a_{12}p_{12} = 0.$$

This equation is independent of p_{i4}.

Theorem 17.] *When the central sphere of a linear hypercomplex is not null, the system consists in the circles meeting this in the same pairs of points as a linear complex of circles orthogonal thereto.*

Theorem 18.] *The assemblage of all circles meeting a not null sphere in the same pairs of points as do the circles of a linear complex orthogonal thereto will be a linear hypercomplex with this as central sphere.**

If we confine ourselves to what we have described as the general linear hypercomplex, we may, as in Ch. XII, reduce to the canonical form

$$a_{01}\,p_{01} + a_{23}\,p_{23} = 0. \qquad (25)$$

Here 11] suffers no exception except for two points on the central sphere, and on a circle of the system.

* Cf. Cosserat, 'Le cercle comme élément générateur de l'espace', *Annales de la Faculté des Sciences de Toulouse*, vol. iii, 1889, p. E. 56.

The axial circles of the linear complex orthogonal to the central sphere shall be called the *axial circles* of the hypercomplex, their cross its *axial cross*. Their coordinates are here

$$\alpha_{01} = 1, \quad \alpha_{ij} = 0; \quad \alpha_{23}' = 1, \quad \alpha_{ij}' = 0.$$

We have thus

$$\frac{I(\alpha p)}{J(\alpha' p)} = \frac{p_{01}}{p_{23}}.$$

For a circle of our hypercomplex this becomes

$$\frac{I(\alpha p)}{I(\alpha' p)} = \frac{\sqrt{I - \sqrt{I^2 - J}}}{\sqrt{I + \sqrt{I^2 - J}}} = \frac{I - \sqrt{I^2 - J}}{\sqrt{J}}. \qquad (26)$$

Theorem 19.] *The ratio of the cosines of the angles which each circle of a general linear complex makes with the two circles of the axial cross is constant, and, conversely, the assemblage of all circles such that the cosines of their angles with the two circles of a proper cross have a constant finite ratio different from zero or unity is a linear hypercomplex with the given cross as axial cross.**

This theorem may be somewhat generalized. Let (q) be any circle orthogonal to the central sphere, and (q') its polar in the linear hypercomplex (25); we see, by XII. (17), that if (p) belong to our hypercomplex,

$$\frac{I(qp)}{I(q'p)} = \text{const.} \qquad (27)$$

Theorem 20.] *The cosines of the angles of the circles of a general linear hypercomplex with any two circles orthogonal to the central sphere and mutually polar in the linear complex of circles of the hypercomplex orthogonal to this sphere will have a constant ratio.*

Let us turn aside for an instant to look at special types of the linear hypercomplex. If

$$I = 0, \quad J \neq 0, \quad I(\alpha p) = \pm i I(\alpha' p).$$

* Cf. the Author's *Circle Cross*, cit., p. 165.

If $I^2 - J = 0$ the circles of the axial cross coalesce in a circle which is in bi-involution with itself, i.e. an isotropic.

Reverting to the general case, we rewrite the canonical equation

$$a_{01} p_{01} + a_{23} p_{23} = 0. \tag{25}$$

Let us take an arbitrary sphere (y). The spheres through the circles of the axial cross which are orthogonal to (y) are

$$(y_1 - y_0\, 0\, 0\, 0)\ (0\, 0\, y_3 - y_2\, 0).$$

The circles of the hypercomplex on the sphere (y) are orthogonal to its intersection with (z) where

$$z_0 = a_{01} y_1,\quad z_1 = - a_{01} y_0,\quad z_2 = a_{23} y_3,\quad z_3 = - a_{23} y_2,\quad z_4 = 0.$$

This sphere is a linear combination of the two spheres above, orthogonal to (y). The ratio of the cosines of the angles of (z) with these last two spheres is found to be

$$\frac{a_{01}\sqrt{y_0^2 + y_1^2}}{a_{23}\sqrt{y_2^2 + y_3^2}},\ \text{hence}$$

Theorem 21.] *In a general linear hypercomplex the circles on an arbitrary sphere are orthogonal to a circle coaxal with the orthogonal projections of the circles of the axial cross thereon, while the cosines of its angles therewith have a ratio which is a constant multiple of the ratio of the cosines of the angles which the circles of the axial cross make with the given sphere.*

This construction for the hypercomplex is simplified if we limit ourselves to those spheres where the ratio of the angles with the circles of the axial cross is

$$\frac{ia_{23}}{a_{01}} = \frac{I - \sqrt{I^2 - J}}{\sqrt{J}}. \tag{28}$$

Here the circles of the complex will be orthogonal to a circle of antisimilitude of the orthogonal projections of the circles of the axial cross. If $I = 0$ we have

Theorem 22.] *A general linear hypercomplex where $I = 0$ is composed of the circles on each sphere tangent to the central* *sphere, orthogonal to one circle of antisimilitude of the orthogonal projections of the circles of the axial cross thereon.*

Suppose that we have a spherical transformation. There is necessarily one fixed sphere which will not be usually null. On this sphere there will be two pairs of fixed points, lying two by two on four isotropics. If we choose our fixed sphere as $x_4 = 0$, and determine our fixed points by the pairs of spheres $x_0 = x_1 = 0$, $x_2 = x_3 = 0$, our transformation, if direct, may be written

$$x_0' = x_0 \cos\theta - x_1 \sin\theta,$$

$$x_1' = x_0 \sin\theta + x_1 \cos\theta,$$

$$x_2' = x_2 \cos\phi - x_3 \sin\phi,$$

$$x_3' = x_2 \sin\phi + x_3 \cos\phi.$$

For an infinitesimal transformation we may write

$$dx_0 = -x_1\, d\theta,$$

$$dx_1 = \quad x_0\, d\theta,$$

$$dx_2 = -x_3\, d\phi,$$

$$dx_3 = x_2\, d\phi.$$

Let (y) be a sphere orthogonal to (x) and $(x+dx)$.

$$(yx) = (ydx) = d\theta\,(x_0 y_1 - x_1 y_0) + d\phi\,(x_2 y_3 - x_3 y_2) = 0.$$

Remembering, lastly, that an infinitesimal transformation must be direct:

Theorem 23.] *If an infinitesimal spherical transformation leave invariant a not null sphere and two distinct isotropics of each set thereon, then the circles on each sphere in space orthogonal to the circle of intersection with the transformed sphere will generate a linear hypercomplex with the given fixed sphere as central sphere, while the axial circles meet this sphere in fixed points for the transformation.*

The general linear hypercomplex contains certain rational systems of circles which are worth mentioning. Let us first define as a *series* of circles a system whose coordinates are proportional to analytic functions of one variable, their ratios being not all constants. When the coordinates are proportional to analytic functions of two variables, the ratios not being functions of one variable, the system shall be called a *congruence*. When the coordinates are proportional to analytic functions of three variables, their ratios not being all functions of two variables, the system is called a *complex*; when they are functions of four variables, their ratios not being functions of three, a *hypercongruence*. We thus find:*

Theorem 24.] *The characteristic circles of every rational series of spheres of order less than six will be contained in a linear hypercomplex.*

Theorem 25.] *Given two rational series of spheres of order less than three; there is a linear hypercomplex containing the circle of intersection of every sphere of one series with every sphere of the other.*

Theorem 26.] *The complex of characteristic circles of every rational congruence of spheres of order less than three lies in a linear hypercomplex.*

The next type of circle system to engage our attention is the hypercongruence. We turn especially to the linear hypercongruence given by

$$\Sigma a_{ij}p_{ij} = \Sigma b_{ij}p_{ij} = 0. \tag{29}$$

The circles of the hypercongruence are common to all linear hypercongruences of the pencil

$$\lambda\,(a) + \mu\,(b).$$

To find the central spheres we write

$$\Omega_i\,(cc) = \lambda^2\Omega_i\,(aa) + 2\lambda\mu\Omega_i\,(ab) + \mu^2\Omega_i\,(bb). \tag{30}$$

Suppose first

$$\rho\Omega_i\,(aa) \equiv \Omega_i\,(bb).$$

* Cf. Mesuret, 'Sur les propriétés infinitésimales des systèmes linéaires de cercles', *Comptes Rendus*, vol. cxxxvi, 1903.

We have a transformation of the twenty-four-parameter linear sphere group which will carry this into $x_4 = 0$. Equations (29) will lack all terms in p_{04}.

Theorem 27.] *If two linear hypercomplexes have the same central sphere, this is also the central sphere of every linear hypercomplex linearly dependent on them. The hypercongruence common to the two hypercomplexes consists in the totality of all circles in involution with two circles, usually distinct, orthogonal to this sphere.*

Theorem 28.] *If two linear hypercomplexes consist in the totality of circles in involution with two non-cospherical circles the pencil of linear hypercomplexes determined by them have all one central sphere, namely, that orthogonal to the two circles; when the circles are cospherical, the hypercomplexes of the pencil consist in the systems of circles in involution with the circles of the coaxal system determined by the given circles.*

Theorem 29.] *If one linear hypercomplex consist in circles in involution with a fixed circle, while a second hypercomplex has a central sphere, the central spheres of their pencil trace a coaxal system.*

Theorem 30.] *If two linear hypercomplexes have different central spheres, the central spheres of their pencil generate a conic series.*

In this case, and this alone, we shall say that the hypercongruence is *general.*

Theorem 31.] *The circles of a linear hypercongruence lying on an arbitrary sphere generate a coaxal system.*

The circles of the hypercongruence on a sphere (y) are those orthogonal to the sphere (z) in (23), and to a second such sphere determined by another hypercomplex of the pencil. It is conceivable that the two spheres orthogonal to (y) should coalesce. Here we should have

$$\lambda \sum_{n=0}^{n=4} a_{in} y_n + \mu \sum_{n=0}^{n=4} b_{in} y_n \equiv 0, \quad i = 0 \ldots 4.$$

These, however, are the equations to determine the central sphere of $\lambda(a)+\mu(b)$.

Theorem 32.] *The only spheres which contain more than a coaxal system of circles of a general linear hypercongruence are the central spheres of the corresponding pencil of linear hypercomplexes.**

Theorem 33.] *Through an arbitrary point-pair will pass but one circle of a linear hypercongruence.*

Theorem 34.] *The circles of a linear hypercongruence orthogonal to an arbitrary sphere generate a linear hypercongruence.*

We leave to the reader the task of noting exceptions to the last few theorems. Let us note that it occasionally happens that a linear hypercongruence splits into two parts. For instance, suppose that we have

$$p_{01}=p_{02}=0.$$

Here, since $\Omega_3(pp)=\Omega_4(pp)=0$, we must either have $p_{03}=p_{04}=0$ or else $p_{12}=0$. In the first case we have the circles orthogonal to a fixed sphere, in the second those cospherical with a fixed circle. We note that in this case

$$\Omega_i(aa)=\Omega_i(ab)=\Omega_i(bb)=0, \quad i=0\ldots 4.$$

Suppose, conversely, that we have

$$\Omega_i(ab)\equiv 0, \quad i=0\ldots 4.$$

From (2) we easily find

$$\sum_{n=0}^{n=4}(\lambda a_{in}+\mu b_{in})(\lambda^2\Omega_n(aa)+2\lambda\mu\Omega_n(ab)+\mu^2\Omega_n(bb))\equiv 0, \qquad i=0\ldots 4.$$

$$\sum_{n=0}^{n=4}a_{in}\Omega_n(bb)\equiv\sum_{n=0}^{n=4}b_{in}\Omega_n(aa).$$

* See an interesting article by Castelnuovo, 'Ricerche nella geometria della retta nello spazio a quattro dimensioni', *Atti del R. Istituto Veneto*, Series 7, vol. ii, Part I, 1890.

Then either

$$\Omega_n(aa) \equiv \rho\,\Omega_n(bb), \quad n = 0 \dots 4.$$

In this case there is but one central sphere for all hypercomplexes of the pencil, or else

$$\Omega_n(aa) = \Omega_n(bb) \equiv 0, \quad n = 0 \dots 4.$$

The hypercongruence consists in circles in involution with the circles of a coaxal system, i.e. either orthogonal to their sphere or cospherical with the circle whose foci are the points common to the coaxal circles.

Two linear hypercomplexes have a simultaneous invariant under the group of spherical transformations which is of interest, namely, the cosine of the angle of their central spheres.

$$\cos\theta = \frac{\sum\limits_{n=0}^{n=4} \Omega_n(aa)\,\Omega_n(bb)}{\sqrt{J}\,\sqrt{J'}}.$$

Among complexes of circles the simplest is the linear one, defined by three equations of the type *

$$\Sigma a_{ij}\,p_{ij} = \Sigma b_{ij}\,p_{ij} = \Sigma c_{ij}\,p_{ij} = 0. \tag{31}$$

The study of this figure is clearly closely connected with that of the linear hypercomplexes linearly dependent on three given hypercomplexes. The complex shall be said to be *general* when no two hypercomplexes of the net have the same central sphere. Remembering that one central sphere can be taken as $x_4 = 0$,

Theorem 35.] *If three linear hypercomplexes have the same central sphere, that is, the central sphere for every linear hypercomplex, linearly dependent on them; if, further, the three be linearly independent, the corresponding linear complex consists in the totality of circles in involution with those of one generation of a cyclide.*

Theorem 36.] *If two linear hypercomplexes have the same central sphere, and a third have a central sphere different*

* Many writers, as Castelnuovo, loc. cit., use the term 'linear complex' for that which we have called 'linear hypercomplex'.

from theirs, the central spheres of the net determined by them will generate a conic series.

Theorem 37.] *An arbitrary sphere will contain but one circle of a general linear complex. The only exceptions to this rule are the central spheres of the linear hypercomplexes of the corresponding net.*

Theorem 38.] *The circles of a linear hypercomplex orthogonal to an arbitrary sphere generate a cyclide.*

This theorem will suffer no exceptions in the general case. Let us look for the equation of the surface enveloped by the central spheres of the net in the general case. If we write

$$y_i = \lambda^2 \Omega_i(aa) + \mu^2 \Omega_i(bb) + \nu^2 \Omega_i(cc) + 2\mu\nu\Omega_i(bc) + 2\nu\lambda\Omega_i(ca) + 2\lambda\mu\Omega_i(ab), \tag{32}$$

we must have

$$\left(x\frac{\partial y}{\partial\lambda}\right) = \left(x\frac{\partial y}{\partial\mu}\right) = \left(x\frac{\partial y}{\partial\nu}\right) = 0.$$

$$\begin{vmatrix} (\Omega(aa)x) & (\Omega(ab)x) & (\Omega(ac)x) \\ (\Omega(ba)x) & (\Omega(bb)x) & (\Omega(bc)x) \\ (\Omega(ca)x) & (\Omega(cb)x) & (\Omega(cc)x) \end{vmatrix} = 0.$$

Theorem 39.] *The central spheres of the linear hypercomplexes which contain a general linear complex envelop a surface of the sixth order which, in cartesian space, has the circle at infinity as a triple curve.*

It will be convenient to call this congruence of central spheres the *central congruence.* Each sphere of the congruence contains a coaxal system of circles of the complex. It will pay to investigate this congruence with some care.*

If (y) be a sphere of the central congruence, we must have five consistent equations.

$$l\sum_{j=0}^{j=4} a_{ij}y_j + m\sum_{j=0}^{j=4} b_{ij}y_j + n\sum_{j=0}^{j=4} c_{ij}y_j = 0, \quad i = 0 \ldots 4. \tag{33}$$

* Cosserat, loc. cit., slights the linear complex surprisingly. The following discussion is based upon Castelnuovo, loc. cit., pp. 867 ff.

The following matrix must therefore have a rank not exceeding two:

$$\left\| \begin{matrix} \sum_{j=0}^{j=4} a_{0j} y_j & \cdot & \cdot & \cdot & \cdot & \cdot & \cdot & \sum_{j=0}^{j=4} a_{4j} y_j \\ \sum_{j=0}^{j=4} b_{0j} y_j & \cdot & \cdot & \cdot & \cdot & \cdot & \cdot & \sum_{j=0}^{j=4} b_{4j} y_j \\ \sum_{j=0}^{j=4} c_{0j} y_j & \cdot & \cdot & \cdot & \cdot & \cdot & \cdot & \sum_{j=0}^{j=4} c_{4j} y_j \end{matrix} \right\|.$$

Again, the spheres of our congruence appear in (32) as depending upon the homogeneous parameters $\lambda : \mu : \nu$, and may be represented by the points of a plane or the points of a rational surface of the fourth order in four-dimensional projective space. The spheres of the congruence orthogonal to an arbitrary sphere will be represented by the points of a conic or by the points of a space quartic in one to one correspondence with the conic, and so, rational, yet without a double point. Such a curve being projected from every one of its points by a rational cubic cone, has one trisecant (line cutting it three times) through each point. Every such trisecant meets the quartic three times, every line meeting the quartic three times lies in ∞^2 hyperplanes of S_4, and so is a trisecant of ∞^2 quartics. The trisecants in any hyperplane will generate an irreducible ruled surface, hence they form an irreducible complex. Let us show that they correspond to the given linear complex of circles. Let us take an arbitrary circle of this complex. We may assume that it is orthogonal to $x_4 = 0$.

$$\sum_{i,j=0}^{i,j=3} a_{ij}\, p_{ij} = \sum_{i,j=0}^{i,j=3} b_{ij}\, p_{ij} = \sum_{i,j=0}^{i,j=3} c_{ij}\, p_{ij} = 0.$$

Let this circle be determined by (x) and (z), while (y) is a sphere of the central congruence through it.

$$y_i = \lambda x_i + \mu z_i, \quad a_{ji} = -a_{ij}, \text{ \&c.}$$

$$z_0 \sum_{j=0}^{j=3} a_{0j} x_j + z_1 \sum_{j=0}^{j=3} a_{1j} x_j + z_2 \sum_{j=0}^{j=3} a_{2j} x_j + z_3 \sum_{j=0}^{j=3} a_{3j} x_j = 0.$$

$$z_0 \sum_{j=0}^{j=3} b_{0j} x_j + z_1 \sum_{j=0}^{j=3} b_{1j} x_j + z_2 \sum_{j=0}^{j=3} b_{2j} x_j + z_3 \sum_{j=0}^{j=3} b_{3j} x_j = 0.$$

$$z_0 \sum_{j=0}^{j=3} c_{0j} x_j + z_1 \sum_{j=0}^{j=3} c_{1j} x_j + z_2 \sum_{j=0}^{j=3} c_{2j} x_j + z_3 \sum_{j=0}^{j=3} c_{3j} x_j = 0.$$

But

$$z_0 \sum_{j=0}^{j=3} a_{0j} z_j + z_1 \sum_{j=0}^{j=3} a_{1j} z_j + z_2 \sum_{j=0}^{j=3} a_{2j} z_j + z_3 \sum_{j=0}^{j=3} a_{3j} z_j = 0, \text{ \&c.}$$

Hence

$$\sum_{i,\,j=0}^{i,\,j=3} a_{ij} z_i (\lambda x_j + \mu z_j) = \sum_{i,\,j=0}^{i,\,j=3} b_{ij} z_i (\lambda x_j + \mu z_j)$$

$$= \sum_{i,\,j=0}^{i,\,j=3} c_{ij} z_i (\lambda x_j + \mu z_j) = 0.$$

Substituting for (y) in (33), we get four equations:

$$l \sum_{j=0}^{j=3} a_{ij} (\lambda x_j + \mu z_j) + m \sum_{j=0}^{j=3} b_{ij} (\lambda x_j + \mu z_j)$$

$$+ n \sum_{j=0}^{j=} c_{ij} (\lambda x_j + \mu z_j) = 0, \quad i = 0 \ldots 3.$$

Every solution of three of these is a solution of the fourth by the equations immediately preceding. Hence we may eliminate $l : m : n$ from three, getting a cubic in λ/μ, or three central spheres go through our circle.

Theorem 40.] *The general linear complex is generated by circles which lie on sets of three spheres of a rational quartic congruence. The congruence will be the central congruence of the complex.*

Certain special forms of the linear complex are worth particular notice. Consider the five equations

$$\lambda^2\Omega_i(aa)+\mu^2\Omega_i(bb)+\nu^2\Omega_i(cc)+2\mu\nu\,\Omega_i(bc)+2\nu\lambda\,\Omega_i(ca) \\ +2\lambda\mu\,\Omega_i(bb)=0.$$

If these equations have a common solution, then all circles of the complex are in involution with a fixed circle. The congruence of spheres orthogonal to this will be a part of the central congruence. A second linear congruence of spheres will split off when the equations have two common solutions. If they have three common solutions the congruence is defined by the circles in involution with three fixed circles. If no two of these are cospherical, and the three are not orthogonal to any common sphere, the central congruence will include all spheres orthogonal to any one of these three circles or to the circle cospherical with all three. We see, in fact, that in four dimensions we wish to find those points through which pass a pencil of lines intersecting each of three planes in general position. The locus will be the three planes and the plane lying with each in a hyperplane.

It is conceivable that four linear hypercomplexes of our net might consist in circles in involution with fixed circles. If three of the fixed circles were (a), (b), (c), the matrix

$$\left\|\begin{matrix} \Omega_0(bc) & \Omega_1(bc) & \Omega_2(bc) & \Omega_3(bc) & \Omega_4(bc) \\ \Omega_0(ca) & \Omega_1(ca) & \Omega_2(ca) & \Omega_3(ca) & \Omega_4(ca) \\ \Omega_0(ab) & \Omega_1(ab) & \Omega_2(ab) & \Omega_3(ab) & \Omega_4(ab) \end{matrix}\right\|$$

would have a rank of two or less. If no row vanish identically, and no two were proportional, the spheres orthogonal to the pairs of our three circles would be coaxal, and there would be a circle in bi-involution with all three. Conversely, when there is such a circle this matrix will have a rank of two or less. But if three circles be in bi-involution with a fourth, every circle cospherical with the latter is in involution with the three former, and we have a hypercongruence, not a complex of circles.

If two rows of the matrix be proportional, but no two identically zero, the three circles are orthogonal to one sphere. If we take this as $x_4 = 0$, the five equations above reduce to

$$\lambda\mu\,\Omega_4(ab) + \mu\nu\,\Omega_4(bc) + \nu\lambda\,\Omega_4(ca) = 0.$$

The circles of our complex are in involution with all those of a series. If all the elements of one row vanish, the circles of our complex will be in involution with the circles of a coaxal system and with one other circle. If all the members of two rows vanish, the circles of the complex are in involution with those of two coaxal systems with a common circle. If the matrix have rank zero, we have no complex, but a hypercongruence.

We next turn to the linear congruence, characterized by the equation

$$\Sigma\, a_{ij}\, p_{ij} = \Sigma\, b_{ij}\, p_{ij} = \Sigma\, c_{ij}\, p_{ij} = \Sigma\, d_{ij}\, p_{ij} = 0. \qquad (34)$$

If we write the conditions that a linear hypercomplex linearly dependent on these four should consist in circles in involution with a given circle, we find exactly the same equations which we encountered in our first analytic treatment of the pentacycle. Defining as *general* the linear congruence where these have five distinct roots, we see

Theorem 41.] *The general linear congruence consists in the circles in involution with those of a pentacycle.**

Theorem 42.] *Two circles of a linear congruence are orthogonal to an arbitrary sphere.*

It appears from this that two circles of the congruence will pass through an arbitrary point. Assuming a theorem to be proved in the last chapter, namely, that the circles of a congruence are, in general, tangent to four surfaces, we see that the circles through a point fall together when it is a focal

* Cf. Koenigs, *Contributions à la théorie*, cit., p. F. 11.

point, i.e. a point of one of these focal surfaces. We find these surfaces by seeking the spheres such that the circles of the congruence orthogonal to them coalesce. The equation of this complex of spheres is

$$\begin{vmatrix} (\Omega\,(aa)\,x) & (\Omega\,(ab)\,x) & (\Omega\,(ac)\,x) & (\Omega\,(ad)\,x) \\ (\Omega\,(ba)\,x) & (\Omega\,(bb)\,x) & (\Omega\,(bc)\,x) & (\Omega\,(bd)\,x) \\ (\Omega\,(ca)\,x) & (\Omega\,(cb)\,x) & (\Omega\,(cc)\,x) & (\Omega\,(cd)\,x) \\ (\Omega\,(da)\,x) & (\Omega\,(db)\,x) & (\Omega\,(dc)\,x) & (\Omega\,(dd)\,x) \end{vmatrix} = 0. \qquad (35)$$

We see, in fact, that this is a covariant form, and when we make a linear transformation of the twenty-four parameter group to carry this sphere (x) to $x_4 = 0$, this equation becomes, in the notation of the last chapter,

$$\begin{vmatrix} (a/a) & (a/b) & (a/c) & (a/d) \\ (b/a) & (b/b) & (b/c) & (b/d) \\ (c/a) & (c/b) & (c/c) & (c/d) \\ (d/a) & (d/b) & (d/c) & (d/d) \end{vmatrix} = 0.$$

This is the necessary and sufficient condition that four linear complexes of circles orthogonal to a sphere should have but one common circle.*

Theorem 43.] *The focal surface of the general linear congruence is of the eighth order and, in cartesian space, has the circle at infinity as a quadruple curve.*

This theorem will admit of special cases when the circles of the pentacycle have special positions. Let us pass over to the consideration of the surface of foci of the circles of the congruence. It will simplify our reckoning if we assume that (a), (b), (c), (d) are circles of the pentacycle. The complex of central spheres will be parametrically expressed by the equations

$$y_i = \lambda\mu\,\Omega_i\,(ab) + \lambda\nu\,\Omega_i\,(ac) + \lambda\rho\,\Omega_i\,(ad) + \mu\nu\,\Omega_i\,(bc) + \mu\rho\,\Omega_i\,(bd) + \nu\rho\,\Omega_i\,(cd). \qquad (36)$$

* Cf. Jessop, *Line Complex*, cit., p. 72.

It is to be noted also that the central spheres are the only ones on which lie circles of the linear congruence. The foci are thus the vertices of the null spheres of this complex. The spheres of this complex are in one to one correspondence with the points of a three-dimensional projective space. The points (1 0 0 0), (0 1 0 0), (0 0 1 0), (0 0 0 1) and one other are exceptional. They correspond to linear hypercomplexes with no central sphere, or rather each has a linear congruence of central spheres orthogonal to a circle of the pentacycle. The vertices of the null spheres of such a congruence are the points of the circle of the pentacycle. The five circles of the pentacycle are thus imbedded in the surface of foci.

Again, if we write the equations

$$(qy) = (ry) = (sy) = 0,$$

we have three quadrics in our three-dimensional space with eight intersections. Five of these must be rejected as they correspond to circles of the pentacycle; there will remain but three, so that the complex of central spheres is of the third order, and has an algebraic equation

$$f^3(x) = 0.$$

The surface of foci is thus a surface of the sixth order.

Suppose, now, that (a') is the circle cospherical with (b), (c), and (d). The spheres through (a) cut (a') in pairs of points which are foci of circles in involution with (a), (b), (c), (d), so that (a') lies on the surface we seek. We thus find another excellent theorem due to Stephanos: *

Theorem 44.] *Fifteen associated circles lie on a surface of the sixth order which is the surface of foci of the six linear congruences of circles in involution with those of the six pentacycles of the associated system.*

This is, of course, the surface of the sixth order previously found.

In the three-dimensional projective space five points correspond to linear congruences of spheres orthogonal to the

* *Quinze cercles*, cit., p. 634.

circles of the pentacycle. Let us call these the notable points. For a point in a plane through three notable points we have such equations as

$$\rho = 0, \quad y_i = \lambda\mu\Omega_i(ab) + \lambda\nu\Omega_i(ac) + \mu\nu\Omega_i(bc).$$

If these notable points correspond to the circles (a), (b), (c), since (d') is cospherical with them, it is orthogonal to the common orthogonal spheres which they determine two by two (cf. XII, theorem 2), so that

$$\sum_{n=0}^{n=4} d_{in}' y_n = 0, \quad i = 0 \ldots 4.$$

Hence these planes will correspond to the congruences of circles orthogonal to the other ten circles of the associated system.

Again, consider a line joining two notable points, say $\nu = \rho = 0$. All points of this line will correspond to a single central sphere, $z_i = \Omega_i(ab)$. It will be a double sphere in our variety. We see, in fact, that the intersections of the variety with a coaxal sphere system including such a sphere will correspond to the intersections of three quadrics through a line of this type. Three such quadrics have but four other common points, whereof three, in the present case, are notable. Every line of this sort contains two notable points, lies in three planes through three notable points, and intersects one plane through three such points, hence

Theorem 45.] *The pairs of foci of fifteen associated circles lie by sixes on ten spheres, four of which pass through each pair of foci. These are the double spheres of the ten cubic complexes of central spheres corresponding to the six pentacycles of the associated system.*

Each double sphere is orthogonal to two circles of our pentacycle; there is thereon a coaxal system of circles in involution with these two and with two other circles of the pentacycle.

Theorem 46.] *Each of the ten double spheres of the complex of central spheres contains a coaxal system of circles of the linear congruence.**

There is almost a transfinite number of special varieties of the linear congruence. The essential facts connected with some are obtained by remembering that such a congruence corresponds to the lines in S_4, which intersect four planes. We leave to the reader the proofs of the following theorems which are easily reached in this way.

Theorem 47.] *The circles in involution with four circles whereof two, and only two, are cospherical, while no three are orthogonal to one sphere, form a reducible congruence composed of a linear congruence of circles orthogonal to the sphere of the two circles, and a congruence of circles cospherical with a fixed circle. The complex of central spheres reduces to that of spheres orthogonal to the sphere of the two circles, and a quadratic complex.*

Theorem 48.] *The congruence of circles in involution with four circles whereof three are orthogonal to one sphere, while no two are cospherical, consists in a linear congruence of circles on the given sphere.*

Theorem 49.] *The circles in involution with four circles which are cospherical in two pairs, but no three orthogonal to one sphere, will consist in a linear congruence orthogonal to each of the given spheres, and the congruence of circles cospherical with the two circles on each of which lie the foci of two of the given circles.*

The last linear family of circles which we shall consider is the *linear series*. This is characterized by such equations as

$$\Sigma\, a_{ij}p_{ij} = \Sigma\, b_{ij}p_{ij} = \Sigma\, c_{ij}p_{ij} = \Sigma\, d_{ij}p_{ij} = \Sigma\, e_{ij}p_{ij} = 0. \quad (37)$$

* This cubic complex of spheres has been quite thoroughly studied. Cf. e.g. Segre, 'Sulla varietà cubica con dieci punti doppii', *Atti della R. Accademia delle Scienze di Torino*, vol. xxii, 1887.

A circle (p) of the series is orthogonal to $x_4 = 0$ if the five equations

$$(a/p) = (b/p) = (c/p) = (d/p) = (e/p) = 0$$

have a common solution which is also a solution of $(p/p) = 0$.

A necessary and sufficient condition for this is

$$\begin{vmatrix} (a/a) & (a/b) & (a/c) & (a/d) & (a/e) \\ (b/a) & (b/b) & (b/c) & (b/d) & (b/e) \\ (c/a) & (c/b) & (c/c) & (c/d) & (c/e) \\ (d/a) & (d/b) & (d/c) & (d/d) & (d/e) \\ (e/a) & (e/b) & (e/c) & (e/d) & (e/e) \end{vmatrix} = 0.$$

This, in turn, may be written in the covariant form

$$\begin{vmatrix} (\Omega(aa)x) & (\Omega(ab)x) & (\Omega(ac)x) & (\Omega(ad)x) & (\Omega(ae)x) \\ (\Omega(ba)x) & (\Omega(bb)x) & (\Omega(bc)x) & (\Omega(bd)x) & (\Omega(be)x) \\ (\Omega(ca)x) & (\Omega(cb)x) & (\Omega(cc)x) & (\Omega(cd)x) & (\Omega(ce)x) \\ (\Omega(da)x) & (\Omega(db)x) & (\Omega(dc)x) & (\Omega(dd)x) & (\Omega(de)x) \\ (\Omega(ea)x) & (\Omega(eb)x) & (\Omega(ec)x) & (\Omega(ed)x) & (\Omega(ee)x) \end{vmatrix} = 0. \quad (38)$$

Theorem 50.] *The circles of a linear series generate a surface of the tenth order which, in cartesian space, has the circle at infinity as a quintuple curve.*

The coordinates of the circles of our series satisfy five linear equations, and are linearly dependent on those of five of their number. Exactly the same is true of the coordinates of the circles in involution with all those of the given series.

Theorem 51.] *The circles of a linear series are in involution with those of a second such series. The relation between the two is reciprocal. Each series will contain every pentacycle whereof it contains four circles.*

Our linear series will correspond to a ruled surface of the tenth order in S_4. In the series of planes which correspond to the foci of the circles of our series ten members will intersect any line. Consider any point on the curve generated by these foci. A circle through it is always cospherical with the other focus, mate to the given one. There will also be two circles through this point in involution with four given circles, as we see from 42].

Theorem 52.] *The foci of the circles of a linear series trace a curve of the tenth order which is a double curve of the surface traced by the linear series of circles in involution with the given ones.**

Suppose that we have six linear equations

$$\Sigma a_{ij} p_{ij} = \Sigma b_{ij} p_{ij} = \Sigma c_{ij} p_{ij} = \Sigma d_{ij} p_{ij} = \Sigma e_{ij} p_{ij} = \Sigma f_{ij} p_{ij} = 0.$$

All solutions of these will be dependent linearly on four such, whence

Theorem 53.] *Six circles of which each four determine a pentacycle not including either of the other two are in involution with the circles of a pentacycle and with no others.*†

§ 3. Other Simple Systems.

Enough has now been said about linear systems of circles in space. We turn to certain other algebraic systems of almost equal simplicity. The first of these shall be the complex of circles lying on the spheres of a general quadratic complex.‡ We shall mean by the *order* of a complex of circles the number of circles on an arbitrary sphere, the linear complex being thus of order one.§

Theorem 54.] *The complex of circles lying on pencils of spheres of a general quadratic complex is of order zero. On*

* Cosserat, loc. cit., p. E. 76.

† Richmond, loc. cit., and Weitzenböck, loc. cit., reach the five-line figure which corresponds to the pentacycle originally in this way.

‡ Cf. Fano, 'Sul sistema di rette di una quadrica dello spazio a quattro dimensioni', *Giornale di Matematica*, Series 2, vol. xii, 1905.

§ For a study of complexes of orders one and two see various articles by Marletta, *Rendiconti del Cercolo Matematico di Palermo*, xxviii, 1909, xxxiv, 1912, xxxviii, 1914; *Atti dell' Accademia di Catania*, Series 5, vol. iii, 1910, and vol. vi, 1913; also *Giornale di Matematica*, Series 2, vol. xix, 1912.

each sphere of the complex will lie a conic series of circles of the circle complex.

Other properties of this circle complex are reached by representing it as the totality of lines on an S_3^2 in S_4. If two lines of such a hypersurface intersect they belong to a cone imbedded in the hypersurface; if they do not intersect their hyperplane cuts the hypersurface in a quadric whereof one system of generators includes the two lines.

Theorem 55.] *Two circles of the complex which are cospherical determine a conic series of the complex which includes them; two which are not cospherical determine likewise one generation of a cyclide.*

We shall mean by the *order* of a congruence of circles the number orthogonal to an arbitrary sphere, that is, the order of the complex of spheres through them; the *class* shall be the number cospherical with an arbitrary circle. Our general linear congruence of circles is thus of the second order and third class. A hyperplane in S_4 tangent to S_3^2 meets it in a cone whereof one generator will meet any chosen line of S_3^2. Thus two lines of S_3^2 meet any chosen line thereof and any arbitrary line. A hyperplane in S_4 meets S_3^2 in a quadric, two of whose generators will intersect an arbitrary line of S_3^2. Three arbitrary lines of S_3^2, whereof no two are cospherical, meet one line in space.

Theorem 56.] *The circles lying on pencils of spheres of a general quadratic complex which are cospherical with one of their number generate a congruence of the second order and class. Three circles of the complex, whereof no two are cospherical, belong to one such congruence and only one. Two such congruences will meet in a conic series, or in one generation of a cyclide.*

An interesting algebraic congruence of circles is reached as follows. Suppose that we have three circles, no two of which are cospherical, nor are all three orthogonal to one sphere. The spheres orthogonal to them will form three linear congruences. Let a projective relation be established between the members of these three congruences, the common orthogonal sphere to two of our given circles being in no case self-corresponding. We propose to study the congruence of circles, each orthogonal to three corresponding spheres in the three projective congruences.* In S_4 we have the lines of intersection of corresponding triads in three projective bundles or nets of hyperplanes. An arbitrary hyperplane will cut these three bundles in three projective bundles of planes. Corresponding planes in the three bundles will be concurrent in the points of a general cubic surface. Analytically, if we express these three bundles in $x_4 = 0$ in the form

$$X_i = \lambda Y_i + \mu Z_i + \nu T_i,$$
$$X_i' = \lambda Y_i' + \mu Z_i' + \nu T_i',$$
$$X_i'' = \lambda Y_i'' + \mu Z_i'' + \nu T_i'',$$
$$S_i = \frac{\partial}{\partial R_i} \mid RYZ'T'' \mid = S_i^{(3)} (\lambda, \mu, \nu), \quad i = 0 \ldots 3.$$

If there be a line of the congruence in this hyperplane, three corresponding planes of the three bundles will pass through it. This will require

$$\lambda (pY_i + qY_i' + rY_i'') + \mu (pZ_i + qZ_i' + rZ_i'') + \nu (pT_i + qT_i' + rT_i'') = 0, \quad i = 0 \ldots 3.$$

Taking two sets of three equations from these four and eliminating p, q, r, we get two homogeneous cubic equations in λ, μ, ν with nine common solutions. Three of these must be rejected, for the equations in λ, μ, ν arose from equating to zero the discriminants of two sets of three linear equations, two equations being the same in both sets. It is easy to see that there will be three values of λ, μ, ν, which will make these two equations identical, yet will not correspond to solutions of all four equations. Six lines of the congruence lie in an arbitrary hyperplane.

Theorem 57.] *If a projective relation be established between the members of three linear congruences of spheres with no*

* Cf. Castelnuovo, 'Una congruenza di terzo ordine', *Atti del R. Istituto Veneto*, Series 6, vol. v, 1887.

common member, where no two of the congruences have more than one common member, and that one not self-corresponding, then the circles orthogonal to corresponding triads of spheres will generate a congruence of the third order and sixth class.

We shall call this the *general congruence of the third order and sixth class.* We may express such a congruence analytically in the form

$$\begin{aligned} x_i &= \lambda y_i + \mu z_i + \nu t_i, \\ x_i' &= \lambda y_i' + \mu z_i' + \nu t_i', \\ x_i'' &= \lambda y_i'' + \mu z_i'' + \nu t_i'', \\ \rho p_{ij} &= | x_k x_l' x_m'' |, \quad i = 0 \dots 4. \end{aligned} \tag{39}$$

The congruence is rational. If we connect λ, μ, ν by a homogeneous linear relation, we get a rational cubic series of circles.

Theorem 58.] *If a projective relation be established among the members of three coaxal systems of spheres, no two having a common sphere, nor are all three orthogonal to one same sphere, the circles orthogonal to corresponding triads of spheres will generate a rational cubic series. Three circles of such a series will be in involution with an arbitrary circle.*

A sphere (x) will be orthogonal to our circle (p) if

$$\sum_{n=0}^{n=4} p_{in} x_n = \sum_{n=0}^{n=4} p_{jn} x_n = 0.$$

If the coordinates (p) be homogeneous binary cubic forms, the resultant of these equations will be of the sixth degree in (x).

Theorem 59.] *A rational cubic series of circles will generate a surface of the twelfth order which, in cartesian space, has the circle at infinity as a sextuple curve.*

Let the reader prove the following:

Theorem 60.] *The spheres orthogonal to pairs of adjacent*

circles of a rational cubic series will generate a rational quartic series of spheres. The circles of curvature of the corresponding annular surface are in bi-involution with the corresponding circles of the cubic series, and will generate a rational sextic series.

The rational cubic series of circles will appear in S_4 as a rational ruled surface of the third order, not lying in a hyperplane. If a hyperplane pass through a line which meets three generators of such a surface in three distinct points, this line will be a part of the cubic curve which the hyperplane cuts from the surface. If, further, the hyperplane include two of the generators meeting this line, the cubic curve must consist in these two generators and the line, hence all the generators meet this line. If three such skew generators could not be found the surface would be a cone. This possibility is ruled out, as otherwise the congruence would be composed of ∞^2 cones, an absurdity.

Theorem 61.] *The circles of a rational cubic series are all cospherical with one fixed circle.*

Reverting to our congruence, we see that two sets of parameter values for λ, μ, ν, which are essentially distinct, will be connected by just one linear relation; hence

Theorem 62.] *Any two circles of the general congruence of the third order and sixth class will belong to just one rational cubic series of the congruence.*

The circles which determine our projective linear congruences of spheres do not belong to our given congruence. They have the coordinates

$$\rho q_{ij} = | y_k z_l t_m |, \quad \rho' q_{ij}' = | y_k' z_l' t_m' |, \quad \rho'' q_{ij}'' = | y_k'' z_l'' t_m'' |. \quad (40)$$

We next write

$$\bar{y}_i = l y_i + m y_i' + n y_i'',$$
$$\bar{z}_i = l z_i + m z_i' + n z_i'',$$
$$\bar{t}_i = l t_i + m t_i' + n t_i''.$$

If we replace (y) (z) (t) by $(\bar{y})$ $(\bar{z})$ $(\bar{t})$ in (39) and (40) the coordinates of (p) are unaltered in value, as are those of (q') and (q''), while (q) becomes any circle of a general congruence of the third order and sixth class which includes (q) (q') (q''). Any three circles of this congruence may be taken to replace (q) (q') (q''). Our new congruence is determined by a projective relation among the linear congruences of spheres respectively including to (y) (y') (y''), (z) (z') (z''), and to (t) (t') (t''). The two congruences of circles bear a reciprocal relation to one another (analogous to that of the two systems of generators of a quadric); each shall be said to be *conjugate* to the other.

Theorem 63.] *There is associated with each general circle congruence of the third order and sixth class a conjugate congruence of like structure. Each congruence is composed of circles orthogonal to the corresponding members of three projective linear congruences of spheres each orthogonal to an arbitrary circle of the other congruence, the three not being members of one same rational cubic series.*

The last restriction amounts in the above case to the inequality $l \neq 0$. We see that otherwise (q) would belong to a rational cubic series with (q') and (q'') and $p_{ij} \equiv 0$. It is thus a characteristic feature of three circles belonging to a rational cubic series that spheres orthogonal to them and to the same circle of the conjugate congruence are coaxal, the common circle which we shall presently find to be that appearing in 61]. Let us call this the *directrix* circle.*

Let us take a cubic series of our congruence, and two circles of the conjugate congruence which are cospherical with the directrix circle. In S_4 we have a rational cubic series of lines, which meet a directrix line d, and two other lines l and l', which also meet d. The lines of the first series will determine projective pencils of hyperplanes through l and l'. Corresponding hyperplanes of these pencils intersect in planes through d (which lies in every hyperplane of each pencil)

* Castelnuovo, *Una congruenza*, cit., calls the circles of the conjugate congruence directrix circles, while these are axial circles.

and the various lines of the series. If, now, l'' be any line of the cubic series of the second congruence which includes l and l', it will intersect every plane which includes d and a line of the first series, for the lines of the second series will determine with any line of the first a pencil of hyperplanes through the plane of the first line and d.

Theorem 64.] *Two conjugate general congruences of the third order and sixth class have the same congruence of directrix circles.*

Each circle of our congruence belongs to ∞^1 cubic series and is cospherical with so many directrix circles. Each two belong to one such series and are cospherical with one same directrix circle. We shall prove presently that each two circles of the congruence are cospherical with but one directrix circle. Since a cubic series arises from a linear equation in the homogeneous variables λ, μ, ν, each two of these series have a common circle, or each two directrix circles are cospherical with (at least) one common circle of the given congruence.

We see that in S_4 the cubic hypersurface corresponding to the complex of spheres through our surface will meet a hyperplane in a cubic surface which will usually be non-singular. This surface contains twenty-seven straight lines. Six of these are lines of the congruence in the hyperplane. We next notice that as each directrix lines meets a line of either congruence at each of its points, and through each point of a line of the congruence will pass one directrix line, the directrix lines generate the same hypersurface as the lines of either congruence. Hence of the twenty-seven straight lines six belong to either congruence, and constitute together a double six of Schläfli.* No two lines of the same double six intersect, but each line of one intersects four of the other. The remaining fifteen lines of the twenty-seven intersect the lines of each double six in pairs, and are the lines of the congruence of directrices which are determined by these pairs. This proves

* 'The twenty-seven lines upon a surface of the third order', *Quarterly Journal of Math.*, vol. ii, 1858.

our statement that two circles of our congruence are cospherical with but one same directrix circle.

Theorem 65.] *The congruence of directrix circles is of the third order and fifteenth class.*

There is one more type of algebraic congruence which is capable of complete and simple discussion. We ask this question, 'What are the possible types of irreducible algebraic congruence of the first class?' We must lay stress on the fact that the congruence is supposed to be irreducible, as otherwise we might adjoin to any such congruence any number of congruences of cospherical circles. In S_4 the analogous question is this: 'What types of line congruence are there such that there are a series of lines in every hyperplane which contains two lines of the congruence?'* Such a congruence will be dual to a congruence of planes whereof but one member goes through an arbitrary point. These will meet an arbitrary hyperplane in a congruence of lines of which but one passes through an arbitrary point. This congruence can have no focal surface but a focal curve, or else consist in concurrent lines. There are but four types of such congruence: †

A) Concurrent lines.

B) Lines meeting two skew lines.

C) Secants to a cubic space curve.

D) Lines intersecting a given line and a space curve of order n which meets the given line $n-1$ times.

Working back to the congruence of planes we see that in the first case each two must be coaxal, i. e. all the planes go through one line. In each other case each plane is coaxal

* Circle congruences of the first class have been studied by Pieri, 'Sopra alcune congruenze di coniche', *Atti della R. Accademia delle Scienze di Torino*, vol. xxviii, 1893. His work contains the vicious assumption that if but one circle goes through an arbitrary point it must be a congruence of the first class, without showing that the complex of null spheres might not be the complex of spheres such that the circles orthogonal to them fall together.

† Cf. Kummer, 'Über algebraische Strahlensysteme', *Berliner Akademie, Abhandlungen*, 1866, pp. 8 ff.

with ∞^1 others. There are thus ∞^1 of these axes, and each is coplanar with one other in each plane of the system. Hence all the axes and all the planes pass through a point, or, in the original case, all the lines of the congruence are in a hyperplane.

Theorem 66.] *There are but four types of irreducible algebraic congruence of the first class:*

A) Circles through two points. Order zero.

B) Circles cospherical with two distinct or adjacent non-cospherical circles. Order one.

C) Circles each containing two members of a rational cubic series of point-pairs anallagmatic in a given sphere. Order three.

D) Circles which contain point-pairs of a series of order n anallagmatic in a fixed sphere and point-pairs of a fixed circle containing $n-1$ point-pairs in the series. Order n.

Let us close our discussion of algebraic circle systems by exhibiting a transformation that is rather different from any that we have yet seen.* We saw frequently in the last chapter that if a circle be given orthogonal to a not null sphere, there is just one other circle orthogonal to that same sphere which is in bi-involution with it. This leads us to a new circle transformation as follows. We start with a fundamental not null circle (q) and transform each circle (p) into the circle (p'), which is in bi-involution with it and orthogonal to the common orthogonal sphere to (p) and (q). If (p) be determined by the spheres (x) and (y),

$$\rho p_{ij}' = \begin{vmatrix} x_k & x_l & x_m \\ y_k & y_l & y_m \\ \Omega_k(pq) & \Omega_l(pq) & \Omega_m(pq) \end{vmatrix}$$

$$= p_{lm}\,\Omega_k(pq) + p_{mk}\,\Omega_l(pq) + p_{kl}\,\Omega_m(pq). \qquad (41)$$

Another way of stating this correspondence is to say that (p') is the circle which passes through the foci of (p) and is

* See the Author's *Circle Cross*, cit., pp. 172 ff.

orthogonal to the sphere connecting these foci with those of (q). The transformation is involutory, and the only singular circles are those which are cospherical with (q). The most striking fact about the transformation is that it does not carry spheres to spheres, i.e. cospherical circles are not carried over into cospherical circles. Suppose, in fact, that we have the circle (p) determined by the spheres (x) and (y), and the circle $(\bar{p})$ determined by $(\bar{x})$ and $(\bar{y})$. We find

$$\Omega_m(p'\bar{p}') = \begin{vmatrix} 0 & x_i & x_j & x_k & x_l & x_m \\ 0 & y_i & y_j & y_k & y_l & y_m \\ \bar{x}_m & \bar{x}_i & \bar{x}_j & \bar{x}_k & \bar{x}_l & 0 \\ \bar{y}_m & \bar{y}_i & \bar{y}_j & \bar{y}_k & \bar{y}_l & 0 \\ 0 & \Omega_i(pq) & \Omega_j(pq) & \Omega_k(pq) & \Omega_l(pq) & \Omega_m(pq) \\ \Omega_m(\bar{p}q) & \Omega_i(\bar{p}q) & \Omega_j(\bar{p}q) & \Omega_k(\bar{p}q) & \Omega_l(\bar{p}q) & 0 \end{vmatrix}.$$

Assuming that (p) and $(\bar{p})$ are cospherical so that

$$\Omega_i(p\bar{p}) \equiv 0, \quad i = 0 \dots 4,$$

$$\Omega_m(p'\bar{p}') = \sum_{l=0}^{l=4} [\Omega_l(pq)\,\Omega_m(\bar{p}q) - \Omega_m(pq)\,\Omega_l(\bar{p}q)] \\ [p_{jk}\bar{p}_{mi} + p_{ki}\bar{p}_{mj} + p_{ij}p_{mk}]$$
$$+ \sum_{i,l=0}^{i,l=4} [\Omega_j(pq)\,\Omega_k(\bar{p}q) - \Omega_k(pq)\,\Omega_j(\bar{p}q)] \\ [\bar{p}_{mi}p_{lm} - p_{mi}\bar{p}_{lm}].$$

This vanishes for all values of m and all pairs of cospherical circles where

$$\Omega_i(\bar{p}q) \equiv \Omega_i(pq).$$

Theorem 67.] *The transformation which carries each circle not cospherical with a fundamental not null circle into the circle in bi-involution with it and orthogonal to the sphere orthogonal to the given and fundamental circle, is single valued and involutory. It will carry cospherical circles into cospherical circles when, and only when, they are orthogonal to the same sphere orthogonal to the fundamental circle.*

It will be found that the relations of involution and bi-involution are not invariant, nor is the transformation usually a contact transformation. Suppose that (p) traces a coaxal system including one cospherical with (q), we have

$$p_{ij}' \equiv \lambda a_{ij} - \mu b_{ij}, \quad i, j = 0 \dots 4.$$

When no member of the coaxal system is cospherical with (q)

$$p_{ij}' \equiv \lambda^2 a_{ij} + 2\lambda\mu b_{ij} + \mu^2 c_{ij}, \quad i, j = 0 \dots 4.$$

All of these circles are orthogonal to the sphere traced by the coaxal circles. We need to impose three linear conditions upon a circle orthogonal to this sphere in order to make it cospherical with all circles of the system, hence:

Theorem 68.] *A coaxal system of circles including one member cospherical with the fundamental circle is transformed into another such coaxal system; a coaxal system which includes no member cospherical with the fundamental circles is transformed into one generation of a cyclide anallagmatic in the sphere generated by the coaxal system.*

It would be pleasant if the coaxal system cospherical and orthogonal to the given one transformed into the conjugate generation of the cyclide; such is not, however, the case. We find the transforms of the circles of a sphere $x_4 = 0$ (which, we may assume, bears no special relation to the fundamental circle) as follows. (p') will have (different from zero) only those coordinates which lack the subscript $_4$, for (p) the only non-vanishing coordinates are those which do involve the subscript $_4$. We find also in the notation of the last chapter $(p'/q) = 0$.

Theorem 69.] *The circles of a sphere not containing the fundamental circle non-orthogonal thereto are transformed into the circles orthogonal to that same sphere and cospherical with that circle which is orthogonal to this sphere and meets in the same points as the fundamental circle.*

Consider the circles of the sphere $x_4 = 0$, which are orthogonal to $x_3 = 0$. The only coordinates which are not zero are p_{04}, p_{14}, and p_{24}. The transformed circles are found to generate a linear congruence.

Theorem 70.] *The circles of a linear congruence on a sphere which does not contain the fundamental circle, and is not orthogonal thereto, are transformed into those of a linear congruence orthogonal to the given sphere.*

Suppose that we have a bundle of circles through two points. Their foci will be any pair of points of the circle whose foci are the given points.

Theorem 71.] *The circles through two points not on the fundamental circle will transform into the circles through each pair of points on the circle with the given points as foci, and orthogonal in each case to the sphere through the pair of points and the foci of the fundamental circle.*

Theorem 72.] *The circles orthogonal to a sphere which does not contain the fundamental circle are transformed into the circles through pairs of points of this sphere orthogonal in each case to the sphere through the pair of points and the foci of the fundamental circle.*

Suppose that we have a fixed circle (r) and that (p) is cospherical therewith. The foci of (p) and (r) are concyclic, and conversely, if a pair of points be concyclic with the foci of (r) they are the foci of a circle cospherical with (r). (p) transforms into a circle through the foci thereof and orthogonal to the sphere which connects them with the foci of (q). If we take any sphere through the foci of (q), the pairs of points thereon concyclic with the foci of (r) are mutually inverse in the sphere through (r) orthogonal to the given sphere.

Theorem 73.] *The circles cospherical with a circle which is not cospherical with the fundamental circle transform into circles orthogonal to the spheres orthogonal to the fundamental circle at pairs of points which are mutually inverse in the sphere through the given circle orthogonal to the sphere in question.*

When (r) and (q) are cospherical, every circle through the foci of (r) lies on a sphere through the foci of (q).

Theorem 74.] *The circles cospherical with a circle which is cospherical with the fundamental circle transform into circles in involution with the given circle, and with the circle which contains the foci of the given and the fundamental circle.*

It is superfluous to state that few parts of our subject offer better opportunities for further study than what we have taken up in the present chapter. There is much that has already been done in the line geometry of four dimensions which gives simple and interesting results in circle geometry, but which we have been forced to omit for lack of space. It seems certain that there is still a good deal left in the linear systems beside what we have taken up. The quadratic hypercomplex is an entirely undeveloped field; no researches whatever seem to have been made there so far, and unless it belies entirely the reputation of the analogous quadratic line complex, there must be a large amount of treasure to be unearthed. Then a study of further congruences and complexes of low order and class would seem advisable. Lastly, the involutory transformation just described is after all of very special type. Is there not an interesting general theory of circle transformations which are not sphere transformations? Truly, the harvest is ripe for the reaper.

CHAPTER XIV

ORIENTED CIRCLES IN SPACE

§ 1. Fundamental Relations.

We saw in Chapters X and XI what important changes were introduced into the geometry of circles in the plane and of spheres in space by orienting the figures, that is to say, by attaching a positive or negative sign to the radius. It is the purpose of the present chapter to study the oriented circle in space.* It will appear that the alterations introduced in this case are much less profound than before. The reason for this is that whereas formerly the non-oriented circle in the plane was treated as a point locus and the oriented one as an envelope, and the two had markedly distinct transformation groups, in the present case it seems impossible to treat the oriented circle in space fruitfully as other than a point locus and under the group of spherical transformations.

We start with two points (x) and $(\bar{x})$ of pentaspherical space. The circle with these as foci has the coordinates

$$\rho p_{ij} \equiv x_i \bar{x}_j - x_j \bar{x}_i.$$

This circle will be null if

$$\Sigma p_{ij}^2 \equiv -(x\bar{x})^2 = 0.$$

We next suppose that the circle has been oriented. This may be done for a real circle by attaching thereto a sense of description, or giving a sign to the radius, or, better still, by

* About half of the results given in the present chapter were first worked out by Dr. David Barrow, and presented in his dissertation for the doctorate of philosophy in Harvard University in 1913. Many were subsequently published in a short article, 'Oriented Circles in Space', *Transactions American Math. Soc.*, vol. xvi, 1915.

establishing an order between the two foci, calling one the *first* and the other the *second*, this latter method holding even in the complex domain. We shall therefore define as an *oriented* circle in pentaspherical space a circle between whose foci there has been established a preference. When the rôles of the two foci have been interchanged, the resulting oriented circle is said to be the *opposite* of the original one. A null circle is its own opposite. An oriented circle shall have twenty-one homogeneous coordinates $p_{ij}\ p$, defined as follows: (p_{ij}) shall be the coordinates of the circle not oriented. The first and second foci being (x) and $(\bar{x})$,

$$\rho\, p_{ij} \equiv x_i \bar{x}_j - x_j \bar{x}_i, \quad \rho\, p \equiv (x\bar{x}). \tag{1}$$

These twenty-six coordinates are connected by the sixteen relations

$$p_{ji} \equiv -p_{ij}, \quad \Omega_i(pp) \equiv 0, \quad i = 0 \ldots 4, \quad \sum_{i=0}^{i=4} \Omega_i^{\,2}(pp) + p^2 \equiv 0. \tag{2}$$

Suppose, conversely, that we have twenty-six homogeneous coordinates, not all zero, which satisfy the equations (2). The circle, not oriented, is uniquely determined. The foci are found as follows. If (x) be the first focus, (z), (s), and (t) three points on the circles, we have

$$(xx) = (xz) = (xs) = (xt) = 0.$$

$$\rho x_i = |\, z_j\, s_k\, t_l\, x_m \,| = \sum_{n=0}^{n=4} p_{in}\, x_n.$$

We have five homogeneous linear equations in (x) which are compatible if

$$\begin{vmatrix} -\rho & p_{01} & p_{02} & p_{03} & p_{04} \\ p_{10} & -\rho & p_{12} & p_{13} & p_{14} \\ p_{20} & p_{21} & -\rho & p_{23} & p_{24} \\ p_{30} & p_{31} & p_{32} & -\rho & p_{34} \\ p_{40} & p_{41} & p_{42} & p_{43} & -\rho \end{vmatrix} = 0.$$

Remembering our equation (2),

$$\rho = \pm p.$$

If we substitute either of these values and find (x) from four of the five equations, we have an unsymmetrical result. A better plan consists in taking an arbitrary sphere (y), multiplying the solution obtained by omitting the equation corresponding to subscript k by y_k, and summing.

$$\sigma x_i = \begin{vmatrix} y_i & p_{ij} & p_{ik} & p_{il} & p_{im} \\ y_j & -\rho & p_{jk} & p_{jl} & p_{jm} \\ y_k & p_{kj} & -\rho & p_{kl} & p_{km} \\ y_l & p_{lj} & p_{lk} & -\rho & p_{lm} \\ y_m & p_{mj} & p_{mk} & p_{ml} & -\rho \end{vmatrix}, \qquad \rho^2 = p^2.$$

If we reintroduce a symbol already used in the last chapter,

$$P_{ij} \equiv \sum_{n=0}^{n=4} p_{in} p_{jn},$$

we find for the first and second foci respectively

$$\lambda x_i = \sum_{j=0}^{j=4} (pp_{ij} - P_{ij}) y_j. \tag{3}$$

$$\mu \bar{x}_i = \sum_{j=0}^{j=4} (pp_{ij} + P_{ij}) y_j.$$

These formulae enable us to write at once the formula for paratactic circles, XIII. (15)

$$p \sum_{n=0}^{n=4} p_{in} P_{in}' + p' \sum_{n=0}^{n=4} p_{in}' P_{in} \equiv pp' \sum_{n=0}^{n=4} p_{in} p_{in}' + \sum_{n=0}^{n=4} P_{in} P_{in}' \equiv 0.$$

$$(4)$$

$$p \sum_{n=0}^{n=4} p_{in} P_{in}' - p' \sum_{n=0}^{n=4} p_{in}' P_{in} \equiv pp' \sum_{n=0}^{n=4} p_{in} p_{in}' - \sum_{n=0}^{n=4} P_{in} P_{in}' \equiv 0,$$

$$i = 0 \ldots 4.$$

When the first equation prevails, the circles shall be said to be *properly paratactic*; in the other case, improperly so. Proper parataxy occurs when the first foci are on an isotropic, and the second foci on another skew to the first. In the case of improper parataxy each first focus is on an isotropic with the second focus of the other circle. We find for the angle of two circles from XII. (12)

$$\sin\frac{\theta}{2} = \frac{\Sigma p_{ij}q_{ij}+pq}{2pq}.$$

Note that the angle of properly tangent circles is zero.

Suppose now that we have a linear transformation of our homogeneous coordinates which leaves invariant our identities (2). Let us show that it must be a spherical transformation or such a transformation followed by reversal of orientation. We first ask, 'When will a linear combination of the coordinates of two oriented circles give the coordinates of another circle?' Let us write

$$p_{ij} \equiv \lambda q_{ij}+\mu r_{ij}, \quad p = \lambda q+\mu r, \quad i,j = 0 \dots 4.$$

Substituting in the identities (2) we see that the circles (q) and (r) must be cospherical and, if their common sphere be not null, they must be properly tangent. If the common sphere be null, the last condition is satisfied automatically for two oriented circles with the same first or second focus. Now a system of circles whereof each two are properly tangent could not depend upon more than two parameters at most, while the circles with a given first or second focus depend on three parameters. Hence a system of circles with a given first or second focus will go into another such system; hence properly tangent circles on a not null sphere will go into other such circles.

If two not null circles be cospherical they will be properly tangent to an infinite number of common circles. Conversely, when this is the case for two not null circles (in cartesian space), every point on the line of intersection of their planes must have the same power with regard to the two. Hence this line meets them in the same two points, and the circles are cospherical. Cospherical circles will, then, go into cospherical circles, or a sphere goes into a sphere, and since none but a null sphere contains two sets of ∞^3 oriented circles with a common first or second focus, a null sphere goes into a null sphere. A null and a not null sphere will have but one common oriented circle when, and only when, the vertex of the null sphere is on the not null one. Hence points of a sphere go into points of a sphere, and we have either a spherical transformation or such a transformation joined with a reversal of orientation.

It is worth while looking a bit more closely at this question of linear dependence. Let us write

$$p_{ij} = \lambda a_{ij} + \mu b_{ij} + \nu c_{ij} + \rho d_{ij} + \sigma e_{ij};$$
$$p = \lambda a + \mu b + \nu c + \rho d + \sigma e, \quad i, j = 0 \dots 4.$$

We then substitute in the equations

$$\Omega_0(pp) = \Omega_1(pp) = \Omega_2(pp) = \Sigma p_{ij}^2 + p^2 = 0.$$

There are usually sixteen different solutions for $\lambda : \mu : \nu : \rho : \sigma$.

The equations

$$\Omega_0(pp) = \Omega_1(pp) = p_{34} = \Sigma p_{ij}^2 + p^2 = 0$$

have eight independent solutions, whereof two are solutions of

$$p_{23} = p_{24} = p_{34} = 0.$$

The remainder are solutions of

$$\Omega_2(pp) = 0.$$

There are thus six solutions for the equations

$$\Omega_0(pp) = \Omega_1(pp) = \Omega_2(pp) = p_{34} = \Sigma p_{ij}^2 + p^2 = 0.$$

The remainder of our sixteen solutions will solve all the equations (2).

Theorem 1.] *The totality of oriented circles of pentaspherical space may be put into one to one correspondence with that of the points of a six-dimensional manifold of the tenth order* S_6^{10} *lying in a space of ten dimensions.*

Theorem 2.] *A necessary and sufficient condition that the coordinates of three distinct oriented circles should be linearly dependent is that they should be properly tangent, or else that they should have a common first or second focus.*

Theorem 3.] *A necessary and sufficient condition that the coordinates of four distinct oriented circles should be linearly dependent is that all should be properly tangent to one another, or else that all should have a common first (second) focus, while their second (first) foci lie on one same sphere through the common focus.*

Theorem 4.] *The necessary and sufficient condition that the coordinates of five distinct oriented circles should be linearly dependent, while no four are dependent, is that they should have a common first (second) focus, while their second (first) foci do not lie on a sphere through the common focus.*

Theorem 5.] *The* S_6^{10} *contains two systems of* ∞^7 *straight lines.* ∞^2 *lines of the first, and two sub-systems each of* ∞^2 *lines of the second, pass through each point thereof.*

Theorem 6.] *The* S_6^{10} *contains a first system of* ∞^5 *planes, and a second system of* ∞^6 *planes. The only lines of the* S_6^{10} *which lie in planes of the first system are lines of the first system, and the same is thus of lines and planes of the second system. Each point of the variety lies in* ∞^1 *planes of the first system, and* $2\infty^3$ *of the second.*

Theorem 7.] *The* S_6^{10} *contains two systems of* ∞^3 *three-way linear spreads. Two spreads of the same system have no common point, two of different systems have one common point. Through each point of the variety will pass one three-way spread of each system. The only lines and planes of the variety lying in these spreads belong to the second system of lines and planes.*

§ 2. Linear Systems.

It is natural for us next to take up linear systems of oriented circles. We shall define *hypercomplexes, hypercongruences, complexes, congruences,* and *series* of oriented circles exactly as was done for non-oriented circles. We begin with the linear hypercomplex. This is given by an equation of the type

$$\Sigma a_{ij} p_{ij} + ap = 0. \tag{5}$$

The hypercomplex shall be said to be *general* if

$$\sum_{i=0}^{i=4} \Omega_i^2(aa) \neq 0, \quad \sum_{i=0}^{i=4} \Omega_i^2(aa) \neq [\Sigma a_{ij}^2]^2.$$

Theorem 8.] *Ten oriented circles will belong to one and, in general, only one linear hypercomplex.*

Theorem 9.] *The totality of oriented circles making a fixed angle with a given not null oriented circle will belong to a linear hypercomplex.*

If in the equation of a linear hypercomplex such as (5) the coefficient a be equal to zero, the hypercomplex is said to be *reduced.* Each general linear hypercomplex is associated with a reduced hypercomplex: the two have the same null circles.

Theorem 10.] *If a linear hypercomplex contain a single pair of opposite circles which are distinct from one another, it contains the opposite of each of its circles and is reduced.*

We shall define as the *central sphere* of a general linear hypercomplex that of the corresponding reduced hypercomplex; the coordinates of this will be

$$\rho z_i = \Omega_i(aa).$$

Substituting in (5) for (p) its value from (1),

$$\sum_{i,j=0}^{i,j=4} a_{ij} x_i \bar{x}_j + a(x\bar{x}) = 0, \quad a_{ji} = -a_{ij}.$$

If (y) be any sphere through the circle, we may put

$$y_i = \lambda x_i + \mu \bar{x}_i.$$

$$(xy) = \mu(x\bar{x}).$$

$$\sum_{i,j=0}^{i,j=4} a_{ij} x_i y_j + a(xy) = 0.$$

If, thus, (y) be fixed, (x) will trace a sphere. The corresponding oriented circles will trace a linear congruence of the type discussed in Ch. X.

Theorem 11.] *The oriented circles of a linear hypercomplex which lie on a not null sphere in general position trace a linear congruence.*

With regard to the words 'in general position' we see that if

$$a\left[a^4 - \Sigma a_{ij}{}^2 a^2 + \sum_{i=0}^{i=0} \frac{\Omega_i{}^2 (aa)}{4}\right] \neq 0.$$

there is no exception. When (y) is the central sphere,

$$a\,(xy) = 0.$$

Theorem 12.] *A general linear hypercomplex which is not reduced will usually share with every not null sphere a linear congruence. In the case of the central sphere this is the congruence of null circles.*

Let us write

$$p_{ij} = \begin{vmatrix} x_k' & x_l' & x_m' \\ r_k & r_l & r_m \\ s_k & s_l & s_m \end{vmatrix}.$$

The equation of our hypercomplex is

$$\left[\Sigma a_{ij} \begin{vmatrix} x_k' & x_l' & x_m' \\ r_k & r_l & r_m \\ s_k & s_l & s_m \end{vmatrix}\right]^2 = a^2 \begin{vmatrix} 0 & (x'r) & (x's) \\ (rx') & 0 & (rs) \\ (sx') & (sr) & 0 \end{vmatrix},$$

(r) and (s) being fixed points; (x') traces some form of cyclide. Let us show that it is a Dupin cyclide. The equation may be written

$$(bx')^2 - (rx')\,(sx') = 0,$$

with the conditions

$$(br) = (bs) = (rr) = (ss) = 0.$$

It is the envelope of the conic series of spheres

$$y_i \equiv t^2 r_i + 2tb_i + s_i.$$

Putting $(yy) = 0$ we get $t^2 = 0$, showing that we have a Dupin series.

Theorem 13.] *The oriented circles of a linear hypercomplex through two points generate a Dupin series.*

We may find a canonical form for the equation of the linear hypercomplex exactly as we did in the last chapter, getting

$$a_{01}p_{01} + a_{23}p_{23} + ap = 0. \tag{6}$$

Theorem 14.] *The locus of the oriented circles, the sines of whose half angles with the two oriented circles of a cross are connected by a linear relation, is a general linear hypercomplex and, conversely, every general linear hypercomplex may be described in this way. The cross is the axial cross of the corresponding reduced hypercomplex.*

Theorem 15.] *The sines of the half angles of each circle of a linear hypercomplex with any two properly oriented circles orthogonal to the central sphere, and mutually polar in the corresponding reduced linear complex of circles orthogonal to this sphere, are connected by a linear relation.*

We pass to the figure next below, the linear hypercongruence, given by two equations of the type

$$\Sigma\, a_{ij}\, p_{ij} + ap = \Sigma\, b_{ij}\, p_{ij} + bp = 0. \tag{7}$$

The oriented circles of the hypercongruence are common to all hypercomplexes of the pencil

$$c_{ij} = \lambda a_{ij} + \mu b_{ij}, \quad c = \lambda a + \mu b, \quad i, j = 0 \ldots 4.$$

Theorem 16.] *In a pencil of linear hypercomplexes there is always one reduced complex. If there be more than one such, all hypercomplexes of the pencil are reduced, and the corresponding linear hypercongruence contains the opposite of each of its circles.*

Theorem 17.] *If two hypercomplexes of a pencil correspond to different reduced hypercomplexes, yet have the same central sphere, this is the central sphere for every hypercomplex of their pencil, and the corresponding linear hypercongruence consists in the totality of oriented circles making fixed angles with two distinct or adjacent circles orthogonal to this sphere. When one or both of these fixed circles are null, the circles of the hypercongruence are invariantly related to them.*

Theorem 18.] *The oriented circles of a linear hypercongruence lying on a not null sphere, which is not a central sphere for the corresponding pencil of linear hypercomplexes, are properly tangent to two distinct or adjacent oriented circles thereof.*

Theorem 19.] *Through two arbitrary points there will pass just two oriented circles of a linear hypercongruence.*

This theorem will suffer exceptions for special positions of the points, but we shall not take the time to determine them. We pass rather to the linear complex given by three equations of the type

$$\begin{aligned} \Sigma a_{ij}p_{ij} + ap &= 0, \\ \Sigma b_{ij}p_{ij} + bp &= 0, \\ \Sigma c_{ij}p_{ij} + cp &= 0. \end{aligned} \tag{8}$$

The circles of the complex are common to the hypercomplexes of a net linearly dependent on three of their number which do not belong to a pencil.

Theorem 20.] *If three linear hypercomplexes not belonging to a pencil, nor corresponding to reduced hypercomplexes of a pencil, have the same central sphere, that is, the central sphere for every linear hypercomplex of their net, and the corresponding linear complex of oriented circles consists in those making fixed angles with each circle of one generation of a cyclide anallagmatic in the fixed sphere.*

The phrasing of this theorem must be slightly varied when the fixed sphere is null or when the fixed circles are. The linear complex shall be said to be *general* when no two hypercomplexes of the net have the same central sphere. The null circles of the linear complex are those of the complex determined by the corresponding reduced hypercomplexes. We thus get from XIII. 40]

Theorem 21.] *The null circles of a general linear complex are those common to sets of three mutually tangent central spheres.*

Consider the circles of our complex which are orthogonal to $x_4 = 0$, which we may assume is not a central sphere. They have seven homogeneous coordinates connected by three linear and two quadratic equations.

If we adjoin one more quadratic equation, to make our circle intersect a chosen circle, we have eight solutions :

Theorem 22.] *The circles of a linear complex orthogonal to an arbitrary not null sphere generate a surface of the eighth order which, in cartesian space, has the circle at infinity as a quadruple curve.*

Let us require (x), the first focus of our oriented circle, to lie on a sphere (z). We get from (3)

$$p \sum_{i,j=0}^{i,j=4} p_{ij} z_i y_i - \sum_{i,j=0}^{i,j=4} P_{ij} z_i y_i.$$

There will be eight solutions common to this and to the equations above. Of these, four must be rejected as extraneous, for this equation is equally well satisfied if the first focus lie on (z) or the second lie on (y).

Theorem 23.] *The first foci of the circles of a general linear complex which are orthogonal to an arbitrary not null sphere generate a cyclic, the second foci generate a second cyclic.*

The *linear congruence* will be given by four equations of the type

$$\begin{aligned} \Sigma a_{ij} p_{ij} + ap &= 0, \\ \Sigma b_{ij} p_{ij} + bp &= 0, \\ \Sigma c_{ij} p_{ij} + cp &= 0, \\ \Sigma d_{ij} p_{ij} + dp &= 0. \end{aligned} \tag{9}$$

If we put

$$e_{ij} = \lambda a_{ij} + \mu b_{ij} + \nu c_{ij} + \rho d_{ij},$$

and write the equations

$$\Omega_i(ee) \equiv 0, \quad i = 0 \ldots 4,$$

we know from XIII. 41] that there are five solutions corresponding to the circles of a pentacycle. We shall say that the congruence is *general* when this pentacycle exists and none of its circles are null. Each circle of the pentacycle can be oriented either way, hence

Theorem 24.] *The general linear congruence of oriented circles may be described in thirty-two ways by an oriented circle making an assigned angle with each oriented circle of a pentacycle.*

Theorem 25.] *The linear congruence of oriented circles is of the fourth class.*

Let us find the locus of the first foci of the oriented circles of our linear congruence. Let us determine our congruence by two circles of the pentacycle, and three other linear hypercomplexes. By a linear combination of the equations connected with the two circles we get a reduced hypercomplex whose equation can be written

$$|\,\bar{x}\,x\,r\,s\,t\,| + |\,\bar{x}\,x\,r'\,s'\,t'\,| \equiv \Delta + \Delta' = 0.$$

The other three linear hypercomplexes may be written

$$\sum_{i,j=0}^{i,j=4} a_{ij}\,x_i\bar{x}_j + a\,(x\bar{x}) = 0,$$

$$\sum_{i,j=0}^{i,j=4} b_{ij}\,x_i\bar{x}_j + b\,(x\bar{x}) = 0,$$

$$\sum_{i,j=0}^{i,j=4} c_{ij}\,(x_i\bar{x}_j) + c\,(x\bar{x}) = 0, \quad (\bar{x}\bar{x}) = 0.$$

Looked upon as five linear equations in $(\bar{x})$, these will be consistent if

$$\left|\begin{array}{ccc}
\dfrac{\partial\Delta}{\partial\bar{x}_0}+\dfrac{\partial\Delta'}{\partial\bar{x}_0} & \cdots\cdots & \dfrac{\partial\Delta}{\partial\bar{x}_4}+\dfrac{\partial\Delta'}{\partial\bar{x}_4} \\
\bar{x}_0 & \cdots\cdots & \bar{x}_4 \\
\sum_{i=0}^{i=4} a_{i0}x_i+a_0x_0 & \cdots\cdots & \sum_{i=0}^{i=4} a_{i4}x_i+a_4x_4 \\
\sum_{i=0}^{i=4} b_{i0}x_i+b_0x_0 & \cdots\cdots & \sum_{i=0}^{i=4} b_{i4}x_i+b_4x_4 \\
\sum_{i=0}^{i=} c_{i0}x_i+c_0x_0 & \cdots\cdots & \sum_{i=0}^{i=4} c_{i4}x_i+c_4x_4
\end{array}\right| = 0.$$

Expanding in terms of the first row and remembering the identities

$$\sum_{i,j=0}^{i,j=4} a_{ij}x_ix_j+a(xx) = \sum_{i,j=0}^{i,j=4} b_{ij}x_ix_j+b(xx)$$

$$= \sum_{i,j=0}^{i,j=4} c_{ij}x_ix_j+c(xx) = 0,$$

we get the equation of our surface

$$\left|\begin{array}{cccc}
(x\bar{x}) & 0 & 0 & 0 \\
(rx) & (dx) & (ex) & (fx) \\
(sx) & (gx) & (hx) & (kx) \\
(tx) & (lx) & (mx) & (nx)
\end{array}\right| + \left|\begin{array}{cccc}
(x\bar{x}) & 0 & 0 & 0 \\
(r'x) & (d'x) & (e'x) & (f'x) \\
(s'x) & (g'x) & (h'x) & (k'x) \\
(t'x) & (l'x) & (m'x) & (n'x)
\end{array}\right| = 0.$$

Theorem 26.] *The first (second) foci of the oriented circles of a general linear congruence generate a surface of the sixth order which, in cartesian space, has the circle at infinity as a triple curve.*

The last linear figure to consider is the linear series given by five equations of the type

$$\begin{aligned} \Sigma a_{ij} p_{ij} + ap &= 0, \\ \Sigma b_{ij} p_{ij} + bp &= 0, \\ \Sigma c_{ij} p_{ij} + cp &= 0, \\ \Sigma d_{ij} p_{ij} + dp &= 0, \\ \Sigma e_{ij} p_{ij} + ep &= 0. \end{aligned} \tag{10}$$

We know from (1) that there are usually ten equations derived linearly from those whose coefficients satisfy the identities (2). When ten such distinct circles can indeed be found and none are null we shall say that our series is *general.* The figure of these ten oriented circles shall be called a *dekacycle.*

Theorem 27.] *The general linear series of oriented circles is composed of the totality of such circles as make null angles with all oriented circles of a dekacycle.*

We may, in an infinite number of ways, find an oriented circle all of whose coordinates but the last are linearly dependent on $a_{ij} b_{ij} c_{ij} d_{ij} e_{ij}$. The circles, unoriented, generate a linear series.

Theorem 28.] *The oriented circles of a general linear series make fixed angles with all not null circles of a series which, when not oriented, is linear.*

Theorem 29.] *If a system of oriented circles make fixed angles with five oriented circles whose coordinates are linearly independent, they make fixed angles with five other oriented circles which complete a dekacycle with the first five.*

We next write the equations

$$q_{ij} = \lambda a_{ij} + \mu b_{ij} + \nu c_{ij} + \rho d_{ij} + \sigma e_{ij},$$

$$q = \lambda \frac{a}{k} + \mu \frac{b}{k} + \nu \frac{c}{k} + \rho \frac{d}{k} + \sigma \frac{e}{k},$$

$$\Omega_i(qq) \equiv \Sigma q_{ij}^2 + q^2 = 0.$$

We get a new dekacycle. If (q) be an oriented circle thereof we have

$$\Sigma q_{ij} p_{ij} + kpq = 0,$$

when (p) belongs to our linear series.

Theorem 30.] *A dekacycle may be found whose ten oriented circles make any chosen angle other than an even integral multiple of π, with every oriented circle of a general linear series.*

The condition that two circles should intersect is quadratic in the coordinates of each; hence

Theorem 31.] *The oriented circles of a linear series will generate a surface of the twentieth order which, in cartesian space, has the circle at infinity as a tenfold curve.*

Theorem 32.] *There are ten null circles in a linear series.*

The condition that the first focus of an oriented circle should lie on a preassigned sphere is quadratic in the coordinates of the circle; it contains, however, extraneous elements which vanish when the second focus lies on another sphere.

Theorem 33.] *The locus of the first (second) foci of the oriented circles of a linear series is a curve of the tenth order.*

§ 3. The Laguerre method for representing Imaginary Points.

One of the most important applications of the study of oriented circles in space is to the representation of imaginary points. The idea goes back to Laguerre.* We represent each real point of pentaspherical space by itself, i.e. by the totality of null circles having that point as vertex; each imaginary point is represented by the real oriented circle whereof it is the first focus. Conjugate imaginary points are thus

* 'Sur l'emploi des imaginaires dans la géométrie de l'espace', *Nouvelles Annales de Math.*, Series 2, vol. xi, 1872. See also Molenbroch, 'Sur la représentation géométrique des points imaginaires dans l'espace', ibid., Series 3, vol. x, 1891. The correct spelling of this author's name seems to be Molenbroek.

represented by opposite oriented circles. Analytically, let us write

$$\rho x_0 = j(x^2+y^2+z^2+t^2),\ \rho x_1 = x^2+y^2+z^2-t^2,$$
$$\rho x_2 = 2xt,\quad \rho x_3 = 2yt,\quad \rho x_4 = 2zt.$$
$$j^2 = -1; \qquad (11)$$

the conjugate imaginary point, which we shall call $(\bar{x})$, will be obtained by replacing x, y, z, t by their conjugate imaginary values, x_1, x_2, x_3, and x_4 will be replaced by their conjugate imaginary values, while x_0 will be replaced by the conjugate imaginary multiple of j.

Theorem 34.] *There is a perfect one to one correspondence between the totality of all complex points of pentaspherical space and that of all real oriented circles of the same space, with the exception that all null circles with the same real vertex represent that vertex.*

The simplest system of points in complex pentaspherical space which depend on a single real parameter is the *chain*. This figure we have already met on p. 202, and defined as the totality of points on a circle such that the cross ratio of any four is real; the definition was there given only for the tetracyclic plane, but may be extended to pentaspherical space, as we see by noticing that the definition of cross ratio there given by means of the angles of circles orthogonal to the given circle may be included in a larger definition based on the angles of spheres orthogonal to a given circle. If the circle whereon lies the chain be not null, we easily find a spherical transformation to carry three points of the chain into three real points. The chain is thus carried into the real domain of a not null circle. In any case it is clear that three points of a chain may be taken entirely at random (when on a null circle they must be on the same isotropic), and that the chain is then completely determined.

The importance of the chain appears very clearly in connexion with the Gauss representation of the complex binary domain. To begin with, we should notice that the Gauss representation is very closely allied to our present representation. For if we take the point $x+iy$ as the first focus of a real oriented circle, the intersections of this circle with the Gauss plane are (x, y) and $(x, -y)$. If four values of the complex variable $x+iy$ have a real cross ratio, the points representing them in the Gauss plane are on a real circle, and vice versa, so that a chain is represented in the Gauss plane by a circle. The corresponding circles in our present system generate a sphere or an anchor ring.

If two points of a chain be (α) and (γ), while the sphere through them is (β), we see from IV. (8) that every chain on this circle connecting (α) and (γ) may be expressed parametrically in the form

$$\rho x_i \equiv t^2\alpha_i + t\beta_i + \gamma_i,$$

where t is real. For the conjugate imaginary chain

$$\bar{\rho}\bar{x}_i \equiv t^2\bar{\alpha}_i + t\bar{\beta}_i + \bar{\gamma}_i.$$

If (x') be a point of the circle whose foci are (x) and $(\bar{x})$

$$(xx') = (\bar{x}x') = 0.$$

Eliminating t,

$$\begin{vmatrix} (\alpha x') & (\beta x') & (\gamma x') & 0 \\ 0 & (\alpha x') & (\beta x') & (\gamma x') \\ (\bar{\alpha} x') & (\bar{\beta} x') & (\bar{\gamma} x') & 0 \\ 0 & (\bar{\alpha} x') & (\bar{\beta} x') & (\bar{\gamma} x') \end{vmatrix} = 0. \tag{12}$$

Theorem 35]. *The general chain on a not null circle is represented by oriented circles generating a surface of the eighth order.*

This surface will be notably simplified in certain cases. If (α) and (γ) be real points we could remove the factors $(\alpha x')$, $(\gamma x')$ from (12), leaving a quadratic equation, i.e. a cyclide, with these points as conical and transformed into itself by $2\infty^1$ inversions, i.e. a Dupin cyclide.

Again, suppose that we have a chain possessing no real points yet lying on a real circle. If (α) and (γ) be two real points of our circle, we may write our chain in the form

$$\sigma x_i \equiv \left(\frac{\lambda t+\mu}{\nu t+\rho}\right)^2 \alpha_i + \left(\frac{\lambda t+\mu}{\nu t+\rho}\right)\beta_i + \gamma_i.$$

A point (x') on the corresponding oriented circle will have coordinates which satisfy the equations

$$\left(\frac{\lambda t+\mu}{\nu t+\rho}\right)^2(\alpha x')+\left(\frac{\lambda t+\mu}{\nu t+\rho}\right)(\beta x')+(\gamma x')$$
$$=\left(\frac{\bar\lambda t+\bar\mu}{\bar\nu t+\bar\rho}\right)^2(\alpha x')+\left(\frac{\bar\lambda t+\bar\mu}{\bar\nu t+\bar\rho}\right)(\beta x')+(\gamma x')=0.$$

Eliminating in turn $(\gamma x')$ and $(\alpha x')$

$$\left[\frac{\lambda t+\mu}{\nu t+\rho}+\frac{\bar\lambda t+\bar\mu}{\bar\nu t+\bar\rho}\right](\alpha x')+(\beta x')=\left[\frac{\nu t+\rho}{\lambda t+\mu}+\frac{\bar\nu t+\bar\rho}{\bar\lambda t+\bar\mu}\right](\gamma x')+(\beta x')=0.$$

On clearing of fractions we find that the coefficients of $(\alpha x')$ and $(\gamma x')$ are the same.

$$[A_0(\alpha x')+B_0(\beta x')]t^2+[A_1(\alpha x')+B_1(\beta x')]t$$
$$+[A_2(\alpha x')+B_2(\beta x')]=0.$$

$$[A_0(\gamma x')+B_0'(\beta x')]t^2+[A_1(\gamma x')+B_1'(\beta x')]t$$
$$+[A_2(\gamma x')+B_2'(\beta x')]=0.$$

$$\begin{vmatrix} A_0(\alpha x')+B_0(\beta x') & A_1(\alpha x')+B_1(\beta x') & A_2(\alpha x')+B_2(\beta x') & 0 \\ 0 & A_0(\alpha x')+B_0(\beta x') & A_1(\alpha x')+B_1(\beta x') & A_2(\alpha x')+B_2(\beta x') \\ A_0(\gamma x')+B_0'(\beta x') & A_1(\gamma x')+B_1'(\beta x') & A_2(\gamma x')+B_2'(\beta x') & 0 \\ 0 & A_0(\gamma x')+B_0'(\beta x') & A_1(\gamma x')+B_1'(\beta x') & A_2(\gamma x')+B_2'(\beta x') \end{vmatrix} =$$

It is easy to see that the second and third columns may be freed from the terms $(\alpha x')$, $(\gamma x')$, so that the factor $(\beta x')^2$ may be struck off. We have left a cyclide which, as before, we see is a cyclide of Dupin. If the chain lie on a self-conjugate imaginary circle it may be put into the form

$$\sigma x_i=\left(\frac{\lambda t+\mu}{\nu t+\rho}\right)^2\alpha_i+\left(\frac{\lambda t+\mu}{\nu t+\rho}\right)\beta_i+\bar\alpha_i.$$

The transformations to be effected in this case are almost exactly like those in the preceding one, and lead to a like result.

Theorem 36.] *Every chain which is not self-conjugate imaginary, but contains two real points, or lies on a real or self-conjugate imaginary circle, will be represented by oriented circles generating a Dupin cyclide.*

A self-conjugate imaginary chain is still easier. Here we have

$$\rho x_i = t^2\alpha_i + t\beta_i + \bar{\alpha}_i.$$

Equation (12) becomes

$$[(\alpha x') - (\bar{\alpha}x')]^2 [\{(\alpha x') + (\bar{\alpha}x')\}^2 - (\beta x')^2] = 0.$$

To decide between the two factors we have but to notice that all points of the circle $(\alpha x) = (\bar{\alpha}x') = 0$ lie on the surface, yet for such points the second factor will not usually vanish. Hence the surface, which is irreducible, is given by the first factor.

Theorem 37.] *The points of a self-conjugate imaginary chain will be represented by coaxal oriented circles.*

The chains which lie on null circles, i.e. in isotropics, are much simpler. Here we have, in general,

$$\rho x_i = \alpha_i t + \gamma_i, \quad \bar{\rho}\bar{x}_i = \bar{\alpha}_i t + \bar{\gamma}_i,$$

$$(\bar{\gamma}x')(\alpha' x') - (\gamma x')(\bar{\alpha}x') = 0.$$

Theorem 38.] *The points of a chain in general position on a null circle with an imaginary vertex are represented by properly paratactic generators of a cyclide; when the vertex of the circle is real, the points are represented by properly tangent oriented circles generating a limiting form of a Dupin cyclide when two conical points fall together.*

Let the reader prove

Theorem 39.] *The points of a chain including one real point and lying on a null circle will be represented by properly tangent oriented circles forming a coaxal system.*

It is time to take up systems of points depending analytically upon two real parameters. An analytic curve of complex space will be an example of such a system, though a peculiar example, as we shall see presently. Suppose that a point traces a non-minimal curve, i.e. such a curve that there is not a tangent isotropic at every point. At each non-singular point of the curve where there is no tangent isotropic, each point in general position, there is a not null osculating circle. Conversely, suppose that we have a system of points in complex space (which we may assume to be cartesian, and in the finite domain) depending on two real parameters, and that the circle connecting a point in general position with two infinitely near points of the variety approaches one definite limiting position as the two points approach the original one in any way in the variety. At each general point in the variety there will be a definite tangent, and there will likewise be a tangent at each point in the projection of the variety on an arbitrary plane. The projection is thus a curve; hence the variety is the total or partial intersection of two cones, and so a curve. Suppose, then, that we take a point P of the variety in general position while P' and P'' are adjacent points of the variety. The conjugate imaginary points shall be $\bar{P}$, $\bar{P}'$, and $\bar{P}''$. The circle through P, P', P'' shall have F_1 and F_2 as first and second foci respectively, that through $\bar{P}$, $\bar{P}'$, $\bar{P}''$ will have the conjugate imaginary points as foci. The three real oriented circles whose first foci are P, P', P'' will be properly paratactic or tangent to the two real circles whose first foci are F_1 and F_2. These latter two are the only two real circles properly paratactic to the three real oriented circles with the first foci P, P', P'', and, if the two circles tend to approach definite limiting positions, the circle through P, P', P'' approaches a definite limiting position.

Theorem 40.] *A necessary and sufficient condition that the oriented circles of a congruence should represent the points of an analytic non-minimal curve is that the two real oriented circles properly paratactic or tangent to a real circle of the congruence in general position, and to two adjacent circles thereof, should tend to approach definite limiting positions no matter how the latter two circles tend to approach the circle in general position as a common limit.*

The words 'in general position' mean that the circle $PP'P''$ is not null, and the circles with the first foci P, P', P'' do not behave in the manner presently described. The state of affairs is somewhat different in the case of a minimal curve. The osculating circle is here an isotropic counted twice, and has ∞^1 foci. Infinitely near circles of the congruence tend to become paratactic, and to be cut twice orthogonally by ∞^1 generators of a cyclide, or else, perhaps, they tend to become properly tangent. This latter case would arise could we find such a minimal curve that the tangent isotropic at each point, and the tangent isotropic to the conjugate curve at the conjugate imaginary point, always intersected. Let us show that there can be no such curve except an isotropic. Suppose, in fact, that we had a curve of this sort in cartesian space. The isotropic tangent lines at the conjugate imaginary points P and $\bar{P}$ intersect, and so lie in a real plane, and $P\bar{P}$ is a real line. There are ∞^2 of these real lines depending analytically on two parameters, and each intersects all lines of the system infinitely near to it. Hence each two lines of the system intersect, and all lie in a plane or pass through a point. If the lines lay in a plane the minimal curves would be plane curves, i.e. isotropics, an excluded case. Hence the real lines and planes pass through a real point. But the real lines project the complex minimal curves, and we cannot have an analytic curve which intersects every real line through a point unless at that point.

Theorem 41.] *A necessary and sufficient condition that the real oriented circles of a congruence should represent the points of a minimal curve not an isotropic is that the circles cutting twice orthogonally a circle of the congruence in general position and neighbouring circles thereof should approach as limiting positions the paratactic generators of a definite cyclide, as the neighbouring circles approach indefinitely near to the circle in general position, or else approach properly tangent oriented circles of a coaxal system.*

The simplest congruence of this sort will arise when we undertake to represent the points of an isotropic.

Theorem 42.] *The points of an isotropic which does not intersect its conjugate imaginary will be represented by the totality of oriented circles properly paratactic to two properly paratactic oriented circles.*

The only exception to theorem 41] is

Theorem 43.] *The points of an isotropic which intersects its conjugate imaginary will be represented by the congruence of real oriented circles properly tangent at a real point.*

All circles of the congruence representing the points of an isotropic skew to its conjugate are cospherical with these two isotropics looked upon as null circles; hence, by XIII. 66]

Theorem 44.] *The congruence of oriented circles whose real members represent the points of an isotropic skew to its conjugate is of the first order and class.*

We pass from the isotropic at once to a null circle.

Theorem 45.] *The points of a null circle with a real vertex will be represented by the totality of real oriented circles properly or improperly tangent to one another at a fixed point.*

Theorem 46.] *The points of a real circle will be represented by the congruence of real oriented circles in bi-involution therewith.*

Theorem 47.] *The points of a self-conjugate imaginary circle will be represented by the totality of real oriented circles through two real points.*

We get from VIII. 27]

Theorem 48.] *The points of a complex circle lying on a real sphere will be represented by real oriented circles orthogonal to this sphere and connecting in a definite order pairs of points which correspond in a real indirect circular transformation on the sphere.*

If we undertake to represent the general complex circle, we see that each representing circle is properly paratactic to the real circle whose first focus is the first focus of the given circle, while its second focus is the conjugate imaginary first focus of the conjugate imaginary circle. It will also be properly paratactic to a second real oriented circle whose first and second foci are the second foci of the given circles.

Theorem 49.] *The points of a general complex circle will be represented by the totality of real oriented circles properly paratactic or tangent to two real non-paratactic oriented circles. When the given oriented circle lies on a self-conjugate imaginary sphere the representing circles are invariant in a real direct involutory spherical transformation.*

This congruence of circles is surely of the second class, for two of its circles will be orthogonal to each real or self-conjugate imaginary sphere. To find the order we notice that the number of circles of the congruence whose coordinates satisfy the equations $p_{01} = p_{02} = 0$ will be the sum of the order and class, for every circle whose coordinates do satisfy these equations is either orthogonal to $x_0 = 0$ or cospherical with the circle whose foci are $(0, 0, 0, 1, i)$. Analytically we write

$$x_i \equiv (r+si)^2\alpha_i + (r+si)\,\beta_i + \gamma_i\,;\;\; \bar{x}_i \equiv (r-si)^2\,\bar{\alpha}_i + (r-si)\,\bar{\beta}_i + \bar{\gamma}_i.$$

We see that these values substituted in the two equations above will give two cartesian cyclides in the $(r,\, s)$ plane whose infinite intersections are unacceptable; hence

Theorem 50.] *The points of a general complex circle will be represented by the real oriented circles of a congruence of the sixth order and second class.*

Theorem 51.] *The points of a real sphere will be represented by the totality of real oriented circles orthogonal thereto.*

Theorem 52.] *The points of a self-conjugate imaginary not null sphere will be represented by the totality of invariant real oriented circles in an involutory direct spherical transformation.*

It is to be noted that in cartesian space we may describe these as the oriented circles with regard to which a real point has a real negative power. The general complex sphere is somewhat harder to grasp. It meets the conjugate imaginary sphere in a circle which is either real, or self-conjugate imaginary. Let us first suppose that the circle is real. We pick out a real point thereon and take all real circles through that point. Our imaginary sphere will be determined by an elliptic involution among the spheres through the first circle, and these will determine an elliptic involution among the points on each circle through the fixed point. Take two pairs of such a point involution, and through each pass a sphere orthogonal to the real circle whereon lie the pairs. The real circle common to these spheres, when properly oriented, will represent the imaginary intersection of the real circle bearing the pairs, with the given imaginary sphere. This construction fails when the circle is not real, and we are compelled to fall back upon the construction of a sphere by means of an isotropic gliding along three skew isotropics.

Theorem 53.] *The points of a general complex sphere will be represented by the real hypercongruence of oriented circles properly paratactic to sets of three properly paratactic oriented circles, each circle of the three being properly paratactic to two given properly paratactic circles.*

Theorem 54.] *Two oriented circles of the hypercongruence whose real members represent the points of a complex sphere will pass through two arbitrary points in space.*

Theorem 55.] *The real oriented circles which represent the totality of points of a null sphere with real vertex will be the totality passing through a real point.*

Theorem 56.] *The oriented circles representing the points of a null sphere with a complex vertex will be the assemblage of all circles properly paratactic or tangent to a real circle.*

The subject of oriented circles in space certainly seems to offer quite as attractive a field for further study as does that of non-oriented circles. There are certain important points which should be cleared up as soon as possible. What is the focal surface of the linear congruence? How do the circles of a dekacycle lie with regard to one another, and how is a dekacycle constructed when five of its members are known? What are the fundamental properties of the general quadratic hypercomplex? Last, but not least, this method of representing imaginaries should be pushed much further than it has ever been before, either here or elsewhere.

CHAPTER XV

DIFFERENTIAL GEOMETRY OF CIRCLE SYSTEMS

§ 1. Differential Geometry of S_6^5.

In this concluding chapter we propose to take up at length the differential geometry of systems of non-oriented circles in space. We begin with a study of the infinitesimal geometry of that point variety in higher space which represents the totality of all pentaspherical circles. We saw in Ch. XIII, in dealing with the pentacycle, that if we put

$$-p_{ji} \equiv p_{ij} \equiv \lambda a_{ij} + \mu b_{ij} + \nu c_{ij} + \rho d_{ij}, \quad i, j = 0 \dots 4,$$

there are five solutions for the five dependent equations

$$\Omega_i(pp) \equiv 0, \quad i = 0 \dots 4.$$

Theorem 1.] *The totality of circles of pentaspherical space can be put into one to one correspondence with that of all points of an S_6^5 in S_9.*

A necessary and sufficient condition that a linear combination of the coordinates of two given circles should always be the coordinates of some circle is that the two should be cospherical. The coordinates of three coaxal circles are linearly dependent, and linear dependence is also a sufficient condition that the circles should be coaxal. Four circles have linearly dependent coordinates, each three being independent when, and only when, the circles pass through a point-pair, or lie on a sphere, and are orthogonal to a circle. The coordinates of five circles are linearly dependent, each set of four being independent when, and only when, the five are on one sphere, but not orthogonal to one circle thereon.

Theorem 2.] *The* S_6^5 *contains* ∞^8 *straight lines,* ∞^3 *passing through each point of the variety. Two lines in general position will not intersect.*

Theorem 3.] *The* S_6^5 *contains* ∞^6 *planes of the first sort, and* ∞^7 *of the second. Through each point of the variety will pass* ∞^2 *planes of the first sort, and* ∞^3 *of the second. Two planes will not usually intersect.*

Theorem 4.] *The* S_6^5 *contains* ∞^4 S_3*'s. Through each point of the variety there will pass* ∞^1 *of these, and each two* S_3*'s will intersect in one point of the variety.*

Two infinitely near points of S_6^5 will determine a tangent, or a direction on the variety. On the other hand, if we have two non-cospherical circles, the spheres through one will meet the other in pairs of points of an involution. Each tangent to S_6^5 at a chosen point will thus correspond to a point involution on the corresponding circle, projectively related to the spheres through the circle. We mean by this, that there is a projective relation between the spheres through the circle and those through any two points, and the pairs of the involution. Suppose, conversely, that we have such a projective relation between the pairs of an involution on a circle and the spheres through it. The circle being (q), consider the linear hypercomplexes

$$\Sigma q_{ij} p_{ij} = \Sigma a_{ij} p_{ij} = 0, \quad \Omega_i(aq) \equiv 0, \quad i = 0 \dots 4.$$

The hypercongruence common to these two hypercomplexes is the limit of that consisting of circles in involution with (q) and $(q+dq)$, where $dq_{ij} = a_{ij}dt$. The ten quantities a_{ij} are subject to five linear equations which, however, are not independent, for it is easy to show that

$$\sum_{n=0}^{n=4} p_{in}\,\Omega_n(ap) \equiv 0, \quad i = 0 \dots 4;$$

hence we have at least five free parameters. The spheres through (q) will cut $(q+dq)$ in pairs of an involution projectively related to these spheres, which will approach a definite limiting position as an involution on (q) when dt

approaches O as a limit. We may make use of our five free parameters to make this projective correspondence between spheres through (q) and pairs of an involution on (q) exactly the correspondence given. Now every linear hypercongruence will correspond to a line in S_9, and if the hypercongruence consist in circles in involution with two given circles, the line will intersect $S_6{}^5$ twice, while a hypercongruence of the present type will correspond to a tangent to $S_6{}^5$. We thus see that there is a one to one correspondence between the tangents to $S_6{}^5$ and the projective relations connecting spheres through circles with pairs of involutions on the same circles.*

Suppose that we have a series of circles. We may express them in the form

$$p_{ij} \equiv p_{ij}(u).$$

There are four different types of circle series:

$$\text{A)} \qquad \sum_{i=0}^{i=4} \Omega_i^2(p'p') \not\equiv 0, \quad p_{ij}' = \frac{dp_{ij}}{du}.$$

This is the *general* case. Adjacent circles have no common point. In S_9 we have a curve of the variety; the tangent at each point corresponds to a projective relation between the spheres through the circle of the series and pairs of an involution thereon.†

Theorem 5.] *If a surface be generated by a general series of circles, each sphere through a circle in general position is tangent in two places to the surface, and the pairs of points of contact trace pairs of an involution projectively related to the system of spheres.*

* This correspondence of tangents and involutions is the fundamental idea in the first part of the article by Cosserat, cit. See also a difficult article by Moore, 'Infinitesimal Properties of Lines in S_4 with applications to circles in S_3', *Proceedings American Academy of Arts and Sciences*, vol. xlvi, 1911.

† Cf. Demartres, 'Sur les surfaces à génératrice circulaire', *Annales de l'École Normale*, Series 3, vol. ii, 1885. For the surfaces contained in linear hypercomplexes, hypercongruences, and complexes see Bompiani, 'Contributo allo studio dei sistemi lineari di rette nello spazio a quattro dimensioni', *Atti del R. Istituto Veneto*, vol. lxxiii, 1914. For an interesting recent article on series of circles see Ranum, 'On the Differential Geometry of Ruled Surfaces in 4 Space, and Cyclic Surfaces in 3 Space', *Transactions American Math. Soc.*, vol. xvi, 1915.

B) $$\sum_{i=0}^{i=4} \Omega_i^2(p'p') \equiv 0, \quad \Omega_j(p'p') \not\equiv 0.$$

Here adjacent circles have one common point, but are not cospherical. This point will trace a locus on the surface generated by the circles, and it seems likely that the circles of this series will be tangent thereto. Let us show that this is verily the fact. Our surface may be expressed by the parametric equations

$$\rho x_i \equiv \alpha_i(u)t^2 + \beta_i(u)t + \gamma_i(u),$$
$$(\alpha\alpha) \equiv (\gamma\gamma) \equiv (\alpha\beta) \equiv (\beta\alpha) \equiv (\beta\beta) + 2(\alpha\gamma) \equiv 0.$$

Let us assume that the locus in question corresponds to $t = 0$. An adjacent circle will contain the point (γ) if dx vanish with t.

$$\beta_i \equiv \lambda\gamma_i'.$$

But (γ') is a sphere through (γ) orthogonal to its line of advance, while (β) is the sphere through (γ) orthogonal to the given circle.

Theorem 6.] *If adjacent circles of a series tend to intersect, all circles of the series are tangent to one curve, and conversely.*

C) $$\Omega_i(p'p') \equiv 0.$$

Here adjacent circles are cospherical, and we have an annular surface. For if (s) be the sphere through our circle and an adjacent one,

$$(s\alpha) = (s\beta) = (s\gamma) = (sd\alpha) = (sd\beta) = (sd\gamma)$$
$$= (ds\alpha) = (ds\beta) = (ds\gamma) = 0,$$

which shows that this circle is a characteristic one for the sphere of x, i.e. the limiting position of its intersection with an adjacent one.

D) $$\Omega_i(p'p') \equiv 0, \quad \Sigma p_{ij}'^2 + p'^2 \equiv 0.$$

Adjacent circles are tangent, and we have the osculating

circles of a curve, or else the surface is generated by null spheres.

We pass to congruences of circles, which we express in the form

$$p_{ij} \equiv p_{ij}(u, v).$$

The condition that adjacent circles should intersect

$$\sum_{i=0}^{i=4} \Omega_i^2(dpdp) = 0$$

is quartic in du/dv. The *general* case being that where the roots are usually distinct

Theorem 7.] *The circles of a general congruence are tangent to four surfaces, some of which may shrink to curves intersecting the circles.*

Consider the equations

$$\Omega_i(dpdp) = 0.$$

These have not, usually, any common solutions. When, however, the five equations are all equivalent to one another, there are two sets of values of du/dv which solve all five (unless they be satisfied identically), and the congruence is said to be *focal*. Such a congruence can be generated in two ways by the circles of curvature of a one-parameter family of annular surfaces. As a matter of fact the largest part of the theory of circle congruences deals with focal congruences.

A complex of circles may be expressed in the form

$$p_{ij} \equiv p_{ij}(u, v, w).$$

How many circles of the complex adjacent to a circle in general position will be cospherical therewith? Let us write

$$\rho x_i \equiv \alpha_i(u, v, w)t^2 + \beta_i(u, v, w)t + \gamma_i(u, v, w),$$

$$\sigma p_{ij} = \begin{vmatrix} \alpha_k & \alpha_l & \alpha_m \\ \beta_k & \beta_l & \beta_m \\ \gamma_k & \gamma_l & \gamma_m \end{vmatrix}.$$

An arbitrary sphere through the circle can be written

$$s_i = \lambda \frac{\partial}{\partial r_i} | r\,\alpha\,\beta\,\gamma\,y | + \mu \frac{\partial}{\partial r_i} | r\,\alpha\,\beta\,\gamma\,z |,$$

where (y) and (z) are arbitrary spheres, not linearly dependent on (α), (β), (γ). This will contain all points of an adjacent circle, if

$$\lambda | d\alpha\,\alpha\,\beta\,\gamma\,y | + \mu | d\alpha\,\alpha\,\beta\,\gamma\,z | = 0,$$
$$\lambda | d\beta\,\alpha\,\beta\,\gamma\,y | + \mu | d\beta\,\alpha\,\beta\,\gamma\,z | = 0,$$
$$\lambda | d\gamma\,\alpha\,\beta\,\gamma\,y | + \mu | d\gamma\,\alpha\,\beta\,\gamma\,z | = 0.$$

Here we have three linear homogeneous equations in $du : dv : dw$. The condition of compatibility will give a cubic in $\lambda : \mu$. When the roots are distinct we shall say that we have the *general* case.

Theorem 8.] *An arbitrary circle of a general complex is cospherical with three adjacent circles thereof. The complex is generated by the circles of curvature of a two-parameter family of annular surfaces.**

We pass to a hypercongruence. Here we have

$$p_{ij} \equiv p_{ij}(u, v, w, \omega).$$

Each circle is cospherical with ∞^1 adjacent circles. In S_4 we have a four-parameter family of lines, and through each will pass ∞^1 planes containing each an adjacent line of the system. In an arbitrary hyperplane there will be a congruence of these lines, and through each will pass two focal planes, the two planes of the sort just mentioned that lie in this hyperplane. The series of planes is thus of the second order and algebraic.

Theorem 9.] *The lines connecting the pairs of points where a circle of a general hypercongruence in cartesian space meets the adjacent circles of the hypercongruence envelop a conic.*†

* Cf. Segre, 'Un' osservazione sui sistemi di rette degli spazi superiori', *Rendiconti del Cercolo Matematico di Palermo*, vol. ii, 1888, p. 148.

† This theorem and the next seem to be due to Moore, 'Some properties of lines in a space of four dimensions', *American Journal of Math.*, vol. xxxiii, p. 151. His excellent article can be consulted with profit in connexion with all that we have done in the present section.

We turn lastly to the hypercomplex. This is written

$$p_{ij} \equiv p_{ij}(u, v, w, \omega, \omega').$$

We may assume that our circle (p) is determined by (x) where $x_i \equiv x_i(u, v, w)$ and (y) where $y_i \equiv y_i(w, \omega, \omega')$. The circle orthogonal to the spheres (x), (y), and $(y+dy)$ has coordinates of the form

$$q_{ij} = \alpha_{ij} d\omega + \beta_{ij} d\omega'.$$

This circle will trace a coaxal system if (x) be fixed. Its foci are on (p) and are the pairs of points where (p) meets adjacent circles of the hypercomplex which lie on (x). But if a circle trace a coaxal system, its foci move on the circle whose foci are the points common to all circles of the coaxal system, and are harmonically separated by the vertices of the null circles of the coaxal system, i.e. they trace an involution.

Theorem 10.] *The circles of a hypercomplex adjacent to a given circle and lying with it on a chosen sphere meet the chosen circle in pairs of points of an involution.*

§ 2. Parametric Method for Circle Congruences.

Of all systems of circles in space indubitably the most interesting is the congruence. There is comparatively little that can be easily reached in this connexion, however, if we stick to our Plücker circle coordinates. In the next two sections we shall develop two other methods which will be found to yield ample returns.* The first of these is called the 'parametric' method, and consists in expressing the coordinates of a point on a circle of a congruence through the fundamental equations

$$x_i \equiv \alpha_i t^2 + \beta_i t + \gamma_i,$$

$$\alpha_i \equiv \alpha_i(u, v), \quad \beta_i \equiv \beta_i(u, v), \quad \gamma_i \equiv \gamma_i(u, v). \tag{1}$$

$$(\alpha\alpha) \equiv (\alpha\beta) \equiv (\beta\gamma) \equiv (\gamma\gamma) \equiv 0, \quad (\beta\beta) \equiv 1, \quad (\gamma\alpha) \equiv -\tfrac{1}{2}.$$

* A good part of all that remains in the present chapter will be found in an article by the Author, 'Congruences and Complexes of Circles', *Transactions of the American Math. Soc.*, vol. xv, 1914.

It appears from these that (α) and (γ) are two arbitrary points of the circle, while (β) is the sphere through them orthogonal to the circle. For the Plücker coordinates of this circle we have

$$\rho p_{ij} = \begin{vmatrix} \alpha_k & \alpha_l & \alpha_m \\ \beta_k & \beta_l & \beta_m \\ \gamma_k & \gamma_l & \gamma_m \end{vmatrix}. \tag{2}$$

If we remember that in S_4 the circles orthogonal to one sphere appear as lines in a hyperplane, and that the lines of a congruence in any S_3 are tangent to two surfaces (or meet a curve or curves), i. e. each intersects two adjacent lines, we reach

Theorem 11.] *If the circles of a congruence be all orthogonal to one sphere the congruence is focal.*

There is a second type of congruence, or rather a sub-type of the focal congruence, which is of special interest. This is called the *normal* congruence, and consists in ∞^2 circles orthogonal to the members of a one-parameter family of surfaces.* In counting the number of surfaces to which a circle is orthogonal we count the number of points where a circle meets an orthogonal surface. Thus the circles orthogonal to a fixed sphere are said to be orthogonal to *two* surfaces. Let us find the analytical conditions for a normal congruence in terms of our various parameters. In (1) we wish t to be such a function of u and v that if dx_n be the corresponding increment for x_n, while δx_n is the increment along the circle,

$$(dx\,\delta x) = 0. \tag{3}$$

From (1)

$$(\alpha d\alpha) \equiv (\beta d\beta) \equiv (\gamma d\gamma) \equiv (\alpha d\beta)+(\beta d\alpha) \equiv (\gamma d\beta)+(\beta d\gamma)$$
$$\equiv (\alpha d\gamma)+(\gamma d\alpha) = 0.$$

$$dx_n \equiv t^2 d\alpha_n + t d\beta_n + d\gamma_n + (2t\alpha_n + \beta_n)dt.$$

$$\delta x_n = (2\,\alpha_n t + \beta_n)\delta t.$$

$$(dx\,\delta x) = \delta t\left[(\alpha d\beta)t^2 + 2\,(\alpha d\gamma)t + (\beta d\gamma) + dt\right].$$

* Some writers, as Eisenhart, *Differential Geometry*, Boston, 1909, call normal congruences 'Cyclic Systems'.

$$-\frac{\partial t}{\partial u}=\left(\alpha\frac{\partial\beta}{\partial u}\right)t^2+2\left(\alpha\frac{\partial\gamma}{\partial u}\right)t+\left(\beta\frac{\partial\gamma}{\partial u}\right),$$

$$-\frac{\partial t}{\partial v}=\left(\alpha\frac{\partial\beta}{\partial v}\right)t^2+2\left(\alpha\frac{\partial\gamma}{\partial v}\right)t+\left(\beta\frac{\partial\gamma}{\partial v}\right).$$

These are compatible if

$$\left\{\left(\frac{\partial\alpha}{\partial v}\frac{\partial\beta}{\partial u}\right)-\left(\frac{\partial\alpha}{\partial u}\frac{\partial\beta}{\partial v}\right)-2\left[\left(\alpha\frac{\partial\beta}{\partial u}\right)\left(\alpha\frac{\partial\gamma}{\partial v}\right)-\left(\alpha\frac{\partial\beta}{\partial v}\right)\left(\alpha\frac{\partial\gamma}{\partial u}\right)\right]\right\}t^2,$$

$$+2\left\{\left(\alpha\frac{\partial\beta}{\partial v}\right)\left(\beta\frac{\partial\gamma}{\partial u}\right)-\left(\alpha\frac{\partial\beta}{\partial u}\right)\left(\beta\frac{\partial\gamma}{\partial v}\right)+\left(\frac{\partial\alpha}{\partial v}\frac{\partial\gamma}{\partial u}\right)-\left(\frac{\partial\alpha}{\partial u}\frac{\partial\gamma}{\partial v}\right)\right\}t,$$

$$+\left\{\left(\frac{\partial\beta}{\partial v}\frac{\partial\gamma}{\partial u}\right)-\left(\frac{\partial\beta}{\partial u}\frac{\partial\gamma}{\partial v}\right)+2\left[\left(\alpha\frac{\partial\gamma}{\partial v}\right)\left(\beta\frac{\partial\gamma}{\partial u}\right)\right.\right.$$

$$\left.\left.-\left(\alpha\frac{\partial\gamma}{\partial u}\right)\left(\beta\frac{\partial\gamma}{\partial v}\right)\right]\right\}=0. \qquad (4)$$

The last equation, being quadratic, is identically satisfied if it have three solutions. We thus get the fundamental theorem of Ribaucour.*

Theorem 12.] *If the circles of a congruence be orthogonal to more than two surfaces, the congruence is a normal one.*

Theorem 13.] *If a congruence of circles orthogonal to a fixed sphere have any other orthogonal surface it is a normal congruence.*

When the fixed sphere is null we imagine that we are in cartesian space, and prove the theorem by inverting into a normal line congruence. It is to be noted that a focal congruence will go into a focal congruence under every transformation of the twenty-four-parameter linear sphere

* The theorems of Ribaucour cited in the present chapter are in the following notes in the *Comptes Rendus*: 'Sur la déformation des surfaces', vol. lxx, 1870, p. 330; 'Sur les systèmes cycliques' and 'Sur les faisceaux de cercles', vol. lxxvi, 1873, pp. 478, 830.

group, while a normal congruence will go into a normal congruence under every spherical transformation.

When we start with the hypothesis that the circles of our congruence are surely orthogonal to two surfaces our equations are vastly simplified. We assume at the outset that (α) and (γ) trace two orthogonal surfaces

$$(\alpha d\beta) \equiv (\beta d\alpha) \equiv (\gamma d\beta) \equiv (\beta d\gamma) = 0. \tag{5}$$

$$-\frac{\partial t}{\partial u} = 2\left(\alpha\frac{\partial \gamma}{\partial u}\right)t, \quad -\frac{\partial t}{\partial v} = 2\left(\alpha\frac{\partial \gamma}{\partial v}\right)t.$$

The condition for a normal congruence is then simply

$$\left(\frac{\partial \alpha}{\partial u}\frac{\partial \gamma}{\partial v}\right) = \left(\frac{\partial \alpha}{\partial v}\frac{\partial \gamma}{\partial u}\right). \tag{6}$$

Two solutions correspond to the values 0 and ∞ for t; if a third solution be $t \equiv t(u, v)$ a fourth will be rt, where r is any constant. But this constant gives the cross ratio of the four corresponding points upon the circle, by IV (9). We thus get another admirable theorem of Ribaucour's.

Theorem 14.] *Any four orthogonal surfaces of the circles of a normal congruence will meet them in sets of points having a fixed cross ratio.*

We next turn back to the more general focal congruence. Let us so choose our parameters u and v that making the one or the other constant, gives us the annular surfaces. If two circles be cospherical, their foci are concyclic and vice versa; if two circles be cospherical with a third, the six foci are cospherical. We thus get Ribaucour's third theorem.

Theorem 15.] *The foci of the circles of a focal congruence will generate the two nappes of the envelope of a congruence of spheres. In cartesian space the planes of the circles will envelop the deferent of the sphere congruence.*

With regard to the last part of the theorem we have merely to notice that the points where a sphere meets two infinitely near spheres are symmetrical with regard to the plane through the centres of the three. Let us proceed to prove the converse of 15]. Let the congruence of spheres be

$$\beta_i \equiv \beta_i(u, v).$$

We may always find six such quantities A, B, C, D, E, F that $\beta_0 \dots \beta_4$ are the six linearly independent solutions of

$$A\frac{\partial^2 \theta}{\partial u^2} + B\frac{\partial^2 \theta}{\partial u\, \partial v} + C\frac{\partial^2 \theta}{\partial v^2} + D\frac{\partial \theta}{\partial u} + E\frac{\partial \theta}{\partial v} + F = 0.$$

Let us now assume explicitly that this differential equation is not parabolic, and define our sphere congruence likewise as *non-parabolic*. Our differential equation can then be reduced to the form

$$\frac{\partial^2 \beta_i}{\partial u\, \partial v} = L\frac{\partial \beta_i}{\partial u} + M\frac{\partial \beta_i}{\partial v} + N\beta_i. \tag{7}$$

The points of contact of the sphere (β) with the envelope lie also on the spheres $\left(\frac{\partial \beta}{\partial u}\right)$ and $\left(\frac{\partial \beta}{\partial v}\right)$. The points where $(\beta) + \left(\frac{\partial \beta}{\partial u}\right)du$ touches the envelope lie on $\left(\frac{\partial \beta}{\partial u}\right) + \left(\frac{\partial^2 \beta}{\partial u^2}\right)du$, $\left(\frac{\partial \beta}{\partial v}\right) + \left(\frac{\partial^2 \beta}{\partial u\, \partial v}\right)dv$. Since the five spheres (β), $\left(\frac{\partial \beta}{\partial u}\right)$, $\left(\frac{\partial \beta}{\partial v}\right)$, $\left(\frac{\partial^2 \beta}{\partial u^2}\right)$, $\left(\frac{\partial^2 \beta}{\partial u\, \partial v}\right)$ are linearly dependent, there is (at least) one sphere orthogonal to all, and the circle whose foci are the points of contact of (β) with the envelope is cospherical with that whose foci are the points of contact of $(\beta) + \left(\frac{\partial \beta}{\partial u}\right)du$. The same will hold for (β) and $(\beta) + \left(\frac{\partial \beta}{dv}\right)dv$; hence

Theorem 16.] *If the foci of the circles of a congruence be the points of contact of the spheres of a congruence with the two nappes of the envelope, the congruence of circles is focal.*

When we surely know that our congruence of circles is focal, we may introduce notable simplifications by choosing as u and v the *focal parameters*, i. e. those which give the annular surfaces. We shall take as (β) the sphere whose points of contact with its envelope give the foci, for such a sphere is surely orthogonal to our circle: (y) and (z) shall be the foci of the circle. We have the equations

$$(y\alpha) = (y\beta) = (y\gamma) = (yd\beta) = 0,$$

$$(z\alpha) = (z\beta) = (z\gamma) = (zd\beta) = 0.$$

The first set of these equations leads to an identity of the type

$$A\alpha_i + B\beta_i + C\gamma_i + D\frac{\partial \beta_i}{\partial u} = 0.$$

Multiplying through by β_i and summing, we find

$$\frac{\partial \beta_i}{\partial u} = a\alpha_i + c\gamma_i, \quad \frac{\partial \beta_i}{\partial v} = a'\alpha_i + c'\gamma_i.$$

Since each of two pairs of points $(\alpha)+(dd)$, $(\gamma)+(d\gamma)$ lies on a sphere through our circle, we find

$$\frac{\partial \gamma_i}{\partial u} = l\alpha_i + m\beta_i + n\gamma_i + r\frac{\partial \alpha_i}{\partial u}.$$

$$\frac{\partial \gamma_i}{\partial v} = \lambda\alpha_i + \mu\beta_i + \nu\gamma_i + \rho\frac{\partial \alpha_i}{\partial v}. \tag{8}$$

If, now, we inquire under what circumstances our focal congruence shall also be a normal one, we must substitute in (4). The coefficient of t^2 and the constant will be found to vanish automatically, and there remains

$$\left[cc' - 4\left(\frac{\partial \alpha}{\partial u}\frac{\partial \alpha}{\partial v}\right)\right](r-\rho) = 0.$$

We first assume $\qquad r = \rho.$

Then every solution of the equations

$$(x\alpha) = (x\beta) = (x\gamma) = (xd\alpha) = 0,$$

where du and dv are variable parameters, will be a solution of

$$(xd\beta) = (xd\gamma) = 0.$$

Hence every series of the congruence will generate an annular surface. Any two infinitely near circles are cospherical, hence any two whatever are cospherical, and all lie on a sphere, or pass through a point-pair. But the circles on a sphere could not certainly be a normal congruence. Hence we must have the circles through a point-pair, distinct or adjacent. We take up now the other hypothesis,

$$cc' - 4\left(\frac{\partial \alpha}{\partial u}\frac{\partial \alpha}{\partial v}\right) = 0;$$

we may take for the *focal spheres*, i.e. the spheres through a circle in general position which contain adjacent circles of the congruence,

$$\sigma s_i = \left| \alpha_j \, \beta_k \, \gamma_l \, \frac{\partial \alpha_m}{\partial u} \right|, \quad \sigma s_i' = \left| \alpha_j \, \beta_k \, \gamma_l \, \frac{\partial \alpha_m}{\partial v} \right|.$$

It then appears that these two are mutually orthogonal in view of the equation above.

Theorem 17.] *A necessary and sufficient condition that a focal congruence should be normal is that it should consist in circles through a point-pair or else that the focal spheres through every circle should be mutually orthogonal.*

Having shown that a focal congruence can be normal, let us proceed conversely to show that a normal congruence must be focal. Starting with (5) we have the additional equations

$$\left(\beta\frac{\partial \alpha}{\partial u}\right) \equiv \left(\beta\frac{\partial \alpha}{\partial v}\right) \equiv (\beta\alpha) \equiv (\beta\gamma) \equiv \left(\beta\frac{\partial \gamma}{\partial u}\right) \equiv \left(\beta\frac{\partial \gamma}{\partial v}\right) = 0,$$

$$\frac{\partial \gamma_i}{\partial u} = l\alpha_i + m\gamma_i + n\,\frac{\partial \alpha_i}{\partial u} + r\,\frac{\partial \alpha_i}{\partial v},$$

$$\frac{\partial \gamma_i}{\partial v} = \lambda\alpha_i + \mu\gamma_i + \nu\,\frac{\partial \alpha_i}{\partial u} + \rho\,\frac{\partial \alpha_i}{\partial v}.$$

We may in two different ways, usually distinct, find such a pair of values, du, dv, that the following equations are compatible:

$$(s\alpha) = (s\beta) = (s\gamma) = (sd\alpha) = (sd\gamma) = 0.$$

We change our parameters u and v so that these shall correspond to the equations $du = 0$ and $dv = 0$. We then have $r = \nu = 0$.

The condition of orthogonality (6) will give

$$n\left(\frac{\partial \alpha}{\partial u}\frac{\partial \alpha}{\partial v}\right) = \rho\left(\frac{\partial \alpha}{\partial u}\frac{\partial \alpha}{\partial v}\right).$$

If $n = \rho$ we should have

$$d\gamma_i \equiv L\alpha_i + M\gamma_i + Nd\alpha_i.$$

If we consider the circles whose foci are (α) and (γ) we see that each is cospherical with every adjacent circle of the same sort. Hence all pass through a point-pair, and (α) and (γ) lie on a fixed circle—an absurdity—or all lie on a fixed sphere, to which the circle through (α) and (γ) is orthogonal. But by 11] a congruence of circles orthogonal to a fixed sphere is focal. There remains the possibility that $n \neq \rho$.

$$\left(\frac{\partial \alpha}{\partial u}\frac{\partial \alpha}{\partial v}\right) = -\left(\alpha\frac{\partial^2 \alpha}{\partial u\,\partial v}\right) = 0.$$

Similarly,

$$\left(\frac{\partial \gamma}{\partial u}\frac{\partial \gamma}{\partial v}\right) = -\left(\gamma\frac{\partial^2 \gamma}{\partial u\,\partial v}\right) = 0.$$

The conditions for compatibility of the equations for $\left(\frac{\partial \gamma}{\partial u}\right)$ and $\left(\frac{\partial \gamma}{\partial v}\right)$ give

$$(\mu-\rho)\frac{\partial^2 \alpha_i}{\partial u\,\partial v} + B\frac{\partial \alpha_i}{\partial u} + C\frac{\partial \alpha_i}{\partial v} + D\alpha_i + E\gamma_i = 0.$$

Multiplying through by α_i and summing, we find $E = 0$. We have, thus, four equations:

$$\frac{\partial^2 \alpha_i}{\partial u\, \partial v} = a\alpha_i + b\frac{\partial \alpha_i}{\partial u} + c\frac{\partial \alpha_i}{\partial v}, \quad \frac{\partial^2 \gamma_i}{\partial u\, \partial v} = a'\alpha_i + b'\frac{\partial \alpha_i}{\partial u} + c'\frac{\partial \gamma_i}{\partial v}.$$

$$\frac{\partial \gamma_i}{\partial u} = l\alpha_i + m\gamma_i + n\frac{\partial \alpha_i}{\partial u}, \quad \frac{\partial \gamma_i}{\partial v} = \lambda\alpha_i + \mu\gamma_i + \nu\frac{\partial \alpha_i}{\partial v}. \tag{9}$$

The equations

$$(x\beta) = \left(x\frac{\partial \beta}{\partial u}\right) = \left(x\frac{\partial \beta}{\partial v}\right) = \left(x\frac{\partial^2 \beta}{\partial u\, \partial v}\right) = 0$$

have two distinct solutions (α) and (γ), so that

$$\frac{\partial^2 \beta_i}{\partial u\, \partial v} = p\frac{\partial \beta_i}{\partial v} + q\frac{\partial \beta_i}{\partial v} + r\beta_i.$$

Again, from the equations

$$\left(\alpha\frac{\partial \beta}{\partial u}\right) = (\alpha\alpha) = (\alpha\beta) = \left(\alpha\frac{\partial \alpha}{\partial u}\right) = \left(\alpha\frac{\partial \alpha}{\partial v}\right),$$

$$\frac{\partial \beta_i}{\partial u} = L\alpha_i + M\beta_i + N\frac{\partial \alpha_i}{\partial u} + R\frac{\partial \alpha_i}{\partial v}.$$

Multiplying through by β_i and summing, $M = 0$.

Multiplying through by $\frac{\partial \alpha_i}{\partial v}$ and summing, $R = 0$.

$$\frac{\partial \beta_i}{\partial u} = L\alpha_i + N\frac{\partial \alpha_i}{\partial u}; \quad \frac{\partial \beta_i}{\partial v} = L'\alpha_i + M'\frac{\partial \alpha_i}{\partial v}.$$

$$\left(\frac{\partial \beta}{\partial u}\frac{\partial \beta}{\partial v}\right) = \left(\beta\frac{\partial^2 \beta}{\partial u\, \partial v}\right) = 0.$$

Hence, in the above partial differential equation for β_i, $r = 0$, and we have

$$\frac{\partial^2 \beta_i}{\partial u\, \partial v} = p\frac{\partial \beta_i}{\partial u} + q\frac{\partial \beta_i}{\partial v}. \tag{10}$$

We see from the equations

$$\left(\frac{\partial \beta}{\partial v} \alpha\right) = \left(\frac{\partial \beta}{\partial v} \beta\right) = \left(\frac{\partial \beta}{\partial v} \gamma\right) = \left(\frac{\partial \beta}{\partial v} \frac{\partial \alpha}{\partial u}\right) = \left(\frac{\partial \beta}{\partial v} \frac{\partial \beta}{\partial u}\right) = \left(\frac{\partial \beta}{\partial v} \frac{\partial \gamma}{\partial u}\right)$$

that $\left(\frac{\partial \beta}{\partial v}\right)$ is a focal sphere, and similarly $\left(\frac{\partial \beta}{\partial u}\right)$ is a focal sphere.

Theorem 18.] *Every normal congruence is focal.*

Again, suppose that we have a focal congruence with two orthogonal surfaces, not consisting in one fixed sphere twice counted. The intersections with these surfaces shall be (α) and (γ), while u and v are the focal parameters.

$$\frac{\partial \gamma_i}{\partial v} = l\alpha_i + m\gamma_i + n\frac{\partial \alpha_i}{\partial u}.$$

$$\frac{\partial \gamma_i}{\partial v} = \lambda\alpha_i + \mu\gamma_i + \rho\frac{\partial \alpha_i}{\partial v}.$$

$$(n-\rho)\frac{\partial^2 \alpha_i}{\partial u\, \partial v} + B\frac{\partial \alpha_i}{\partial u} + C\frac{\partial \alpha_i}{\partial v} + D\alpha_i + E\gamma_i = 0.$$

$$\left(\beta\frac{\partial^2 \alpha}{\partial u\, \partial v}\right) = -\left(\frac{\partial \beta}{\partial u}\frac{\partial \alpha}{\partial v}\right) = -\left(\frac{\partial \beta}{\partial v}\frac{\partial \alpha}{\partial u}\right) = 0.$$

$$\frac{\partial \beta_i}{\partial u} = L\alpha_i + M\frac{\partial \alpha_i}{\partial u}, \quad \frac{\partial \beta_i}{\partial v} = L\alpha_i' + M'\frac{\partial \alpha_i}{\partial v}.$$

$$\left(\frac{\partial \alpha}{\partial u}\frac{\partial \alpha}{\partial v}\right) = 0.$$

$$\left(\frac{\partial \alpha}{\partial u}\frac{\partial \gamma}{\partial v}\right) = \left(\frac{\partial \alpha}{\partial v}\frac{\partial \gamma}{\partial u}\right).$$

Theorem 19.] *If the circles of a congruence have two orthogonal trajectories, not a fixed sphere counted twice, the congruence will be focal if it be normal, and vice versa.*

We easily see from (9) that the parameters u and v give curves on the (α) and (γ) surfaces which are both orthogonal and conjugate. We thus reach another important theorem, also due to Ribaucour.

Theorem 20]. *In a normal congruence the lines of curvature correspond to one another on all orthogonal surfaces, and give the annular surfaces of the congruence.*

If we take the two focal spheres through a circle of a normal congruence they are orthogonal to one another, and each touches one annular surface all along that circle, whence

Theorem 21.] *The orthogonal surfaces of the circles of a normal congruence and the annular surfaces determine a triply orthogonal system.*

Theorem 22.] *The lines of curvature of the orthogonal surfaces of a normal congruence of circles in cartesian space correspond to the focal developables in the congruence of axes of these circles.**

We next seek a converse to 20]. Suppose that we have such a congruence of not null spheres (β) that the lines of curvature correspond to one another in the two nappes of the envelope. We take these to determine the parameters u and v. The first two equations (9) will hold, (α) and (γ) being the points of contact of (β) with its envelope. We have also the equations

$$\frac{\partial \gamma_i}{\partial u} = l\alpha_i + m\gamma_i + n\frac{\partial \alpha_i}{\partial u} + r\frac{\partial \alpha_i}{\partial v},$$

$$\frac{\partial \gamma_i}{\partial v} = \lambda\alpha_i + \mu\gamma_i + \nu\frac{\partial \alpha_i}{\partial u} + \rho\frac{\partial \alpha_i}{\partial v}.$$

Differentiating the first equation to v, and substituting in the second, we have, with the aid of (9),

$$r\frac{\partial^2 \alpha_i}{\partial v^2} + A\frac{\partial \alpha_i}{\partial u} + B\frac{\partial \alpha_i}{\partial v} + C\alpha_i + D\gamma_i = 0.$$

Multiplying through by β_i and summing,

$$r\left(\beta\frac{\partial^2 \alpha}{\partial v^2}\right) = 0.$$

* The congruence of axes of the circles of a normal congruence has been extensively studied under the name of 'cyclic congruence'. Vide, *inter alia*, Eisenhart, loc. cit., pp. 431 ff.

If the second factor vanish, (β) would be the osculating sphere for one line of curvature, and so generate a one-nappe envelope, which is not the case. Hence $r = 0$ and, similarly, $\nu = 0$. Under these circumstances, however, we easily see that our congruence is focal, u and v being focal parameters, whence, from 19],

Theorem 23.] *If a congruence of spheres establish such a point to point correspondence between the two distinct nappes of the envelope that the lines of curvature correspond in the two, then the circles orthogonal to the various spheres at their points of contact with the envelope will generate a normal congruence.*

We next vary our hypotheses by assuming that the circles of the congruence have two orthogonal surfaces, traced by the points (α) and (γ), and that the surfaces $v =$ const. are annular. We have at our disposal equations (1) and (5) as well as

$$\frac{\partial \gamma_i}{\partial u} = l\alpha_i + m\gamma_i + n\frac{\partial \alpha_i}{\partial u}, \quad \frac{\partial \gamma_i}{\partial v} = \lambda\alpha_i + \mu\gamma_i + \nu\frac{\partial \alpha_i}{\partial u} + \rho\frac{\partial \alpha_i}{\partial v}.$$

Since (α), (β), (γ), $\left(\frac{\partial \alpha}{\partial u}\right)$, $\left(\frac{\partial \beta}{\partial u}\right)$ are orthogonal to one sphere

$$A\alpha_i + B\beta_i + C\gamma_i + D\frac{\partial \alpha_i}{\partial u} + E\frac{\partial \beta_i}{\partial u} = 0.$$

Multiplying through by β_i and summing, $B = 0$. Multiplying through by α_i and summing, $C = 0$. It is easy to see that we could not have $E = 0$, hence we may take $E = 1$. Our parameter u shall now be so chosen as to give with v an orthogonal system of curves on the surface (α). We have the equations

$$\left(\frac{\partial \alpha}{\partial u}\frac{\partial \alpha}{\partial v}\right) = \left(\frac{\partial \beta}{\partial u}\frac{\partial \alpha}{\partial v}\right) = -\left(\alpha\frac{\partial^2 \beta}{\partial u\,\partial v}\right) = \left(\frac{\partial \beta}{\partial v}\frac{\partial \alpha}{\partial u}\right)$$

$$\equiv -\left(\beta\frac{\partial^2 \alpha}{\partial u\,\partial v}\right) = 0.$$

We next observe that (α) and (β) are independent solutions of

$$\left(x \frac{\partial^2 \alpha}{\partial u\,\partial v}\right) = \left(x \frac{\partial \alpha}{\partial v}\right) = \left(x \frac{\partial \alpha}{\partial v}\right) = (x\alpha) = 0.$$

Hence

$$\frac{\partial^2 \alpha_i}{\partial u\,\partial v} = H\frac{\partial \alpha_i}{\partial u} + K\frac{\partial \alpha_i}{\partial v} + L\alpha_i.$$

This, however, combined with the condition of compatibility of the equation for $\frac{\partial \gamma_i}{\partial u}$ and $\frac{\partial \gamma_i}{\partial v}$, gives $\nu = 0$, and our congruence is focal.

Theorem 24.] *If the circles of a congruence be orthogonal to two surfaces other than a sphere counted twice, and if they constitute the circles of curvature of a one-parameter family of annular surfaces, the congruence is normal.**

Let us see if we can find the condition for a normal congruence in terms of (β) alone. We begin with a slight change of notation, writing

$$\beta_i = \rho z_i, \quad 1 = \sqrt{(\beta\beta)} = \rho\sqrt{(zz)}.$$

Differentiating, and substituting in (10), we see that our quantities z_i are solutions of a differential equation of the type

$$B\frac{\partial^2 \theta}{\partial u\,\partial v} + D\frac{\partial \theta}{\partial u} + E\frac{\partial \theta}{\partial v} + F\theta = 0.$$

Since the expression $1 = \sqrt{(\beta\beta)}$ is a solution of (10), so will $\sqrt{(zz)}$ be a solution of the last equation. In other words, the six coordinates of the *oriented* sphere (z) are solutions of one same non-parabolic partial differential equation of the

* This theorem is sometimes stated without the restriction upon the two orthogonal trajectories, but a moment's reflection shows that this restriction is necessary.

second order. Suppose, conversely, that we have a non-parabolic equation

$$A\frac{\partial^2\theta}{\partial u^2}+B\frac{\partial^2\theta}{\partial u\,\partial v}+C\frac{\partial^2\theta}{\partial v^2}+D\frac{\partial\theta}{\partial u}+E\frac{\partial\theta}{\partial v}+F\theta=0,$$

$$B^2-4AC\not\equiv 0,$$

whereof six solutions are given by the coordinates of an oriented sphere (z). If we change variables so as to write

$$\beta_i=\frac{z_i}{\sqrt{(zz)}},$$

we see that β_i and $\sqrt{(\beta\beta)}=1$ are solutions of an equation of like type, so that by a suitable change of the parameter u and v we find β_i is a solution of (10). If (α) and (γ) be the points of contact of (β) with its envelope,

$$\left(\alpha\frac{\partial^2\beta}{\partial u\,\partial v}\right)=\left(\beta\frac{\partial^2\beta}{\partial u\,\partial v}\right)=\left(\gamma\frac{\partial^2\beta}{\partial u\,\partial v}\right)=\left(\frac{\partial\beta}{\partial u}\frac{\partial\beta}{\partial v}\right)=0.$$

$$\left(\alpha\frac{\partial\beta}{\partial u}\right)=(\alpha\alpha)=(\alpha\beta)=\left(\alpha\frac{\partial\alpha}{\partial u}\right)=\left(\alpha\frac{\partial\alpha}{\partial v}\right).$$

$$\frac{\partial\beta_i}{\partial u}=A\alpha_i+B\beta_i+C\frac{\partial\alpha_i}{\partial u}+D\frac{\partial\alpha_i}{\partial v}.$$

Multiplying through by β_i and summing, $B=0$.

Multiplying through by $\frac{\partial\beta_i}{\partial v}$ and summing, $D=0$.

Differentiating to v and substituting in (10) we get to the first equation (9). The second equation (9) comes similarly.

Theorem 25.] *A necessary and sufficient condition that a congruence of oriented spheres should establish such a point to point correspondence between the two nappes of the envelope that the lines of curvature correspond to one another is that their coordinates should be the solutions of a non-parabolic partial differential equation of the second order.**

* Cf. Darboux, *Théorie générale des surfaces*, vol. ii, Paris, 1889, p. 332.

We derive an interesting corollary from this by means of the line sphere transformation of Ch. XI. A congruence of lines which establishes a correspondence between the asymptotic curves on the two nappes of the focal surface is called a W congruence.

The Plücker coordinates of the lines of a W congruence and of no other are solutions of a partial differential equation of the second order and non-parabolic type.

We saw that in the case of a focal congruence the foci of the various circles generated the two nappes of the envelope of a congruence of spheres whose deferent is the envelope of the planes of the circles. This suggests that perhaps, in certain cases, the locus of the centres of the spheres might also be the locus of the centres of the circles. Suppose that this is the case. The annular surfaces of the congruence must correspond to the lines of curvature of the deferent, and the centres of the focal spheres must be the centres of curvature for the deferent. Since the locus of the centres of the circles is the envelope of their planes, the distances from the centre of a focal sphere to the centres of the adjacent circles lying thereon will differ by an infinitesimal of the second order, as will the radii of the two circles. The circles of the congruence have thus a constant radius.

Theorem 26.] *If the envelope of the planes of a congruence of circles in cartesian space be the locus of their centres, a necessary and sufficient condition that the congruence should be focal is that the circles should have a constant radius.*

We now suppose that the congruence is normal. The focal spheres are mutually orthogonal, by 17]. If ρ_1 and ρ_2 be the radii of curvature of the surface, while r is the constant radius of our circles,

$$\rho_1 \rho_2 = r^2.$$

Theorem 27.] *If with each real point of a real surface in cartesian space as centre a real circle be drawn in the tangent plane, a necessary and sufficient condition that these circles* *should generate a normal congruence is that the surface should be pseudospherical, and that the radius of the circle should be equal to the square root of the negative of the reciprocal of the measure of total curvature.**

Suppose that we have two points A and B and a not null sphere. It is easy to show that a necessary and sufficient condition that the two spheres with centres A and B orthogonal to the given sphere should be orthogonal to one another is that the sphere on (AB) as diameter should be orthogonal thereto. Secondly, if two lines intersect, the circles orthogonal to a non-planar sphere and having these lines as axes are cospherical, whence

Theorem 28.] *If the spheres on the focal segments of a line congruence as diameters intersect orthogonally a fixed sphere which is non-planar, the lines are the axes of the circles of a normal congruence orthogonal to the fixed sphere.*†

Theorem 29.] *If the circles of a normal congruence be orthogonal to a fixed non-planar sphere, the spheres whose diameters are the focal segments on their respective axes are orthogonal to the fixed sphere.*

Every focal congruence is associated with a congruence of spheres. We start with the equation (7). There will be a similar equation with solution z_i where

$$z_0 = i\beta_0 + \beta_1,\ z_1 = i\beta_0 - \beta_1,\ z_2 = -\beta_2,\ z_3 = -\beta_3,\ z_4 = -\beta_4.$$

If, thus, we assume that we are in cartesian space, and that our points are determined by special pentaspherical coordinates, while $z_0 = 1$, then z_1, z_2, z_3 will be the rectangular cartesian coordinates of the centre of the sphere.

Theorem 30.] *In a focal congruence in cartesian space the focal parameters give conjugate systems of curves on the*

* Cf. Bianchi, 'Sopra alcuni casi di sistemi tripli ciclici', *Giornale di Matematiche*, vol. xxi, 1883, p. 278.

† Eisenhart, loc. cit., p. 443.

deferent of the system of spheres whose envelope is generated by the foci of the given circles.

An interesting and difficult question connected with normal congruence is the following: given a surface, to determine a normal congruence having that surface as an orthogonal surface.* Let us suppose that we have a non-developable surface of cartesian space, and that it is expressed parametrically in terms of the lines of curvature. Then, by the formulae of Olinde Rodrigues, the two equations

$$\frac{\partial\theta}{\partial u}+\rho_1\frac{\partial\Theta}{\partial u}=0,\quad \frac{\partial\theta}{\partial v}+\rho_2\frac{\partial\Theta}{\partial v}=0$$

have the four pairs of solutions

$$(x, X),\ (y, Y),\ (z, Z),\ \left(\frac{x^2+y^2+z^2}{2},\ xX+yY+zZ\right).$$

Eliminating θ,

$$\frac{\partial}{\partial v}\left(\rho_1\frac{\partial\Theta}{\partial u}\right)=\frac{\partial}{\partial u}\left(\rho_2\frac{\partial\Theta}{\partial v}\right). \tag{11}$$

This has the solutions X, Y, Z, $-(xX+yY+zZ)$.

Similarly, the equation

$$\frac{\partial}{\partial v}\left(\frac{1}{\rho_1}\frac{\partial\theta}{\partial u}\right)=\frac{\partial}{\partial u}\left(\frac{1}{\rho_2}\frac{\partial\theta}{\partial v}\right) \tag{12}$$

will have the solutions $-2x$, $-2y$, $-2z$, $x^2+y^2+z^2-r^2$, and these are the coordinates of a sphere with (x, y, z) as centre. There is but one equation of the type

$$\frac{\partial^2\theta}{\partial u\,\partial v}+a\frac{\partial\theta}{\partial u}+b\frac{\partial\theta}{\partial v}=0$$

which has the solutions X, Y, Z; hence we get all surfaces, with the same spherical representation as our given surface, by taking the envelope of all planes with the coordinates X, Y, Z, Θ, where the latter is a solution of (11). It appears

* This development is taken direct from Darboux, *Surfaces*, cit., vol. iv, pp. 137 ff.

also from the Codazzi equations that the lines of curvature will correspond in any two such surfaces. Suppose, then, that we have a surface enveloped by the planes X, Y, Z, Θ', and write

$$\Theta = \Theta' - (Xx + Yy + Zz).$$

Here Θ and Θ' are both solutions of (11). The equation of the tangent plane will be

$$\Sigma(\xi - x)X = \Theta.$$

To find the envelope we differentiate to u and v respectively, remembering

$$\Sigma X\frac{\partial x}{\partial u} = \Sigma X\frac{\partial x}{\partial v} = 0,$$

$$\Sigma(\xi - x)\frac{\partial X}{\partial u} = \frac{\partial \Theta}{\partial u}, \qquad \Sigma(\xi - x)\frac{\partial X}{\partial v} = \frac{\partial \Theta}{\partial v}.$$

We transform these by means of (11) and the Codazzi equations, getting

$$\Sigma(\xi - x)\frac{\partial x}{\partial u} = \frac{\partial \theta}{\partial u}, \qquad \Sigma(\xi - x)\frac{\partial x}{\partial v} = \frac{\partial \theta}{\partial v},$$

where θ is a solution of (12). These equations give us also the secant of contact of the points of contact with its envelope of the sphere

$$(x-\xi)^2 + (y-\eta)^2 + (z-\zeta)^2 = -2\theta.$$

If, thus, two non-developable surfaces have the same spherical representation, the normals to the one are secants of contact with the envelope of spheres whose centres lie on the other. Conversely, the orthogonal trajectories of such secants of contact will be a surface of the form required, provided the square of the diameter is a solution of (12).

We next suppose that P and P' are two infinitely near points on a line of curvature of that surface which is the locus of the centre of the moving sphere; HK and $H'K'$ the corresponding pairs of points of contact of the sphere with its envelope. Since the lines of curvature correspond in the two surfaces, the lines HK and $H'K'$ meet (to the fourth order of infinitesimals), let us say, in R, while the normals at P and P' meet, let us say, in Q. The four points H, K, H', K' are concyclic, by 16]; hence R has the same power with regard to the circles PHK and $P'H'K'$. Again, PQ is tangent to the first of these circles at P, and $P'Q$ touches the second at P' while PP' is orthogonal to PQ, so that Q also has the same power with regard to both. It follows that the two circles are cospherical, their common sphere being orthogonal to the other line of curvature at P. A similar state of affairs will hold if we proceed infinitesimally along this other line of curvature. Hence the congruence of circles P, H, K is focal, and, since the focal spheres are mutually orthogonal, it is normal, the given surface being one orthogonal surface.

Theorem 31.] *If a non-developable surface be given in cartesian space, and a second having the same spherical representation as the first, then the normals to the latter are secants of contact with the two nappes of the envelope of spheres whose centres lie at the corresponding points of the former. The circles, each inverse to one of these secants, with regard to the corresponding sphere, generate a normal congruence, having as one orthogonal trajectory the given surface.*

It is not at all clear that all normal congruences with a given orthogonal surface can be obtained in this way.

Any triple orthogonal system of surfaces will lead to a normal congruence in the following simple way. Let the parameters giving the various surfaces of the system be u, v, w.

$$x_i = x_i(u, v, w), \quad i = 0 \ldots 4.$$

$$\frac{\partial^2 x_i}{\partial v\, \partial w} = a\frac{\partial x_i}{\partial v} + b\frac{\partial x_i}{\partial w} + c x_i,$$

$$\frac{\partial^2 x_i}{\partial w\, \partial u} = a'\frac{\partial x_i}{\partial w} + b'\frac{\partial x_i}{\partial u} + c' x_i,$$

$$\frac{\partial^2 x_i}{\partial u\, \partial v} = a''\frac{\partial x_i}{\partial u} + b''\frac{\partial x_i}{\partial v} + c'' x_i,$$

$$\frac{\partial^3 x_i}{\partial w^2\, \partial u} = a'''\frac{\partial x_i}{\partial w} + b'''\frac{\partial x_i}{\partial u} + c''' x_i + d'''\frac{\partial^2 x_i}{\partial w^2}.$$

The sphere (s), where

$$\rho s_i = \left| x_j \frac{\partial x_k}{\partial u} \frac{\partial x_l}{\partial w} \frac{\partial^2 x_m}{\partial w^2} \right|,$$

contains all points of the osculating circle to the w curve at (x) and at $(x)+\left(\frac{\partial x}{\partial w}\right)du$. It is orthogonal to the u, v surface, and to the v curve thereon. A similar sphere is found by interchanging the parameters u and v; and these two spheres are mutually orthogonal, whence

Theorem 32.] *If a triply orthogonal system of surfaces be given, the osculating circles to one system of trajectory curves at the points where they meet one orthogonal surface form a normal congruence.*

§ 3. The Kummer Method.

There is a totally different method of analysis which may be profitably applied to congruences of circles, and which leads to theorems of a different sort from the classical ones which we have just proved. We have frequently had occasion to point out that the circles of a pentaspherical space could be treated as lines in an S_4 of elliptic measurement. If, then, we are interested, not only in the points on a circle, but also in the spheres through it, there is much to be gained by copying the standard methods of line geometry adapted to a space of elliptic measurement, and of four dimensions.*

A circle of a given congruence shall be determined by two mutually orthogonal not null spheres (y) and (z), whose coordinates are analytic functions of two independent parameters u, v,

$$y_i \equiv y_i(u, v), \quad z_i \equiv z_i(u, v), \quad i = 0 \ldots 4,$$
$$(yy) = (zz) = 1, \quad (yz) = 0. \tag{13}$$

* For a discussion of the corresponding formulae for line geometry in elliptic geometry of three dimensions, see the Author's *Non-Euclidean Geometry*, cit., Ch. XVI.

We have three fundamental equations

$$
\begin{aligned}
(dy\,dy)-(z\,dy)^2 &= E du^2+2\,F du\,dv+G dv^2,\\
(dz\,dz)-(y\,dz)^2 &= E' du^2+2\,F' du\,dv+G' dv^2,\\
(dy\,dz) &= e du^2+(f+f')du\,dv+g dv^2.
\end{aligned}
\tag{14}
$$

More specifically

$$
\left(\frac{\partial y}{\partial u}\frac{\partial y}{\partial u}\right)-\left(z\frac{\partial y}{\partial u}\right)^2\equiv E,\quad \left(\frac{\partial y}{\partial u}\frac{\partial y}{\partial v}\right)-\left(z\frac{\partial y}{\partial u}\right)\left(z\frac{\partial y}{\partial v}\right)\equiv F,
$$

$$
\left(\frac{\partial y}{\partial v}\frac{\partial y}{\partial v}\right)-\left(z\frac{\partial y}{\partial v}\right)^2\equiv G;
$$

$$
\left(\frac{\partial z}{\partial u}\frac{\partial z}{\partial u}\right)-\left(y\frac{\partial z}{\partial u}\right)^2\equiv E',\quad \left(\frac{\partial z}{\partial u}\frac{\partial z}{\partial v}\right)-\left(y\frac{\partial z}{\partial u}\right)\left(y\frac{\partial z}{\partial u}\right)\equiv F',
$$

$$
\left(\frac{\partial z}{\partial v}\frac{\partial z}{\partial v}\right)-\left(y\frac{\partial z}{\partial v}\right)^2\equiv G'; \tag{15}
$$

$$
\left(\frac{\partial y}{\partial u}\frac{\partial z}{\partial u}\right)\equiv e,\quad \left(\frac{\partial y}{\partial u}\frac{\partial z}{\partial v}\right)\equiv f',\quad \left(\frac{\partial y}{\partial v}\frac{\partial z}{\partial u}\right)\equiv f,\quad \left(\frac{\partial y}{\partial v}\frac{\partial z}{\partial v}\right)\equiv y.
$$

These various coefficients are connected by a symmetrical syzygy. We apply the Frobenius identity to the six spheres

$$
(y),\quad (z),\quad \left(\frac{\partial y}{\partial u}\right),\quad \left(\frac{\partial y}{\partial v}\right),\quad \left(\frac{\partial z}{\partial u}\right),\quad \left(\frac{\partial z}{\partial v}\right).
$$

$$
\begin{vmatrix}
E & F & e & f'\\
F & G & f & g\\
e & f & E' & F'\\
f' & g & F' & G'
\end{vmatrix}=0. \tag{16}
$$

We saw in Ch. XII that if two circles be given in general position, i.e. no focus of one lying on an isotropic with a focus of the other, there are two circles of a cross cospherical and orthogonal to both. Let us find this cross for two adjacent circles of our congruence. Let the spheres through the first circle be $\cos\phi\,(y)+\sin\phi\,(z)$, $-\sin\phi\,(y)+\cos\phi\,(z)$, while those through the adjacent circle are $l\,(y+dy)+m\,(z+dz)$,

$-m(y+dy)+l(z+dz)$. Writing out the conditions for criss-cross orthogonality, and remembering

$$2(ydy) = -(dy\,dy),\ 2(zdz) = -(dz\,dz),$$
$$(ydz)+(zdy) = -(dy\,dz),$$

we have two linear homogeneous equations in l and m. Equating the discriminant to zero, and neglecting infinitesimals of a higher order,

$$\begin{vmatrix} [1-\frac{1}{2}(dy\,dy)]\sin\phi-(zdy)\cos\phi, & (ydz)\sin\phi-[1-\frac{1}{2}(dz\,dz)]\cos\phi \\ -[1-\frac{1}{2}(dz\,dz)]\sin\phi-(ydz)\cos\phi, & (zdy)\sin\phi+[1-\frac{1}{2}(dy\,dy)]\cos\phi \end{vmatrix} = 0.$$

Expanding, and casting aside higher infinitesimals,

$$[edu^2+(f+f')\,du\,dv+gdv^2]\,(\sin^2\phi-\cos^2\phi)+[(E-E')\,du^2$$
$$+2\,(F-F')\,du\,dv+(G-G')\,dv^2]\sin\phi\cos\phi = 0. \quad (17)$$

In order to discuss this equation, we write two others:

$$(E-E')\,du^2+2\,(F-F')\,du\,dv+(G-G')\,dv^2 = 0,$$
$$edu^2+(f+f')\,du\,dv+gdv^2 = 0.$$

These are the expanded forms of

$$(dy\,dy)-(dz\,dz) = 0, \quad (dy\,dz) = 0.$$

Let the foci of our circle be (α) and (γ). We may write

$$\alpha_i \equiv \frac{-iy_i+z_i}{2}, \quad \gamma_i \equiv -\frac{iy_i+z_i}{2}. \quad (18)$$
$$y_i \equiv i(\alpha_i+\gamma_i), \quad z_i \equiv (\alpha_i-\gamma_i), \quad i = 0 \ldots 4.$$

Our differential equations amount to the pair

$$(d\alpha\,d\alpha) = (d\gamma\,d\gamma) = 0.$$

With regard to these two we have the following possibilities:

A) They have, in general, no common root. The isotropic

curves do not correspond to one another on the two surfaces of foci. We shall say that the congruence is *non-conformal.*

B) The equations have, in general, one common root. Then the surfaces of foci (which must not on any account be confused with the *focal surfaces*) are so related that one system of isotropic curves on the one surface corresponds to one such system on the other. These congruences shall be called *semi-conformal.*

C) The equations are equivalent to one another. The surfaces of foci are conformally related, and the congruence shall be called *conformal.*

Let us begin with the non-conformal congruence. Here, if du/dv be given, we usually get a unique value for $\tan 2\phi$, that is to say, two mutually orthogonal spheres, and so the cross required. On the other hand, if ϕ be given, we have a quadratic equation in du/dv, so that on each sphere through a circle of a non-conformal congruence will lie two circles orthogonal thereto, orthogonal and cospherical also to an adjacent circle of the congruence. These two circles will fall together if this equation in du/dv have equal roots, i.e. if

$$\begin{gathered}[eg-\tfrac{1}{4}(f+f')^2](\tan^2\phi-1)^2\\ +[e(G-G')-(F-F')(f+f')+g(E-E')](\tan^2\phi-1)\tan\phi \qquad (19)\\ +[(E-E')(G-G')-(F-F')^2]\tan^2\phi=0.\end{gathered}$$

This equation is unaltered if we replace $\tan\phi$ by $-\operatorname{ctn}\phi$.

Theorem 33.] *In a non-conformal congruence, through each circle in general position will pass two pairs of mutually orthogonal spheres, on each of which there is but a single circle orthogonal to the given circle, cospherical and orthogonal to an adjacent circle of the congruence.*

The words 'in general position' mean that the two quadratic differential equations just written have no common solution. We shall call these the *limiting* spheres through the circle. They correspond to maximum or minimum values for $\tan\phi$ in (17). We see, in fact, that if we equate to zero the partial derivatives to du and dv, we get

$$\left[edu+\frac{f+f'}{2}dv\right](\tan^2\phi-1)+[(E-E')\,du + (F-F')\,dv]\tan\phi = 0.$$

$$\left[\frac{f+f'}{2}du+gdv\right](\tan^2\phi-1)+[(F-F')du + (G-G')dv]\tan\phi = 0.$$

Eliminating du/dv we fall back upon (19). In the case of a real congruence the real spheres containing real circles orthogonal to the given circle and a next neighbour will lie in specific angular openings determined by the limiting spheres.

We next suppose that our congruence is focal. We have for the focal sphere

$$\cos\phi y_i+\sin\phi z_i \equiv \cos(\phi+d\phi)\,(y_i+dy_i)+\sin(\phi+d\phi)\,(z_i+dz_i).$$

$$dy_i\cos\phi+dz_i\sin\phi-(y_i\sin\phi-z_i\cos\phi)\,d\phi = 0.$$

Multiplying through by y_i, summing, and neglecting higher infinitesimals,

$$d\phi = (ydz) = -(zdy).$$

$$\left[\left(\frac{\partial y_i}{\partial u}\cos\phi+\frac{\partial z_i}{\partial u}\sin\phi\right)-(y_i\sin\phi-z_i\cos\phi)\left(y\frac{\partial z}{\partial u}\right)\right]du$$

$$+\left[\left(\frac{\partial y_i}{\partial v}\cos\phi+\frac{\partial z_i}{\partial v}\sin\phi\right)-(y_i\sin\phi-z_i\cos\phi)\left(y\frac{\partial z}{\partial v}\right)\right]dv = 0.$$

Multiplying through by $\frac{\partial z_i}{\partial u}$ and summing, then doing the same for $\frac{\partial z_i}{\partial v}$,

$$[edu+fdv]\cos\phi+[E'du+F'dv]\sin\phi = 0.$$

$$[f'du+gdv]\cos\phi+[F'du+G'dv]\sin\phi = 0.$$

Similarly

$$[edu+f'dv]\sin\phi+[Edu+Fdv]\cos\phi = 0.$$

$$[fdu+gdv]\sin\phi+[Fdu+Gdv]\cos\phi = 0.$$

Eliminating ϕ,

$$(E'f'-F'e)\,du^2+[E'g-F'(f-f')-G'e]\,du\,dv+(F'g-G'f)\,dv^2=0. \tag{20}$$

$$(Ef-Fe)\,du^2+[Eg-F(f'-f)-Ge]\,du\,dv+(Fg-Gf')\,dv^2=0.$$

Eliminating du/dv,

$$(E'G'-F'^2)\tan^2\phi+[E'g-F'(f+f')+G'e)]\tan\phi+(eg-ff')=0. \tag{21}$$

$$(eg-ff')\tan^2\phi+[Eg-F(f+f')+Ge]\tan\phi+(EG-F^2)=0.$$

$$[(eg-ff')-(E'G'-F'^2)]\tan^2\phi+[(E-E')g-(F-F')(f+f')+(G-G')e]\tan\phi+[(EG-F^2)-(eg-ff')]=0. \tag{22}$$

In order that a congruence should be focal, it is necessary that the two equations (20) should be equivalent, and the same for the two equations (21). It should, further, be noted that the middle coefficient is the same in (19) and (22). This vanishes when, and only when, the pairs of solutions make equal angles with (y) and (z).

Theorem 34.] *In a focal congruence the spheres of antisimilitude of the focal spheres are also spheres of antisimilitude of the limiting spheres in pairs.*

The conditions that a congruence should be focal can be written in better shape. The foci of the circle being (α) and (γ), which coordinates we find from (18), the condition for a focal congruence is found from (15),

$$\left|\frac{\partial\gamma}{\partial u}\alpha\gamma\frac{\partial\alpha}{\partial u}\frac{\partial\alpha}{\partial v}\right|=\left|\frac{\partial\gamma}{\partial v}\alpha\gamma\frac{\partial\alpha}{\partial u}\frac{\partial\alpha}{\partial v}\right|=0. \tag{23}$$

This gives

$$\left|\frac{\partial z}{\partial u}yz\frac{\partial y}{\partial u}\frac{\partial y}{\partial v}\right|=\left|\frac{\partial z}{\partial v}yz\frac{\partial y}{\partial u}\frac{\partial y}{\partial v}\right|=0.$$

Squaring

$$\begin{vmatrix} E & F & e \\ F & G & f \\ e & f & E' \end{vmatrix}=\begin{vmatrix} E & F & f' \\ F & G & g \\ f' & g & G' \end{vmatrix}=0. \tag{24}$$

What additional requirements must be fulfilled if our focal congruence is to be a normal one? We take as u and v the focal parameters, then

$$\frac{\partial z_i}{\partial u} = ay_i + acz_i + c\frac{\partial y_i}{\partial u}, \quad \frac{\partial z_i}{\partial v} = a'y_i + a'c'z_i + c'\frac{\partial y_i}{\partial v}. \qquad (25)$$

$$E' = ce, \quad F' = cf', \quad = c'f, \quad G' = c'g; \quad cE = e, \quad cF = f,$$
$$c'F = f', \quad c'G = g.$$

Then, since the expression

$$c\left[(y) + \left(\frac{\partial y}{\partial u}\right)du\right] - \left[(z) + \left(\frac{\partial z}{\partial u}\right)du\right]$$

is a linear combination of (y) and (z), the focal spheres are

$$c(y) - (z) \text{ and } c'(y) - (z). \qquad (26)$$

The condition for a normal congruence is thus, by (17),

$$cc' + 1 = 0.$$

This gives

$$(EG - F^2) = (E'G' - F'^2) = -(eg - ff'). \qquad (27)$$

These equations are invariant, and our reasoning is reversible. They will, therefore, give necessary and sufficient conditions that a focal congruence should be a normal one.

The normal congruence of circles seems, at first sight, the most natural extension of the normal congruence of straight lines, which is after all but a special case of the other. There is, however, another extension of the normal line congruence which possesses not a little interest for us. Interpreting our circles as lines in S_4, what sort of a congruence of circles will correspond to a normal line congruence in this space? We mean by a *normal* congruence of lines in S_4 a two-parameter family orthogonal to an analytic surface. What will such a line congruence give us in circle geometry? The measurement in S_4 being elliptic, we see by a little reflection that in S_3 we must have such a circle congruence that through each circle we may pass a sphere whose pairs of points of contact with the two nappes of the envelope are mutually inverse in that circle. Suppose that such a sphere has the coordinates

$$y_i' \equiv y_i \cos\phi + z_i \sin\phi, \quad i = 0 \ldots 4.$$

Then every sphere through the points of contact of (y') with its envelope cuts (y') in a circle orthogonal to the given circle, i.e. such a sphere is orthogonal to $(-y)\sin\phi+(z)\cos\phi$, the sphere through our circle orthogonal to (y'). We thus get the equations

$$-\left(y\frac{\partial z}{\partial u}\right)\sin^2\phi+\left(z\frac{\partial y}{\partial u}\right)\cos^2\phi+\frac{\partial\phi}{\partial u}=-\left(y\frac{\partial z}{\partial v}\right)\sin^2\phi$$

$$+\left(z\frac{\partial y}{\partial u}\right)\cos^2\phi+\frac{\partial\phi}{\partial v}=0.$$

$$\frac{\partial\phi}{\partial u}=\left(y\frac{\partial z}{\partial u}\right),\quad \frac{\partial\phi}{\partial v}=\left(y\frac{\partial z}{\partial v}\right). \tag{28}$$

$$f=f'. \tag{29}$$

The differential equations (28) are equally well satisfied if we replace ϕ by $\phi+k$, where k is any finite constant.

Theorem 35.] *If through each circle of a congruence it be possible to pass a sphere whose points of contact with its envelope are mutually inverse in that sphere, then an infinite number of such spheres may be passed through each circle. These spheres will generate a one-parameter family of congruences; corresponding spheres of any two congruences pass through the same circle of the circle congruence and make a fixed angle.*

A congruence of circles which possesses this property is said to be *pseudo-normal.*

Suppose that we have a congruence of circles which is both focal and pseudo-normal. We find from (25) and (29) that

$$(c-c')f=0.$$

We could not have $c=c'$, for then would

$$dz_i\equiv py_i+qz_i+rdz_i,$$

and we should have a set of circles through two points, or one sphere, an uninteresting set. Hence

$$f=f'=0.$$

Substituting in (17) we get an equation whereof (26) gives two solutions.

Theorem 36.] *A necessary and sufficient condition that a focal congruence which does not consist in circles through a point-pair or on a sphere should be pseudo-normal is that the focal spheres should coincide with a pair of limiting spheres.*

Let us see what relations subsist among the foci of the circles of a pseudo-normal congruence. We return to (18). If, then,

$$\left(\frac{\partial y}{\partial u}\frac{\partial z}{\partial v}\right)=\left(\frac{\partial y}{\partial v}\frac{\partial z}{\partial u}\right),$$

we have also

$$\left(\frac{\partial \alpha}{\partial u}\frac{\partial \gamma}{\partial v}\right)=\left(\frac{\partial \alpha}{\partial v}\frac{\partial \gamma}{\partial u}\right),$$

and vice versa. If, now, (β) be a sphere touching its envelope at (α) and (γ), we see, by (6), that the circle orthogonal to (β) at (α) and (γ) (in bi-involution with the given circle) will generate a normal congruence, and vice versa. Moreover, u and v will be the focal parameters for this normal congruence, i.e. give the lines of curvature of the surface (α) and (γ).

Theorem 37.] *A necessary and sufficient condition that a focal congruence of circles should be pseudo-normal is that the lines of curvature should correspond in the two surfaces of foci.*

Let us follow further the relation between the normal and pseudo-normal congruence. The focal spheres are

$$c(y)-(z)\,;\ c'(y)-(z).$$

The spheres orthogonal to our circle through the pairs of focal points are

$$\left(\frac{\partial y}{\partial u}z\right)(z)-\left(\frac{\partial y}{\partial u}\right),\quad \left(\frac{\partial y}{\partial v}z\right)(z)-\left(\frac{\partial y}{\partial v}\right).$$

These will be mutually orthogonal if $F=f=f'=0$.

When, however, these spheres are mutually orthogonal, the pairs of focal points separate one another harmonically, and conversely, whence

Theorem 38.] *A necessary and sufficient condition that a focal congruence should be pseudo-normal is that the pairs of focal points on each circle should separate one another harmonically.*

We get at once from (28)

Theorem 39.] *If a normal congruence of circles be given, the congruence of circles whose foci are the pairs of intersections of the circles of the normal congruence with any two chosen orthogonal surfaces is focal and pseudo-normal.*

Theorem 40.] *If a congruence of circles be both focal and pseudo-normal, the foci are the pairs of intersections of the circles of a normal congruence with two orthogonal surfaces.*

When a normal and a pseudo-normal congruence are related in the fashions described in the last two theorems, we shall speak of them as *associated.* The normal congruence has the parametric form

$$x_i \equiv t^2\alpha_i + t\beta_i + \gamma_i.$$

$$\beta_i \equiv \left| y_j \; z_k \; \frac{\partial y_l}{\partial u} \; \frac{\partial y_m}{\partial v} \right|, \quad i = 0 \dots 4.$$

Here we suppose that u and v are the common focal parameters of the two congruences. Let one of the spheres whose points of contact with its envelope are mutually inverse in the (y), (z) circle be (y'), where

$$y_i' \equiv \cos\phi y_i + \sin\phi z_i, \quad i = 0 \dots 4.$$

$$\frac{\partial y_i'}{\partial u} \equiv \cos\phi \frac{\partial y_i}{\partial u} + \sin\phi \frac{\partial z_i}{\partial u} + \left(y \frac{\partial z}{\partial u}\right)(-y_i \sin\phi + z_i \cos\phi);$$

$$\frac{\partial y_i'}{\partial v} \equiv \cos\phi \frac{\partial y_i}{\partial v} + \sin\phi \frac{\partial z_i}{\partial v} + \left(y \frac{\partial z}{\partial v}\right)(-y_i \sin\phi + z_i \cos\phi);$$

as we see by the aid of (28). We also find from (18) and the above value of (β) that all points of the associated circle of the normal congruence lie on the spheres $\left(\frac{\partial y'}{\partial u}\right)$, $\left(\frac{\partial y'}{\partial v}\right)$, so that, in particular, the points of contact of (y') with its envelope lie on this associated circle, and this circle is orthogonal to (y'), since $\left(y'\frac{\partial y'}{\partial u}\right)=\left(y'\frac{\partial y'}{\partial v}\right)=0.$

Theorem 41.] *If a focal and pseudo-normal congruence be given, the spheres whose pairs of points of contact with their envelopes are mutually inverse in the circles of the given congruence will envelop the orthogonal surfaces of the circles of the associated normal congruence.*

Since u and v give the focal parameters for both congruences,

Theorem 42.] *If a normal and pseudo-normal congruence be associated, the annular surfaces correspond in the two.*

Theorem 43.] *If a normal congruence be given, not consisting in the circles through a point-pair, pairs of intersections with any two orthogonal surfaces may be taken as the foci of the circles of an associated pseudo-normal congruence. The other orthogonal surfaces will then be paired in such a way that the intersections of each circle with a pair of surfaces are mutually inverse in the corresponding circle of the associated congruence.*

The pseudo-normal congruence enjoys a sort of indestructability akin to that of the normal line congruence. Let one such congruence be given by the spheres (y) and (z). We may then determine (z') so that, θ being fixed,

$$(z'z')=1,\quad (z'y)=0,\quad (z'z)=\cos\theta,\quad \left(\frac{\partial z'}{\partial u}\frac{\partial y}{\partial v}\right)=\left(\frac{\partial z'}{\partial v}\frac{\partial y}{\partial u}\right).$$

The spheres (y) and (z') will determine a second pseudo-normal congruence of such sort that each of its circles is cospherical with one of the given circles, and makes therewith a fixed angle. The sphere through the circle of (z) and (z') making an angle ϕ with the former is (z''), where

$$z_i'' = \frac{\sin(\theta-\phi)}{\sin\theta} z_i + \frac{\sin\phi}{\sin\theta} z_i'.$$

This cuts (y) in a circle coaxal with the previous circles, and making an angle ϕ with the first of these. If, then, ϕ be constant,

$$\left(\frac{\partial z''}{\partial u}\frac{\partial y}{\partial v}\right) = \left(\frac{\partial z''}{\partial v}\frac{\partial y}{\partial u}\right).$$

Theorem 44.] *If two pseudo-normal congruences be so related that corresponding circles are cospherical and make a constant angle, then each circle coaxal with both and making constant angles with them will generate a pseudo-normal congruence.*

We reach another theorem of the same sort in the following manner. Suppose that we have two correlative complexes of spheres, given by the equations

$$\rho x_i = x_i(u, v, w), \quad \sigma x_i' = \left| x_j \frac{\partial x_k}{\partial u}\frac{\partial x_l}{\partial v}\frac{\partial x_m}{\partial w} \right|.$$

We assume also that we have a pseudo-normal congruence of circles. Through each circle will pass at least one sphere of the first complex, and this we shall take as (y). The correlative sphere is (t) where

$$t_i = \left| y_j \frac{\partial y_k}{\partial u}\frac{\partial y_l}{\partial v} s_m \right|.$$

s_m is a function of u, v, and w. Further, let

$$z_i' = \frac{\sin(\phi-\theta)}{\sin\theta} t_i + \frac{\sin\phi}{\sin\theta} z_i,$$

so that ϕ is the angle of (z') and (t), while θ is the angle of (z) and (t), and let us assume that $\dfrac{\sin\phi}{\sin\theta} = k$, a constant. We have

$$\left(\frac{\partial z'}{\partial u}\frac{\partial y}{\partial v}\right) - \left(\frac{\partial z'}{\partial v}\frac{\partial y}{\partial u}\right) = k\left[\left(\frac{\partial z}{\partial u}\frac{\partial y}{\partial v}\right) - \left(\frac{\partial z}{\partial v}\frac{\partial y}{\partial u}\right)\right].$$

Theorem 45.] *Through each circle c of a pseudo-normal congruence a sphere is passed belonging to a given non-developable complex in such a way as to generate a congruence of that complex. This sphere meets the corresponding sphere of the correlative complex in a circle c′, and a circle c″ is so taken as to be coaxal with the circles c and c′ while the sines of the angles which c and c″ make with c′ have a constant ratio. Then the circles c″ will also generate a pseudo-normal congruence.**

It is now time to take up some of the hitherto excluded types of congruence. The semi-conformal type has the property that each circle in general position is paratactic or tangent to one adjacent circle. The circles are, however, necessarily imaginary, and we prefer to pass to the more interesting type of conformal congruences. Here we have

$$E-E':F-F':G-G'=e:\frac{f+f'}{2}=g. \tag{30}$$

Equations (17) take the form

$$[edu^2+(f+f')\,du\,dv+gdv^2]\,[\sin^2\phi-\cos^2\phi+k\sin\phi\cos\phi]=0.$$

The roots of the second factor give two mutually orthogonal spheres, which we may take for (y) and (z). A circle cospherical and orthogonal to our given circle, and to one infinitely near, must lie on (y) or (z). The sphere (y) is, then, orthogonal to (z) and $(z+dz)$.

$$(ydz)\equiv(zdy)\equiv-\tfrac{1}{2}(dy\,dz)\equiv 0.$$

$$e\equiv f+f'\equiv g\equiv 0.$$

But if we return now to our equations (18) we find

$$\left(\frac{\partial\alpha}{\partial u}\frac{\partial\alpha}{\partial u}\right)\equiv\left(\frac{\partial\gamma}{\partial u}\frac{\partial\gamma}{\partial u}\right),\quad \left(\frac{\partial\alpha}{\partial u}\frac{\partial\alpha}{\partial v}\right)\equiv\left(\frac{\partial\gamma}{\partial u}\frac{\partial\gamma}{\partial v}\right),$$

$$\left(\frac{\partial\alpha}{\partial v}\frac{\partial\alpha}{\partial v}\right)\equiv\left(\frac{\partial\gamma}{\partial v}\frac{\partial\gamma}{\partial v}\right).$$

$$(d\alpha\,d\alpha)\equiv(d\gamma\,d\gamma).$$

* This is the interpretation in terms of circles of the four-dimensional extension of the Malus-Dupin theorem.

If these various expressions do not all vanish identically, the two surfaces of foci are conformally related. If they do vanish identically, (α) and (γ) trace two minimal curves, given respectively by the parameters u and v. Suppose, conversely, that (α) and (γ) are corresponding points on two conformally related surfaces,

$$(d\alpha d\alpha) \equiv \rho(d\gamma d\gamma), \quad 2(\alpha\gamma) \equiv -1.$$

Replacing α_i by $\frac{\alpha_i}{\sqrt[4]{\rho}}$ and γ_i by $\sqrt[4]{\rho}\gamma_i$,

$$(d\alpha d\alpha) \equiv (d\gamma d\gamma), \quad 2(\alpha\gamma) \equiv -1, \quad e \equiv f + f' \equiv g = 0.$$

Equation (17) reduces to

$$[(E-E')du^2 + 2(E-F')du dv + (G-G')dv^2] \sin\phi \cos\phi = 0,$$

a circle orthogonal and cospherical with our given circles, and one of its next neighbours must lie either on (y) or (z). We next write

$$y_i' = y_i \cos\phi + z_i \sin\phi, \quad y_i'' = y_i \cos\phi - z_i \sin\phi, \quad \phi = \text{const}.$$

$$(dy'dy') \equiv (dy''dy'').$$

Conversely, if this equation hold, and if ϕ be constant,

$$(dy dz) \equiv 0.$$

We shall say that two congruences of spheres are *conformally related* if they be in one to one analytic correspondence (at least in some continuous region), and if the angle of two adjacent spheres of the one be equal to the corresponding angle in the other. We thus get

Theorem 46.] *If the foci of the circles of a congruence be corresponding points on two conformally related surfaces, or trace two minimal curves, then, through each circle of the congruence we may, in an infinite number of ways, pass two spheres which shall make a constant angle with one another, and describe conformally related congruences.*

Theorem 47.] *If corresponding spheres in two conformally related congruences make a constant angle, then will their circle of intersection generate a conformal congruence.*

Let us now make the additional assumption that our congruence is focal. The isotropic curves will correspond on the two surfaces of foci. If we take these to determine our parameters u and v, we have two conceivable cases

$$\text{(A)}\quad \frac{\partial \gamma_i}{\partial u} = p\alpha_i + q\gamma_i + r\frac{\partial \alpha_i}{\partial v};\quad \frac{\partial \gamma_i}{\partial v} = p'\alpha_i + q'\gamma_i + r'\frac{\partial \alpha_i}{\partial u}.$$

$$\text{(B)}\quad \frac{\partial \gamma_i}{\partial u} = p\alpha_i + q\gamma_i + r\frac{\partial \alpha_i}{\partial u};\quad \frac{\partial \gamma_i}{\partial v} = p'\alpha_i + q'\gamma_i + r'\frac{\partial \alpha_i}{\partial v}.$$

Leaving aside for the moment the question of whether both cases are possible, let us consider them in turn. In (A) we have

$$\left(\frac{\partial \alpha}{\partial u}\frac{\partial \gamma}{\partial v}\right) = \left(\frac{\partial \alpha}{\partial v}\frac{\partial \gamma}{\partial u}\right).$$

Our congruence is pseudo-normal; the focal parameters u' and v' will, by 42], give the focal directions for the associated normal congruence, so that they give mutually orthogonal lines of advance for (α) and (γ). We have the partial differential equations, analogous to those previously found for a normal congruence,

$$\frac{\partial \gamma_i}{\partial u'} = bc\alpha_i + b\gamma_i + c\frac{\partial \alpha_i}{\partial u'},\quad \frac{\partial \gamma_i}{\partial v'}\, b'c'\alpha_i + b'\gamma_i + c'\frac{\partial \alpha_i}{\partial v'}.$$

The relation between the coefficients of (α) and (γ) comes from the equations

$$\left(\alpha\frac{\partial \gamma}{\partial u'}\right) + \left(\gamma\frac{\partial \alpha}{\partial u'}\right) = \left(\alpha\frac{\partial \gamma}{\partial v'}\right) + \left(\gamma\frac{\partial \alpha}{\partial v'}\right) = 0.$$

The condition that our surfaces (α) and (γ) should be conformally related is

$$\left(\frac{\partial \alpha}{\partial u'}\frac{\partial \alpha}{\partial u'}\right)du'^2 + \left(\frac{\partial \alpha}{\partial v'}\frac{\partial \alpha}{\partial v'}\right)dv'^2 = \left(\frac{\partial \gamma}{\partial u'}\frac{\partial \gamma}{\partial u'}\right)du'^2 + \left(\frac{\partial \gamma}{\partial v'}\frac{\partial \gamma}{\partial v'}\right)dv'^2.$$

This gives $c^2 = c'^2$. Now, if $c = c'$, we have at once

$$d\gamma_i = P\alpha_i + Q\gamma_i + Rd\alpha_i.$$

Each two adjacent circles are cospherical, and we have circles through two points or one sphere, cases which we may exclude. Hence

$$c+c'=0.$$

This shows that the focal spheres $c(\alpha)-(\gamma)$, $c'(\alpha)-(\gamma)$ are mutually orthogonal, and so, by 17],

Theorem 48.] *If a congruence be both conformal and pseudo-normal it is normal.*

Theorem 49.] *If a congruence be both normal and conformal it is pseudo-normal, or consists in circles touching a given circle at a given point.*

Congruences of this type are well known. On the other hand, in congruences of type (B) the focal parameters are also the isotropic ones. We thus see that each circle is tangent to two adjacent ones, or the circles of the congruence are the osculating circles to two different one-parameter families of curves. This much is true if such congruences exist; unfortunately, the present writer has signally failed in all his attempts to find an example of such a congruence.*

§ 4. Complexes of Circles.

Our leading object in attempting the Kummer method for the study of circle congruences was to follow the methods which are fruitful in line geometry, or, rather, to study line

* The problem of finding a conformal focal congruence is sometimes called the problem of Ribaucour; it amounts to finding a congruence of spheres which establishes a conformal relation between the two nappes of the envelope. For an interesting discussion of congruences of type (A) see Darboux, 'Sur les surfaces isothermiques', *Annales de l'École Normale*, Series 3, vol. xvi, 1899, pp. 498 ff. Darboux here proves that this is the only type of conformal congruence, but his proof is erroneous, as he subsequently acknowledged in a letter to the Author. He doubted, however, whether any congruences of type (B) really existed. The theorem that congruences of type (A) are normal was casually mentioned by Cosserat in a short article, 'Sur le problème de Ribaucour', *Bulletin de l'Académie des Sciences de Toulouse*, vol. iii, 1900, pp. 267 ff.

systems in S_4. The idea lies close at hand that if in our line geometry we move up one in the number of dimensions, we should do well to allow ourselves an extra parameter. In other words, will not these same methods yield interesting results when applied to the study of circle complexes? We start with the equations

$$y_i \equiv y_i(u_1 u_2 u_3), \quad z_i \equiv z_i(u_1 u_2 u_3),$$
$$(yy) \equiv (zz) \equiv 1, \quad (yz) \equiv 0. \tag{33}$$

$$(dy\,dy) - (z\,dy)^2 \equiv \sum_{i,j=1}^{i,j=3} a_{ij}\,du_i\,du_j, \quad a_{ij} = a_{ji}.$$

$$(dz\,dz) - (y\,dz)^2 \equiv \sum_{i,j=1}^{i,j=3} b_{ij}\,du_i\,du_j, \quad b_{ij} = b_{ji}. \tag{34}$$

$$(dy\,dz) \equiv \sum_{i,j=1}^{i,j=3} c_{ij}\,du_i\,du_j.$$

$$\left(\frac{\partial y}{\partial u_i}\frac{\partial y}{\partial u_j}\right) - \left(z\frac{\partial y}{\partial u_i}\right)^2 \equiv a_{ii};$$

$$\left(\frac{\partial y}{\partial u_i}\frac{\partial y}{\partial u_j}\right) - \left(z\frac{\partial y}{\partial u_i}\right)\left(z\frac{\partial y}{\partial u_j}\right) \equiv a_{ij} + a_{ji} \equiv 2a_{ij} \equiv 2a_{ji}, \quad j \neq i.$$

$$\left(\frac{\partial z}{\partial u_i}\frac{\partial z}{\partial u_i}\right) - \left(y\frac{\partial z}{\partial u_i}\right)^2 \equiv b_{ii};$$

$$\left(\frac{\partial z}{\partial u_i}\frac{\partial z}{\partial u_j}\right) - \left(y\frac{\partial z}{\partial u_i}\right)\left(y\frac{\partial z}{\partial u_j}\right) \equiv b_{ij} + b_{ji} \equiv 2b_{ij} \equiv 2b_{ji}, \; j \neq i. \tag{35}$$

$$\left(\frac{\partial y}{\partial u_i}\frac{\partial z}{\partial u_j}\right) \equiv c_{ij}.$$

These coefficients are connected by the following identical relations

$$\sum_{i,\,m=0}^{i,\,m=4} \begin{vmatrix} \frac{\partial y_j}{\partial u_1} & \frac{\partial y_k}{\partial u_1} & \frac{\partial y_l}{\partial u_1} \\ \frac{\partial y_j}{\partial u_2} & \frac{\partial y_k}{\partial u_2} & \frac{\partial y_l}{\partial u_2} \\ \frac{\partial y_j}{\partial u_3} & \frac{\partial y_k}{\partial u_3} & \frac{\partial y_l}{\partial u_3} \end{vmatrix} \begin{vmatrix} \frac{\partial z_j}{\partial u_1} & \frac{\partial z_k}{\partial u_1} & \frac{\partial z_l}{\partial u_1} \\ \frac{\partial z_j}{\partial u_2} & \frac{\partial z_k}{\partial u_2} & \frac{\partial z_l}{\partial u_2} \\ \frac{\partial z_j}{\partial u_3} & \frac{\partial z_k}{\partial u_3} & \frac{\partial z_l}{\partial u_3} \end{vmatrix}$$

$$\equiv \begin{vmatrix} c_{11} & c_{12} & c_{13} \\ c_{21} & c_{22} & c_{23} \\ c_{31} & c_{32} & c_{33} \end{vmatrix} \equiv |\, c_{ij} \,|. \tag{36}$$

$$\left| zy \frac{\partial y}{\partial u_1} \frac{\partial y}{\partial u_2} \frac{\partial y}{\partial u_3} \right|^2 = |\, a_{ij} \,|, \quad \left| yz \frac{\partial z}{\partial u_1} \frac{\partial z}{\partial u_2} \frac{\partial z}{\partial u_3} \right|^2 = |\, b_{ij} \,|.$$

From

$$(yz) \equiv \left(y \frac{\partial y}{\partial u_1}\right) \equiv \left(y \frac{\partial y}{\partial u_2}\right) \equiv \left(y \frac{\partial y}{\partial u_3}\right) \equiv 0.$$

$$(zy) \equiv \left(z \frac{\partial z}{\partial u_1}\right) \equiv \left(z \frac{\partial z}{\partial u_2}\right) \equiv \left(z \frac{\partial z}{\partial u_3}\right) \equiv 0.$$

$$y_i \equiv \frac{\left| z_j \frac{\partial y_k}{\partial u_1} \frac{\partial y_l}{\partial u_2} \frac{\partial y_m}{\partial u_3} \right|}{\sqrt{|\, a_{ij} \,|}}, \quad z_i \equiv \frac{\left| y_j \frac{\partial z_k}{\partial u_1} \frac{\partial z_l}{\partial u_2} \frac{\partial z_m}{\partial u_3} \right|}{\sqrt{|\, b_{ij} \,|}}.$$

If we write

$$A_{kl} \equiv \frac{\partial |\, a_{ij} \,|}{\partial a_{kl}}, \quad B_{kl} \equiv \frac{\partial |\, b_{ij} \,|}{\partial b_{kl}},$$

$$a_{ij} = -\frac{\sum_{k,\,l=1}^{k,\,l=3} c_{ik} c_{jl} B_{kl}}{|\, b_{ij} \,|}, \quad b_{ij} = -\frac{\sum_{k,\,l=1}^{k,\,l=3} c_{ik} c_{jl} A_{kl}}{|\, a_{ij} \,|}. \tag{37}$$

$$|\, a_{ij} \,|\,|\, b_{ij} \,| = |\, c_{ij} \,|^2. \tag{38}$$

Let us find the circles cospherical and orthogonal to a circle in general position and to one immediately next. We pursue exactly the same calculation which led to (17), and reach the equation

$$\sum_{i,j=1}^{i,j=3} [c_{ij}(\sin^2\phi - \cos^2\phi) + (a_{ij} - b_{ij})\sin\phi\cos\phi]\,du_i du_j = 0. \quad (39)$$

As in the case of a congruence we are thrown back upon the equations

$$(d\alpha\, d\alpha) = 0, \quad (d\gamma\, d\gamma) = 0.$$

When these equations are not equivalent the relation between (α) and (γ) does not give a conformal transformation of space, and the complex shall be said to be *non-conformal.* When the two equations are equivalent we have a conformal transformation of space established by (α) and (γ), and the complex shall be called *conformal.* There is no intermediate case corresponding to the semi-conformal congruence, unless (α) or (γ) be restricted to lie on a surface, since the equations are irreducible. We start with the non-conformal complex, and notice that there will be two circles cospherical and orthogonal to each circle and each of its next neighbours, unless the two be cospherical or paratactic. Analytically, we have trouble in finding these orthogonal circles when

$$\sum_{i,j=1}^{i,j=3} c_{ij}\,du_i du_j = 0, \quad \sum_{i,j=1}^{i,j=3} (a_{ij} - b_{ij})\,du_i du_j = 0.$$

Let us see whether the solutions of these equations give us cospherical or paratactic circles. We know from 8] that each circle of a complex is (usually) cospherical with three adjacent circles. On the other hand, the equations as they stand are irreducible quadratics, with four solutions, distinct or coincident, and in the case of each solution ϕ is indeterminate; hence

Theorem 50.] *A circle in general position in a non-conformal complex is paratactic with four distinct or coincident adjacent circles of the complex.*

The qualification *pseudo-normal* may be applied to complexes as well as to congruences. We wish to find those complexes of circles which correspond to the normals to a hypersurface in S_4. A little reflection shows that such circles must be the intersections of corresponding spheres in two correlative complexes. Let the corresponding spheres be (y') and (z'),

$$y_i' \equiv y_i \cos\phi + z_i \sin\phi, \quad z_i' \equiv -y_i \sin\phi + z_i \cos\phi.$$

We wish to have

$$(y'dz') \equiv (z'dy') \equiv 0.$$

Proceeding exactly as before, we find

$$(ydz) = d\phi. \tag{40}$$

$$c_{ij} \equiv c_{ji}, \quad i, j = 1, 2, 3. \tag{41}$$

We note also that if ϕ be one solution of the differential equation, so is $\phi + \text{const.}$

Theorem 51.] *If through each circle of a complex it be possible to pass a pair of spheres which correspond in two correlative complexes, then we may pass an infinite number of such pairs of spheres generating as many pairs of correlative complexes. The spheres of any two of these complexes through each circle make a fixed angle.*

We shall define circle complexes of this type as *pseudo-normal.* We next choose for u_1, u_2, u_3 such parameters that by holding two constant we get the annular surfaces of the complex. We can repeat almost word for word what we did in the previous case of pseudo-normal congruences, and find

$$c_{12} = c_{21} = c_{23} = c_{32} = c_{31} = c_{13} = 0.$$

The spheres $\left(\frac{\partial y}{\partial u_i} z\right)(z) - \left(\frac{\partial y}{\partial u_i}\right)$, $\left(\frac{\partial y}{\partial u_j} z\right)(z) - \left(\frac{\partial y}{\partial u_j}\right)$ are mutually orthogonal, hence

Theorem 52.] *A necessary and sufficient condition that a complex of circles should be pseudo-normal is that the pairs of focal points should separate one another harmonically.*

On an arbitrary sphere there will lie a finite number of circles of a given complex of order greater than zero. Suppose that two pseudo-normal complexes are determined respectively by $(y)(z)$, and $(y)(z')$, and that (z) and (z') make a constant angle θ. Then, if

$$z_i'' \equiv \frac{\sin(\phi-\alpha)}{\sin\theta} z_i + \frac{\sin\phi}{\sin\theta} z_i', \quad i = 0 \ldots 4,$$

where ϕ is constant, we see that (y) and (z'') determine a pseudo-normal complex.

Theorem 53.] *If two pseudo-normal complexes be so related that corresponding circles are cospherical and make a constant angle, then the complex of circles, each coaxal with two corresponding circles of the given complexes and making with them a fixed angle, is also pseudo-normal.*

Suppose next that (s) corresponds to (y) in the complex correlative to that generated by the latter. If, then, we write

$$z_i' = \frac{\sin(\phi-\theta)}{\sin\theta} s_i + \frac{\sin\phi}{\sin\theta} z_i,$$

and if $\frac{\sin\phi}{\sin\theta}$ be constant, we find, as before, that (y) and (z') generate a pseudo-normal complex.

Theorem 54.] *If through each circle of a pseudo-normal complex a sphere be passed belonging to a given non-developable complex, and if the original circle be replaced by such a circle coaxal with it, and the circle cut by the corresponding sphere of the correlative complex, that the sine of the angles of the original and the replacing circles with the circle on the correlative sphere have a fixed ratio, then will the replacing circles also generate a pseudo-normal complex.*

The consideration of circles adjacent and cospherical with a given circle presents certain special features in the case where the complex is conformal. Let the foci of a circle in general position be P and P', while those of an adjacent sphere cospherical with the first are $P+\Delta P$, $P'+\Delta P'$. If we follow the conformal transformation of space established by our complex by an inversion which interchange P and P', we have a spherical transformation which leaves invariant P and the circle through our four points. We are thus led to the consideration of the invariant line-elements in a spherical transformation which leaves P invariant. We have four possibilities :

1) One proper and two isotropic line elements fixed.

2) One proper line element fixed, and all line elements orthogonal thereto fixed.

3) All line elements orthogonal to an isotropic fixed.

4) All line elements fixed.

Remembering that a necessary and sufficient condition that two circles on a not-null sphere should touch is that the circle through their foci should be null, we have

Theorem 55.] *If a circle of a conformal complex be cospherical with but three adjacent circles of the complex, the common sphere being in no case null, it will touch two of them.*

Theorem 56.] *If a circle of a conformal complex be cospherical with three adjacent circles, and be not tangent to any one, nor lie with one on a null sphere, it will be cospherical with a series of adjacent circles.*

Theorem 57.] *If a circle of a conformal complex be neither tangent nor on a null sphere with more than one adjacent circle, it will be cospherical with a series of adjacent circles.*

Theorem 58.] *If a circle in general position in a conformal complex touch three adjacent circles, the complex will consist in the totality of circles on one sphere.*

Of course these statements in terms of adjacent circles could be translated into terms of annular surfaces and osculating circles if we chose.

The following theorems are deduced by methods exactly analogous to those used in the case of circle congruences. Two complexes of spheres shall be said to be *conformally related* when their members are in a continuous one to one correspondence, and the angle of two adjacent spheres in one is equal to the corresponding angle in the other.

Theorem 59.] *If a conformal circle complex be given, we may, in an infinite number of ways, pass two spheres through each circle which shall make a constant angle with one another, and describe conformally related complexes.*

Theorem 60.] *If two sphere complexes be conformally related, and two corresponding spheres meet always at a fixed angle, then will their circles of intersection generate a conformal complex.*

The subject-matter of the present chapter offers quite as much opportunity for fruitful further study as did that which preceded it. We only managed to prove known theorems with the aid of the parametric method, but it is a method of great power and, skilfully handled, seems likely to furnish new and interesting results. What will be found if we attempt to extend this method to the study of complexes of circles? On the other hand, the Kummer method which has done so much for the differential geometry of the straight line is certainly capable of much larger development than it has here received. Other writers will prefer to study directly the infinitesimal properties of the S_6^5 by the methods which are proving so fruitful in modern projective differential geometry. What is certain is that the circle has been diligently studied for two thousand years, and that it will be similarly studied for many thousands more. The methods of attack here exhibited are no more in advance of those known to Euclid and Apollonius than will be those of future geometers in comparison with the best that we have been able to show. This, at least, is what we have a right to hope and expect. For ourselves, 'Let us shut up the box and the puppets, for our play is played out.'

SUBJECT INDEX

INDEX OF PROPER NAMES